14th International Symposium on Advanced Materials 2015

AA006930

IOP Conference Series: Materials Science and Engineering
Volume 146

Islamabad, Pakistan
12 – 16 October 2015

Editors:

Sara Qaisar
A. Nusair Khan
Ejaz Ahmed Mukhtar

ISBN: 978-1-5108-2998-5
ISSN: 1757-8981

Printed from e-media with permission by:

Curran Associates, Inc.
57 Morehouse Lane
Red Hook, NY 12571

CURRAN ASSOCIATES INC.
proceedings
.com

Some format issues inherent in the e-media version may also appear in this print version.

Copyright© (2015) by the Institute of Physics
All rights reserved. The material featured in this book is subject to
IOP copyright protection, unless otherwise indicated.

Printed by Curran Associates, Inc. (2016)

For permission requests, please contact the Institute of Physics
at the address below.

Institute of Physics
Dirac House, Temple Back
Bristol BS1 6BE UK

Phone: 44 1 17 929 7481
Fax: 44 1 17 920 0979

techtracking@iop.org

Additional copies of this publication are available from:

Curran Associates, Inc.
57 Morehouse Lane
Red Hook, NY 12571 USA
Phone: 845-758-0400
Fax: 845-758-2633
Email: curran@proceedings.com
Web: www.proceedings.com

Table of contents

Volume 146

14th International Symposium on Advanced Materials
12–16 October 2015, National Centre for Physics, Islamabad, Pakistan

Accepted papers received: 28 July 2016
Published online: 6 September 2016

Preface

Papers

IOP Conf. Series: Materials Science and Engineering **146** (2016) 011001 doi:10.1088/1757-899X/146/1/011001

14th International Symposium on Advanced Materials (ISAM 2015)

Islamabad, Pakistan
12-16 October 2015

Editors: Sara Qaisar, A. Nusair Khan and Ejaz Ahmed Mukhtar

IOP Publishing

Content from this work may be used under the terms of the Creative Commons Attribution 3.0 licence. Any further distribution of this work must maintain attribution to the author(s) and the title of the work, journal citation and DOI.
Published under licence by IOP Publishing Ltd

IOP Conf. Series: Materials Science and Engineering **146** (2016) 011001 doi:10.1088/1757-899X/146/1/011001

Selected peer reviewed papers from the

14th International Symposium on Advanced Materials (ISAM 2015)

12-16 October 2015, Islamabad, Pakistan

Editors
Sara Qaisar
A. Nusair Khan
Ejaz Ahmed Mukhtar

Organized by : International Symposium on Advanced Materials, Islamabad, Pakistan
Co-Organizers : National Centre for Physics, Islamabad, Pakistan
Institute of Space Technology, Islamabad, Pakistan

PREFACE

Advanced Materials XIV is a compilation of peer reviewed papers presented at the 14th International Symposium on Advanced Materials (ISAM 2015) held from October 12- 16th 2015 at the National Centre for Physics, Islamabad, Pakistan. In the role of Secretary ISAM-2015, I feel honored that the symposium ended on a positive note. The challenges of the present day industry entail an in-depth knowledge of the science behind the processes involving advanced materials. It is this need that ISAM, along with similar other forums, endeavors to address.

The five day deliberations of ISAM 2015, consisted of 29 technical sessions and 2 poster sessions, inclusive of a poster competition on nanostructures which displayed 150 research papers depicting various aspects of nanomaterials. In all 307 papers were presented, inclusive of 61 contributory, invited and oral presentations. The symposium also hosted 3 parallel panel discussions led by renowned scientists and eminent researchers from foreign as well as local institutes.

The ultimate aim of this proceeding is to archive various findings in the field of advanced materials. I hope that the technical data available in this publication proves its worth to scientists and researchers working in this area of science.

I wish to acknowledge the Institute of Physics (IOP) Publishing UK, for accepting the research papers from ISAM-2015 for publication in the IOP Conference Series: Materials Science and Engineering. The proceeding will be available on the IOP website as an online open access document.

I am profoundly thankful to the Symposium Chairman for his firm support and for the wisdom with which he guided the Organizing Committee of ISAM 2015 to make it a successful event. On the same note, I must sincerely acknowledge the truly commendable efforts of the organizing committee. My gratitude to all our distinguished participants, session chairs/co-chairs, and reviewers for their active role in the symposium. Last yet not the least, my thankfulness goes to all our sponsors for financing the event.

Dr. Sara Qaisar
Symposium Secretary

IOP Conf. Series: Materials Science and Engineering **146** (2016) 011001 doi:10.1088/1757-899X/146/1/011001

ORGANIZING COMMITTEE

Ejaz Ahmed Mukhtar	:	(Chairman)
Sara Qaisar	:	(Secretary)
M. Muneeb Asim	:	(Treasurer)
S. Wilayat Husain		
Anjum Tauqir		
M. Farooque		
Shaheed Khan		
Iftikhar us Salam		
Zubair Ahmed		
Shabbar A. Rizvi		
A. Nusair Khan		

ADVISORY BOARD

Amir Azam Khan	:	(Pakistan)
F. Ahmad Khalid	:	(Pakistan)
Mohamad M. Ayad	:	(Egypt)
Mi Yan	:	(China)
Muhammad Mujahid	:	(Pakistan)
S. Wilayat Husain	:	(Pakistan)
Tahir I. Khan	:	(Canada)
Yosry Mashal	:	(Saudi Arabia)
Zhongwu Liu	:	(China)

SECRETARIAT

Zubair Ahmed	:	Incharge
Shahab Bhatti		
Junaid Shinwari		
Majid Nizami		
Mudassar Rauf		
Khalid Mehmood		
Yasir Raza		
Qadeer Ahmad		
M. Nasir		
Amar Ajaz Kazmi		

TECHNICAL COMMITTEE

A. Nusair Khan	:	Incharge
Amjad Ali		
Tayyab Subhani		
M.B. Bhatty		
M. Mansoor Malik		
M. Adnan Alvi		
Shams Iqbal		

SPONSORS

Askari Bank Ltd., Pakistan

EPRF, Pakistan

Higher Education Commission, Pakistan

Islamic Development Bank, Saudi Arabia

National Bank of Pakistan

National Talent Pool, Pakistan

Peoples Steel Mills Ltd., Pakistan

IOP Conf. Series: Materials Science and Engineering **146** (2016) 011001 doi:10.1088/1757-899X/146/1/011001

AUTHOR INDEX

M

M. A. A. Al-Matari
M. A. A. Qattous
M. A. Aleem
M. A. Gondal
M. A. H. Nazzal
M. A. Munot
M. A. Noshad
M. A. Rafiq
M. A. Raza
M. A. Rehman
M. Abas
M. Awais
M. B. Bhatty
M. B. Khan
M. Bashir
M. Çakmakkaya
M. F. Ahmed
M. F. Bodla
M. Farooque
M. Jalil1
M. M. Ahmed
M. M. Asim
M. Mansoor
M. Maqsood
M. Mehmood
M. Mujahid
M. R. Saleem
M. Rahim
M. Rosemal
M. S. Awan
M. S. Khan
M. Saleem
M. Shah
M. Shahid
M. Shakeel
M. Shamraiz
M. Shuaib
M. Siddique
M. Sohail
M. Waseem
M. M. Rauf
M.R. Usman
M. S. Awan
M. S. Mehmood
M-H. Kim

N

N. Ejaz
N. Fatima
N. Glandut
N. Iqbal
N. Iqbal
N. S. Al-Hunetti
N. S. Khattak

Q

Q. Shahzad
Q. F. Gillani

R

R. A. Pasha
R. Ahmad
R. Ali
R. Khan
R. Waheed
R. A. Malik
Raghad
Ramiza

S

S. A. Rizvi
S. Abbas
S. Ahmed
S. Akbar
S. Akbar
S. Akhtar
S. Amira
S. Bashir
S. Hememat
S. Honkanen
S. K. Ajmal
S. Khan
S. Mehmood
S. S. Iqbal
S. Talaş
S. W. Husain
S. Z. Ramadan
S. Zaman
S. M. A. Al-Qawabah

IOP Conf. Series: Materials Science and Engineering **146** (2016) 011001 doi:10.1088/1757-899X/146/1/011001

T

T. Ghani
T. Hussain
T. Jamil
T. Rashid
T. Yasin
T. Zaman
T. Hussain
T-K. Song

U

U. A. Khan
U. Sultan
U. Zulfiqar

W

W. A. Khan
W. A. Tarar
W. Muhammad
W. Tarar
W.A. Khan
W-J. Kim

Y

Y. Jamil
Y.F. Joya

Z

Z. Ahmad
Z. Butt

Peer review statement

All papers published in this volume of *IOP Conference Series: Materials Science and Engineering* have been peer reviewed through processes administered by the proceedings Editors. Reviews were conducted by expert referees to the professional and scientific standards expected of a proceedings journal published by IOP Publishing.

Content from this work may be used under the terms of the Creative Commons Attribution 3.0 licence. Any further distribution of this work must maintain attribution to the author(s) and the title of the work, journal citation and DOI.
Published under licence by IOP Publishing Ltd

Characterization of nanomaterials with transmission electron microscopy

D. H. Anjum
Imaging and Characterization (IAC) Lab, King Abdullah University of Science and Technology (KAUST), Thuwal, Makkah 23955, Kingdom of Saudi Arabia (KSA)

E-mail: dalaver.anjum@kaust.edu.sa

Abstract. The field of nanotechnology is about research and development on materials whose at least one dimension is in the range of 1 to 100 nanometers. In recent years, the research activity for developing nano-materials has grown exponentially owing to the fact that they offer better solutions to the challenges faced by various fields such as energy, food, and environment. In this paper, the importance of transmission electron microscopy (TEM) based techniques is demonstrated for investigating the properties of nano-materials. Specifically the nano-materials that are investigated in this report include gold nano-particles (Au-NPs), silver atom-clusters (Ag-ACs), tantalum single-atoms (Ta-SAs), carbon materials functionalized with iron cobalt (Fe-Co) NPs and titania (TiO_2) NPs, and platinum loaded Ceria (Pt-CeO_2) Nano composite. TEM techniques that are employed to investigate nano-materials include aberration corrected bright-field TEM (BF-TEM), high-angle dark-field scanning TEM (HAADF-STEM), electron energy-loss spectroscopy (EELS), and BF-TEM electron tomography (ET). With the help presented of results in this report, it is proved herein that as many TEM techniques as available in a given instrument are essential for a comprehensive nano-scale analysis of nano-materials.

1. Introduction

The field of nanotechnology deals with the science, engineering, and technology of materials whose at least one of three dimensions is in the range of less than one hundred nanometers [1]. The modern concept of nanotechnology was introduced by 20th century celebrated Physicist Richard P. Feynman in which he proposed the use of 1 nm^3 size clusters as "computer bits" [2]. It is to be noted that TEM was most elaborate and powerful technique that was available for the imaging of materials. However due to inadequate spatial resolution, the TEM instruments was unable to meet the challenges of imaging such underlined one nm size atom-clusters. This is why Feynman stressed upon improving the spatial resolution of TEM instruments [2]. Albeit not related but there had been consistent efforts in improving the resolution and capability of TEM instruments [3-4]. One of most important developments in this regard is arguably spherical aberration correctors for both conventional TEM and scanning TEM (STEM) modes of TEM instruments [5-6]. With the help of such correctors, the contemporary TEMs are capable of providing atomic scale imaging, elemental and chemical analyses of samples [7-9].

A complete introduction about a TEM instrument is beyond the scope of this report. As there are number of textbooks are currently available on this technique [10-13]. Nevertheless for the sake of completeness a brief introduction is given in this report to highlight the most important parts of TEM instruments. As an aside, it is imperative to note that the thickness of materials in the direction of

Content from this work may be used under the terms of the Creative Commons Attribution 3.0 licence. Any further distribution of this work must maintain attribution to the author(s) and the title of the work, journal citation and DOI.
Published under licence by IOP Publishing Ltd

penetrating electron beam is crucial for their TEM analysis. In fact this thickness has to be below the value of about 500 nm in order to have the enough fraction of electron beam transmitted through sample elastically which is a prerequisite to the formation of TEM-images. It is also essential that the electrons, provided by a thermionic or a field-emission gun, are collimated nicely with the help of electromagnet condenser lenses prior to their passing through the samples. A number of interactions take place between the sample atoms and penetrating electron beam when it passes through the samples. The transmitted electrons are then focused by using another electromagnetic lens called the "objective" lens which forms the first image of samples. The resolution of images formed in TEMs depends upon the wavelength of electrons used as well as the value of spherical aberration co-efficient (Cs) of the objective lens. As a practical fact, a typical resolution of an image in contemporary TEM instruments is in the range of 0.2 nm without using a spherical aberration corrector (Cs-corrector). Whereas the spatial resolution of 0.05 nm can be reached for both TEM and STEM modes by employing Cs-correctors after and before the samples, respectively [10,12]. Likewise the image-plane of the objective lens in TEM, its conjugate focal plane or the diffraction plane also possesses the same information on samples but presents in the form of reciprocal space. This is also quite useful to the researcher and allows them to study diffraction of electrons from samples to acquire the information on the phases present in materials. Another important optical element called "electron prism" which can be attached to a TEM and forms the basis for performing EELS in TEM instruments [13]. It can be attached either in column or post column of TEMs.

The object of this study presented in this report is to make list the advantages and limitations of a TEM analysis performed on nano-materials. This is accomplished by performing main TEM analyses of materials used in the field of nanotechnology. Such examples were deliberately taken from representative nano-materials having NPs below the size of 100 nm.

2. Experimental Methods

Altogether there are three microscopes of models Titan 60-300 ST from FEI Company (Hillsboro, OR) were employed to perform the TEM analyses of nano-materials. One of the utilized microscopes was equipped with an image corrector, the other one was equipped with a probe corrector, and while the third one was having no corrector of either type. Moreover all the TEMs were equipped with post-column energy filters or electron prisms of model GIF Tridiem 865 from Gatan Inc. (Pleasanton, CA). The most of datasets were acquired and processed in Gatan Microscopy Suite (GMS) of version GMS v1.85. Variety of TEM characterizations were performed on nano-materials suitable for a certain type of TEM analysis. For instance NPs with having sizes from 10 nm to single atoms were chosen for aberration corrected TEM (AC-TEM) and aberration corrected STEM (AC-STEM) analysis experiments. In both cases, the correctors were tuned prior to the experiments so that the co-efficient of spherical aberration (Cs) was measured to be below the value of negative one micron. AC-STEM and AC-STEM techniques were also applied to the imaging of silver NPs (Ag-NPs) below the size of 2 nm and tantalum (Ta) single atoms loaded on to nickel oxide (NiO) catalyst support, respectively. A carbon (C) nano composite containing oxidized iron cobalt NPs (FeCo-NPs) was included to perform the spectrum imaging (SI) in the STEM-EELS mode to generate the Fe and Co elemental maps. SI technique in the energy-filtered TEM (EFTEM) mode was applied to another material made up of C nano-fibers. Nano-fibers contained about 10 at.% of nitrogen (N) and were functionalized with TiO_2 NPs. EFTEM-SI analysis was carried out by acquiring energy loss images by setting slit width of 5 eV and with a step of 5 eV as well. At the end a nano-composite material containing 2-3 nm size platinum (Pt) NPs loaded on to 20-25 nm size CeO_2 NPs was selected for the ET analysis in the bright-field TEM (BF-TEM) mode. The main objective of the ET analysis was to investigate the loading of Pt NPs onto the CeO_2 NPs. The ET analysis was carried out by tilting the specimens from -70° to +70° with an increment of 1° during the acquisition of datasets. The acquired ET datasets were then aligned and reconstructed with freely available software package IMOD to determine the loading of Pt-NPs on CeO_2 NPs in three dimensions.

3. Results and discussion

3.1. Imaging of nano-materials

The focus was first turned to compare and contrast AC-STEM and AC-STEM techniques by carrying out the imaging of ~10 nm size Au-NPs. The corresponding results compiled by using these techniques are shown in Figure 1. It can be seen from there that in the low magnification range, the both techniques provide about similar quality information on the morphology of NPs. Furthermore both types of images can be used to determine the NP size distribution as well as for the determination of crystal structure. However, in higher magnification range, the situation changes a bit between AC-TEM and AC-STEM analyses of those 10 nm size Au-NPs. It can be seen by high resolution images of couple of NPs shown as inserts in Figure 1a and Figure 1b which were acquired with HR-TEM and HR-STEM techniques, respectively. There are a couple of important differences exist between those two images. First, it is the presence of no background in AC-STEM image as compared to BF-TEM. While the second difference is on quality of lattice fringe pattern, especially in the middle, exhibited by the NPs. It is apparent that mentioned lattice pattern in the HR-STEM image is of better quality as compared to image acquired with HR-TEM. The underlined differences between these techniques basically stem from different principles governing their image contrasts. It is well known that image-contrast in AC-TEM image is formed by the interference of transmitting electron waves through the NP and the observed interference pattern depends upon thickness of NP and difference of phases in the electron waves [10-11]. As a consequence the areas with different thickness can have a different interference pattern. It can be destructive on one place but can be constructive o another place. In spite of these difficulties, HR-TEM analysis of nano materials is a preferred way because of its ease and flexibility. On the other hand, the image-contrast in the STEM image of NP is not formed by the interference of transmitting electron waves and hence is less prone to changes in the thickness of NP. The lattice pattern in the image is a direct result of convolution of scanned sub-nanometer probe size with lattice columns.

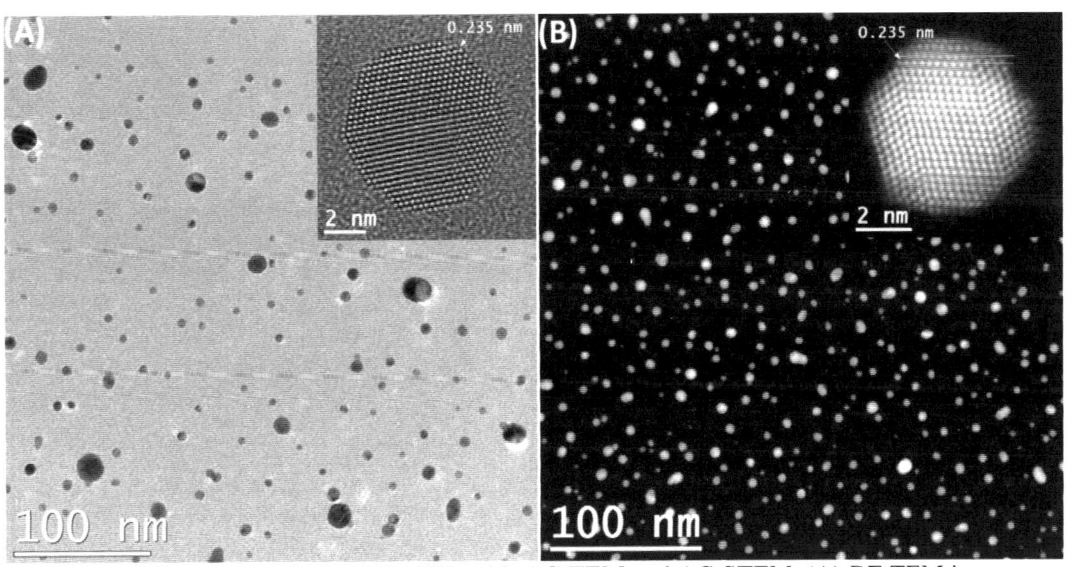

Figure 1. Imaging analysis of Au-NPs with AC-TEM and AC-STEM. (A) BF-TEM image showing a uniform size Au-NPs and an inset of HR-TEM image of a single Au-NP. (B) DF-STEM image showing a uniform size Au-NPs and an inset of HR-STEM image of a single Au-NP

The situation for realizing a successful TEM analysis changes dramatically for the imaging of NPs below the size of 2 nm to going down to single atoms. This is because during of the imaging of such small size NPs, the electron beam can also modify them. So special imaging conditions need to be determined or established before achieving a successful imaging-analysis of such small size NPs. In this regard, the electron beam energy and mode of imaging are the most important parameters. Generally there is a specific range of parameters for certain type of samples. For instance, the lowering the accelerating voltage below 100 kV is a must for imaging of 2 dimensional materials as well as the imaging of NPs below the size of 2 nm in order to keep the beam-induced sample modification at the minimum level. Figure 2a contains an example of such analysis which is an aberration corrected electron micrograph of about 2 nm size Ag-NPs. The micrograph has been acquired by setting the image corrected microscope in BF-TEM mode at the accelerating voltage of 60 kV. It can be seen from there the Ag-NPs exhibit an excellent quality image contrast and further no beam damage was observed at this accelerating voltage. So it be stated that the underlined analysis enabled to determine the particle size distribution and crystal structure without causing any modification to NP size and shapes. The imaging of single atoms sitting on a 20-50 nm support is not that straightforward and requires completely different imaging conditions. This is because the convoluted image contrast of a single atom in the BF-TEM image is difficult to interpret and hence is prone to mistakes. Such subtleties can be avoided in certain cases by using the microscope in DF-STEM mode. This is particularly true for the cases when the atomic number (Z) of single atoms is quite different than that of support NPs or film. Figure 2b contains an aberration corrected DF-STEM micrograph of such a sample that was having Ta single atoms loaded on to ~10 nm size NiO support. This analysis not only allowed the imaging of Ta single atoms but it also enabled determining the population of Ta single atoms on NiO crystals.

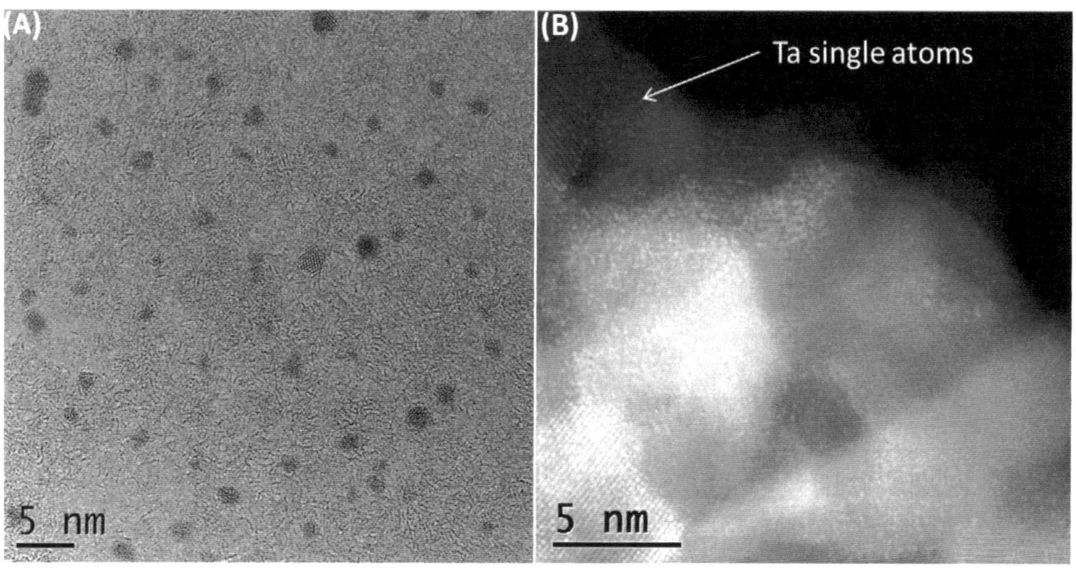

Figure 2. Imaging of 2nm size NPs and single atoms. (a) Aberration corrected BF-TEM image showing a uniform size Ag-NPs. (b) Aberration corrected DF-STEM image of Ta single atoms on NiO support revealing a non-uniform decoration of NiO crystals with Ta atoms

3.2. Analytical electron microscopy of nano-materials

Another advantage of a TEM analysis is to enable having the spectroscopy analysis of samples at the spatial resolutions of nanometer to sub-nanometer range. The spectroscopy analysis can be accomplished in a couple of ways with a TEM namely the x-ray energy dispersive spectroscopy (EDS) and EELS. Both EDS and EELS spectroscopy methods have advantages and disadvantages over each other and are normally used as complementary techniques in a given TEM analysis. In spite of this, the EELS spectroscopy analysis is a preferred one when there exists the availability of both techniques because of its ability to provide both elemental and chemical information at the nanometer scale when utilized along with DF-STEM mode. Figure 3 contains an example of such a STEM-EELS analysis that has been performed on a sample which was having a porous carbon (C) material functionalized with oxidized bimetallic FeCo-NPs. Such materials are promising candidates for the energy related applications [14]. It can be seen from Figure 3a that the oxidized FeCo-NPs were uniformly embedded in porous C material and while the corresponding EELS spectrum not only revealed the presence of all the expected elements in the sample but it also allowed the analysis of Fe-L23 and Co-L23 white lines to determine the oxidation states of both Fe and Co elements [13].

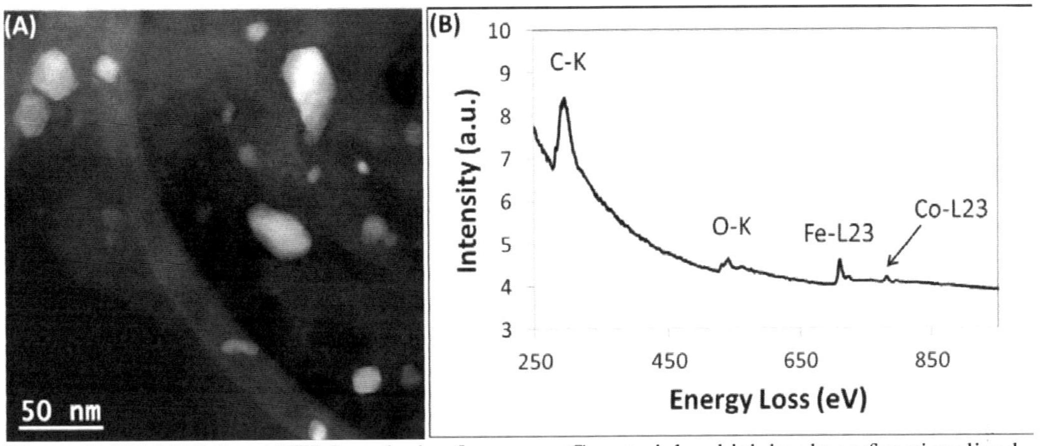

Figure 3. DF-STEM and EELS analysis of a porous C materials which has been functionalized with oxidized FeCo bimetallic NPs. (a) DF-STEM micrograph showing the average size of NPs was in the range of 30 nm. (b) Corresponding EELS spectrum revealing the presence of all four elements. Further, the analysis of Fe-L23 and Co-L23 white lines allowed determining the oxidation states of Fe and Co

In addition to getting above mentioned useful information about the samples with STEM-EELS technique, high spatial resolution mapping of elements can also be accomplished by making the use of spectrum imaging (SI) method in this mode. The SI data sets were acquired in STEM-EELS mode first and then the acquired data sets were processed to generate the elemental maps having sub-nanometer spatial resolutions. Fe and Co maps of a couple of oxidized FeCo-NPs in Figure 3a are shown in Figure 4. It can be noticed from there that one of the NPs was a uniform alloy of Fe and Co and while the second NP was a having a core shell morphology. So it can be inferred from the presented results that such type of morphological information on NPs is essential for establishing an accurate structure-property relationship. Otherwise, by using results of conventional STEM-EELS analysis of Figure 3, it was possible to state that the FeCo-NPs were uniform in composition. It is to be noted that the elemental maps of several NPs of Figure 3 were generated in order to gather a statistics on the presence of uniform alloy and core shell NPs. In this way it was found out that there were about 40% NPs were possessing core shell morphology.

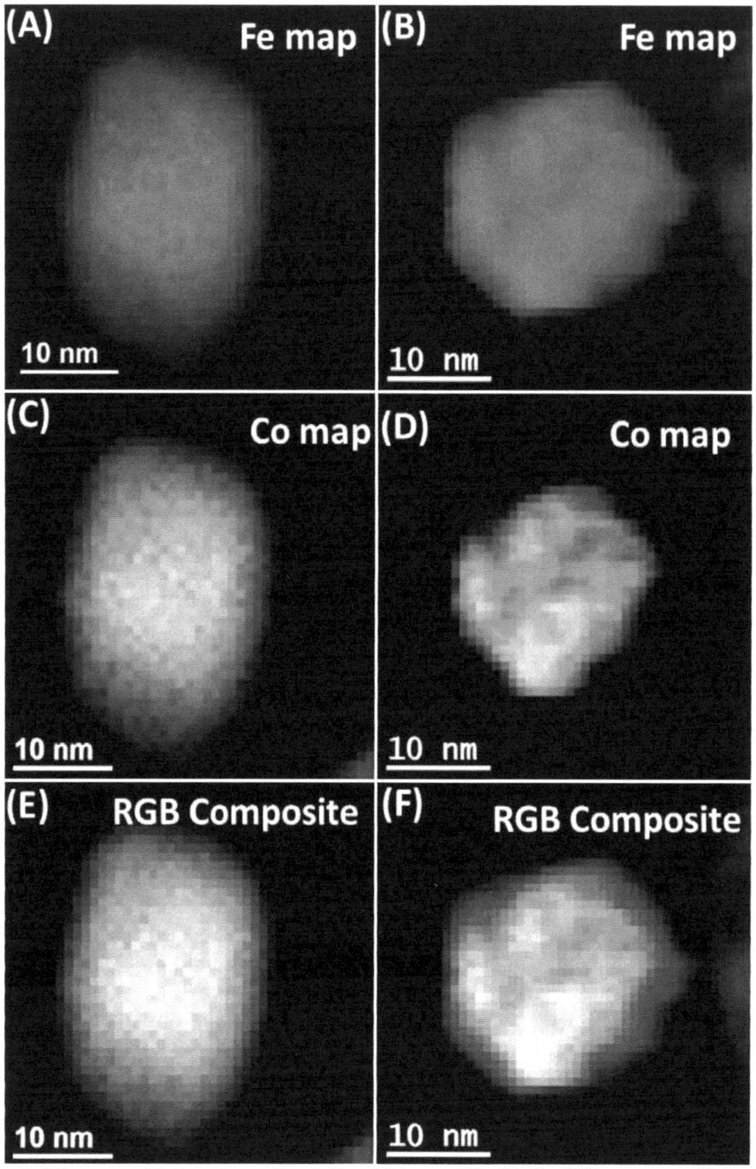

Figure 4. Elemental Mapping of oxidized bimetallic FeCo NPs accomplished with STEM-EELS SI datasets. (a) & (b) Fe maps. (c) & (d) Co maps. (e) & (f) Resultant composite maps of Fe and Co to reveal the nature of elemental distribution in NPs

Carbon based nano fiber like materials are increasingly being utilized as a support for photo catalysis applications. Especially when these are functionalized with high band gap semiconductor oxides such as TiO_2. It is shown here next that such materials can be characterized at nano scales quite easily and accurately by using the TEM based techniques. For instance Figure 5 contains a BF-TEM micrograph along with a core-loss EELS spectrum from an empty region (or hole). It can be seen that the C nano fibers were functionalized with TiO_2 NPs quite uniformly as confirmed by a core-loss EELS spectrum of Figure 5(b) which has been from a region enclosed by a circle-1 in Figure 5(a).

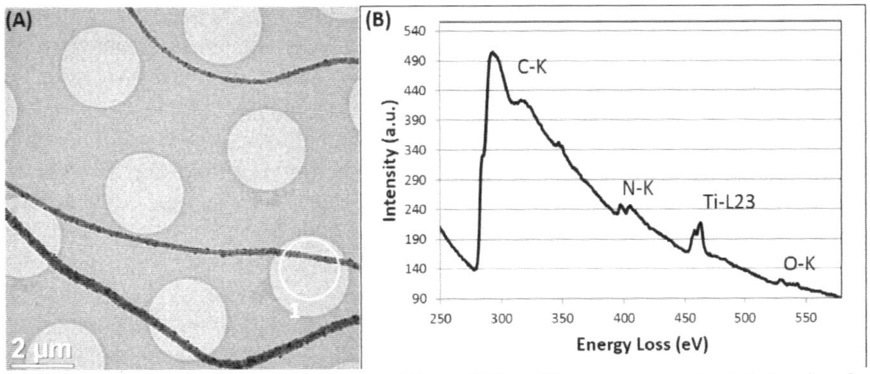

Figure 5. Imaging and elemental composition of fiber-like carbon material that has been functionalized with TiO_2 NPs. (a) BF-TEM of TiO2 functionalized C-fibers that are supported by holey-carbon film. (b) A core-loss EELS spectrum from an area enclosed with circle-1 in (a) and it gives the elemental composition of functionalized fibers

The intermixing of functionalized NPs with host materials like shown in Figure 5a is always a matter of concern and hence it is addressed next by using the EFTEM-SI technique. Figure 6 contains the results acquired with this technique. It was applied to a part of functionalized C nano fiber shown in Figure 6a which has been imaged with EFTEM mode to exhibit higher image contrast. The EFTEM generated elemental maps of all four elements namely C, N, Ti, and O are given in Figure 6b-e and then fat the end resultant Red-Green-Blue (RGB) composite of maps is given in Figure 6f. It can be clearly seen that the intermixing of TiO_2 NPs with C fiber was virtually non-existing and thus these fibers behave as only the carrier of TiO_2 NPs. In this way, the properties of TiO_2 NPs are not expected to change on these fibers and hence will retain their high efficiency for photo catalysis applications.

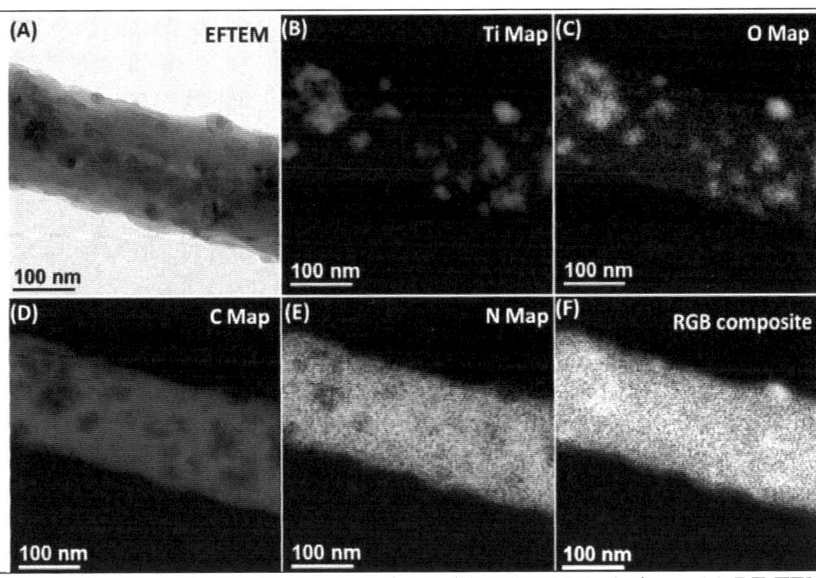

Figure 6. Elemental analysis of nano materials with EFTEM technique. (a) BF-TEM image of a TiO_2 functionalized nano fiber. (b) Elemental map of Ti generated with Ti-L23 edge. (c) Elemental map of O generated with O-K edge. (d) Elemental map of C generated with C-K edge. (e) Elemental map of N generated with N-K edge. (f) RGB Composite of all four elemental maps

3.3. Three-dimensional imaging of nano materials with a TEM

It is important to remember that the TEM images are essentially two dimensional projections of three dimensional objects. As a consequence, the information about the third-dimension or along the penetrating electron-beam direction is suppressed and hence becomes difficult to find out about the sample properties in this direction. However it is becoming increasingly important in the field of nanotechnology to have information about NPs in all three dimensions in order to fully understand their properties [10-12]. Electron tomography (ET) in a TEM provides primarily provides just such type of three dimensional analysis at the nanometer scale lengths [15]. In ET method, first of all, the images of a sample at various angles are acquired by tilting it in small angular increments. The acquired stack of images is then aligned and back projected to have a three-dimensional volume or tomogram of the sample capable of revealing the three dimensional information about samples. Such analysis is particularly important for the characterization of catalytic materials which are sometimes functionalized with small size NPs [16]. Figure 5 contains the results acquired from the ET method applied to a material which was composed of CeO_2-NPs functionalized with Pt-NPs. The conventional BF-TEM image presented in Figure 7a is basically used as a control for the ET analysis. It is a projection image and hence contains some places where the presence of Pt-NPs along with CeO_2-NPs cannot be easily established. For instance, the four places marked with four circles have been selected in Figure 7a to underline this point. Interestingly but expectedly the three volume-slices from the different region of tomogram of Figure 7a revealed that not only all those four encircled places have Pt-NPs but were found to be lying at different height levels. The presence of Pt-NPs at different height levels in tomogram is corroborated by the given volume-slices in Figure 7b-d where it can be seen that the Pt-NPs appeared in encircled areas in different slices. So it is suffice to say that the ET analysis is essential to fully characterize the loading of small size NP onto large size support NPs.

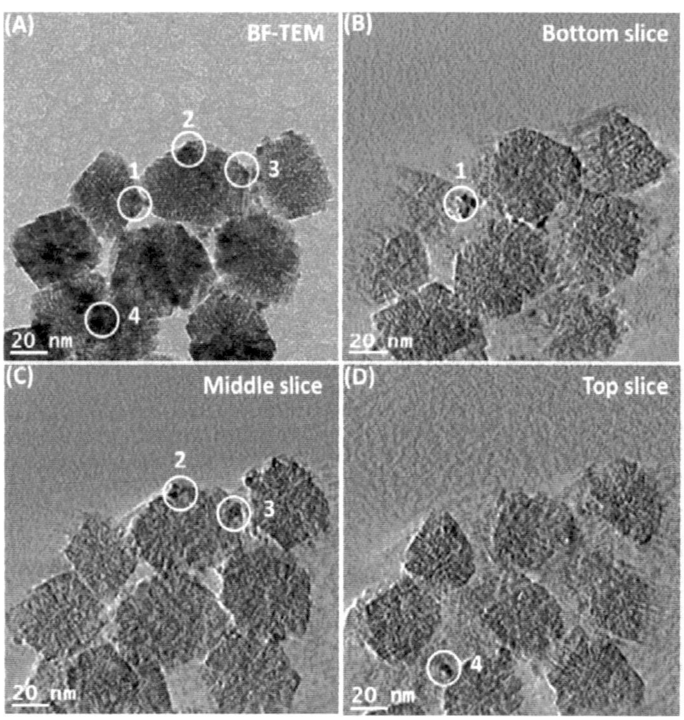

Figure 7. ET analysis of CeO_2 and Pt NPs. (a) BF-TEM projection image of sample at 0° angle. There are few ambiguous places in it where the presence of Pt NPs does not seem to be an obvious thing. (b) The volume slice from the bottom part of the area revealing that the place enclosed with circle 1 indeed had Pt NPs. (c) & (d) Similarly the middle and top slices revealing the presence of Pt NPs at the remaining three places highlighted in (a).

4. Conclusions

In the light of presented TEM analyses of different nano materials, it can be stated with confidence that TEM offers a comprehensive recipe of their characterization. Specifically both the AC-TEM and AC-STEM techniques can be employed to image materials comprised of single atoms to small size NPs. Though a suitable electron beam conditions should be adapted for the imaging of materials having NPs of size 5 nm or close. In other words for the imaging of atom clusters, the electron energy of 100 keV or less should be used during the analysis. While AC-STEM mode is quite suitable for the imaging of high atomic number single atoms sitting on different support NPs. For the analytical characterization of nano materials, both STEM-EELS and EFTEMSI techniques can be employed in determining the elemental distributions in NPs. STEM EELS technique offers composition and distribution of elements in NPS with spatial resolution below the size of one nanometer. However its datasets are acquired in the sequential manor and thus it takes generally several minutes in completing the acquisition of datasets. On the other hand, EFTEM SI technique offer the same elemental analysis of nano materials but its datasets generally possess spatial resolution worse than one nanometer. However, unlike STEM EELS, EFTEM SI it a parallel data acquisition and hence offers the completion of datasets normally in just few seconds. At last but certainly not least, the ET analysis allows determining three dimensional analysis of nano materials at the nanometer scales and can be useful in investigating the loading of small size NPs onto larger size host NPs.

5. Acknowledgements

KAUST is acknowledged for the financial support. We are also thankful to different people who provided the samples for the TEM analyses.

6. References

[1] K.D. Sattler, *Handbook of Nanophysics, Principles and Methods* (CRC, New York, 2010)
[2] R. P. Feynman, *Engineering and Science,* 22 (1960).
[3] P. W. HAWKES, Aberration correction past and present, *Phil. Trans. R. Soc. A*,367, 3637 (2009)
[4] S.J. Pennycook, Seeing the atoms more clearly: STEM imaging from the Crewe era to today, *Ultramicroscopy*, 123, 28 (2012)
[5] M. Hairder, P. Hartel, H. Muller, S. Uhlemann, and J. Zach, Current and future aberration correctors for the improvement of resolution in electron microscopy, *Phil. Trans. R. Soc. A,* 367, 3665 (2009)
[6] O. L. Krivanek, M. F. Chisholm, V.Nicolosi, T. J. Pennycook, G. J. Corbin, N. Dellby, M. F. Murfitt, C. S. Own, Z. S. Szilagyi, M. P. Oxley, S. T. Pantelides, and S. J. Pennycook, Atom-by-atom structural and chemical analysis by annular dark-field electron microscopy, *Nature*, 464, 25 (2010)
[7] Knut W. Urban, Is science prepared for atomic-resolution electron microscopy? *Nature materials*, 8, 260 (2009)
[8] Ping Lu, Lin Zhou, M. J. Kramer and David J. Smith, Atomic-scale Chemical Imaging and Quantification of Metallic Alloy Structures by Energy-Dispersive X-ray Spectroscopy, *Scientific Reports*,4, 3945 (2013)
[9] Stuart Turner, Sorin Lazar, Bert Freitag, Ricardo Egoavil, Johan Verbeeck, Stijn Put, YvanStrauvend and Gustaaf Van Tendeloo, High resolution mapping of surface reduction in ceria nanoparticles, *Nanoscale*, 3, 3385 (2011)
[10] David D. Williams and C. Barry Carter, Transmission Electron Microscopy: A Textbook for Materials Science, 2nd Edition, Springer-Verlag Berlin,(2009)
[11] Brent Fultz and James Howe, Transmission Electron Microscopy and Diffractometry of Materials, 4th Edition, Springer-Verlag Berlin, (2013)

[12] Stephen J. Pennycook and Peter D. Nellist, Scanning Transmission Electron Microscopy: Imaging and Analysis, Springer-Verlag Berlin, (2011)

[13] Ray F. Egerton, Electron Energy Loss Spectroscopy in an Electron Microscope, 3rd Edition, Springer-Verlag Berlin, (2011)

[14] Guang-Hui Wang, JakobHilgert, Felix Herrmann Richter, FengWang, Hans-Josef Bongard, Bernd Splietho, ClaudiaWeidenthaler, and FerdiSchüth, Platinum–cobalt bimetallic nanoparticles in hollow carbon nanospheres for hydrogenolysis of 5-hydroxymethylfurfural, *Nature Materials*, 13, 293(2014)

[15] Chao Wan, Dang-guo Cheng, Fengqiu Chen and Xiaoli Zhan, Fabrication of CeO_2 nanotube supported Pt catalyst encapsulated with silica for high and stable performance, *Chem. Commun.*, 51, 9785 (2015)

IOP Conf. Series: Materials Science and Engineering **146** (2016) 012002 doi:10.1088/1757-899X/146/1/012002

Laser surface treatment of porous ceramic substrate for application in solid oxide fuel cells

D. S. A. Mahmod[1], A. A. Khan[1], M. A. Munot[1], N. Glandut[2] and J. C. Labbe[2]
[1]Department of Mechanical and Manufacturing Engineering, Faculty of Engineering, Universiti Malaysia Sarawak (UNIMAS), 94300 Kota Samarahan, Sarawak, Malaysia.
[2]SPCTS, UMR 7315, CNRS, Centre Européen de la Céramique, 12 Rue Atlantis, Université de Limoges, 87068 Limoges, France

E-mail: akamir@unimas.my

Abstract. Laser has offered a large number of benefits for surface treatment of ceramics due to possibility of localized heating, very high heating/cooling rates and possibility of growth of structural configurations only produced under non-equilibrium high temperature conditions. The present work investigates oxidation of porous ZrB_2-SiC sintered ceramic substrates through treatment by a 1072 ± 10 nm ytterbium fiber laser. A multi-layer structure is hence produced showing successively oxygen rich distinct layers. The porous bulk beneath these layers remained unaffected as this laser-formed oxide scale and protected the substrate from oxidation. A glassy SiO_2 structure thus obtained on the surface of the substrate becomes subject of interest for further research, specifically for its utilization as solid protonic conductor in Solid Oxide Fuel Cells (SOFCs).

1. Introduction

Nowadays, laser has become an alter native tool for materials processing, in particular for surface modification of refractory ceramics. However, the laser involvement on the treatment of structural ceramics for the purpose of protonic conducting fuel cell electrodes is still in initial stage.

As known, typical physical phenomenon occurs during laser-ceramic interaction such as heating, surface melting, surface vaporization, plasma formation and ablation [1]. These phenomena provide various treatments on the samples. One of the desirable lasers for surface treatment is ytterbium fiber laser (Yb-fiber laser).According to M. N. Zervas [2], high power Yb-fiber laser has been drawing attention to the material processing and other industrial applications. This due to their attractive features such as higher energy conversion efficiency (> 30 %), which is more than yttrium aluminium garnet (YAG) or CO_2 lasers [3–5] and finally the most highlighted feature is high heat localization that lead to higher precision of customizing product.

Several studies have reported Yb-fiber laser as a major tool for various properties modification such as electrical, mechanical, chemical and thermal. Study by Boutinguiza et al. [6] has reported that pulsed Yb-fiber laser ablation can be used to synthesize titanium oxide (TiO_2) at controllable size nano particles. In addition, they have also reported that the laser can be employed for formation of nano particles in solutions with reduced contamination, high particle collection, ease of preparation, low costs of processing and eventually many kind of colloidal nano particles can be obtained [6].

Meanwhile, other researchers [3–5] have employed laser particularly for the modification of physical properties of ceramics and morphology modification. The most recent study is on

Content from this work may be used under the terms of the Creative Commons Attribution 3.0 licence. Any further distribution of this work must maintain attribution to the author(s) and the title of the work, journal citation and DOI.
Published under licence by IOP Publishing Ltd

modification of electrical properties which have been conducted via Yb-fiber laser [4,5]of such ceramics such as zinc oxide (ZnO) and TiO_2.Kido et al. [3] have also conducted patterns of the thermoelectric iron silicide on alumina substrates using the Yb-fiber laser. Apart from that, the same laser type also has been employed for the fabrication of ion conductive lines on glass surface, preparation of MgB_2 superconductive phase lines, crystal growth in CuO-doped lithium disilicate glasses[7] and surface densification of porous ZrC [8].

The idea to expand the usage of laser in favour of multi-layer formation is considered to be compelling. In aiming to produce protonic conducting layer which serves as electrolyte and undisturbed sub-layer as electrode could be a solution towards lowering operating temperature of solid oxide fuel cells. In near future it is expected that the weakness of anion conducting electrolyte could be resolved.

Yb-fiber laser have shown numerous promising future prospects for material processing to be further developed therefore, the current study will focus on the surface characterizations of ZrB_2-SiC ceramic substrates that had been treated via laser specifically for protonic conduction. Additionally, it may lead to greater understanding of laser treatment.

2. Materials and experimental procedures

The substrates consist of mixed raw powders which purchased from H.C Starck, USA; zirconium diboride (ZrB_2; Grade B; 97 %+) and silicon carbide (SiC; Grade UF-25-α-SiC; 98 %+). Average particle size for ZrB_2 and SiC was 1.5 μm and 0.45 μm, respectively. The average particle size was identified by laser diffraction particle size analyzer (Horiba; LA-950V2). They were used without further treatment.

2.1. Powder mixture preparation and pelletizing

With the intention of simple mixing, both powders were immersed together in ligroin also known as petroleum ether.ZrB_2 was mixed with 30 vol.% SiC. The mixtures were stirred for approximately 5 min in a 30 kHz ultrasonic bath. Then, the slurry was left overnight for solvent evaporation and put in the oven at 80 °C to remove excess moisture for the pelletizing. The mixed powders were pelletized by uniaxial cold pressing at 40 MPa using stainless steel die 10 mm in diameter. The dimension of pellets produced from 1.15g was roughly 10 mm in diameter and 4 mm in thickness.

2.2. Sintering

Ambient sintering was conducted in the Nabertherm (VHT-08-22/GR) furnace under argon flow. The set-up temperature and holding time were 1900 °C and 2.5 hr, respectively. Both heating and cooling rate was kept at10 °C/min.

2.3. Laser oxidation conditions

The surface oxidation was employed via ytterbium fiber laser (IPG, Oxford, MA, USA; model LCF 100).It was conducted under ambient environment. As shown in Figure 1, a round spiral laser pattern was scanned onto the ZrB_2-SiC ceramic substrates. The laser specifications were as tabulated in Table 1. The beam diameter was reduced to 1.25 mm with beam reducer (LINOS) and the pattern dimension was 8 mm × 8 mm in diameter. An optimum laser power was identified at 70 W.

Table 1. Typical specification of ytterbium fibre laser and treatment condition

Laser Parameters	Condition
Mode of operation	Continuous wave
Central emission wavelength	1072 ± 10 nm
Nominal output power	100 W
Typical beam quality	< 1.1
Laser beam diameter	1.25 mm
Environment	Ambient

3. Characterization techniques

The surface view and fractured morphological structure of the cross section of ZrB_2-SiC ceramic composites were observed through scanning electron microscope SEM (Philips XL30) equipped with EDS (Oxford Instruments-INCA, UK). X-ray diffraction was performed using Siemens D5000 diffractometer equipped with a Cu anticathode and a back monochromator. These characterizations were conducted for as sintered and laser-treated pellets.

4. Results and discussion

4.1. Untreated ZrB_2-SiC ceramic substrates

Figure 3 represents micrograph of sintered SiC-ZrB_2 polished surface in BSE mode after pressure less sintering at 1900 °C. As can be observed, light phase indicates ZrB_2 and grey phase as SiC. Approximately 30 % remaining porosity can be seen very well in between the grains. The pores were expected due to the usage of pressure less sintering and absence of the sintering aids other than a major contribution from fine SiC powder which acts as a second phase and also helps reduce the sintering time and temperature. The SiC grains were distributed uniformly and played a positive role in sintering as can be observed in the micrograph. The microstructure obtained is consistent with the other previous studies on ZrB_2-SiC ceramic composites[9–25].

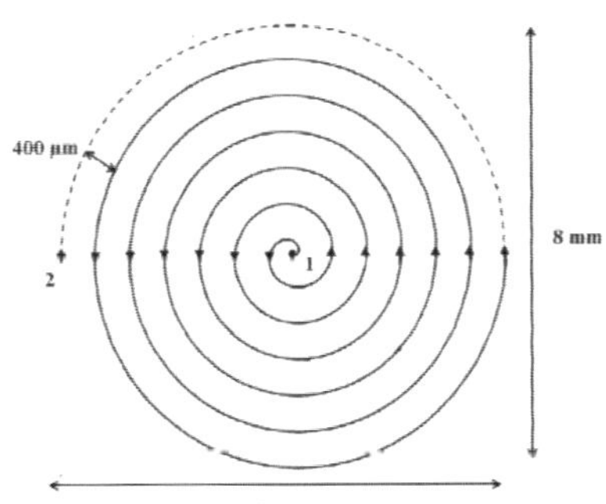

Figure 1. A round spiral laser pattern was designed for laser surface treatment[26]

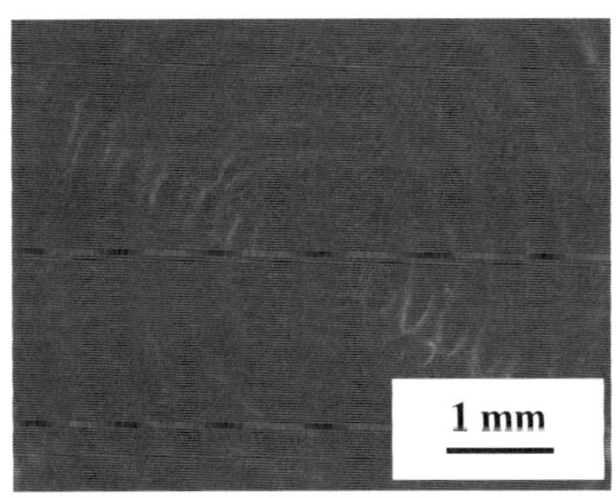

Figure 2. Micrograph of laser-treated surface of ZrB_2-SiCat 70 W in secondary electrons mode (SE)[26]

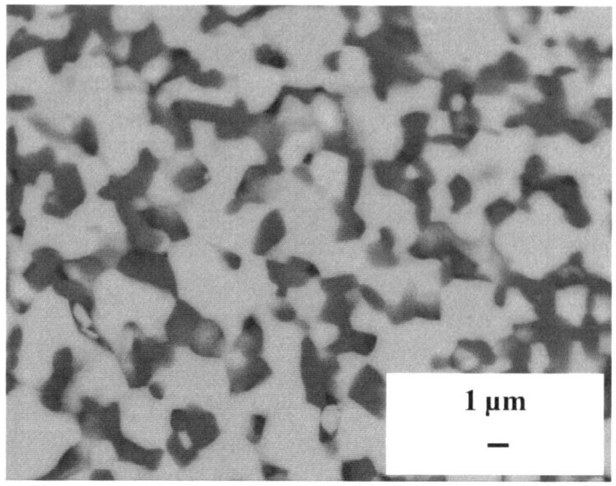

Figure 3. Micrograph of sintered ZrB_2-SiC polished surface

4.2. Laser-treated ZrB_2-SiC ceramic substrates

Micrograph of laser-treated surface of ZrB_2-SiC substrate is presented in Figure 2. The laser treatment on the surface of the substrate has resulted in surface melting. This glassy surface layer is expected to be rich in oxide. The overlapping laser track is still visible due to high heat localization from the laser as well as the spiral pattern. If observed closely a network of fine white line scan be observed which is also known as crazing. This similar structure is discussed Mahmod et al. (2015) studies [26].

The SEM study of the cross section reveals a multilayer structure. On Figure 4 it is revealed just below the-treated ZrB_2-SiC surface. Beneath the heat affected multi-layer structure the bulk of the substrate remained unaffected. Here, a multi-layer structure formation could be divided into four main layers, although these are not very clearly divided by a boundary but are diffused into each other. A descriptive micrograph in Figure 4 displays the predicted compounds that are present in every layer. In Figure 5, micrograph of higher magnification of silica rich layer is presented in order to highly focus on the surface layer fractured cross-section. Apparently, the silica thickness is approximately 8 µm. From our point of view, spherical particles can also be observed and identified as ZrO_2 particles. EDS analysis was conducted to verify the presence of each element. Figure 6 (a), (b), (c), and (d) present the composition and elemental distribution within each layer.

As seen at layer 1, EDS pattern in Figure 6(a) exhibited higher peak of Siand also Zr together with O element. These peaks give indication of silica rich layer with some presence of ZrO_2. However, C element could also be observed. In our opinion, it could be due to the decomposition of SiC. The following reaction could have possibly occurred:

$$ZrB_2 + SiC + 4O_2 \rightarrow SiO_2 + B_2O_3 + ZrO_2 + CO \qquad (1)$$

Figure 6(b) displayed EDS analysis at layer 2, sharp peaks of Zr, Si and O elements were present and the region is depleted of B and C elements. The high peak of Zr indicates a higher concentration of ZrO_2 and lower peak of Si indicates that small amount of $SiO_2$is also present. The depleted B and C elements could be assumed as a result of partial decomposition ZrB_2 and SiC. Figure 6(c) represents a very sharp peak of Zr and B. These peaks place in evidence that layer 3is rich with ZrB_2 and apparent porous SiC. Finally, Figure 6(d) represents the unaffected region just above the bulk. The depletion of O element also confirmed the effective glassy layer which is pronounced as resistant towards the oxidation.

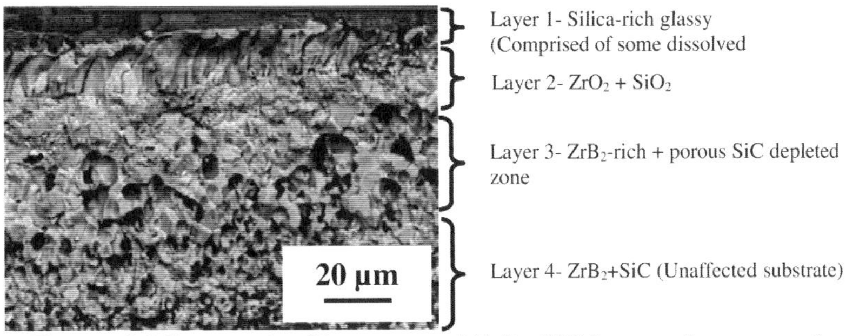

Figure 4. Micrograph of laser-treated ZrB_2-SiC fractured cross-section backscattered electrons mode (BSE)

Figure 5. Micrograph of fractured silica rich layer at higher magnification

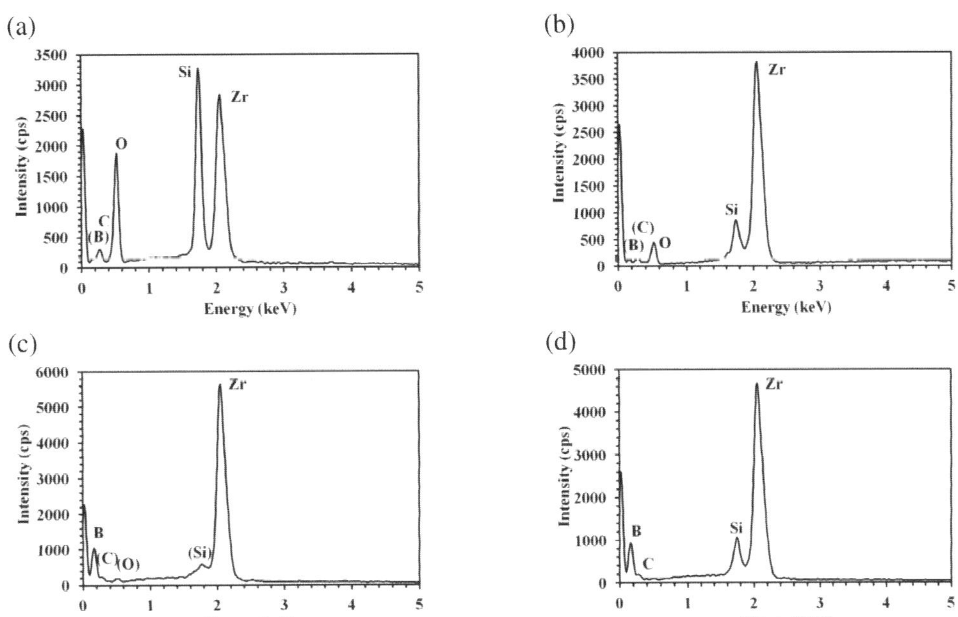

Figure 6. EDS of the multi-layer structure. (a) Layer 1. (b) Layer 2. (c) Layer 3. (d) Layer 4.[26]

The use of laser surface treatment in this particular case is conducted to cover the surface of a porous mixed boride carbide ceramic composite substrate with a glassy layer. The treatment was conducted in ambient atmosphere hence sufficient oxygen was available for oxidation of the surface layer. Such interesting results have been reported for the first time [26] whereby the laser heat is localized and the oxygen rich surface layer actually protects further oxygen penetration into the bulk of the substrate. Future investigation of the thermal and electrical conductivity of glassy layer would allow understanding and forecasting the use of such structures for protonic conductivity within solid oxide fuel cells.

5. Conclusion and outlooks

Conclusively, the present study has shown that laser surface treatment can be employed on ZrB_2-SiC ceramic composites. The localized heat treatment due to intense laser results in a multi-layer structure. The silica-rich glassy layer formation signifies that the oxidation has been highly localized. This formation protects the beneath layer. Furthermore, the round spiral laser pattern allowed a uniform heat distribution but it can be also customized according to various applications. Further study is required to be carried out on the performance of glassy layer as a protonic conductor. Other type of ceramic materials such as TiB_2 could be suggested to be employed for laser surface treatment that might offer solutions for the crazing pattern that occurred in the present study.

6. Acknowledgments

The authors would like to acknowledge the French Government through their Embassy in Kuala Lumpur and the Ministry of Education Malaysia for their support and also colleagues at the Science of Ceramic Processing and Surface Treatment (SPCTS) in Limoges, for helpful advice and fruitful discussion.

7. References

[1] Samant A N and Dahotre N B 2009 Laser machining of structural ceramics—A review *J. Eur. Ceram. Soc.*29 969–93

[2] Zervas M N 2014 High power ytterbium-doped fiber lasers — fundamentals and applications *Int. J. Mod. Phys. B*28 1442009

[3] Kido H, Takahashi M, Tani J, Sawairi Y and Iwasaki M 2014 Laser patterning of thermoelectric iron silicide on alumina substrates by continuous-wave ytterbium fi ber laser irradiation *J. Ceram. Soc. Japan*122 802–5

[4] Kido H, Takahashi M, Tani J, Abe N and Tsukamoto M 2011 Modification of electrical properties of zinc oxide by continuous wave ytterbium fiber laser irradiation *IOP Conf. Ser. Mater. Sci. Eng.*18 1–4

[5] Kido H, Takahashi M, Tani J, Chigane M, Abe N and Tsukamoto M 2014 Modification of electrical properties of TiO2 by continuous-wave ytterbium fiber laser irradiation *J. Ceram. Soc. Japan*1 21–4

[6] Boutinguiza M, del Val J, Riveiro A, Lusquiños F, Quintero F, Comesaña R and Pou J 2013 Synthesis of Titanium Oxide Nanoparticles by Ytterbium Fiber Laser Ablation *Phys. Procedia*41 787–93

[7] Honma T, Nguyen P T and Komatsu T 2008 Crystal growth behavior in CuO-doped lithium disilicate glasses by continuous-wave fiber laser irradiation *J. Ceram. Soc. Japan*116 1314–8

[8] Bacciochini A, Glandut N and Lefort P 2009 Surface densification of porous ZrC by a laser process *J. Eur. Ceram. Soc.*29 1507–11

[9] Rezaie A, Fahrenholtz W G and Hilmas G E 2007 Evolution of structure during the oxidation of zirconium diboride–silicon carbide in air up to 1500°C *J. Eur. Ceram. Soc.*27 2495–501

[10] Lonné Q, Glandut N, Labbe J-C and Lefort P 2011 Fabrication and characterization of ZrB2–SiC ceramic electrodes coated with a proton conducting, SiO2-rich glass layer *Electrochim. Acta*56 7212–9

[11] Han J, Hu P, Zhang X and Meng S 2007 Oxidation behavior of zirconium diboride–silicon carbide at 1800°C *Scr. Mater.*57 825–8

[12] Karlsdottir S N and Halloran J W 2009 Oxidation of ZrB 2 -SiC: Influence of SiC Content on Solid and Liquid Oxide Phase Formation *J. Am. Ceram. Soc.*92 481–6

[13] Li N, Hu P, Zhang X, Liu Y and Han W 2013 Effects of oxygen partial pressure and atomic oxygen on the microstructure of oxide scale of ZrB2–SiC composites at 1500°C *Corros. Sci.*73 44–53

[14] Patel M, Reddy J J, Bhanu Prasad V V and Jayaram V 2012 Strength of hot pressed ZrB2–SiC composite after exposure to high temperatures (1000–1700°C) *J. Eur. Ceram. Soc.*32 4455–67

[15] E. Wuchina, Opila E, Opeka M, Fahrenholtz W and Talmy I 2007 UHTCs HighUltra-Temperature Ceramic Materials for Extreme Environment Applications *Electrochem. Soc. Interface* 30–5

[16] Zimmermann J W, Hilmas G E, Fahrenholtz W G, Dinwiddie R B, Porter W D and Wang H 2008 Thermophysical Properties of ZrB 2 and ZrB 2 –SiC Ceramics *J. Am. Ceram. Soc.*91 1405–11

[17] Simonenko E P, Sevast'yanov D V., Simonenko N P, Sevast'yanov V G and Kuznetsov N T 2013 Promising ultra-high-temperature ceramic materials for aerospace applications *Russ. J. Inorg. Chem.*58 1669–93

[18] Karlsdottir S N and Halloran J W 2007 Rapid Oxidation Characterization of Ultra-High Temperature Ceramics *J. Am. Ceram. Soc.*90 3233–8

[19] Kim S, Chae J-M, Lee S-M, Oh Y-S, Kim H-T and Jang B-K 2014 Change in microstructures and physical properties of ZrB2–SiC ceramics hot-pressed with a variety of SiC sources *Ceram. Int.*40 3477–83

[20] Lonné Q, Glandut N and Lefort P 2012 Laser densification of porous ZrB2-SiC composites *Process. Prop. Adv. Ceram. Compos. IV Ceram. Trans. Vol. 234*234 199–209

[21] Fahrenholtz W G 2007 Thermodynamic Analysis of ZrB2-SiC Oxidation: Formation of a SiC-Depleted Region *J. Am. Ceram. Soc.*90 143–8

[22] Tian C, Gao D, Zhang Y, Xu C, Song Y and Shi X 2011 Oxidation behaviour of zirconium diboride–silicon carbide ceramic composites under low oxygen partial pressure *Corros. Sci.*53 3742–6

[23] Karlsdottir S N, Halloran J W and Grundy A N 2007 Zirconia Transport by Liquid Convection during Oxidation of Zirconium Diboride-Silicon Carbide *J. Am. Ceram. Soc.*91 272–7

[24] Hu P, Guolin W and Wang Z 2009 Oxidation mechanism and resistance of ZrB2–SiC composites *Corros. Sci.*51 2724–32

[25] Zhu S, Fahrenholtz W G and Hilmas G E 2007 Influence of silicon carbide particle size on the microstructure and mechanical properties of zirconium diboride–silicon carbide ceramics *J. Eur. Ceram. Soc.*27 2077–83

[26] Mahmod D S A, Glandut N, Khan A A and Labbe J-C 2015 Surface oxidation of porous ZrB2-SiC ceramic composites by continuous-wave ytterbium fibre laser *Appl. Surf. Sci.*357 1982–90

IOP Conf. Series: Materials Science and Engineering **146** (2016) 012003 doi:10.1088/1757-899X/146/1/012003

Study on structure, electrical and dielectric properties of $Eu_{0.65}Sr_{0.35}Fe_{0.3}Mn_{0.7}O_3$

I. A. Abdel-Latif
Physics Department, College of Science & Arts, Najran University,
P. O. 1988 Najran, Kingdom of Saudi Arabia
Reactor Physics Department, NRC, Atomic Energy Authority,
Abou Zabaal P. O. 13759, Cairo, Egypt

E-mail: ihab_abdellatif@yahoo.co.uk

Abstract. Polycrystalline $Eu_{0.65}Sr_{0.35}Fe_{0.3}Mn_{0.7}O_3$ was synthesized using a standard solid state reaction technique. The preliminary microstructure of this compound at room temperature was studied using a field emission scanning electron microscope (FESEM), and the crystal structure was investigated using X-ray diffraction (XRD). The considered compound was found to crystallize in a single-phase orthorhombic structure in the P_{bnm} space group (62). The Raman spectrum shows three Raman-active vibrational modes at approximately 208, 476 and 616cm^{-1}. The temperature variation of the resistivity shows that these compounds have semiconductor behaviour with an activation energy 0.498 eV. The frequency dependence of the dielectric constant in these materials indicates that space charge polarization contributes significantly to their observed dielectric parameters.

1. Introduction

Over the last few decades, the ABO_3perovskite structure has attracted scientific interest due to its remarkable properties. The technologically useful properties exhibited by these perovskite-type oxide materials are electrical, magnetic, dielectric, and optical. Among these materials, the doped rare-earth manganites $R_yA_{1-y}MnO_3$ have been extensively investigated by several researchers [1-4] because of their scientific significance and potential applications [5]. In particular, these systems exhibit the novel phenomena of colossal magnetoresistance (CMR), metal-insulation (MI) and charge ordering (CO), in manganites [6-8]. The underlying physics of those phenomena has been interpreted in terms of the double exchange (DE) model, in which a strong exchange interaction occurs between Mn^{3+} and Mn^{4+} ions through intervening filled oxygen 2p states [9], the Jahn-Teller (JT) polaron, and a tolerance factor [10]. Knowledge of the crystal structure and the chemical bonding of these compounds is of a great importance to the understanding of the peculiar electro transport and magneto transport properties in these perovskites. It is known that the *A*-site substitution primarily changes the carrier density and strongly affects the Mn^{3+}-O-Mn^{4+} angle (lattice distortion), thus transforming the parent compound *R*MnO from an insulating antiferromagnet into a metallic ferromagnet [11]. Alternatively, the *B*-site substitution directly modifies the crucial Mn^{3+}-O-Mn^{4+} network. The important correlation between the crystal structure of the materials and their transport properties are reported in [12-14], where the tilt in the octahedra plays an important role in defining the magnetic and electric exchange interaction between transition metal element e_g orbital and oxygen *2p* orbital. The composition and microstructure of ABO_3 may be modified by doping in the A site or the B site or by simultaneous

Content from this work may be used under the terms of the Creative Commons Attribution 3.0 licence. Any further distribution of this work must maintain attribution to the author(s) and the title of the work, journal citation and DOI.
Published under licence by IOP Publishing Ltd

doping to tailor the properties to specific applications [15-20]. To describe the ion match between A and B ions in this type of compound, a geometrical quantity, the tolerance factor t, is usually introduced; t is defined as [17]:

$$t = (\langle r_A\rangle + \langle r_O\rangle)/\sqrt{2}(\langle r_B\rangle + \langle r_O\rangle)$$

where $\langle r_A\rangle$ is the average ionic radius of the A-site ions (Eu^{3+} and Sr^{2+}) and $\langle r_B\rangle$ is the average ionic radius of the B-site ions (Mn and Fe) in our studied samples. In this case, the average ionic radius of the A-site is fixed. Thus, the lattice distortion observed is caused by the partial replacement of Mn by Fe. Therefore, the structural studies of manganites doped by Fe with systematic compositional variation can be advantageous for gaining important insight into the structure–property relationships for these compounds.$Eu_{0.65}Sr_{0.35}Fe_xMn_{1-x}O_3$ (x=0.1 and x=0.5 compounds were studied in ref. [21], so that it is very important to complete the series and see what the difference and the effect of iron substitution. The valance of manganese in $Eu_{0.65}Sr_{0.35}Mn_{0.9}Fe_{01}O_3$ is ($0.35Mn^{+3} \Leftrightarrow 0.35Mn^{+4} + 0.2Mn^{+3}$) where the double exchange phenomenon is complete. The valence of manganese is ($0.35Mn^{+3} \Leftrightarrow 0.15Mn^{+4}$) in $Eu_{0.65}Sr_{0.35}Mn_{0.5}Fe_{0.5}O_3$, whereas, double exchange is not complete. As it is clear that the valence of manganese in $Eu_{0.65}Sr_{0.35}Fe_{0.3}Mn_{0.7}O_3$ is the critical case.
The purpose of the present work is to study the structural, electrical and dielectric properties of iron-doped europium ferrimanganites with a critical concentration of iron of x=0.3.

2. Experimental details

The sample synthesis and preparation are performed using a conventional solid-state reaction. The proper stiochiometric amounts of pure oxides were carefully mixed, and then the mixed powder was pressed at 2 tons. The pellet was fired at 1100°C for 12 h in air. The pre-sintered sample was ground again into a fine powder and then pressed and fired at 1200°C for 12 h. After repeating the grinding and pressing of the powder, it was finally fired at 1350°C for 72 h, followed by natural furnace cooling. The phase purity and crystallographic information of the synthesized samples were tested using powder X- ray diffraction (XRD). The XRD analysis was performed using a PANalyticalX' PertPRO diffractometer at room temperature with Cu-K_α radiation (λ=1.5406 Å). Micro-structural observations and chemical analysis were performed on the samples using a field emission scanning electron microscope (FESEM) (JEOL JSM-7600F) combined with energy dispersive spectroscopy (EDS). The Raman spectra were obtained using a (Perkin Elmer) Raman spectrometer in the 3000-60 cm^{-1} region at 4 cm^{-1} resolution in a continuously purged sample compartment. The DC resistivity–temperature dependence measurements were performed using the four-point probe technique (Scientific Equipment & Services) in the temperature range from room temperature to 573 K. The dielectric characteristics at room temperature were determined using an inductance–capacitance–resistance (LCR) meter (HP4284A).

3. Results and discussions

The surface morphology and microstructure analysis of the samples have been investigated using FESEM and EDS analysis. Figure 1 shows the morphology and the crystalline size of the synthesized sample. Upon examining the morphology of the $Eu_{0.65}Sr_{0.35}Mn_{1-x}Fe_xO_3$ samples (X=0.1, 0.3 and 0.5), we find that there is a fairly uniform grain size distribution throughout the surface that becomes non-uniform as the iron concentration increases. Our morphological investigation reveals the absence of high-density cracks. This result is an indication that we have a good-quality material with a minimum number of defects. These micrographs show that doping with Fe^{3+} strongly reduces the grain size, which decreases from 2.5-3.5 μm for x = 0.1 to 0.5-1μm for x = 0.5, i.e., substitution of Fe for Mn leads to a reduction in the grain size. Accordingly [22], the specimens become magnetically harder

IOP Conf. Series: Materials Science and Engineering **146** (2016) 012003 doi:10.1088/1757-899X/146/1/012003

(a)

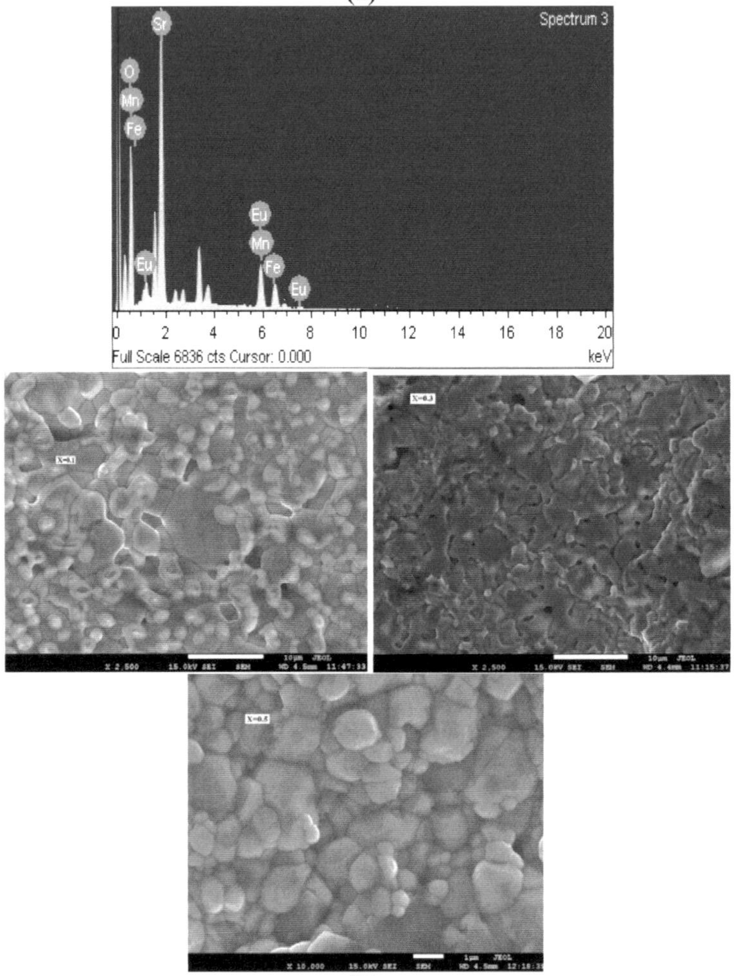

(b)

Figure 1. (a) EDS spectrum of $Eu_{0.65}Sr_{0.35}Mn_{0.7}Fe_{0.3}O_3$ and (b) SEM micrograph of $Eu_{0.65}Sr_{0.35}Mn_{1-x}Fe_xO_3$ samples (X=0.1, 0.3 and 0.5)

when reducing the particle size and exhibit a large magnetic anisotropy upon reducing the grain size. Micro-qualitative analysis using EDS was performed for our sample; the results indicate that all of the elements of our compound are in the proper ratio. From the EDS result of $Eu_{0.65}Sr_{0.35}Mn_{1-x}Fe_xO_3$, no impurities were found in the content of the proposed compounds, indicating the quality of the synthesis. The experimentally observed percentages of elements (which constitute the proposed composites) are clearly in good agreement with those predicted theoretically.

X-ray powder diffraction investigations at room temperature indicated that the samples were single phase with no detectable secondary phases, see Figure 2. XRD data for the system $Eu_{0.65}Sr_{0.35}Mn_{0.7}Fe_{0.3}O_3$ are given in Table 1. All of the reflection peaks in the measured XRD pattern were indexed according to the orthorhombic symmetry of the space group *Pbnm* (No. 62). However, a minor shift is observed in the peak positions, indicating a small change in the lattice parameters. The orthorhombic unit cell volume increases slightly due to differences in ionic radii of Mn^{3+} and Fe^{3+} ions [23]. Due to the increase in the iron concentration, both the full-width at half

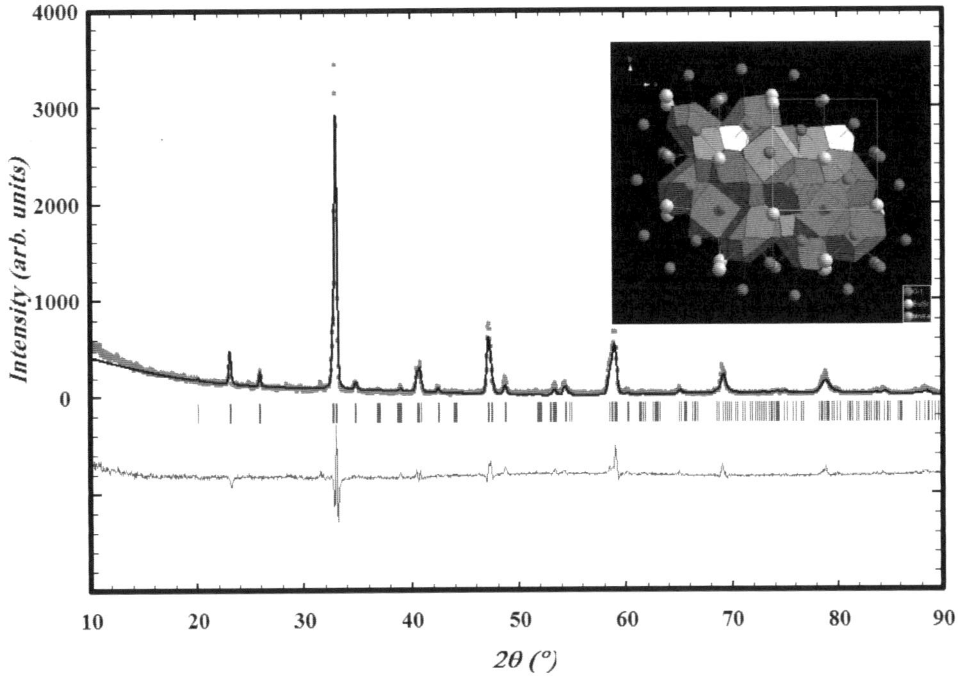

Figure 2. Refined XRD pattern of $Eu_{0.65}Sr_{0.35}Mn_{0.7}Fe_{0.3}O_3$.

maximum (FWHM) (broadness) and the intensities of the peaks increased. The broadness of the peaks indicates either particles of very small crystalline size or particles that are semicrystalline in nature [24]. The maximum reflection intensity corresponds to the 112 plane; see Figure 2. The lattice parameters resulting from fitting of the experimental XRD patterns using the Fullprof program [25] are listed in Table 1. The simple formula for the [b] and [c] tilt are shown in ref. [14]. The first -order approximation of the [b] and [c] tilt are given as following; [b]tilt~(180-α)/2 and [c]tilt~(180-β)/2where α and βareMn-O_1-Mn andMn-O_2-Mn angles, respectively. From the results of XRD refinement, we found that the value of Mn-O_1-Mn and Mn-O_2-Mn angles equal 168.40° and 153.87°, respectively. So the resulting tilt angles were calculated for $Eu_{0.65}Sr_{0.35}Fe_{0.3}Mn_{0.7}O_3$ to be; [b]tilt~5.8° and [c]tilt~13.1°. Comparing with the tilt angles in $Sm_{0.6}Sr_{0.4}MnO_3$ our sample has more distortion than it. looking at the lattice parameters of $Eu_{0.65}Sr_{0.35}Fe_xMn_{1-x}O_3$ (where x= 0.1, 0.3 and 0.5) previously reported for the same concentration[26], it is clear that no change occurred in the crystal structure of the present system over time, indicating the stability of this composition. The XRD results are consistent with the FESEM results.

The scattering density inside the unit cell of a crystal, regardless of the symmetry, is calculated using Fourier as a subprogram in Full Prof. A *F*ast *F*ourier *T*ransform (*FFT*) is applied as a subroutine to accelerate the calculation of the following expression:

$$\rho(\mathbf{r}) = \frac{1}{V}\sum_{\mathbf{H}} F(\mathbf{H})\exp\left\{-2\pi i(\mathbf{H}\cdot\mathbf{r})\right\}$$

Where *V* is the volume of the unit cell. *H* is the reciprocal lattice vector, *r* is the vector position inside the unit cell, and *F*(*H*) are complex Fourier coefficients used to perform different types of Fourier

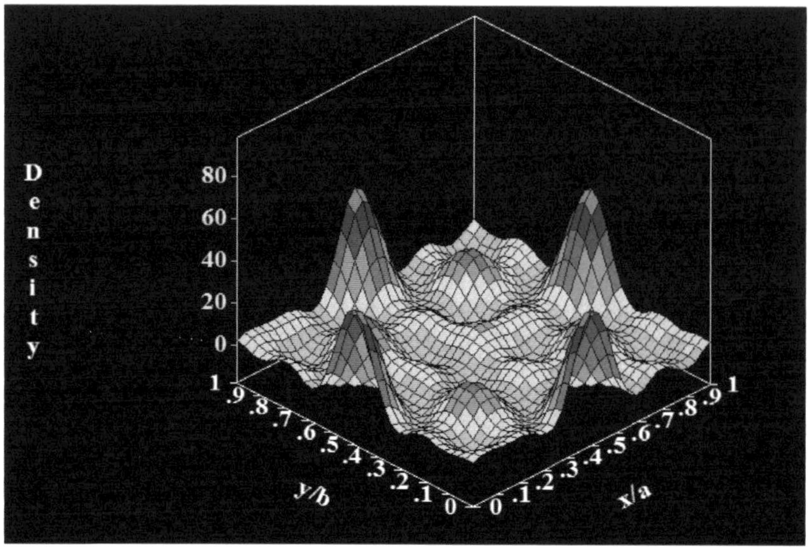

Figure3. Electron density of $Eu_{0.65}Sr_{0.35}Mn_{0.7}Fe_{0.3}O_3$ inside the unit cell

syntheses. The units of ρ (*r*) are those of *F*(*H*) divided by those of *V*. The density of electrons inside the unit cell is calculated as shown in Figure 3.

Raman spectroscopy is a powerful technique for studying and characterizing the lattice-vibrational properties. Raman spectroscopy is not only used to probe the spin–phonon interaction but also accounts for the electron–phonon interaction and is effective in manganites. Thus, the symmetry information regarding phonon, electron, and spin excitations can be simultaneously explored using this technique. The optical mode frequencies in the vibrational spectra of manganites are shown in the literature, either by theoretical or experimental studies, and are interpreted on the basis of lattice distortions.

Table 1. Cell parameters and atomic parameters

a(Å)	*b(Å)*	*c(Å)*	V (Å^3)	*a*/b	b/c=	c/*a*
5.4183(5)	5.4725(5)	7.6630(7)	227.22(4)	0.9901	0.7141	1.4143
Atom	**Ox.**	**S.O.F.**	**x/a**	**y/b**	**z/c**	**U [Å^2]**
O2	-2	1	-0.28229	0.27634	0.05029	0.0013
O1	-2	1	0.02896	0.47896	0.25000	0.0132
Sr	+2	0.35	-0.00481	0.0372(5)	0.25000	0.0086
Eu	+3	0.65	-0.00481	0.0372(5)	0.25000	0.0086
Mn	+3	0.7	0.50000	0.00000	0.00000	0.0011
Fe	+3	0.3	0.50000	0.00000	0.00000	0.0011

It is well known that the Raman-active mode of lattice vibrations is absent in the ideal cubic perovskite, and the Raman-active modes of the orthorhombic structure are as follows: A_g+B_{1g} symmetric and $2B_{2g}+2B_{3g}$antisymmetric stretching modes; $A_g+2B_{1g}+B_{3g}$ bending modes and $2A_g+2B_{2g}+B_{1g}+B_{3g}$ rotation and tilt modes of the octahedral structure; and $3A_g+B_{2g}+3B_{1g}+B_{3g}$ modes related to the rare earth movement. Therefore, the vibrational bands reported in [26-31] for various families of the manganite compounds were taken as a reference for the discussion of the present observed spectrum. Thus, we can use Raman spectroscopy to study the lattice distortions in $Eu_{0.65}Sr_{0.35}Fe_{0.3}Mn_{0.7}O_3$ (Figure 4) and compare the data with the data induced by the doping effect. Themain features observed in the spectrum at room temperature are the three bands centered at approximately 210, 490 and 610 cm^{-1}. The assignment of these modes is based on the lattice dynamic calculation [28] and is similar to the results of Choi et al. [31]. In our case, the spectrum at room temperature three bands centered at approximately 208, 476 and 616 cm^{-1}. The 208 cm^{-1} (low-frequency mode) band is attributed to the A_{1g} mode, which is the in-phase rotation of the oxygen cage about the y-axis to the adjacent MnO_6octahedra. The mode near 476 cm^{-1} corresponds to the out-of-phase bending (asymmetric stretching) mode ($B_{2g}(3)$), which is associated with the JT distortion and is

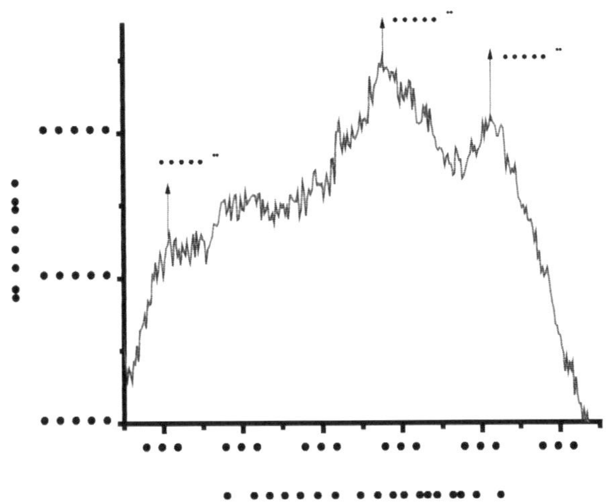

Figure 4. Raman spectrum of $Eu_{0.65}Sr_{0.35}Mn_{0.7}Fe_{0.3}O_3$.

sensitive to a change in the Mn-O-Mn bond angle. The mode observed at 616 cm^{-1} is due to symmetric stretching of the basal oxygen ion (B_{2g} (1)). By comparing these results with those published by I A Abdel-Latif and S A Saleh [21], it is clear that no major change in the shape and position of the characteristic vibration bands with the increase in Fe content has been observed. With an increase in the iron content in $Eu_{0.65}Sr_{0.35}Mn_{0.5}Fe_{0.5}O_3$, the A_{1g} mode is slightly shifted toward higher frequency (216 cm^{-1}). The bending mode ($B_{2g}(3)$) is observed to be shifted to lower frequency (476 cm^{-1}). Because the $B_{2g}(3)$ modes are sensitive to the change in the Mn-O-Mn bond angle, the variation of this mode directly reflects the change of the bond angle. The peak for the stretching mode (B_{2g} (1)) gradually becomes stronger and moves slightly to higher frequency as a result of the increasing in Fe^{3+} concentration. This stretching mode is sensitive to a change in the bond length of the Mn-O bond. From the lattice dynamic calculations, the most intense modes, that is, those at 210 - 216, 488 - 491, and 610 - 632 cm^{-1}, are associated with rotational-, bending-, and stretching-like vibrations of the $Mn(Fe)O_6$octahedra, respectively. The peak that corresponds to the $B_{2g}(1)$ mode is correlated with the JT distortion in the compounds with large ionic radii [33]. The distortion of $Eu_{0.65}Sr_{0.35}Fe_xMn_{1-x}O_3$ increases as the concentration of iron increases and may lead to an increase in the frequency; this phenomenon is the same as that reported in [26].

The electrical properties of any material are an important factor that reveal important and reliable information about the transport phenomenon and other physical properties of the material. To determine the general behavior of the electrical resistivity and to gather information about the current transport mechanisms, the resistivity data were analyzed to distinguish between the possible mechanisms in the tested samples, namely, band transfer, thermionic emission, thermally assisted tunneling and variable range hopping [34]. The variation in the DC electrical resistivity of

IOP Conf. Series: Materials Science and Engineering **146** (2016) 012003 doi:10.1088/1757-899X/146/1/012003

$Eu_{0.65}Sr_{0.35}Mn_{0.7}Fe_{0.3}O_3$ as a function of temperature is shown in Figure 5. It is clear that the temperature dependence of the resistivity curve shows that the semiconductor behavior of the resistivity of $Eu_{0.65}Sr_{0.35}Fe_xMn_{1-x}O_3$ decreases with increasing temperature. Attempts were made to fit the semiconducting behaviors to better understand the conduction of the electrons between localized states. The fit of the resistivity curves could clarify the type of conduction, discriminating between pure thermally activated conductivity and variable range hopping (VRH) between localized electronic

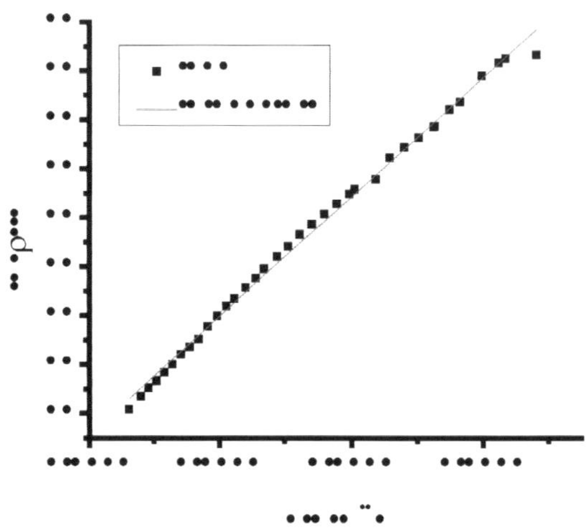

Figure 5. DC resistivity – temperature dependence of $Eu_{0.65}Sr_{0.35}Mn_{0.7}Fe_{0.3}O_3$

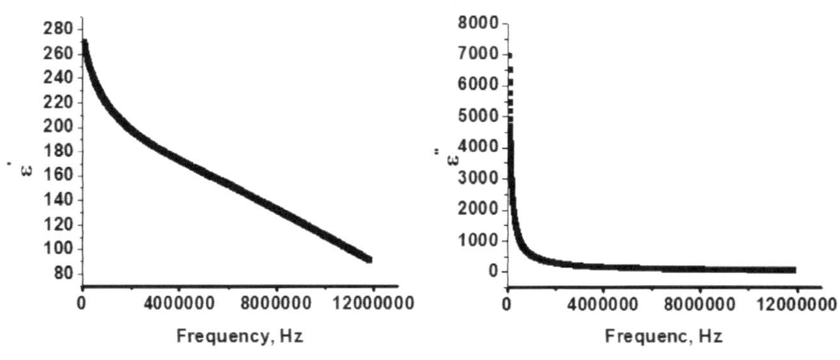

Figure 6. Dielectric constant of $Eu_{0.65}Sr_{0.35}Mn_{0.7}Fe_{0.3}O_3$

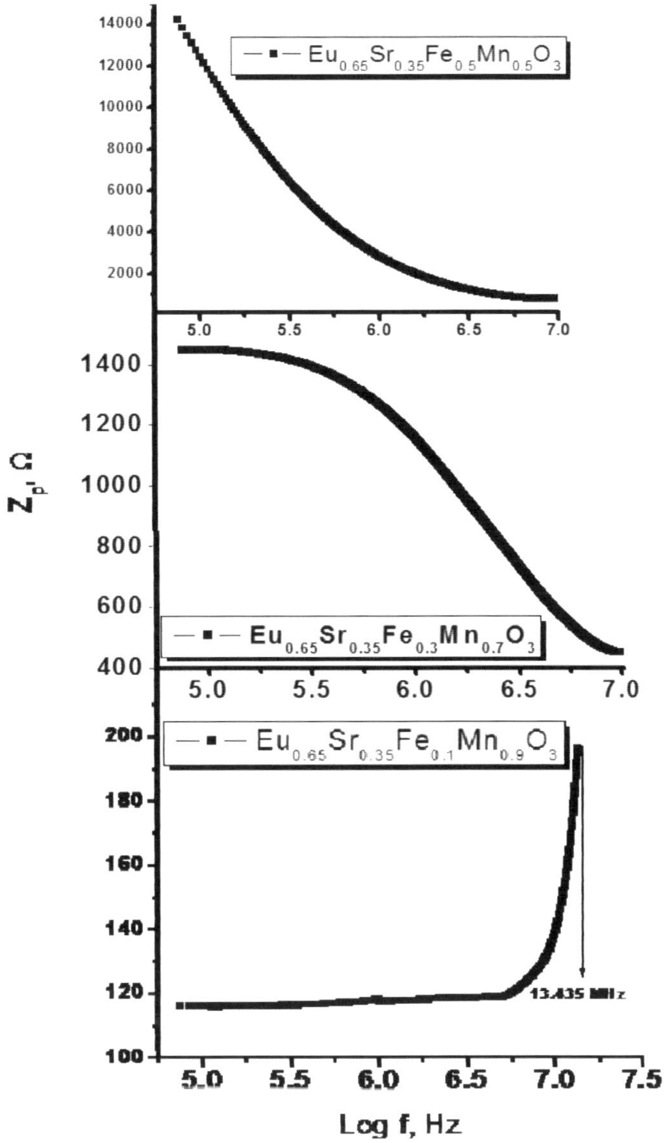

Figure 7. Resistivity - Frequency dependence plots of $Eu_{0.65}Sr_{0.35}Mn_{0.7}Fe_{0.3}O_3$

states. For thermally activated conductivity, the electrical resistivity can be written as a function of temperature, T, as follows: $\rho = \rho_o \exp(-E_\rho / K_B T)$, where ρ_o is a pre-exponential coefficient, E_A is the activation energy, k is the Boltzmann constant and T is the absolute temperature [34]. According to this formula, one can calculate the activation energy, which has a value of 0.498 eV.

From the SEM, XRD and Raman data, one can correlate the increase in the bond length and the activation energy of $Eu_{0.65}Sr_{0.35}Fe_xMn_{1-x}O_3$ with the increasing iron content, which is similar to Fe substitution at the Mn site in $Nd_{0.65}Sr_{0.35}Fe_xMn_{1-x}O_3$ [36] and $SmFexMn_{1-x}O_3$ [37]. Because Fe doping is the direct replacement of Mn^{3+} by Fe^{3+}, the sites that are now occupied by Fe^{3+} can no longer effectively participate in the DE process because only the Mn $e_{g\uparrow}$ band is electronically active, within which electron hopping can occur between Mn^{3+} and Mn^{4+}. Thus, DE is suppressed, resulting in an increase in the resistivity [38].

Using the measured capacitance values and the dimensions of the pellet (real part) of the materials, one can calculateδ $\varepsilon' = \frac{Cd}{\varepsilon_o A}$

where C is the capacitance of the pellet in farad, d is the thickness of the pellet in meters, $\delta\varepsilon_o = 8.85x10^{-12} Fm^{-1}$, is the permittivity of free space and A is the cross-sectional area of the cylindrical pellet in m^2. The dielectric constant (real part) in the investigated system is contributed by several structural and microstructural factors. The imaginary part of the dielectric constant ε'' was calculated using the relation $\varepsilon'' = \acute{\varepsilon}\tan\delta$, where $\tan\delta = \tan(90 - \phi)$ is the dispersion of the dielectric loss tangent. Structural and various electrical properties of polycrystalline $Ca_{0.5+}xNd_{0.5-}x(Ti_{0.5}Fe_{0.5})O_3$ is studied in ref [39] and it is found that that the composition dependence of lattice parameters, density and microstructural vary significantly with Ca content. The dielectric frequency dependent measurements in [40] showed that the dielectric constant (ε') increases with increasing Ca content and similarly the dielectric loss (tan δ) and ac conductivity (σ_{ac}). The high ε' observed in [41] samples are similar to our case and these results are suitable for fabrication of devices. The plots of the dielectric constant ρ' with frequency at room temperature for the system $Eu_{0.65}Sr_{0.35}Mn_{0.7}Fe_{0.3}O_3$ are shown in Figure 7. At lower frequencies, the compound shows different types of polarization (i.e., interfacial, dipolar, atomic, ionic and electronic). A decrease in $\acute{\varepsilon}$ was observed with an increase in frequency; this variation can be explained on the basis of space charge polarization. At higher iron concentrations, e.g., $Eu_{0.65}Sr_{0.35}Mn_{0.5}Fe_{0.5}O_3$, the dielectric constant of the sample depends strongly on frequency, as explained in ref. [20]. The high value of $\acute{\varepsilon}$ observed in the $Eu_{0.65}Sr_{0.35}Fe_xMn_{1-x}O_3$ system shows the ferroelectric nature of this system. Voids, dislocations and other defects appear at low frequencies [40-41]. The decrease in the penetration depth of the electromagnetic waves at a high dielectric constant is due to the increase in the skin effect. Figure 7 depicts the frequency variation of loss tangent (tan δ) of $Eu_{0.65}Sr_{0.35}Mn_{0.5}Fe_{0.5}O_3$ at room temperature. Note that at relatively lower frequencies, the values of tan δ are high and decrease rapidly with increasing frequency. The decrease in tan δ with increasing frequency is attributed to the fact that the hopping frequency of the charge carrier cannot follow the changes of polarity of the external field beyond a certain frequency [41]. Both $\acute{\varepsilon}$ and saturated polarizations depend on the grain size of the materials. However, the SEM micrograph shows a granular structure with a dense morphology and little porosity, which offers an abundance of insulating grain boundaries, leading to an increase in the dielectric constant with decreasing grain size. The valance of manganese in $Eu_{0.65}Sr_{0.35}Mn_{0.9}Fe_{0.1}O_3$ is ($0.55Mn^{+3}$ + 0. $35Mn^{+4}$) and the valence of manganese in $Eu_{0.65}Sr_{0.35}Mn_{0.7}Fe_{0.3}O_3$ is ($0.35Mn^{+3}$+0. $35Mn^{+4}$), where the double exchange phenomenon is complete, whereas in $Eu_{0.65}Sr_{0.35}Mn_{0.5}Fe_{0.5}O_3$, double exchange is not complete, as indicated by ($0.35Mn^{+3}$+0. $15Mn^{+4}$); thus, we found that the resistivity is much higher for concentrations of iron greater than 0.3; see Figure 7.

4. Conclusions

The conduction mechanism in $Eu_{0.65}Sr_{0.35}Mn_{0.3}Fe_{0.7}O_3$ is thermally activated, where the valance of manganese is considered as half trivalent and tetravalent; ($0.35Mn^{3+}$— O — 0. $35Mn^{4+}$). This makes the double exchange phenomenon is complete, whereas in $Eu_{0.65}Sr_{0.35}Mn_{0.5}Fe_{0.5}O_3$ [33], double exchange is not complete, as indicated by ($0.35Mn^{3+}$— O — 0. $15Mn^{4+}$); thus, we found that the resistivity is much higher for concentrations of iron greater than 0.3. The frequency dependence of the dielectric constant of $Eu_{0.65}Sr_{0.35}Mn_{0.3}Fe_{0.7}O_3$ indicates that space charge polarization contributes significantly to their observed dielectric parameters.

The first -order approximation of the [b] and [c] tilt are calculated to be ; [b]tilt~5.8° and [c]tilt~13.1° that is mean that we have more distorted structure than other compounds.

5. Acknowledgement
The author is thankful to the Deanship of Scientific Research for Grant Research code NU/ESCI/14/019 to Dr. I. A. Abdel-Latif, Najran University, Najran, Saudi Arabia. Advanced Materials and Nano Research Center in Najran University is gratefully acknowledged for providing access to the instruments at their facility.

6. References
[1] Coey JMD, Viret M, von Molnar S1999 *Adv. Phys.*48 167
[2] Von Helmolt R, Wecker J, Holzapfel B, Schultz L, Samwer K 1993*Phys. Rev. Lett.*71 2331
[3] Mandal K D and Behera L 2008 *J. Alloys & Comp.* 448 313
[4] Ahmed AM, Saleh S A, Ibrahim EMM, Mohamed H F 2006*Journal of Magnetism and Magnetic Materials*301 452
[5] Jin S, Tiefel T H, Cormack M Mc, Fastnacht P A, Ramesh R, and Chen L H 1994*Science*264 413
[6] Zemni S, DhahriJa, Cherif K, Dhahri J, Oumezzine M, GhediraMand Vincent H 2005*J. Alloys & Compd.*392 55
[7] Hwang H Y, Cheong S-W and Ballogg B 1996 *Appl. Phys. Lett.*68 3494
[8] Shankar K S, Kar S, Raychaudhuri .K andSubbanna G N 2004 *Appl. Phys. Lett.* 84 (6), 993
[9] Wang L M, et al.,2007 *J. Appl. Phys.*102 023915
[10] Millis A J, *Nature* 1998 392 147.
[11] Urushibara A, Moritomo Y, Arima T, Kito G, and Tokura Y 1995*Phys. Rev.* B51 14103
[12] Bashkirov Sh Sh, Parfenov V V, Valiullin A A, Khramov A S, Trounov V A, Smirnov A P, Abdel-Latif I A2003 *Izv. RAS, Physical Series*67 No 7, , p 1072, (in Russian)
[13] Aguadero A, Perez-Coll D, de la Calle C, Alonso J A, Escudero M J, Daza L 2009 *J. Power Sources*192 132
[14] Abdel-Latif I A, HassenA,Zybill C, Abdel-Hafiez M , Allam S, El-SherbiniTh 2008*J. Alloys & Comp*452 245
[15] Bhide VG, Rajoria DS, Rama Rao G, Rao CN 1972 *Phys. Rev.* B6 1021
[16] Raccah P, Goodenough J B 1967*Phys. Rev.*155 932
[17] Ganguli P, Rao C N R 1973*Mate. Res. Bull.*8 405
[18] Zheng G H, Sun Y P, Zhu X B, Song W H2006 *Solid Commun.* 137 326
[19] Ahmed M A, ImamN G, AbdelmaksoudM K, SaeidYA2015*Journal of Rare Earths,*33 965–971
[20] Abdel-Latif I A, Ismail A A, Bouzid H, Al-Hajry A2015 *Journal of Magnetism and Magnetic Materials* 393 233
[21] Abdel-Latif I A, Saleh SA2012*Journal of Alloys and Compounds*530116
[22] Martínez B, Balcells L, Fontcuberta J, CohencaC H and Jardim R F 1998*J. Appl. Phys,*83 7058
[23] Uusi-Esko K, Malm J, Imamura N, Yamauchi H, Karppinen M 2008*Mater. Chem. and Phys.*112 1029
[24] Shannon R D and Prewitt C T 1969*ActaCrystallogr.* B25 925
[25] Channu VSR, Holze R, Wicker Sr S A, Walker Jr E H, Williams QL, Kalluru R R 2011*Mater. Sci. Appl.*2 1175
[26] Rodriguez-Carvajal J, 1993 *Physica* B192 55
[27] Ahmed Farag I S, MostafaA M and Abdel-Latif I A 2007*Egypt. J. Solids*, 30 149
[28] Iliev M N, et al., (1998) *Phys. Rev.* B57 2872

[29] Yoon S, Liu H L, Schollerer G, Cooper S L, Han P D, Payne D A, Cheong S-W, and Fisk Z1998 *Phys. Rev.* B58 2795

[30] Liarokapis E, Leventouri Th, Lampakis D, Palles D, Neumeier J J, and Goodwin D H1999 *Phys. Rev.* B 60 12758

[31] Granado E Sanjurjo J A, Rettori C, Neumeier J J,Oseroff S B2000 *Phys. Rev.* B62 11304

[32] Choi K-Y et al.,2003 *J. Phys. : Condens. Matter*15 3333

[33] Martin-Carron L de Andrés A 2001 *J. Alloys & Comp.*323-324 417

[34] Abdullah H Halim S A and Jannah AN2012*Journal of Composite Materials,* 463103-3112

[35] Abdel-Latif I A2011 *AIP Conf. Proc.*1370 108

[36] Abdel-Latif I A et al 2006 *Egypt. J. Solids*, 29 341

[37] Bashkirov Sh Sh Parfenov VV Abdel-Latif IAZaripova LD 2005 *J. Alloy. & Compd.*,387 70–73

[38] Li Y., et al, 2007 *J. Alloys & Comp.* 441 1

[39] Hussain S., et al, 2006*J. Crys. Grow.*297 403

[40] Nikola I. Ilić, Adis S. Džunuzović, Jelena D. Bobić, Bojan S. Stojadinović, Peter Hammer, Mirjana M. VijatovićPetrović, Zorana D. Dohčević-Mitrović, Biljana D. Stojanović, 2015 *Ceramics International*41 69

[41] Shah M R, Akther Hossain A KM 2013 *Journal of Materials Science &Technolog,* 323–329

Optical design of a rotating eyepiece telescope

M. Siddique, F. Nasim, A. N. Khan and A. Gul
Institute of Industrial Control Systems, P. O. Box 1398, Rawalpindi, Pakistan

E-mail: drsiddikbit@gmail.com

Abstract. Flexible eyepiece telescope has been designed and verified. The rotating eyepiece of telescope will facilitate viewing of objects in a remote or out of sight target. Eyepiece arm of telescope can be rotated upto 360° keeping objective and reticule unchanged and ensuring zero deviation in reticule inclination. Main application of this scope is off axis viewing of objects. Image inversion has been carried out by using pair of mirrors and length of telescope is controlled by using relay lenses. The optical design, simulation and image analysis has been carried out by using ZEMAX®. Magnification of telescope is between 10~12 times with FOV of 6^0. Experiment has been carried out using uncoated Edmund Optics and optical tool box of Micro Series Kit, NEWPORT.

1. Introduction

A periscope is an optical instrument that facilitates viewing of objects which are beyond line of sight [1]. It is an optical instrument that facilitates visual imagery through bent geometries and can be described as a telescope having a bent optical path and a limited field of view [2]. The requirement of brightness, proper image orientation, field of view and maneuverability has led the periscope optics to consist of several image forming, image rotating and image scanning elements.

Periscope is an important tool of submarines. It is a necessity in searching and identifying the targets before attack. Due to the powerful and vital threat, submarine technologies such as periscope design, sonar sensors, engines, communications, and weaponry are confidential by designers. A complete periscope design includes the head piece, the relay optics in the mast, and the observation eyepiece.

An endoscope is a miniature periscope used to examine the inside of a cavity through a small orifice; they are widely used in medical applications. The size of the optics in a medical endoscope is on the order of 2 or 3mm in diameter. In an endoscope or periscope, the number of relay stages is determined by the length of the instrument. The relay lenses are typically made as cemented doublets, with the flint (negative) element made thick enough to fill the space. The outer surface of the flint is made convex so that it functions as the field lens. This is often referred to as a rod-lens endoscope. The reduction in the number of relay components both reduces the cost of the endoscope and improves the image quality (especially by reducing the secondary spectrum and the Petzval field curvature) [3].

Design presented can also be utilized in these applications when realized at miniature scale. Further CCD cameras can also be used directly to view image on mounts.

Content from this work may be used under the terms of the Creative Commons Attribution 3.0 licence. Any further distribution of this work must maintain attribution to the author(s) and the title of the work, journal citation and DOI.
Published under licence by IOP Publishing Ltd

2. Design parameters

The eyepiece of rotating eyepiece telescope is rotated up to 360° keeping objective and reticule unchanged and ensuring zero deviation in reticule inclination. The design parameters are listed below:

I.	Total length of the telescope :	470mm
II.	Magnification:	12^x
III.	Field of view :	6.1°
IV.	Viewing object scan:	360°
V.	Exit pupil diameter:	5mm
VI.	Exit pupil position:	25mm
VII.	Objective diameter:	65mm

3. Optical design

The basic design concept is to image the object field by an objective lens and then to relay this image to the eye-piece at the required distance by a set of relay lenses. The object field is scanned by rotating the entire rotating eyepiece about the optical axis. The overall design ensures that the image is finally seen in a proper perspective and no lateral inversion is introduced.

Taking 1arc-min as the angular resolution of the human eye [4], the minimum magnification needed is 6^x. The clear aperture of the objective is 65mm limiting the maximum useful magnification to 12^x. Our system is designed with a maximum magnification of 12^x.

The first order Gauss optics parameters can be determined from the design specification of FOV and angular magnification of the telescope system as shown in Figure 1. Firstly, according to the need of high slope angle and long eye-relief, the Erfle-type of eyepiece is chosen. While the focal length of the eyepiece is about 25mm and the magnification of telescope is set as 12^x, the focal length of objective is 340mm.

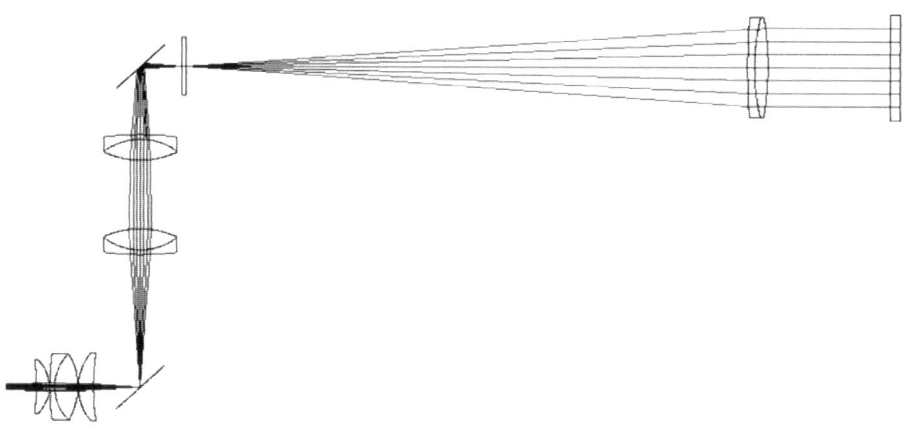

Figure 1. The 3D Optical Layout of Telescope

In order to control light for the sake of brighter illumination on the observer's eye, the exit pupil diameter of telescope can be set at 4mm and the diameter of the objective at 65mm. The full field-of-view (FOV) at the pressure window of the telescope is approximately 6° as usually required, and the full FOV exit angle to the observer's eye will be 72°. Secondly, the overall length is often (often more than one hundred times greater) than its diameter. This can be implemented by adding more than two sets of relay lenses. We use the achromatic doublet with a focal length of 97mm, and diameter of 43mm. In this report, we use one pair of relay to represent the image forming system. Finally, a set of mirrors with the size of 50mm X 50mm at the top and button of the mast are used to redirect the sight line.

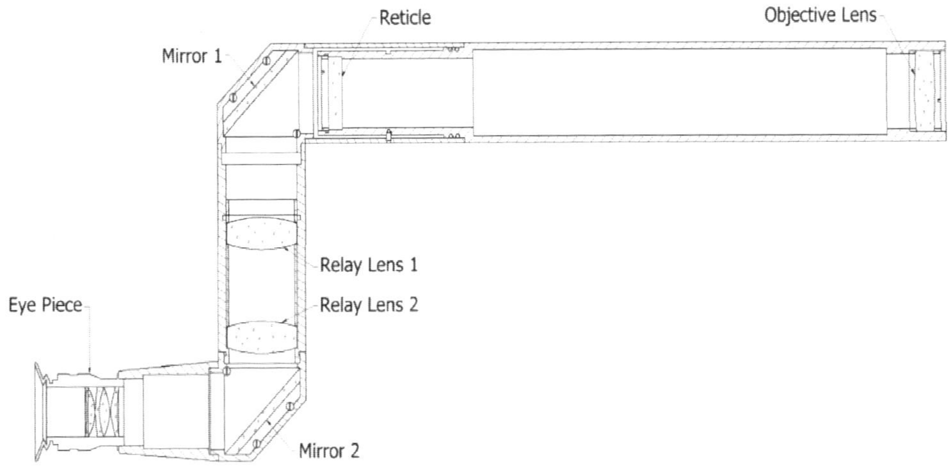

Figure 2. Cross-sectional View of Opto-mechanical Assembly of Telescope

The telescope is comprised by components as shown in Figure 2, W: pressure window, O: objective lens, R:Reticle, M_1 and M_2 are mirrors; L_1 and L_2 are relay lenses, Erfle: Eye-relief eyepiece.

3.1 Paraxial working layouts

The basic layout of the system was worked using paraxial ray tracing [5]. An axial ray is traced through the system for determining the lens positions for the required image positions in the system. A principal paraxial ray is traced for determining the clear apertures needed.

The periscopes are formed by relay lenses to transfer the image from one plane to the next with a magnification equal to one, and field lenses located at the image planes, to form the image of the exit pupil of the preceding system, on the entrance pupil of the next system (Hopkins, 1949, 1962b) [6]. A good aberration correction may be obtained if each relay systems formed by a pair of doublets in a symmetrical configuration.

The primary image is formed by the objective at the fixed reticule. Then this image is shifted on the pair of relay lenses. This is relayed in the relay module and focused at the final image plane in the eyepiece module.

The relay system is designed with images forming at the focal planes of the lenses so that beams are parallel between the lenses. The total number of reflections in the system is two with one reflection at mirror M_1 and second reflection at mirror M_2 is not laterally inverted at the eye-piece.

The prescription of rotating eyepiece telescopic system and general lens data editor are given in Table 1 and Table 2.

Table 1. The Prescription of the Rotating Eye-piece Periscopic Telescope

Lens Data Editor

Edit Solves Options Help

Surf	Type	Comment	Radius	Thickness	Glass	Semi-Diameter
OBJ	Standard		Infinity	Infinity		Infinity
1*	Standard	PRESSURE WINDOW	Infinity	6.000000	F_SILICA	35.000000
2*	Standard		Infinity	72.000000		35.000000
3*	Standard	OBJECTIVE LENS	214.800000	8.000000	H-BAK2	32.750000
4*	Standard		-153.460000	4.000000	ZF2	32.750000
5*	Standard		-515.450000	333.000000		32.750000
6*	Standard	RETICLE	Infinity	3.000000	H-K9L	18.990000
7*	Standard		Infinity	24.960000		18.990000
8	Coordinate B..	Element Tilt		0.000000	-	0.000000
9*	Standard		Infinity	0.000000	MIRROR	20.000000
10	Coordinate B..	Element Tilt		-45.000000	-	0.000000
STO*	Standard	RELAY LENS 1	-162.930000	-3.000000	ZF5	21.750000
12*	Standard		-32.500000	-14.000000	BAK4	21.750000
13*	Standard		47.420000	-45.000000		21.750000
14*	Standard	RELAY LENS 2	-47.420000	-14.000000	BAK4	21.750000
15*	Standard		32.500000	-3.000000	ZF5	21.750000
16*	Standard		162.930000	-87.250000		21.750000
17	Coordinate B..	Element Tilt		0.000000	-	0.000000
18*	Standard		Infinity	0.000000	MIRROR	20.000000
19	Coordinate B..	Element Tilt		28.050000	-	0.000000
20*	Standard	EYE-PIECE LENS	-134.000000	9.000000	H-BAK3	22.500000
21*	Standard		-36.980000	0.300000		22.500000
22*	Standard		47.420000	14.000000	H-BAK3	21.750000
23*	Standard		-32.500000	3.000000	ZF5	21.750000
24*	Standard		-162.930000	0.300000		21.750000
25*	Standard		26.300000	8.000000	H-K9L	17.500000
26*	Standard		267.900000	26.800000		17.500000
27	Standard		Infinity	-33.000000		2.000000
IMA	Standard		Infinity	-		3.000000

Table 2. The general lens data of Rotating Eye-piece Periscopic Telescope

```
GENERAL LENS DATA:

Surfaces                  :             28
Stop                      :             11
System Aperture           : Entrance Pupil Diameter = 70
Glass Catalogs            : SCHOTT SUMITA INFRARED CHENGDU
Ray Aiming                : Off
Apodization               : Uniform, factor =    0.00000E+000
Temperature (C)           :    2.00000E+001
Pressure (ATM)            :    1.00000E+000
Effective Focal Length    :       831.1741 (in air at system temperature and pressure)
Effective Focal Length    :       831.1741 (in image space)
Back Focal Length         :       76.82691
Total Track               :            451
Image Space F/#           :       11.87392
Paraxial Working F/#      :       11.87392
Working F/#               :         13.318
Image Space NA            :     0.04207183
Object Space NA           :   3.500001e-009
Stop Radius               :      -7.310051
Paraxial Image Height     :       44.28735
Paraxial Magnification    :              0
Entrance Pupil Diameter   :             70
Entrance Pupil Position   :      -1888.753
Exit Pupil Diameter       :       4.852227
Exit Pupil Position       :       25.41197
Field Type                : Angle in degrees
Maximum Field             :           3.05
Primary Wave              :      0.5875618
Lens Units                :   Millimeters
Angular Magnification     :       14.42636
```

3.2 Objective of telescope

An objective lens of achromatic doublet of 340mm focal length is used as shown in Figure 3. An achromatic lens reduces the achromatic aberrations. An achromatic can be bent like a singlet; this means that the system power is maintained during the bending as shown in Figure 4. Color correction is also maintained since this depends only on the power of the individual elements [7]. At the focal plane of objective lens a cross reticule with circle at centre is fixed for aiming purpose. The general data of objective and eyepiece lenses are shown in Tables 3 and 4.

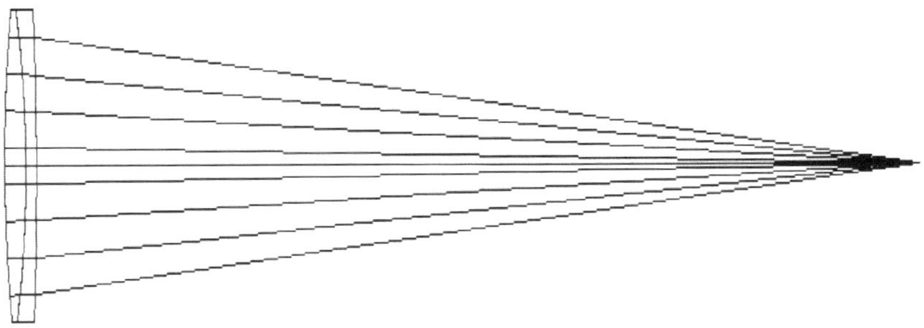

Figure 3. The Objective lens 3D layout

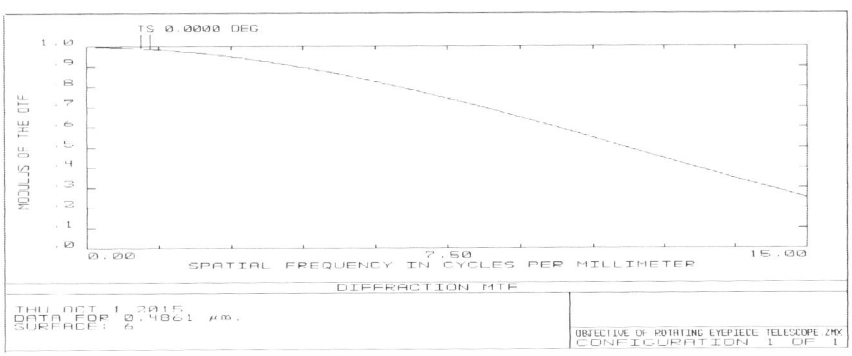

Figure 4. The MTF of Objective lens of Rotating Eyepiece Telescope

Table 3. The general Lens data of the Objective lens of Rotating Eyepiece Telescope

Lens Data Editor

Edit Solves Options Help

Surf:Type		Comment	Radius	Thickness	Glass	Semi-Diameter
OBJ	Standard		Infinity	Infinity		Infinity
1*	Standard		Infinity	6.000000	F_SILICA	35.000000 U
STO*	Standard		Infinity	72.000000		35.000000 U
3*	Standard		214.800000	8.000000	H-BAK2	32.750000 U
4*	Standard		-153.460000	4.000000	ZF2	32.750000 U
5*	Standard		-515.450000	334.008735 M		32.750000 U
IMA	Standard		Infinity	-		0.000000 U

3.3 Erfle eyepiece of telescope

A rotating eyepiece of Erfle type has been used in our telescope. It consists of three optical elements including one doublet lens, which is used to reduce the chromatic aberrations. The eyepiece collects the image made by the objective lens of the telescope and provides parallel rays for viewing directly through the eye. The eyepiece is assembled in such a way that the focal planes of objective lens and eyepiece are coincident. The optical lenses lay out, shaded view of eyepiece lenses, general lens data and MTF are shown in Figures 5-6.

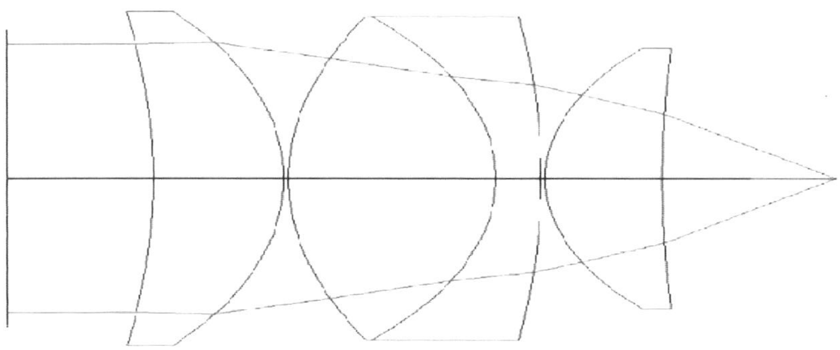

Figure 5. The Eyepiece lenses layout

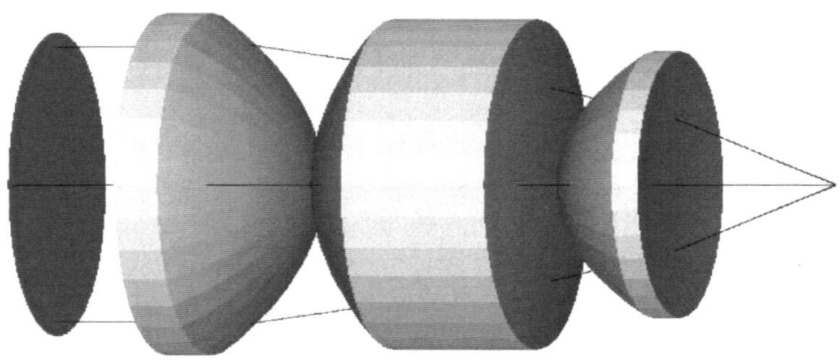

Figure 6. The 3D shaded view of Eyepiece lenses layout

Table 4. The general lens data of Rotating Eye-piece Periscopic Telescope

Lens Data Editor

Edit Solves Options Help

	Surf:Type	Radius	Thickness	Glass	Semi-Diameter	
OBJ	Standard	Infinity	Infinity		0.000000	
1	Standard	Infinity	10.000000		20.000000	U
STO*	Standard	-134.000000	9.000000	H-BAK3	22.500000	U
3*	Standard	-36.980000	0.300000		22.500000	U
4*	Standard	47.420000	14.000000	H-BAK3	21.750000	U
5*	Standard	-32.600000	3.000000	ZF5	21.750000	U
6*	Standard	-162.930000	0.300000		21.750000	U
7*	Standard	26.300000	8.000000	H-K9L	17.500000	U
8*	Standard	267.900000	11.750000		17.500000	U
IMA	Standard	Infinity	-		0.061329	

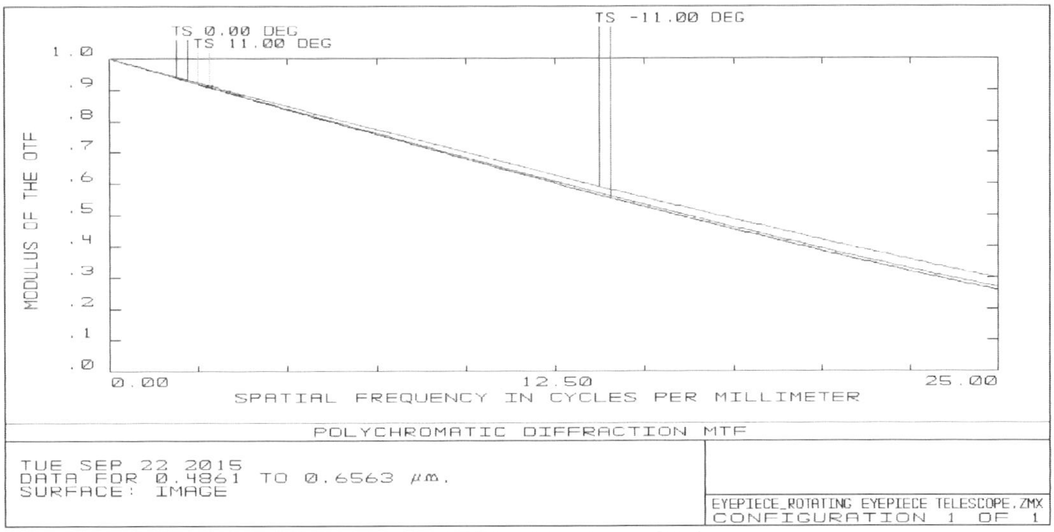

Figure 6. The MTF of Eyepiece of Telescope

4 Conclusions

I. The applications of such type of telescopes are found in military and civil projects. Main application of this telescope is in submarines for off axis viewing of objects.

II. Aberrations are balanced by adding a pair of doublets as relay lenses in a symmetrical configuration.

III. Image inversion has been carried out by using pair of mirrors and relay lens system systems and also the length of telescope is extended by using relay lenses.

IV. It is concluded that the MTF of objective and eyepiece lenses of the telescope is good enough to see the target clearly up to 4km. The image is shifted from focal plane of objective lens to focal plane of eyepiece by using a set of relay lenses of focal length 98mm.

V. The rotating eyepiece will facilitate viewing of objects in a remote or out of sight objects. Eyepiece arm of periscope can be rotated up to 360^0 keeping objective and reticule unchanged and ensuring zero deviation in reticule inclination.

VI. The optical design simulation and image analysis has been carried out by using ZEMAX® software. Magnification of telescope has been achieved between 10^x~12^x with FOV of 6^0.

VII. Prototype telescope was fabricated by using uncoated Edmund optics and optical tool box of Micro Series Kit, NEWPORT.

5 References

[1] Francis B Patrick, *"Military Optical Instruments" in Applied Optics and Optical Engineering*, Vol. Part II, R. Kingslake, ed. Academic Press, New York, pp. 209-214 (1969)

[2] M Born and E Wolf, *"Principles of Optics", Pergamon Press London*, New York, 6th Ed., pp. 243–245, (1980)

[3] W J Smith, *"Modern Optical Engineering", MC Graw Hill Book Company, London*, pp. 257-257, (2000)

[4] W J Smith, *"Modern Optical Engineering", MC Graw Hill Book Company, London, pp.* 214-215, (1966)

[5] Kingslake R, *"Lens Design Fundamentals", Academic Press, London*, pp. 63-66, (1978)

[6] Danial Malacara & Zacarias Malacara, *"Hand Book of Optical Design", Marcel Dekker Inc*, 2004, p-455

[7] Joseph M. Geary, *"Introduction to Lens Design with Practical Examples"*, p-187

ESD coating of copper with TiC and TiB_2 based ceramic matrix composites

S. Talas[1,2], **E. Mertgenç**[3] **and B. Gökçe**[2]
[1,2]Afyon Kocatepe University, Faculty of Technology, Afyon, Turkey
[3]Afyon Kocatepe University, Vocational High School, Afyon, Turkey

E-mail: stalas@aku.edu.tr

Abstract. In automotive industry, the spot welding is a general practice to join smaller sections of a car. This welding is specifically carried out in short time and in an elevated number with certain pressure applied on copper electrodes. In addition, copper electrodes are expected to endure against cyclic mechanical pressure and temperature that is released during the passage of the current. The deformation and oxidation behaviour of copper electrodes during service appear with increasing temperature of medium and they also need to be cleaned and cooled or replaced for the continuation of joining process. The coating of copper electrodes with ceramic matrix composites can provide alternative excellent high temperature strength and ensures both economic and efficient use of resources. This study shows that the ESD coating of copper electrodes with a continuous film of ceramic phase ensures an improved resistance to thermal effects during the service and the change in content of film may be critical for cyclic alloying.

1. Introduction

Ultra-high-strength steels (UHSS), dual phase (DP) steels and transformation induced plasticity steels (TRIP) make car bodies stronger and lighter and also contain complex microstructures from which their strength is obtained and is critical for durability of welded sections. The production of car parts from these steels is relatively difficult compared to conventional steels since the formability is limited and short moulding life and high forming force are important problems due to spring back at room temperature [1-2]. The high strength of steels used for the production may be affected by heat treatment occurring during the spot welding of which its quality is strongly dependent on the thermal history. This is, alone, a reason for a fail in mechanical tests, in addition to coating on the surface of these automotive steels. Such coatings have an effect on the quality of welds [2-3] by reacting with copper spot weld electrodes, resulting in either excessive softening or hardening of the tip and then breaking off in the form of flake. The coating of Zn or AlSi on these steel sheets causes Cu electrodes degrade rapidly and these electrodes more often require maintenance than uncoated sheets.

The resistance welding of hot-dip galvanized zinc coated steel is particularly difficult as Zn in coating forms intermetallic compound with Cu on the surface electrodes, degrading the quality of nugget rapidly. The degradation of electrode and alloying at the tip can be prevented by a ceramic layer by forming a metallurgical barrier [4-5]. The loss of production in welding process and in weld nugget properties such as its service life affected by the production method of parts such as hot shaping causes the formation of martensite and harder second phase constituent as a result of fast

Content from this work may be used under the terms of the Creative Commons Attribution 3.0 licence. Any further distribution of this work must maintain attribution to the author(s) and the title of the work, journal citation and DOI.
Published under licence by IOP Publishing Ltd

cooling and transformation temperature [5]. In any welding, these two parameters i.e. hard phases and microstructures make the weld zone sensitive to cracking leading to interfacial fracture failure.

Electrospark deposition (ESD) is a microwelding technique that is a low heat input process with very short weld duration which may be used on many different materials. Thermal distortion on substrate is low and a heat sensitive materials can be well managed for a successful joining. The heat input of the process is just due to the localised electrical energy from the short circuiting from which heat is extracted. In this process cycle, the electrode travels along the surface of substrate to deposit a uniform layer. Some combinations of materials as a substrate and a coating are said to be successful and the variety of materials are similar to diffusion welding. In many cases, the coating is composed of hard phases or its corrosion resistance is higher compared to substrate [6]. Due to low heat generated in the process, many electrically conductive materials such as metals, alloys, or cermets are possible candidates for ESD coating and electrode content can be deposited on metal substrates by the help of arc effect [6-8]. It is questionable to assume that rapid solidification is responsible for the success of this method but the rapid formation of compounds on the surface of substrate is more plausible with the help of arc plasma effect [5]. Coatings can be made with and without binder phase on the substrate [9] and cracking in the coating may be prevented with ESD method [10]. TiCp/Ni metal matrix composite and copper substrate combinations provided strong bonds with Ni diffusing into both Cu substrate and TiCp composite without the formation of excessive intermetallic phase [11].

Various improvements have been achieved with this method, such that, 70% and 57% improvements have been observed with TiC [12] and multilayer coating [13], respectively. TiB_2/Ni coating have also been attempted for Zn coated steel sheets and successfully demonstrated by Luo et al. [14]. The results indicated that deformation of the electrode was reduced and its spot welding lifespan was much longer than that of an uncoated electrode on steel sheet without TiB_2/Ni coating.

In this study, TiB_2 and TiC based composites were utilized to form a layer of ceramic on Cu substrate with ESD method. The composite coated Cu surfaces were characterized by SEM (Scanning Electron Microscope) and XRD (X-Ray Diffraction) method while hardness and resistivity were also measured.

2. Experimental procedures

The ESD process is relatively simple and does not require any advanced equipment design, which normally consists of the power source and the electrode holder torch which is sometimes called the applicator. TiC+Co mixture with trace elements of Ni and Fe i.e. TiC+CoNiFe and TiB_2+Co mixture with trace elements of Fe i.e. TiB_2+CoFe composites have been produced and used to coat resistance welding electrode tips made from copper (See Table 1 for compositions), using ESD (Electro Spark Deposition) machine Model SZ3000, (Shangai Shengzao Mechanical and Electrical Co). A voltage of 100 and a frequency of approx. 1.8 kHz were selected for the deposition. A standard metallographic study was carried out on all specimens using optical microscope and SEM. The hardness measurements were made using Shimadzu HMV 2 Microhardness tester with a load of 9.8N. XRD study was carried out using Shimadzu XRD- 6000 model machine for the 2-theta range of 20-100° at a scanning rate of 2° per minute. Electrical resistance was measured by GWINSTEK GOM-802 milliohm meter with a sensitivity of 1 microohm. All the resistance measurements were carried out using an apparatus with two coppers pins between which the distance is 11.5 mm. The device was set for a constant value hold measurement which is obtained after a long duration until a stabilized value is displayed.

Table 1. Nominal compositions of electrodes and copper substrate.

	Cu	Ni	Fe	Other	TiC	TiB_2	Co
TiC+Co	-	-	-	-	90	-	10
TiB_2+Co	-	-	-	-	-	90	10
Substrate	97.9	0.86	0.8	0.44	-	-	-

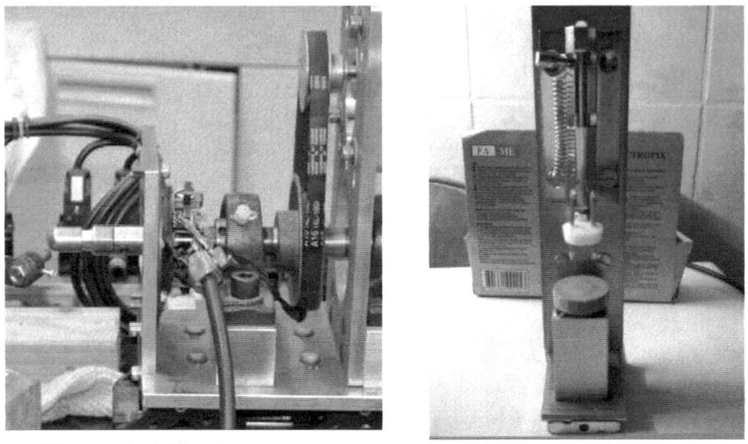

Figure 1.a) Coating apparatus b) resistance testing apparatus

3. Results and discussion

3.1. Characterization of specimen surfaces

Coating electrodes were prepared according to composition given in Table 1 and mixed for 24 hrs then compacted using 700MPa pressure and sintered at a temperature of 1650 °C for 2 hrs in a tube furnace. Metallographic characterization of coating was carried out following ESD coating of resistance weld tips (Figure 2) with SEM. Coated surface was examined under microscope with BSE (Back Scattered Electron) mode and the results showed that a continuous layer of ceramic phase was formed on the surface after ESD process. The optimisation of the process was made prior to coating as to produce an even surface as shown in the upper most specimen in Figure 2. In general, the surface that is affected by the heat is subjected to coating, as in the specimen in the most left in Figure 2. This ensures that the surface of copper tips for the following coating is not affected by low resistance ceramic layer.

Figure 2. TiC and TiB2 coated specimens; specimens A and D have an uneven surface or large area of coating; and specimen B and C are ideally coated for the resistance welding

3.2. TiC+Co composite coated specimens

XRD spectrograms from TiC+Co electrode and the surface of TiC+Co coated Cu specimens are given in Figure 3a,b, respectively. As can be seen in Figure 3, that Cu and TiC are successfully detected on the top surface of copper tips. Diffraction peaks of TiC and Cu are clearly visible; however, Co peaks overlap with first two peaks of Cu and can be assumed to be present since they exist in EDX results in Figure 3b. Co concentration is very low in coating layer, and therefore, it is not even observed in the XRD patterns of the coating, as seen in Figure 3c and d. A possible reason for that is Co and Cu may have evaporated at high temperatures during the ESD arc process.

The integrity of TiC in the coating is very important for the performance of electrode as it is the hardest phase. The bonding between the copper matrix surface and TiC layer is continuous and low level of roughness of deposited compound can also be differentiated. It is an indication of successful coating for the service, however, coating was carried out in air and a possible oxidation of copper surface or alloying constituents is expected where the arc is occurred. Indeed, Figure 3c and d show that oxygen peak is detected in EDX analysis with possible compounds of Cu or Ti elements. In addition, TiC particles are also visible as dark features in Figure 3c as a result of detaching from the electrode and it unreacted but was engulfed with remaining layer. This may be a possible imperfection which will cause the layer fail. Cracking is the main problem in coatings deposited on softer substrates such as Cu and Ni and it may fail prior to service as a result of incompatibility due to linear expansion coefficients [15]. This was not the case in this study and the surface layer was intact and no flaking was also observed.

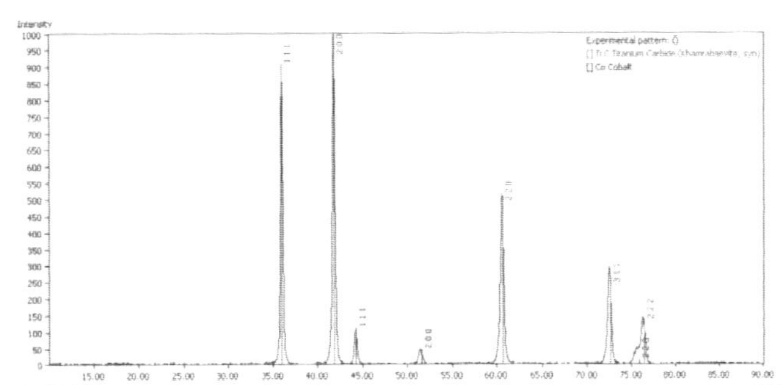

(a)

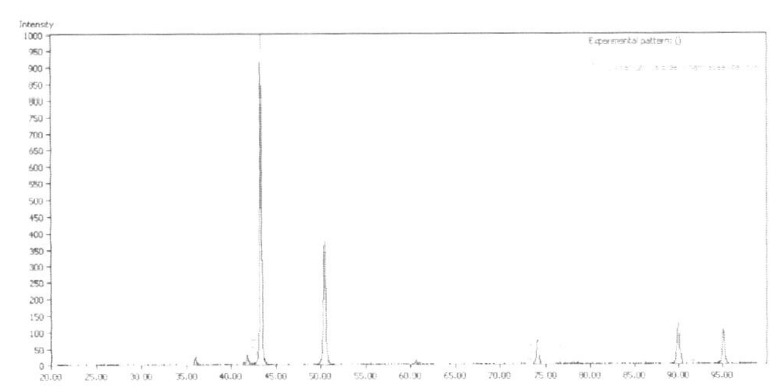

(b)

Figure 3. continue

(c)

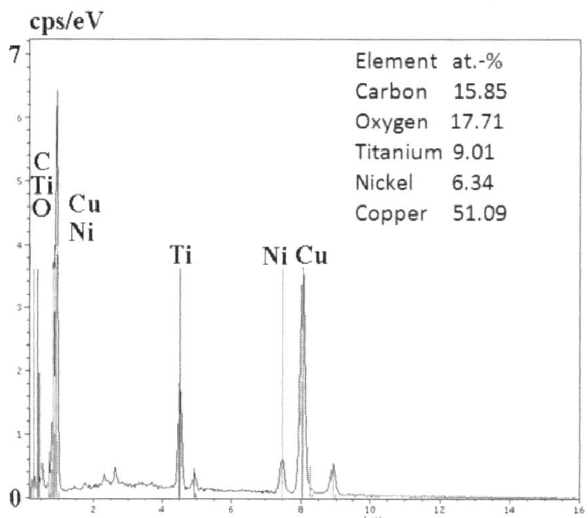

(d)

Figure 3. XRD spectrograms of a) TiC+Co electrode, b) coated surface, c) SEM image of coated section of electrode caps with d) EDX area analysis of TiC+Co composite coated Cu surface

A group of large arc craters can also be seen on the surface which is ideally removed when polishing prior to service in Figure 4. It is also interesting to note that particles of small size are embedded in the matrix and possibly contributes towards bulk hardness. The composition of these particles is unknown.

Hardness of Cu substrate was in average 118 HV while the hardness of coated Cu substrate was approximately 594±75 HV. This is similar to observations made by Tang et al. [11]. The increase in hardness is mostly result of hard phase/particle hardening from TiC interacting with Cu matrix. Since the hardness of TiC is higher than 594 HV, the formation of oxides or the thickness of layer may also be contributing factors for the overall hardness of the surface. Another parameter is the electrode's self resistance which is relatively low compared with conventional ceramic electrodes. It should be as low as possible to pass high currents and to cause an increase in the production rate. For TiC+Co electrode, the self resistance is 153.4 microohm/cm and the resistance of the surface layer is slightly higher than this value which 202.1 microohm/cm. This may be explained by the fact that oxidation and alloying with the substrate Cu are two important parameters for the conductivity of layers. Metallic additions are mostly responsible for the loss of resistance as they are oxidised or evaporated easily during the high temperature arc processes.

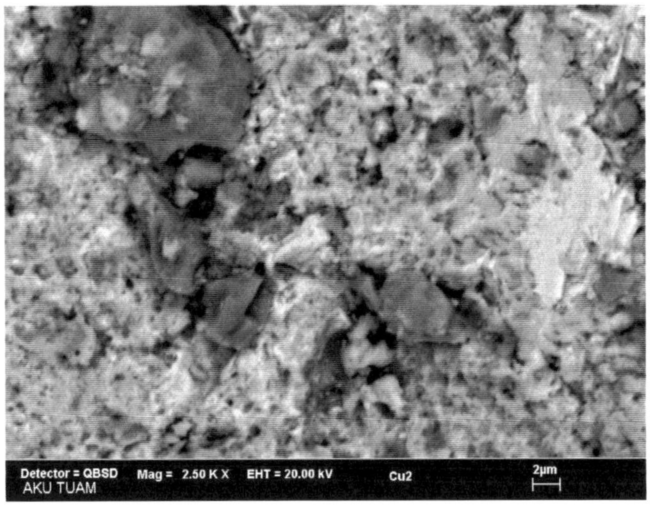

Figure 4. Detailed view Cu surface after coating with TiC+Co composite

3.3 TiB_2+Co composite coated specimens

ESD process may generate high energy mass transfer defined by surface spot localization, electrode and substrate overheating. The compound formation is,therefore, affected by excessive heat and mass transfer characteristics. As seen in XRD spectrogram of TiB_2+Co electrode given in Figure 4a, TiB_2, CoO and B_2Co_3 compound peaks are clearly observed. CoO oxidation product during the sintering are assumed to be responsible for the absence of Co peaks as stated in previous section; they are present in the form of oxide. The formation of B_2Co_3 phase is expected at 1650°C sintering temperature. The compound types on the surface of Cu tips coated by TiB_2+Co electrode were determined by XRD study with a result of TiB_2, Cu and Co_3Ti compounds. Co_3Ti is believed to form as a result of decomposition of constituents; some studies suggest that the additions of Co and Ni are important in the kinetics of decomposition process of TiB_2 and WC [16]. In addition to $CoTi_3$, there is a nitride of Cu primarily peaked at 2 theta of 48 degrees. The reaction of evaporated/heated alloying elements and Cu with air in arc plasma region is highly possible at elevated temperatures, forming oxides and nitrides of alloying and matrix elements. Oxides of Cu may easily be dissociated contrary to nitride and therefore may not be present in the spectrogram but EDX results show that a continuous layer of oxide is present which may be just limited to surface due to washing procedure prior to SEM analysis. The amount of TiB_2 is weaker than Cu, which increases the possibility of detection of phases resulted from reaction with Cu substrate. SEM images from TiB_2 coated specimens are given in Figure 5c with EDX analysis in Figure 5d. It should be noted that the EDX detector is blind to B because of its low radiation energy and therefore there is no B shown in EDX results.

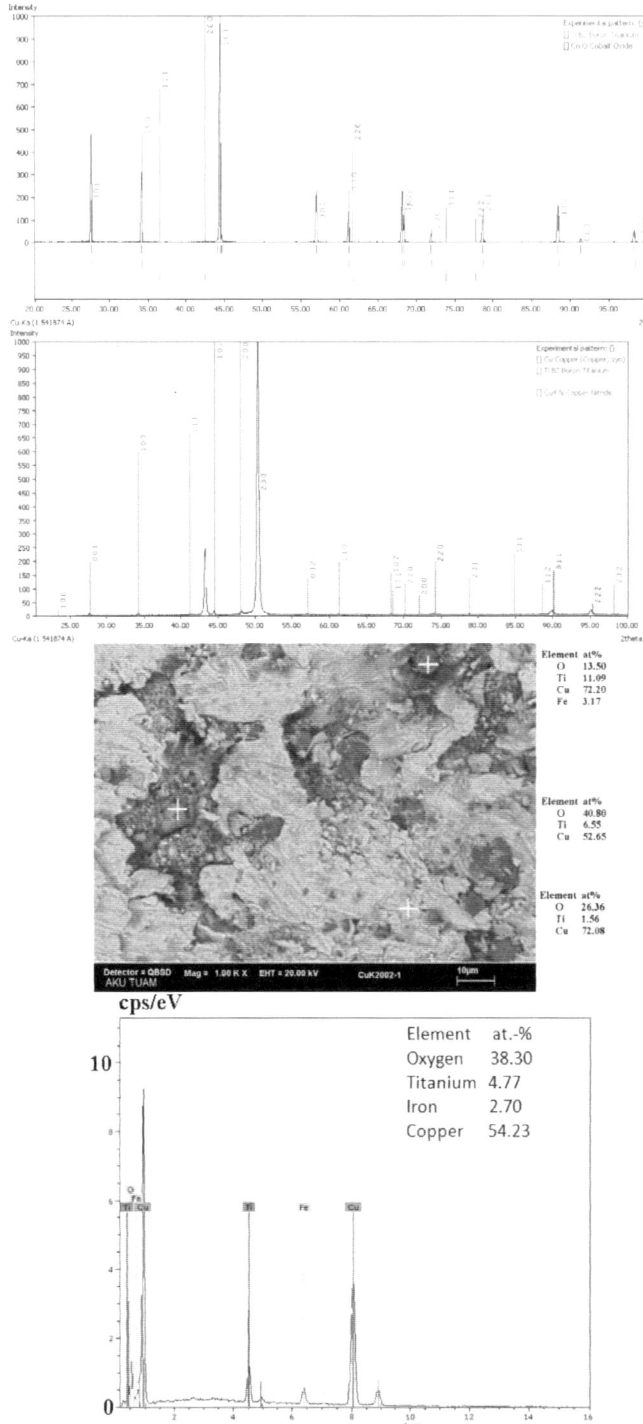

(a)
(b)
(c)
(d)

Figure 5. XRD spectrograms of a) TiB2+Co electrode, b) coated surface, c) SEM image of coated section of electrode caps with d) EDX area analysis of TiB_2+Co composite coated Cu surface

As seen in Figure 6 that, the integrity of surface layer is comparatively better than TiC coating because of its smooth appearance. There are no cracks observed on the surface of ceramic layer. The presence of crater and large areas similar to spatter suggest that the mass transfer was in the form of droplet, which may eventually be solidified slower than anticipated. Pores and large craters are mainly distributed well within molten regions. Large mass transfer together with arc force may give rise to such features.

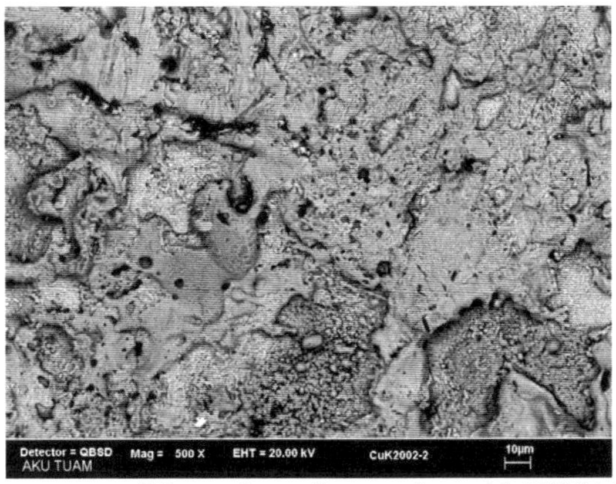

Figure 6. Appearance of Cu surface after coating with TiB2+Co composite

The hardness of Cu substrate was, as given above, 118 HV and the hardness of TiB_2+Co coated Cu substrate was in average 513± 68 HV. The formation of a continuous layer of hard phase contributed towards the increase in hardness which is a result of hardening effect from TiB_2particles/alloying with Cu matrix and Co. The nitride of Cu is distributed amongst large patches of Cu and Ti rich areas. The presence of air with high content of N forms arc plasma with thermal conductivity exceeding the melting point of Co and Cu. The temperature and arc forces inside arc plasma help local melting of constituents and cause immediate reactions with surrounding gaseous elements which take place rapidly. Fast solidification rates and gaseous phases evolving from arc plasma are responsible for porous surface as seen in Figure 4c.

The measured self resistance of TiB_2+CoNi electrode is comparatively higher than that of TiC, similar trend is observed in the literature. The high self resistivity improves the consumption rate of the electrode but the thickness of layer is compromised, hence, the alloying additions should be promoting the passage of current by providing free electrons to pass high currents and increase the deposition rate. For TiB_2+Co electrode, the self resistance is measured to be 238.4 microohm/cm and the resistance of the surface layer is 728 microohm/cm which is higher than that of TiC+Co electrode. Excessive oxidation and nitride formation on the Cu surface can be a plausible explanation for high resistance. Alloying with Cu substrate is also important parameter for the electrical conductivity of coating layer. High temperature of plasma may not even be achieved due to high resistance of TiB_2+Co electrodes and hence the decomposition of compounds did not take place, resulting in high self resistance.

4. Conclusions

The appearance TiC coatings are smoother than that of TiB2 coatings, which is possibly a result of the temperature of arc plasma and decomposition characteristics of constituents. The low self resistance of TiC electrodes also affects the hardness of both coatings. TiC coating without additional compounds possess higher hardness than TiB_2 composite coating. The intrinsic high resistance of TiB_2 affects the surface quality of the coating and its composition.

5. Acknowledgment

This study was financially supported by TUBITAK (Project Number 213M547).

6. References

[1] An JH, Ko DC, Lee CJ and Kim DM 2008 J. Korean Society of Mechanical Engineering A, 32 905
[2] Choi HS, Park G, Lim Wand Kim B2011 Journal of Mechanical Science and Technology 25 1543
[3] Gould J 2008Welding Journal 90 191s
[4] Chen Z, Zhou Y2006 Surf. Coat. Technol. 201 1503
[5] Marya M and Gayden X Q 2005 Welding Journal 84 172s
[6] Johnson RN2002 ElectroSpark Deposition: Principles and Applications Society of Vacuum Coaters 45th Annual Technical Conference, (Lake Buena Vista, FL April 13-18) pp 87-92
[7] Johnson RN and Sheldon G L 1986 J. Vac. Sci. Technol. 4 2740
[8] Parkansky N, Boxman RL and Goldsmith S 1993 Surf. Coat. Technol. 61 268
[9] Johnson RN and Sheldon GL 1986 J. Vac. Sci. Technol. A4 2740
[10] Reynolds JL, Holdren RL and Brown LE 2003 Advanced Materials and Process161 35
[11] Tang SK, Nguyen TC, Zhou Y2010 Welding Journal, 89 172s
[12] Dong SJ and Zhou Y2003 Metall. Mater. Trans. 34A 1501
[13] Chen Z andZhou Y2006 Surf. Coat. Technol. 201 2419
[14] Luo C, Dong S, Xiong X and Zhou N2009 Surf. Coat. Technol. 203 3333
[15] Agarwal A, Dahotre NB and Sudarshan TS 1999 Surf. Eng. 15 27
[16] Shibuya M, Yoneda T, Yamamoto Y, and Ohyanagi M2003 J. Am. Ceram. Soc.86 354

IOP Conf. Series: Materials Science and Engineering **146** (2016) 012006 doi:10.1088/1757-899X/146/1/012006

Structural and electromechanical properties of $Na_{0.5}Bi_{0.5}TiO_3$ ceramics produced by different synthesis routes

A. Hussain, A. Maqbool, R. A. Malik[1], M. H. Kim, T. K. Song and W. J. Kim[1]
School of Advanced Materials Engineering, Changwon National University, Gyeongnam 641-773, Republic of Korea
[1]Department of Physics, Changwon National University, Gyeongnam 641-773, Republic of Korea

E-mail: hussain@changwon.ac.kr

Abstract. Sodium bismuth titanate, $Na_{0.5}Bi_{0.5}TiO_3$ (NBT) ceramics were produced by three different methods; conventional mixed-oxide (CMO) route, molten salt synthesis (MSS) and topochemical microcrystal conversion (TMC) and then sintered at 1150 °C for 2 h in air atmosphere. The crystal structure, dielectric, ferroelectric and field-induced strain properties were investigated for all samples. All samples showed a single phase perovskite structure without any evidences of unwanted secondary phases. The NBT ceramics synthesized by the TMC method show slightly better dielectric, ferroelectric and field induced strain response as compared with CMO and MSS synthesized ceramics. The room temperature dielectric constant measured at 1 kHz increased from 218 for NBT ceramics synthesized by MSS method to 271 and 330 for CMO and TMC synthesized ceramics, respectively. Similarly, the dynamic piezoelectric coefficient (d_{33}^*) enhanced from 91 pm/V for CMO synthesized to 97 pm/V and 107 pm/V for MSS and TMC synthesized ceramics, respectively.

1. Introduction

Oxide piezoelectric ceramics are used in numerous devices such as sensors, actuators, filters, resonators, multilayer capacitor, transducers, micro electromechanical systems (MEMS) and memory devices [1-3]. Most of these devices are fabricated from the lead-based ceramics such as lead zirconate titanate (PZT) and its solid solutions because of their high performance [4]. Nevertheless, the pollution of lethal lead during the synthesis and waste products cause serious environmental problems. Therefore, there is an increasing demand for the development of lead-free piezoelectric ceramics to minimize pollution produced by lead-based materials [5].

Sodium bismuth titanate, $Na_{0.5}Bi_{0.5}TiO_3$(NBT) is considered a promising substitute for PZT because of its high Currie temperature (T_c) and large polarization [6]. Moreover, it is well established that the covalency of unoccupied states of the *Pb6d* in the perovskite structure and oxygen p-states favor ferroelectric ground states [7]. Pb^{2+}, Bi^{3+}ions are in an isoelectronic state which show a longer pair effect that stimulate research on NBT as an alternative to PZT ceramics. In lead-based ceramics, Pb^{2+} is responsible for high polarization due to its long pair effect of 6s valence shell electron. It may be pointed out that Bi^{3+}in NBT has similar electronic configuration to that of lead in PZT which motivates the piezoelectric community to consider the NBT system as replacement of PZT. NBT has a perovskite structure and shows a sequence of phase transitions as a function of temperature. There are

Content from this work may be used under the terms of the Creative Commons Attribution 3.0 licence . Any further distribution of this work must maintain attribution to the author(s) and the title of the work, journal citation and DOI.
Published under licence by IOP Publishing Ltd

numerous research reports from different research groups about the structural phases, types of electric ordering and its relaxor behaviorof NBT that exits at different temperatures [8-12].

Beside all these, little efforts have been concentrated on the development and comparison of electromechanical properties of NBT ceramics produced by different routes. In this work, an effort has been made to prepare NBT ceramics by three different routes CMO, MSS and TMC followed by investigation and comparison of their crystal structural, dielectric, ferroelectric and field-induced strain properties to get insight into their structure and properties relationship.

2. Experimental method

Na_2CO_3 (99.95%),Bi_2O_3 (99.95%) and TiO_2 (99.99%) were used as starting raw materials for preparation of NBT ceramics by CMO and MSS methods. First, the starting raw materials were weighed using a physical balance according to stoichiometric formula of $Na_{0.5}Bi_{05}TiO_3$. For synthesis of NBT by MSS process, NaCl (99.95%) was added to oxide powder in the weight ratio of 1.5:1. Both batches were separately ball-milled in ethanol with yettria stabilized zirconia balls as a medium for 24 h and then dried. The dried slurries were placed in a tightly covered alumina crucibles and heated for 4 h at 950 °C. After heat treatment, the MSS synthesized NBT ceramics were thoroughly washed with hot de-ionized water to eliminate residual salt and then dried at 80 °C.

Aurivlliustype bismuth layered structure ferroelectric, $Na_{0.5}Bi_{4.5}Ti_4O_{15}$ (NBT4) was used as precursor material for the synthesis of simple NBT ceramics by TMC method. Reagent grade Na_2CO_3,Bi_2O_3 and TiO_2 powders of purity more than (99.90%) were initially mixed following the stoichiometry of $Na_{0.5}Bi_{4.5}Ti_4O_{15}$ (NBT4) and then NaCl (99.95%) was added to NBT4 mixture in weight ratio (1.5:1). Detail procedure for the preparation of the NBT ceramics by TMC method can found in reports [13-14]. All powder produced by CMO, MSS and TMC methods were pressed into green disks of diameter of 10 mm and thicknesses of 2 mm. The green disks were embedded in the same compositional powder to minimize the evaporation of the volatile elements (Bi, Na) and then sintered at 1150 °C for 2 h in air atmosphere.

The crystalline phase and purity information of the NBT ceramics synthesized by different routes were checked by X-ray diffraction machine (XRD, RAD III, Rigaku, Japan) using CuKα radiation (λ = 1.541 Å) at room temperature. The upper and the lower extensive surfaces of the samples were polished and coated with a silver-palladium paste to form electrodes for electrical properties measurements. The dielectric constant and loss response were measured through an impedance analyzer (HP4194A, Agilent Technologies, Palo Alto, CA). The ferroelectric hysteresis loops were measured using a Precision Premier II device (Radiant Technology, Inc.) at 50 Hz. Field-induced strain response was measured using a contact-type displacement sensor (Model 1240; Mahr GmbH, Göttingen, Germany) at 50 mHz.

3. Results and discussion

Figure 1 shows the XRD pattern of sintered NBT ceramics synthesized by different routes (CMO, MSS and TMC). All samples crystallized into a single-phase perovskite structure, which is consistent with JCPDS card No. 36-0340 of the $Na_{0.5}Bi_{0.5}TiO_3$ ceramics. Neglecting a small rhombohedral distortion, all diffraction peaks were indexed on the basis of pseudocubic perovskite unit cell. The NBT ceramics synthesized by CMO and MSS methods have similar diffraction pattern to that of TMC, however, different heights of the intensity peaks evident from the (110) and (200) peaks. NBT ceramics synthesized by CMO and MSS have strong (110) peaks, suggesting random orientation while that synthesized by TMC method have (100) and (200) intense peaks indicating (*h*00) preferred orientation. The dissimilarity of NBT ceramics synthesized by TMC can be attributed to the difference in its synthesis routes.

The variation of dielectric constant and loss of NBT ceramics synthesized by different routes as a function of temperature at different frequencies (1, 10 and 100 kHz) are shown in Figure 2. It can be observed from the dielectric constant vs temperature graph that dielectric constant increases with increase in temperature up to certain point and shows two broad dielectric maximum at specific temperature, commonly referred as depolarization temperature (T_d) and maximum permittivity temperature (T_m), The dielectric constant decreases slowly when temperature is raised above T_m. This variation of dielectric constant as a function of temperature is consistent with previously published reports [15-17]. The dielectric loss curves show high dispersion for the CMO sample as compared with MSS and TMC synthesized samples. This high loss at higher temperature may be due to transport of ions with higher thermal energy. The broad transition peaks and frequency dependent response confirm the relaxor-like characteristics of the samples. However, probably due to difference in fabrication processes, the depolarization temperature (T_d) and T_m appears at different temperatures. It can be seen in Table 1 that T_d and T_m values of MSS and TMC synthesized NBT ceramics are slightly higher than TMC synthesized ceramics. The frequency dispersion at T_m is quite high for the CMO synthesized ceramics as compared with the MSS and TMC processed ceramics. This is probably due to the elements Na^+ and Bi^{3+}, which are randomly distributed in the 12-fold coordination sites of NBT, resulting in the disordering of *A*-site cations and the compositional fluctuation [17]. So, the observed diffused phase transition at T_m is attributing to inhomogeneous distribution of *A*-site cations, which appears to be responsible for relaxor behavior. All the samples show the typical character of a ferroelectric relaxor because of diffused phase transition. The dielectric constant has high frequency dependence especially at low frequency and high temperature, which is called as low frequency dielectric dispersion. This may be due to increase in dielectric loss at low frequency and high temperature.

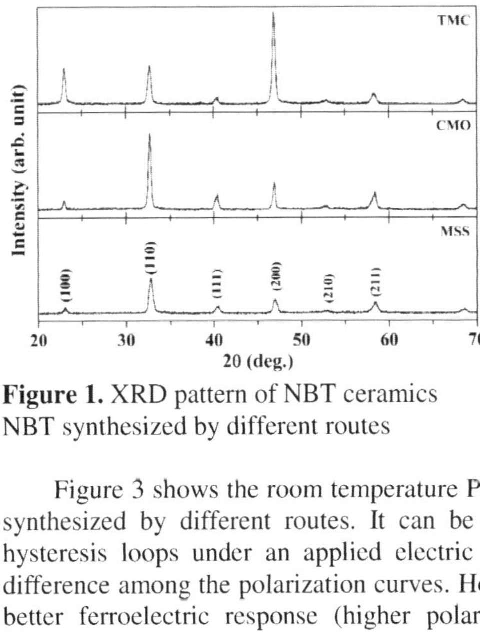

Figure 1. XRD pattern of NBT ceramics NBT synthesized by different routes

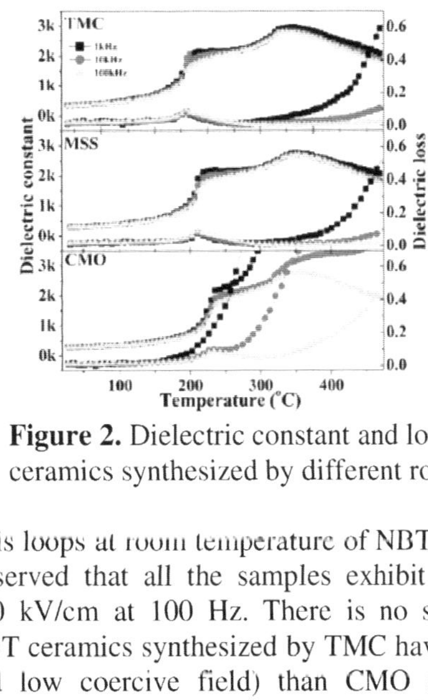

Figure 2. Dielectric constant and loss of ceramics synthesized by different routes

Figure 3 shows the room temperature P-E hysteresis loops at room temperature of NBT ceramics synthesized by different routes. It can be clearly observed that all the samples exhibit saturated hysteresis loops under an applied electric field of 90 kV/cm at 100 Hz. There is no significant difference among the polarization curves. However, NBT ceramics synthesized by TMC have slightly better ferroelectric response (higher polarization and low coercive field) than CMO and MSS synthesized samples. As TMC synthesized ceramics exhibit single crystal type nature and for single crystal it is known that their coercive field is usually lower than that of random ceramics of the same composition [18]. Detailed values of the Remnantpolarization (P_r), and coercive (E_c) of these samples are given in Table 1.

Table1. Dielectric, ferroelectric and field-induced strain response of NBT synthesized by different routes

Technique	T_d (°C)	ε_r	P_r(pC/N)	E_c (kV/mm)	S_{max} (%)	d^*_{33} (pm/V)
CMO	233	271	30.88	57.80	0.068	97
MSS	218	290	30.01	61.02	0.072	102
TMC	195	330	31.50	55.44	0.075	107

Figure 3 shows the room temperature P-E hysteresis loops at room temperature of NBT ceramics synthesized by different routes. It can be clearly observed that all the samples exhibit saturated-hysteresis loops under an applied electric field of 90 kV/cm at 100 Hz. There is no significant difference among the polarization curves. However, NBT ceramics synthesized by TMC have slightly better ferroelectric response (higher polarization and low coercive field) than CMO and MSS synthesized samples. As TMC synthesized ceramics exhibit single crystal type nature and for single crystal it is known that their coercive field is usually lower than that of random ceramics of the same composition [18]. Detailed values of the Remnantpolarization (P_r), and coercive (E_c) of these samples are given in Table 1.

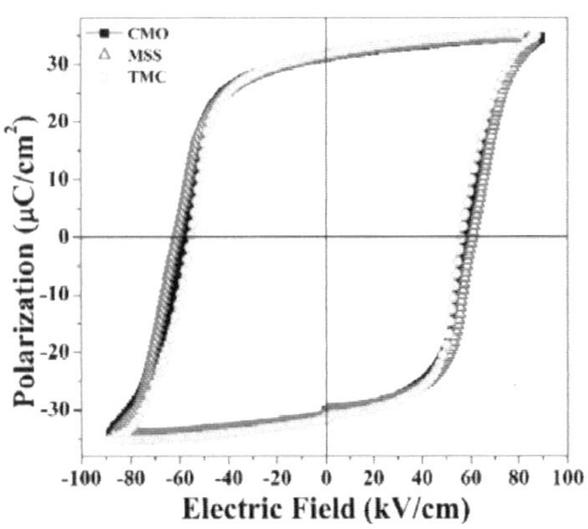

Figure 3. P-E hysteresis loops of NBT ceramics synthesized by CMO, MSS and TMC methods

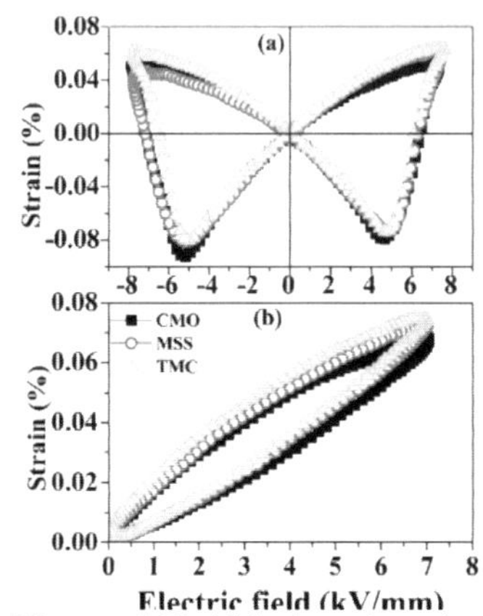

Figure 4. Field-induced strain curves of NBT ceramics synthesized by (a) Unipolar strain (b) Bipolar strain

The electric field-induced bipolar and unipolar strain curves of NBT ceramics synthesized by different routes are shown in Figure 4. The bipolar strain curves (Figure 4a) present that all samples have negative strain (S_{neg}; which is associated with domain back switching, i.e., the difference between zero field and lowest strain) as well as maximum strain (S_{max}; which is the difference between positive and negative maximum strain) similar to that of the typical for ferroelectric materials. The variation in the bipolar curves is consistent with the observed P-E hysteresis loops response shown in Figure 3. The grain oriented NBT ceramics synthesized by TMC results in better ferroelectric and field-induced strain response as compared with randomly oriented CMO and MSS synthesized samples. The unipolar field-induced strain response of the NBT ceramics synthesized by different routes were measured at 7 kV/mm and is shown in Figure4b. The room temperature unipolar field induced-strain increased from 0.068 for CMO synthesized sample to 0.075for TMC synthesized sample. Detail information of field-induced strain response along with their corresponding dynamic piezoelectric coefficient (d^*_{33} = S_{max}/E_{max}) values obtained for these specimens provided in Table 1.The better ferroelectric and field induced strain response of the NBT ceramics synthesized by the TMC method can be attributed to their single crystal type nature.

4. Conclusions

NBT ceramics were successfully synthesized by three different routes and their electromechanical properties were investigated. All sample have same crystal structure, however different grain orientations. NBT ceramics synthesized by TMC show (100) strong intensity peaks indicating (100) grain orientation, while those by CMO and MSS exhibit strong (110) peak signifying random orientation. At measuring frequency of 100 kHz, room temperature dielectric constant increase from 271 for CMO sample to 290 and 330 for MSS and TMC samples. Enhancement in unipolar strain from

0.068% for CMO synthesized sample to 0.072% and 0.075% were observed for MSS and TMC synthesized samples.

5. Acknowledgements
This work is supported by the National Research Foundation of Korea (NRF) grant funded by the Korean government (MOE) (2013R1A1A2058345) and Basic Research program through the National Research Foundation of Korea (NRF) funded by Ministry, Science and Technology (MEST) (2011-0030058).

6. References
[1] Uchino K1997 *Piezoelectric Actuators and Ultrasonic Motors*(Boston, Dordrecht, London: Kluwer Academic Publishers)
[2] Haertling G H1999 Ferroelectric ceramics: history and technology*J. Am. Ceram. Soc.* 82 797-818
[3] Scott JF2000 *Ferroelectric Memories* (Heidelberg, Germany: Springer-Verlag)
[4] Jaffe B, Cook W R and Jaffe H 1971 Piezoelectric ceramics vol 1 (New York: Academic Press)
[5] EU-Commision Directive 2013/28/EU of 17 May 2013 amending annexII to directive 2000/53/EC of the European Parliament and of the Councilon end-of-life vehicles *Off. J. Eur. Union*L135 14–8.3
[6] Smolenskii G A, Isupov V A, Agranovskaya A I and Krainik N N 1961 New ferroelectrics of complex composition Sov. Phys. Solid State 2 2651-4
[7] Cohen R E 1992 Origin of ferroelectricity in Perovskite Oxides*Nature*358136-138
[8] Jones G O and Thomas P A2000 The tetragonal phase of $Na_{0.5}Bi_{0.5}TiO_3$ a new variant of the perovskite structure*Acta Crystallogr. Sect. B: Struct. Sci.* 56426-430
[9] Jones G O and Thomas P A 2002 Investigation of the structure and phase transitions in the novel A-site substituted distorted perovskite compound $Na_{0.5}Bi_{0.5}TiO_3$*Acta Crystallogr. Sect. B: Struct. Sci.* 58 168-178
[10] Suchanicz J 1998 Behaviour of $Na_{0.5}Bi_{0.5}TiO_3$ ceramics in the a.c. electric field*Ferroelectrics* 209 561-568
[11] JZvirgzds J A, Kapostis P and Zvirgzde J 1982 X-ray study of phase transitions in ferroelectric $Na_{0.5}Bi_{0.5}$TiO3*Ferroelectrics* 40 75-77
[12] Suchanicz J, Roleder K, Kania A and Handerek J1988 Electrostrictive strain and pyroeffect in the region of phase coexistence in $Na_{0.5}Bi_{0.5}TiO_3$*Ferroelectrics*77 107-110
[13] Hussain A, Rahman JU, Ahmed F, Kim JS, Kim MH, Song T K and Kim W J 2015Plate-like $Na_{0.5}Bi_{0.5}TiO_3$ particles synthesized by topochemical microcrystal conversion method*J. Eu. Ceram. Soc.* 35 919-925
[14] Hussain A, Maqbool A, Malik R A, Rahman J U, Song T K, Kim W J and Kim M H 2015 $Na_{0.5}Bi_{0.5}TiO_3$–$BaZrO_3$ textured ceramics prepared by reactive template grain growth method*Ceram. Int.*141 S26-S30
[15] Rahman JU, Hussain A, Maqbool A, Song TK, Kim WJ, Kim SS and Kim M H 2014 Dielectric, ferroelectric and field-induced strain response of lead-free $BaZrO_3$-modified $Bi_{0.5}Na_{0.5}TiO_3$ ceramics*Curr. Appl. Phys.* 14 331-336
[16] Maqbool A, Hussain A, Rahman JU, Jong KP, Park TG, Song JS, Kim MH 2014 Ferroelectric and piezoelectric properties of $SrZrO_3$-modified $Bi_{0.5}Na_{0.5}TiO_3$ lead-free ceramics *Trans. Nonferrous. Met. Soc. China* 24 s146-s151
[17] Li Y, Chen W, Xu Q, Zhou J, Gu X and Fang S 2005 Electromechanical and dielectric properties of $Na_{0.5}Bi_{0.5}TiO_3$-$K_{0.5}Bi_{0.5}TiO_3$-$BaTiO_3$ lead-free ceramics*Mater. Chem. Phys.* 94328-332
[18] Su S and Zuo R 2012 Fabrication and electrical properties of $0.94Na_{0.5}Bi_{0.5}TiO_3$-$0.06BaTiO_3$ textured ceramics by RTGG method using micrometer sized $BaTiO_3$ plate-like templates *J. Alloys compd.* 525 133-136

Synthesis and characterization of mesoporous hydrocracking catalysts

D. Munir and M. R. Usman
Institute of Chemical Engineering and Technology, University of the Punjab, Lahore 54590, Pakistan

E-mail: dureem83@gmail.com

Abstract. Mesoporous catalysts have shown great prospective for catalytic reactions due to their high surface area that aids better distribution of impregnated metal. They have been found to contain more adsorption sites and controlled pore diameter. Hydrocracking, in the presence of mesoporous catalyst is considered more efficient and higher conversion of larger molecules is observed as compared to the cracking reactions in smaller microporous cavities of traditional zeolites. In the present study, a number of silica-alumina based mesoporous catalysts are synthesized in the laboratory. The concentration and type of surfactants and quantities of silica and alumina sources are the variables studied in the preparation of catalyst supports. The supports prepared are well characterized using SEM, EDX, and N_2-BET techniques. Finally, the catalysts are tested in a high pressure autoclave reactor to study the activity and selectivity of the catalysts for the hydrocracking of a model mixture of plastics comprising of LDPE, HDPE, PP, and PS.

1. Introduction

In the past few decades, there has been a massive increase in the demand of plastics worldwide, in all sectors of industry and also in daily life. The production of plastic products are increasing day by day due to its light weight, durability, low cost, low degradation rates, and good insulating properties. This huge plastic production results in the generation of gigantic plastic waste. About 25% of all the solid waste comprised of plastics. These plastic wastes are treated by the popular waste management techniques that include landfilling, incineration, and recycling. Landfilling of plastics is not a valued method of waste disposal. Plastic waste occupies a huge volume of landfill for a longer time due to its slow degradation rate. Moreover, the plastic waste in a landfill is a source of many environmental hazards. Incineration of plastic waste, for getting energy, results in the emission of poisonous gases that are unacceptable from the environmental point of view. Feedstock recycling or chemical recycling for waste plastics is highly encouraged all over the world. It includes chemical depolymerization to monomers, gasification, pyrolysis, catalytic cracking and reforming, and hydrocracking [1]. Compared to the other chemical recycling methods, hydrocracking of waste plastic into liquid fuels is highly advantageous process. The fuel obtained as the result of pyrolysis and catalytic cracking is highly unsaturated and found unfit to be used as engine fuel [2]. However, high quality liquid fuel is obtained as the result of waste plastic hydrocracking. Hydrocracking of waste plastic requires a dual-functional catalyst that has a hydrogenation-dehydrogenation function as well as acidic function [3]. Microporous acidic zeolite supports such as HZSM-5 [1, 4-12]and HY [4, 12, 13] have been used by many researchers and resulted in appreciable conversions. However, most of the converted product

Content from this work may be used under the terms of the Creative Commons Attribution 3.0 licence. Any further distribution of this work must maintain attribution to the author(s) and the title of the work, journal citation and DOI.
Published under licence by IOP Publishing Ltd

comprised of gases and a rather small quantity of liquid was obtained due to the microporous cavities of zeolites that hinder diffusion of larger molecules of plastics. On the other hand, the role of mesoporous supports for waste plastic hydrocracking has been studied by only a few investigators. Mesoporous catalysts are expected to demonstrate great prospective for hydrocracking reactions due to their high surface area, that aids better distribution of the impregnated metal, and larger pore distribution. Mesopores in these catalysts allow the diffusion of heavy molecules and result in the production of more liquid fuel as compared to gaseous product. The only shortcoming of mesoporous catalysts is their weak acidic character. In the present study, mesoporouscatalysts that contain strong acid sites are synthesized by novel methods. These catalysts contain the acidity along with mesoporosityrequired for the hydrocracking of heavy molecules like polymers. The aluminum is incorporated during the gel formation in a direct synthesis approach. Different Si/Al ratios are used and their effect on the hydrocracking of a model waste plastic mixture is investigated.

2. Experimental

2.1. Materials

Model waste plastic feed used for the hydrocracking reactions is a mixture of virgin high density polyethylene(HDPE, d=0.952g/cm^3), low density polyethylene(LDPE, d=0.918g/ cm^3), polypropylene (PP, MW = 250000,d=0.9g/cm^3), and polystyrene (PS, MW = 192000). Tetraethylor-thosilicate (TEOS, 98% purity), sodium hydroxide, aluminum is opropoxide (≥98% purity), HCl (35.4% purity), F127, cetyltrimethylammoniumbromide (CTAB, ≥99% purity), and tetrapropyl ammonium bromide(TPAB) are used in the synthesis of catalysts. Apart from HCl (35.4% purity) and NaOH, all the chemicals were purchased from Sigma-Aldrich and used without any further purification. All the solutions are made in a double distilled water.

2.2. Catalyst preparation

2.2.1 Al-SBA16(10)and Al-SBA16(20)

In a typical procedure F127 was dissolved in 2.0 M HCl with continuous stirring for 1.0 h at room temperature. Afterwards, 6.0 g of TEOS was added into it. The mixture was hydrolyzed with vigorous stirring for 4 h at 40°C. Then, a calculated amount of aluminum isopropoxide was added for achieving Si/Al wt. ratio of 10. This solution was allowed to stir vigorously, overnight, to form ordered mesoporous structure. For hydrothermal treatment, the mixture was transferred into Teflon lined stainless steel autoclave and crystallized at 100°C in an oven for a complete day. The final product was centrifuged and washed several times with double distilled water. The catalyst was dried in oven for 20 h at 100°C. The catalyst was calcined for 6 h in a muffle furnace at 600°C where the temperature increased from room temperature to the calcination temperature at the rate of 2°C/min. The catalyst was named as Al-SBA16(10). Al-SBA16(20) was prepared with the same method with Si/Al wt. ratio of 20.

2.2.2 Mesoporous zeolite

Mesoporous zeolite was prepared by dissolving TPAB in double distilled water. After complete dissolution, TEOS was added into it. The mixture was hydrolyzed with vigorous stirring at room temperature. Aluminum isopropoxide was added in calculated amount in 0.1 M NaOH solution for achieving Si/Al wt. ratio of 20. The two solutions were mixed together and stirred vigorously. After this, CTAB was dissolved into the solution at room temperature. The mixture resulted was again subjected to stirring. For hydrothermal treatment, the mixture was transferred into Teflon lined stainless steel autoclave and crystallized at 120°C in an oven for 2 days. The final product was centrifuged and washed several times with double distilled water. The catalyst was dried in oven for 20 h at 100°C. The catalyst was calcined for 5.0 h in a muffle furnace at 550°C. The catalyst was ion exchanged in order to achieve proton form. The catalyst was then filtered and washed with double distilled water and dried in an oven at 100°C for 12 h. The dried catalyst was calcined again at 550°C for 4.0 h in a muffle furnace. The catalyst was named as MZ-20.

2.3 Catalyst characterization

Scanning electron microscopy (SEM) of samples were obtained by using Field Emission Scanning Electron microscope. All images were obtained using samples coated with carbon (sputtering) to reduce the effects of charging. Energy Dispersive X-ray (EDX) spectroscopy was performed for elemental analysis to find out the elemental composition of the catalysts. Potential silica to alumina ratio was evaluated by using this technique. Micromeritics TriStar II-3020 (Surface Area and Porosity Analyzer) was used for N_2 adsorption/desorption isotherms of the catalysts at 77.3 K. All the samples were degassed under vacuum at 200°C for 2 h, by using Micromeritics Smart Prep (Programmable Degas System) before the measurement. The total pore volume was estimated from the analysis of nitrogen amount adsorbed at a relative pressure (P/P_0) of 0.9933 by assuming negligible adsorption on the external surface as compared to adsorption in the pores. For all the analysis, correlation coefficients of 1.0 were obtained. N_2-BET surface area was calculated by using Brunauer-Emmett-Teller (BET) method. Pore surface area and pore volume were estimated by t-plot method. Barrett-Joyner-Halenda (BJH) model was used for the calculation of pore size distribution using adsorption branch of the isotherm. The estimation of the size of a mesopore was carried out by using BJH procedure to the adsorption branch of the isotherm.

2.3 Activity tests

Catalysts activity is evaluated by using them for the plastic hydrocracking. A waste plastic model mixture is prepared with HDPE (40%), LDPE (10%), PS (20%), and PP (30%). Feed to catalyst ratio used for all the reactions is 10:5 by wt. The reactions are performed at 400°C with reaction time of 50 min. The reactions are carried out in a high pressure 600 mL SS autoclave reactor (Parr Instrument Co.) under initial H_2 pressure of 20 bar at room temperature. After the reaction, the reactor is allowed to cool down. The gases are collected at room temperature. The remaining product obtained is solvent extracted in order to evaluate the yield of oil (n-heptane solubles), liquid (THF solubles), and solids (the THF insolubles containing catalyst, coke, and unreacted plastic remained after drying in an oven).

3. Results and discussion

Figure 1 (a-c) are representative SEM photographs of Al-SBA16(10) consisting of mesoporous SBA16 with Si/Al ratio of 10. The images show some irregular shaped particles with rough surfaces along with some randomly shaped particles, as is observed by the other researchers [15]. The surface is rough with marble like texture which seems to exhibit amorphous mesoporous phase.

Figure1 (d-f) are the SEM photographs of Al-SBA16(20). The surface of the catalyst seems irregular. Aggregated spherical particles of small size could be seen along with the presence of some macrospheres. The loose aggregates of spherical particles may indicate the presence of mesoporous phase [14]. However, macrospheres are formed due to the higher silica content [15] in Al-SBA16(20) as compared to Al-SBA16(10). Figure1 (g-i) are representative SEM photographs of MZ-20. The images show some loose aggregates along with some spherical as well as cubical particles. The loose aggregates found may indicate the presence of mesoporous MCM-48 phase [14]. The cubic particles seen, may demonstrate the presence of ordered zeolite structure. It is observed by the images that the cubic particles are interconnected with the mesoporous spherical aggregates, thus strongly support the concept of mesoporous zeolite structure formation.

Table 1 shows the Si/Al ratio of the gel and actual Si/Al ratio in the catalysts, as obtained from EDX analysis. It is found that the Si/Al ratios obtained from the EDX analysis are nearly the same as that of the gel. This could be attributed to their formation in neutral medium that inhibits the possible aluminium leaching

Structural properties of the two composite catalysts are summarized in Table2.BET surface area of Al-SBA16(10) and MZ-20 is higher than that of Al-SBA16(20). The mesoporous area, as is evaluated from BJH area of all the catalysts, is much higher than that of their corresponding microporous area. It shows the presence of large area with mesopores. The microporous area of MZ-

20 is higher than the microporous area of the other two catalysts. It shows that the catalyst is composite with micropores and mesopores.

Pore size distribution of catalysts is demonstrated in Figure 2. Al-SBA16(10) shows a narrow pore size distribution cantered at 4 nm. Pore sizes are in the range of 2-32 nm. Al-SBA16(20) demonstrates bimodal pore size distribution at 4.5 nm and 11.5 nm. Both are in the mesoporous range. Pore size distribution of MZ-20 also shows a bimodal pore size distribution with most of the pore sizes are 5 nm in diameter. Pore sizes are in the range of 2-24 nm in that catalyst sample. Micropores of 2 nm are present and mesopores of up to 24 nm show their existence. Pore size distribution of MZ-20 confirms its composite structure with micropores and mesopores.

Table 1. Si/Al weight ratios in gel and that obtained by EDXanalysis of different catalysts.

Name of catalyst	Si/Al (initial)	Si/Al (EDX)
Al-SBA16(10)	10	10.84
Al-SBA16(20)	20	19.52
MZ-20	20	20.36

Table 2. Structural properties of different catalyst samples

Samples	**Surface area**[a] **(m^2/g)**	**Microporearea**[b] **(m^2/g)**	**BJH area**[c] **(m^2/g)**	**Micropore volume**[d] **(cm^3/g)**	**BJH volume**[e] **(cm^3/g)**	**BET Pore diameter**[f] **(nm)**	**BJH Pore diameter**[g] **(nm)**
Al-SBA16(10)	220.98	5.81	219.5	0.0030	0.4866	9.389	8.868
Al-SBA16(20)	110.01	5.65	101.5	0.0033	0.3426	13.17	13.51
MZ-20	239.50	17.6	205.9	0.0057	0.6350	11.32	12.33

a). Surface area by using BET method
b). Micropore area obtained by t-plot
c). BJH adsorption cumulative surface area of the pore between 17Å and 3000Å width
d). Micropore volume obtained by t-plot
e). BJH adsorption cumulative volume of the pore between 17Å and 3000Å width
f). Adsorption average pore dia (4V/A) by BET
g). BJH Adsorption average pore dia (4V/A)

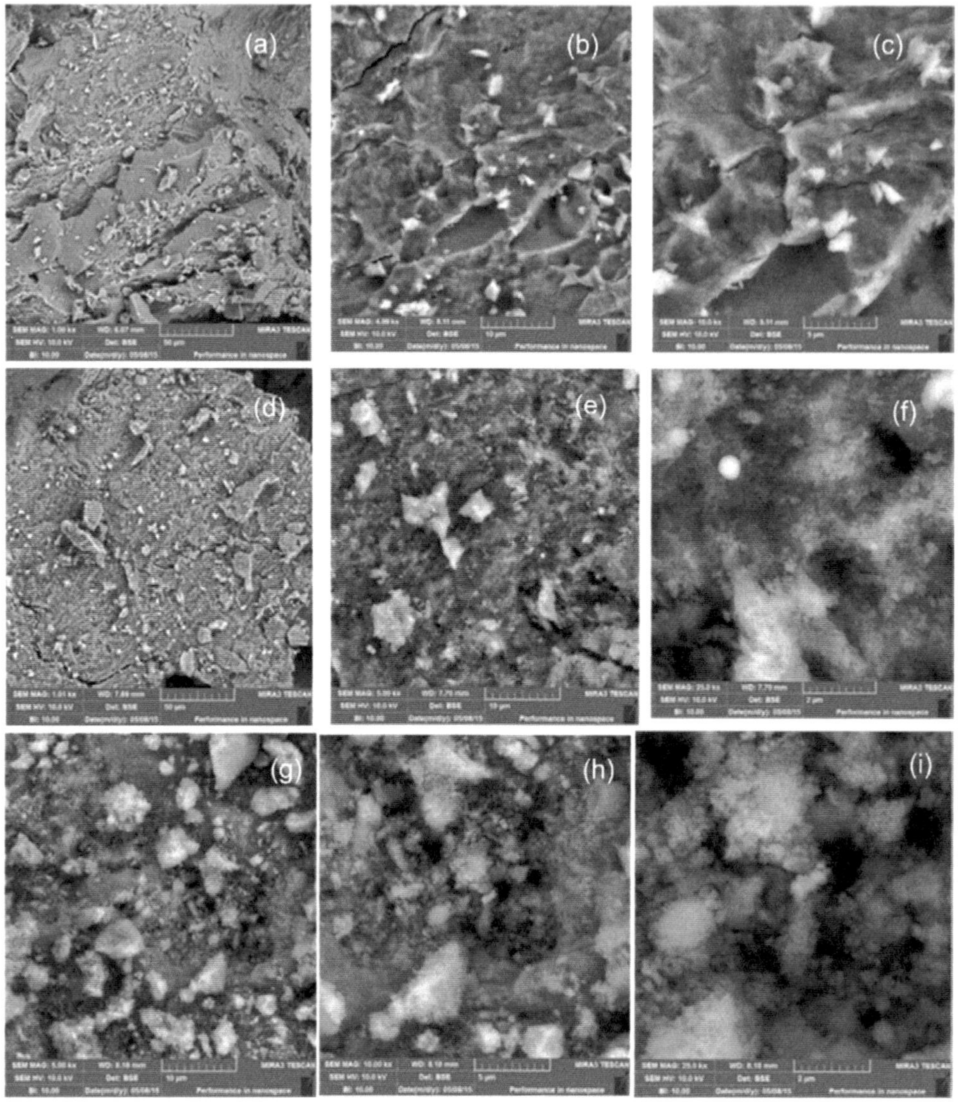

Figure 1. SEM images of catalysts: (a-c) Al-SBA16(10), (d–f) Al-SBA16(20), (g–i) MZ-20

The results of catalyst activity tests are shown in Table 3. The catalyst activity tests in high pressure autoclave reactor with model waste plastic mixture show that improved results of activity and selectivity are obtained with Al-containing SBA catalysts compared to micro-mesoporouscatalyst, MZ-20, and no catalyst conditions. The minimum conversion and lower yield of oil are obtained using MZ-20 catalyst where the THF insolubles are the highest. For the two Al-SBA catalysts,the conversions are virtually similar but the higher oil is produced with Al-SBA16(20), a mesoporous catalyst having higher silica-alumina ratio and broader range of mesopores that may aid in the formation of higher yield of liquid fuel. Important to note is that the surface area of the MZ-20 is the highest among the catalysts, shown in Table 2, however, it has given the minimum oil yield while the surface area of Al-SBA16(20) is the minimum but it yields the maximum oil product though the Si/Al ratios of the two catalysts are similar. It may be concluded that pore size distribution is also a critical factor for obtaining the desired product yield.

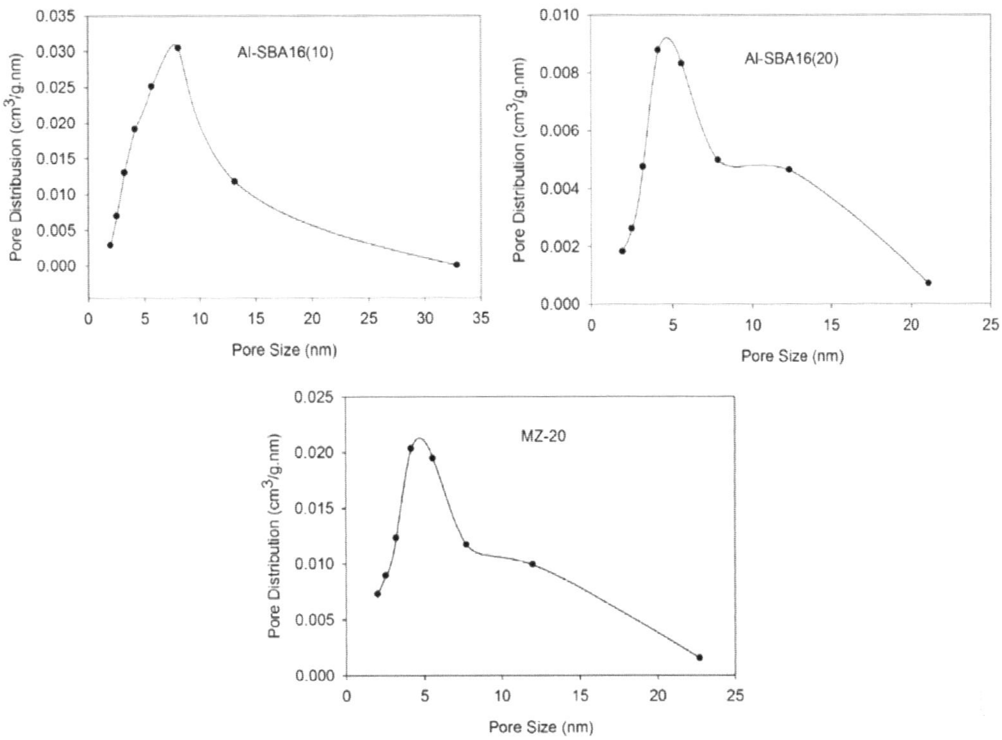

Figure 2. Pore size distribution of Al-SBA16(10), Al-SBA16(20), and MZ-20 catalysts

Table 3. Results of hydrocracking of model plastic mixture

Run	Catalyst used	Gas yield (wt%)	Oil yield (wt%)	Total liquid yield (wt%)	Solid yield (wt%)	Conversion (wt%)
1	No catalyst	12.9	13.0	22.2	61.1	38.9
2	Al-SBA16(10)	8.34	21.0	27.4	59.7	40.2
3	Al-SBA16(20)	3.89	26.3	31.0	60.3	39.7
4	MZ-20	4.97	12.0	21.0	69.5	30.5

4. Conclusion

The characterization results of the catalysts have revealed the presence of mesopores in the catalysts and confirm the formation of composite structure of micro-mesoporous zeolites. A higher silica-alumina ratio results in the formation of macrospheres which is a characteristic feature of high silica containing mesoporous catalysts. Higher silica-alumina ratio catalyst, Al-SBA(16)20, shows larger pore diameter, slightly larger pore volume, and lower surface area and shows a bimodal pore size distribution compared to lower silica-alumina ratio catalyst, Al-SBA(16)10, which shows a unimodal pore size distribution. Al-SBA(16)10 gives the highest gas yield showing it has higher cracking activity characteristics whereas highest liquid yield is obtained over Al-SBA-16(20) with highest pore diameter. It is therefore concluded that silica-alumina ratio and pore size distribution are the key factors in determining the required activity and selectivity of a catalyst. Surface area solely does not seem to play decisive role in the cracking ability of a catalyst.

IOP Conf. Series: Materials Science and Engineering **146** (2016) 012007 doi:10.1088/1757-899X/146/1/012007

5. Acknowledgments

We acknowledge Higher Education Commission of Pakistan for funding the project NRPU-12470. We also acknowledge Mr. Abdullah and Mr. Hassaan for assisting us in different ways.

6. References

[1] Aguado J, Serrano D P, Escola J M, Garagorri E and Fernández J A 2000 *PolymDegrad Stab* 69 11
[2] Walendziewski J and Steininger M 2001 *Catal. Today* 65 323
[3] Venkatesh K R, Hu J, Wang W, Holder G D, Tierney J W and Wender I 1996 *Energy Fuels* 10 1163
[4] Ochoa R,Woert HV, Lee W H, Subramanian R, Kugler E andEklund PC 1996*Fuel Process Technol*49 119
[5] Liu K andMeuzelaar HL C 1996 *Fuel Process Technol* 49 1
[6] Luo M and Curtis CW 1996 *Fuel Process Technol* 49 91
[7] Ding W, Liang J and Anderson L L 1997 *Fuel Process Technol* 51 47
[8] Ding W, Liang J and Anderson L L 1997 *Energy Fuels* 11 1219
[9] Shah N, Rockwell J and Huffman GP 1999*Energy Fuels* 13832
[10] Karagöz S,Yanik J, Ucüar S and Song C 2002*Energy Fuels* 16 1301
[11] Ali MF, Siddiqui M N andRedhwi SHH2004*J Mater Cyc Waste Manag* 6 27
[12] Hesse ND and White RL 2004 *JAppl Pol Sci* 921293
[13] Akah A, Hernandez-Martinez J, Rallanb C andGarforth AA2015 *ChemEng Trans*43
[14] Guo W, Xiong C, Huanga L and Li Q 2001 *J MaterChem* 11 1886
[15] Wesley J J S, Kurt L, Myrjam M, Gustaaf V T, Pegie C, and Etienne F V 2006 *J. Phys. Chem. B*110 9183

IOP Conf. Series: Materials Science and Engineering 146 (2016) 012008 doi:10.1088/1757-899X/146/1/012008

Development of material model for assessment of brittle cracking behavior of plexiglas

A. J. Khan, N. Iqbal, H. A. Saeed and W. A. Tarar
Department of Mechanical Engineering, College of E & ME,
National University of Science & Technology, Rawalpindi, Pakistan

E-mail: ali_khanpk@hotmail.com

Abstract. The objective of this study is to investigate the brittle cracking behavior of Plexiglas material when subjected to indentation loading. Indentation tests were conducted on Modified Vickers testing machine to acquire the experimental data in the form of load-displacement curve. Several mechanical properties such as hardness, yielding stress and fracture toughness have been ascertained from the analysis of the experimental data. The experimental data then used to create a mathematical model of Plexiglas for its brittle cracking behavior with indentation loading. Furthermore, a numerical simulation based study was carried out to simulate the brittle cracking in Plexiglas plate when subjected to indentation loading. The simulations were performed in the FE solver Abaqus. The brittle cracking model in Abaqus/Explicit is used which determines the required force and displacement to produce crack in Plexiglas. Finally a comparison of simulation results was made to the experimental data to validate the FEA procedures and accuracy of predictions. The numerical predictions of load-displacement curve found remarkably consistent with experimental results.

1. Introduction

Indentation tests have been commonly employed because they do not require elaborate sample preparation and are straightforward to execute. Initially they were used for determining the hardness of a material. Indentation techniques have also been used to find fracture toughness and other mechanical properties of material.

Finite element method (FEM) is the most powerful numerical technique for understanding mechanical behavior of materials and has become more common day by day because it's a time saving and cost effective alternative. It has been successfully used to solve various indentation problems with sufficient accuracy. Numerical simulations using finite element methods have also been employed to investigate the indentation fracture characteristics of brittle materials. Experiments and simulations have indicated that when brittle materials are subjected to indentation, various forms of cracking are detected [1-4] including median, radial, lateral cracking etc. The nature of cracking generally depends on the shape of the indenter and on the properties of the material.

Plexiglas is the common name for a type of polymer that is a cross between plastic and glass. It is a commercial name of PMMA (poly methyl methacrylate). It is also sometimes called acrylic glass and has many properties of glass. It has been widely used for residential, military, medical and commercial purposes as it possesses high toughness and transparency.

Various researchers and investigators have studied the behaviour of a Plexiglas under different mechanical and environmental conditions. R. Rikards et al. [5] developed a numerical and

Content from this work may be used under the terms of the Creative Commons Attribution 3.0 licence. Any further distribution of this work must maintain attribution to the author(s) and the title of the work, journal citation and DOI.
Published under licence by IOP Publishing Ltd

experimental method to determine elasto-plastic properties i.e. yield stress, tangential modulus and hardness of different polymers when subjected to Vickers indentation. Zhouhua Li et al. [6] investigated the thermo-mechanical behavior of polymethyl methacrylate (PMMA) when subjected to compressive dynamic loading. SaiSarva et al. [7]carried out experimental and analytical investigation to understand the high rate deformation and projectile impact behavior of polycarbonate and poly-mathylmethacrylate and develop a new constitutive model. Based on a chain of experiments and by using theoretical model (ZWT), Tao Suo et al. [8] investigate the temperature and strain rate effect on acrylic polymer during quasi-static and dynamic loading. N.K. Naik et al. [9] have carried out experimental studies and presented an analytical method to understand the effect of strain rate on tensile properties of acrylic. By using both experimental and simulation through commercial software Abaqus, Arjun Tekalur et al. [10] have studied the failure mechanism of PMMA plate when subjected to low velocity impact response. From the literature, it is ascertained that different analytical, experimental and numerical investigations are carried out in order to study the behavior of Plexiglas and other polymers under different mechanical and environmental conditions. However, in order to simulate the brittle cracking behaviour of Plexiglas, no appropriate model is available in literature. The study of brittle cracking in Plexiglas is of great interest due to its wide range of application.

In this paper, the failure mechanism of PMMA plate subjected to indentation loading is investigated. For this purpose indentation experiments were conducted on Vickers testing machine to acquire the data. This data is then used to derive various mechanical properties as well as to develop a mathematical model for Plexiglas. Furthermore a numerical simulation based study is made to predict the brittle cracking in Plexiglas plate when subjected to indentation loading. Finally a comparison of simulation result with the experimental result is given to validate the FEA procedures and accuracy of predictions.

2. Experimental

2.1 Procedure

Experiments of hybrid nature are performed in which a sharp tipindenter is pressed into a flat surface of Plexiglas under a specific load for a definite period of time. The indenter employed in the test is conical with self-similar geometry. The Plexiglas specimen has 25.4 mm diameter and 3 mm thickness. The time was few seconds to reach the maximum load regardless of its volume. These tests are based on technology that allows for either load-controlled or depth-controlled experiments. The applied load, penetration depth and cycle time are continuously measured during these tests. The mechanical properties are determined by analysing the geometrical dimensions of the indent and from the analysis of a load-depth curve. All the experiments were performed at room temperature. Our resulting data is in the form of load displacement curves. In our final output, we have both the plots and the pictures of cracking. The same tests were repeated thrice, and similar results were obtained repeatedly.

2.2 Indentation cracking in plexiglas plate

Figure 1 shows an image after a conical indenter was forced into the surface of Plexiglas. Two cracks are initiated from the edge of the indent. In the literature, these kinds of cracks are referred to as radial cracks [11]. Therefore it can be seen in these experiments that when a conical indenter is indented into Plexiglas materials, radial cracks develop normal to the surface after a critical load has been reached.

2.3 Analysis of indentation load displacement curve

Primary information about an investigated Plexiglas material in this test is described by the load-displacement curve. The load-displacement data obtained from the indentation process contain the information about hardness, fracture toughness, maximum crack load etc. Figure 2 displays the average load–displacement curve acquired from these indentation experiments on Plexiglas material. As shown, when the indentation force reaches to a value of 400 N, the crack is initiated in Plexiglas material. This P-h curve can be used to derive different mechanical properties of materials such as the hardness and fracture toughness of material.

Figure 1. Indentation in Plexiglas with the conical indenter showing well defined radial cracks

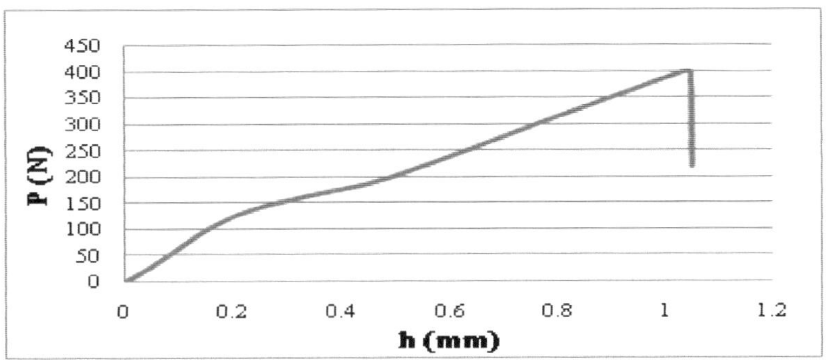

Figure 2. Average load–deflection curve

3. Data Processing and material modelling

3.1 Measurement of hardness and yield stress

Hardness is an important property of a material that enables it to resist deformation, usually by penetration of a hard indenter. The indentation hardness is calculated from the peak indentation load, P_{max}, divided by the projected contact area A_p at the contact depth h_c:

$$H = \frac{P_{max}}{A_P} \qquad (1)$$

For conical indenter as shown in Figure 3, the projected contact area under the maximum load is determined by

$$A_p = \pi {h_c}^2 tan^2\theta \qquad (2)$$

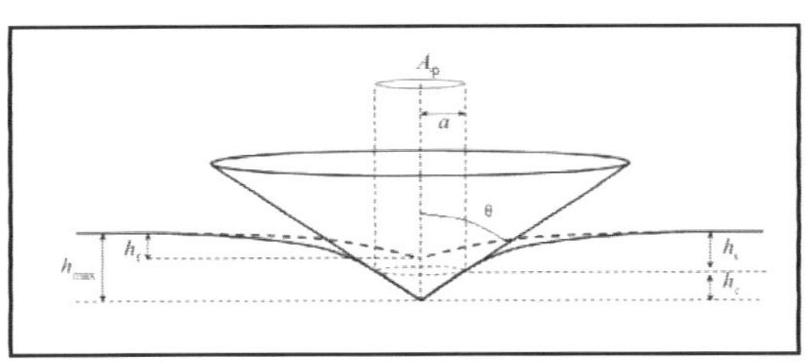

Figure 3. Geometry to characterize Indentation

The specimen's yield stress, σ_Y, is directly proportional to the hardness and can be expressed as

$$H \approx C \times \sigma_Y \quad (3)$$

Where C ≈ 1.5 for materials with a low ratio of $\frac{E}{\sigma_Y}$ such as polymers, glass etc. By substituting values in equations 1 and 3, the indentation hardness and yield stress can be calculated as shown in Table 1.

Table 1. Properties of Plexiglas material used in measurement of indentation hardness and yield stress.

Material	h_c(mm)	Ap(mm^2)	P(N)	H(MPa)	σ_Y(MPa)
Plexiglas	0.664	4.155	400	96	64

3.2 Measurement of fracture toughness

Accurately measuring the fracture toughness of brittle materials can often be challenging. Different indentation techniques could be used to determine the fracture toughness of brittle materials. Assessing fracture toughness by making direct measurements of cracks created using a sharp indenter can appear to be an attractive alternative to traditional fracture toughness testing techniques. As is evident from Figure 1, when a sharp tip such as conical indenter is indented into brittle materials, radial cracking usually occurs after a critical load is reached, which can be used to calculate fracture toughness based on the maximum indentation load and the crack length. The equation used to calculate the material fracture toughness Kc by indentation is

$$K_C = \alpha \left[\frac{E}{H}\right]^{1/2} + \left[\frac{P}{C_0^{3/2}}\right] \quad (4)$$

Where P is the indentation load, H is the hardness, and c is the average radial crack length as shown in Figure 4. α is a calibration constant and its is considered 0.04 for sharp tip indentation geometries.

By substituting the values of elastic modulus, E, hardness, H, crack length and maximum indentation load in equation 4, the fracture toughness value is obtained as shown in Table 2.

Table 2. Properties of Plexiglas material used in indentation cracking measurement of fracture toughness.

Material	E(MPa)	H(MPa)	P(N)	C_o(mm)	K_c (MPa.mm$^{1/2}$)
Plexiglas	3100	96	400	2.4	0.773

3.3 Strain graph for Conical Indentation

The stress and strain values are determined from the force displacement data by following calculation. The average stress can be calculated from the indentation force, P, divided by the true contact area A, as follow

Stress = Force/Contact Area

$$\sigma = \frac{P}{A} \quad (5)$$

For simple conical indenter as shown in Figure 5, the true contact area A at indentation depth h can be calculated as

$$A = \pi h^2 tan^2 \alpha \quad (6)$$

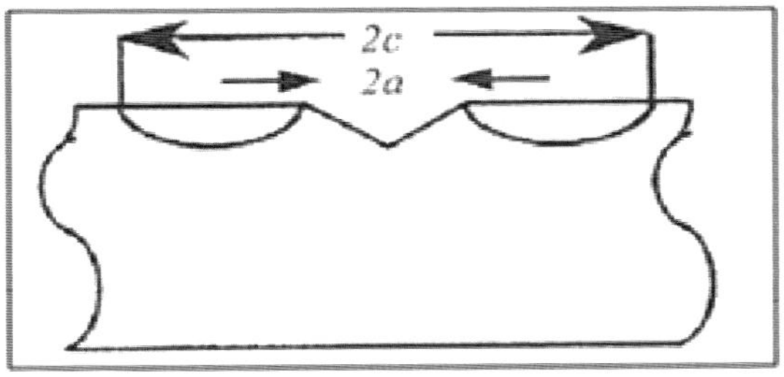

Figure 4. Radial crack (side view)

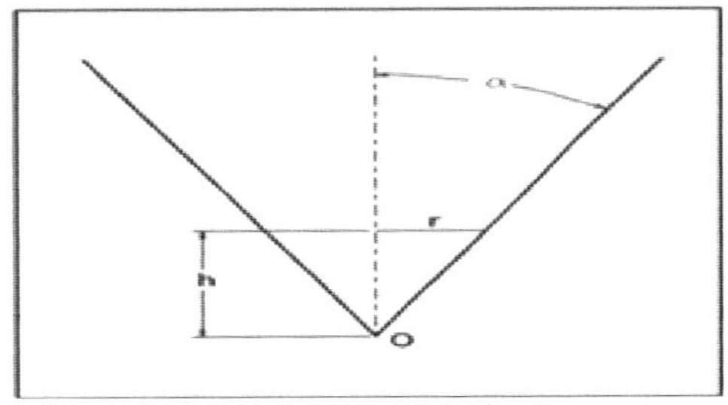

Figure 5. Cone indenter

The average strain in the direction of the indentation load can be calculated by dividing the change in depth, h, by the initial depth h_o:

Strain = change in depth/ initial depth

$$\varepsilon = \frac{h}{h_0} \quad (7)$$

After calculating the stress strain values from the load displacement data, the stress strain graph is generated as shown in Figure 6.

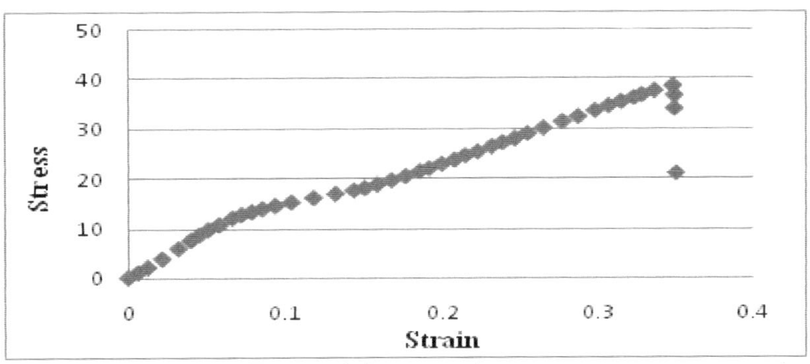

Figure 6. Stress strain curve

3.4 Mathematical model created using curve fitting technique

After getting the stress strain graph, a mathematical model is developed for this graph by curve fitting of the data. This model can be used for data visualization to infer values of a function where no data are available and to summarize the relationships among the variables. The model is created by using Fourier fit to the data points in curve fitting toolbox of Matlab as shown in Figure 7. The mathematical model is expressed as

$$\sigma = a_0 + a_1 \times cos(\varepsilon \times \omega) + b_1 \times sin(\varepsilon \times \omega) + a_2 \times cos(2 \times \varepsilon \times \omega) + b_2 \times sin(2 \times \varepsilon \times \omega) + a_3 \times cos(3 \times \varepsilon \times \omega) + b_3 \times sin(3 \times \varepsilon \times \omega) + a_4 \times cos(4 \times \varepsilon \times \omega) + b_4 \times sin(4 \times \varepsilon \times \omega) \quad (8)$$

The values of the model coefficients are shown in Table 3.

Table 3. Coefficients of mathematical model values

Coefficients	a_0	a_1	b_1	a_2	b_2	a_3	b_3	a_4	b_4	w
Values	1.491e6	-1.78e6	-2.294e6	2.101e4	2.055e6	3.703e5	-7.152e5	-1.025e5	8.268e4	1.664

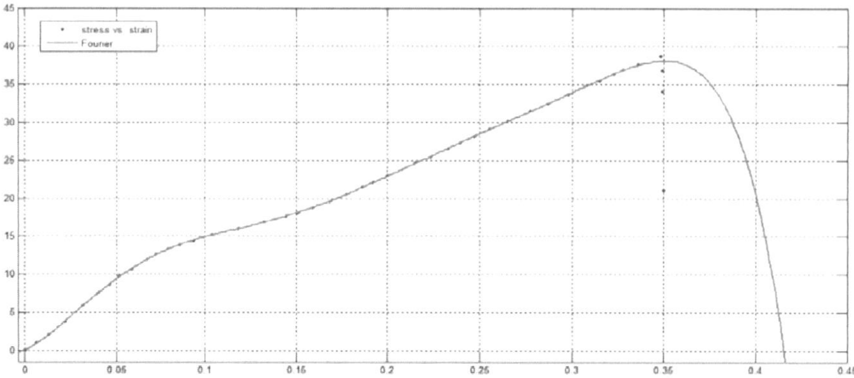

Figure 7. Stress strain curve with Fourier fit

4. Numerical simulation

The use of numerical simulations for understanding mechanical behavior of materials is becoming more popular due to its advantage of time saving and cost effectiveness. In this study acommercial FE package Abaqus 6.13 is used to simulate the brittle cracking in Plexiglas. Modeling of a conical indentation experiment can be accomplished by mean of an axi-symmetric model as the system is symmetric about an axis. The numerical model consists of two parts created separately: a plate and an indenter. Since the indenter is much stiffer than the plate, the indenter is considered to be perfectly rigid and is modeled as analytical rigid surface. The reference point is assigned to the indenter, which manipulates the rigid body translation of the indenter. The plate has been created as 2D axi-symmetric deformable with 25.4 mm diameter and 3 mm thickness. The plate was modeled using four noded bilinear axi-symmetric quadrilateral elements. The element type "CAX4R" is used. The contact interactions were defined as surface to surface contact, which can describe the contact between a deformable surface and a rigid body. The indenter surface is defined as the "master" surface since the indenter is rigid body. The top of the specimen is the "slave" surface. In this work it is assumed that the friction effect is negligible and $\mu=0$ is defined in all the models assuming there is no slip between the surfaces in contact.

The indenter is constrained in all directions except in the vertical direction normal to the plate surface (y-axis). The bottom of the plate is constrained axially and YASYMM boundary conditions are applied to the left side of the plate. The boundary conditions of plate and indenter are shown in Figure 8. Slightly finer mesh is created around the indenter and the element size is increased making the mesh denser away from the indenter towards the boundary of the model as shown in Figure 9.The indentation test was simulated by applying a downward displacement of 1.045 mm to the indenter which causes the indenter to push into the material to be indented.

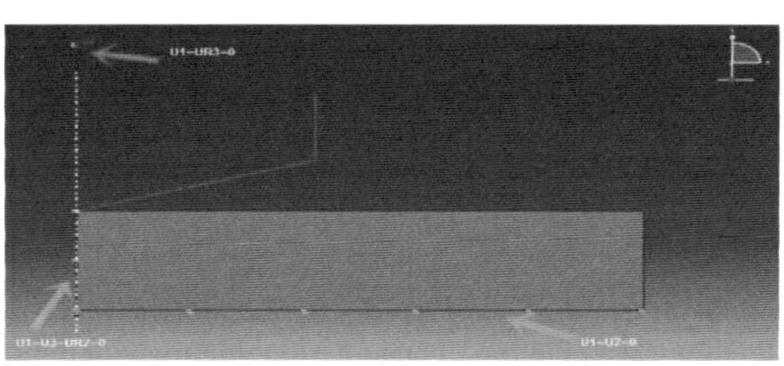

Figure 8. Boundary conditions applied in the model.

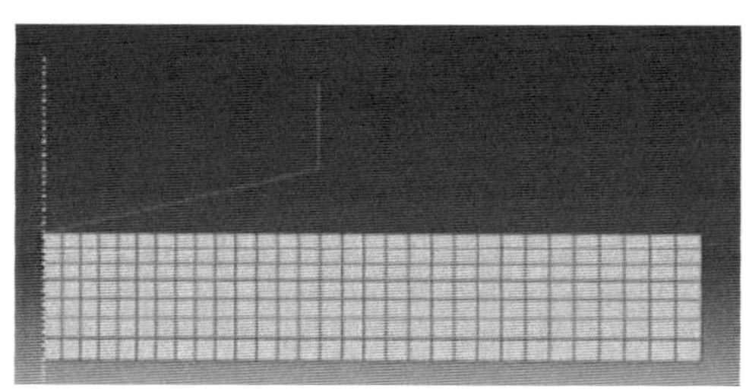

Figure 9. Meshed model of plate.

1.1. Brittle cracking model for plexiglas

The brittle cracking material model in Abaqus/Explicit is used to predict brittle cracking behavior of Plexiglas material when subjected to indentation loading. The brittle cracking parameters are extracted from the stress strain data obtained earlier and final values are fixed after an admissible result is obtained. The values of Plexiglas properties and the brittle cracking parameters are given in Table 4.

Table 4. Plexiglas mechanical and brittle cracking properties

Plexiglas Properties	**Values**	
Density	1.2x10-6 (kg/mm^3)	
Elasticity Modulus	3100 (MPa)	
Poisson ratio	0.37	
	Direct stress after cracking	**Direct cracking strain**
Brittle Cracking	36.8	0
	34.1	0.000333
	21.1	0.000667
Direct cracking failure strain	1x10-6	
	Shear retention factor	**Crack opening strain**
Brittle shear	1	0
	0.5	0.001
	0.25	0.002
	0.125	0.003

5. Results and discussion

5.1 Load displacement curve obtained from simulation

For brittle cracking material model, finite element simulation results gives a set of "P" and displacement "h" data for each loading step and then these data points are plotted on a graph. The load-depth curve obtained from simulation is shown in Figure 10.

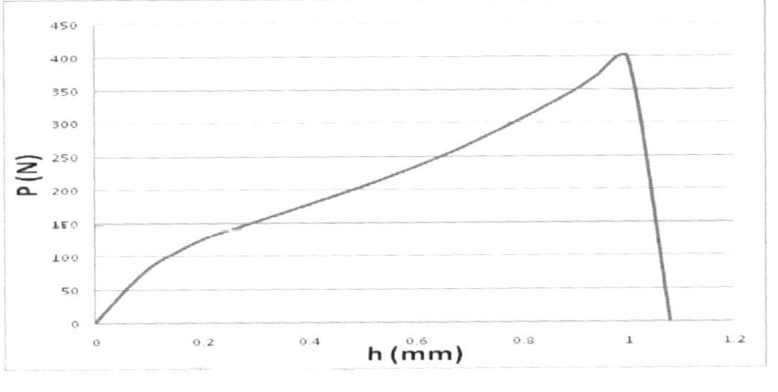

Figure 10. Load depth curve using brittle cracking model.

5.2 Comparison between FEM results and experimental results

Load displacement curve obtain by using brittle cracking model in current numerical simulation is compared with previously extracted experimental load displacement curve is shown in Figure 11. The Figure indicates that the model predictions of P-h relationship agree well with the experimental results.

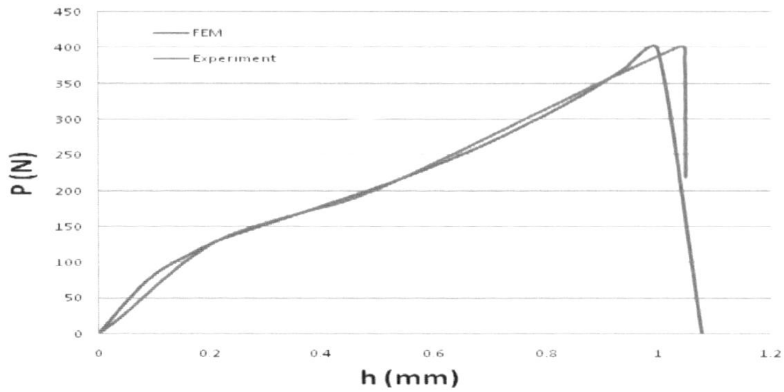

Figure 11. Load depth curve comparison between simulation and experimental results

As seen in Figure 11, the force displacement curve is linear when the indention force is small, and after that the curve shows non-linear behaviour as the magnitude of the indentation force increases. When the force reaches to critical valve of 400 N, the curve has a sudden drop which indicates that cracking started at this point, relieving the load. Therefore both experimental and numerical results provide evidence that radial cracks are initiated during the loading stage, when the indentation force reaches to a value of 400 N. Hence by using brittle cracking material model the crack initiation in Plexiglas plate during indentation process can be predicted.

6. Conclusion

The brittle cracking behavior of poly methyl methacrylate (PMMA) material, subjected to sharp indentation loading is investigated in this study. ModifiedVickers testing apparatus is used to acquire the experimental data in the form of load displacement curve. The mechanical properties such as hardness, yield stress and fracture toughness are determined by analyzing the geometrical dimensions of the indent and from the analysis of a load-displacement curve. In addition stress strain graph is generated from the force displacement data. After getting the stress strain graph, a mathematical model is created for this graph by doing curve fitting.

Furthermore, a numerical simulation based study is made and a brittle cracking model in Abaqus/Explicit is proposed in order to predict the crack initiation in Plexiglas material when subjected to indentation loading. The crack prediction in Plexiglas plate by simulation of the experimental data is achieved successfully and a good correlation is observed between experiments and simulations results.Therefore by using brittle cracking model the crack initiation in Plexiglas plate when subjected to sharp indentation loading can be predicted without doing series of experiments.

7. References

[1] Jewan I, Fahmi Z, Moussa NA and Zitouni A 2008 Computational modeling of static indentation-induced damage in glass *Computational Materials Science* 407-415

[2] Ghatu S, Spandan M, Philippe H G and Dipankar G 2008Recent Advances in Dynamic Indentation Fracture, Impact Damage and Fragmentation of Ceramics *J. Am. Ceram. Soc.*91 (9) 2777-2791

[3] Alessandro L, Franco F, Robert J K Wood and Stavros S 2010Numerical analysis of brittle materials fractured by sharp indenters *Engineering Fracture Mechanics* 77 264-274

[4] Ji-Peng L and Gan-Yun H 2012Mechanical modeling of Lateral Cracking in brittle materials under Indentation*Proc. Int. Conf. on Mechanical Engineering and Material Science*

[5] R Rikards, A Flores, F Ania, V Kushnevski and F J Calleja 1998 Numerical-experimental method for the identification of plastic properties of polymers from micro hardness tests *Computational Material science* 11 233-244

[6] Zhouhua L and John L 2001 Strain rate effects on the thermo mechanical behaviour of polymers *Int. J. of solid and structures* 38 3549-3562
[7] Sai S, Adam D M and Mary C B 2004 Mechanics of transparent polymeric material assemblies under projectile impact: simulations and experiments*Cambridge: Massachusetts Institute of Technology*
[8] Tao S, Yulong L, Hong Y, Fei X, Zhongbin T and Lei L 2005 Temperature effect on the mechanical behavior of acrylic polymers under quasi-static and dynamic loading *Xi'an: Northwestern Polytechnical university*
[9] NK Naik and Yernamma P 2009 Mechanical behaviour of acrylic under high strain rate tensile loading*Mumbai: Indian institute of technology*
[10] Srinivasan A T, Wei Z and Luan H 2010 Dynamic failure of monolithic and layered PMMA and PC plates*Proc. of the Implast Conference Rhode Island USA*: *Society of experimental mechanics*
[11] LawnB R, Evans A G and Marshall D B1980Elastic/plastic indentation in ceramics. The median/radial crack system*J. Am. Ceram. Soc.* 63 574-581

IOP Conf. Series: Materials Science and Engineering **146** (2016) 012009 doi:10.1088/1757-899X/146/1/012009

Deep drawing mechanism, parameters, defects and recent results: state of the art

Adnan I. O. Zaid
Mechanical and Industrial Engineering Department, Applied Science University, Amman 11931, Jordan

E-mail: 1adnan_kilani@yahoo.com

Abstract. Deep drawing is a sheet metal forming process which is widely used in manufacturing parts for automobile and air craft industries. In this paper, the mechanism of deformation, effect of the geometrical parameters involved and the defects encountered in the process are presented and discussed. These include: the radial clearance percentage, punch and die profile radii. Despite the number of publications on the subject, there is still a great demand for further research. Recent experimental investigation on the effect of radial clearance percentage, punch and die profile radii on their autographic records (punch load- punch displacement curves) and on quality of the produced blanks is carried out and the results are presented and discussed. It was found that the maximum drawing force decreases with increase of the die profile radius and increases by increase of the punch profile radius. Furthermore, the liability of the produced cups to wrinkle increases by the increase of both punch and die profile radii being more influenced by the die profile radius.

1. Introduction

Deep drawing is a secondary forming process which in its simplest form a cylindrical shape or alike (for example a cone or frustum) is produced from a thin disc of sheet metal by subjecting it to a compressive force (while it is held between a die and blank holder) through a circular punch which mainly on the blank thickness as illustrated in Figure 1.

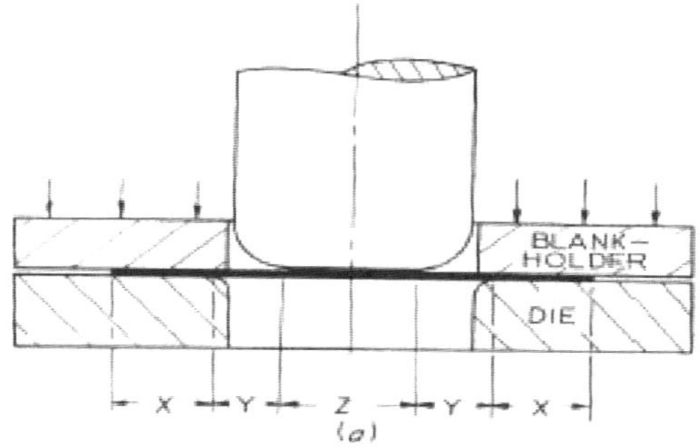

Figure 1. Deep drawing setup at start

Content from this work may be used under the terms of the Creative Commons Attribution 3.0 licence. Any further distribution of this work must maintain attribution to the author(s) and the title of the work, journal citation and DOI.
Published under licence by IOP Publishing Ltd

1.1 Mechanism of deformation

The deep drawing mechanism is a complicated process particularly the different types of stresses on each region in the blank. To facilitate the understanding of the mechanism, the blank is divided into three regions X, Y and Z. The outer annual region X is sandwiched between the die at its bottom part and the blank holder at its top part. Region Y, the inner annular region is not in contact with either the punch or the die and Z the central region of the blank is only in contact with the punch as illustrated in Figure 1(a).When the compressive force is applied to the punch, the draw proceeds the material in region X starts to draw progressively inwards towards the die profile under the effect of the applied tensile stress resulting in continuously decreasing the radii in this region which causes induced compressive hoop stress which causes an increase in the material thickness at the outer part of region X. Unless holding down pressure is applied, the induced hoop stress will cause the blank to fold causing wrinkling. When the material in region X passes over the die profile it is thinned by plastic bending under the effect of the tensile stress. The net effect of the outer part of region X is increase in thickness of the material. Regarding the material in region Y, it can be readily seen that it is subjected to bending and sliding over the die profile; part to stretching in tension in the clearance region, part to stretching between the die and punch in the clearance zone and part to bending and sliding over the punch profile. Finally, zone Z is subjected only to stretching and sliding over the punch head. The above mechanism can be summarized in accordance with the above division of the blank and the type of stresses which each region is subjected to:

i. Pure radial drawing between the die and the blank holder.
ii. Bending and sliding over the die profile.
iii. Stretching between the die and the punch in the clearance zone.
iv. Bending and sliding over the punch profile radius.
v. Stretching and sliding over the punch head.

Various parts of region X may go through some or all of the processes i, ii and iii; while parts of region Y may go through some or all of processes ii, iii and iv; finally, parts of region Z may go through some or all of processes iii, iv and v.
It should be noted that process i causes thickening of the blank whereas processes ii, iii, iv and v causes its thinning.
Figure 2 shows the variation of the thickness along the wall of a drawn cylindrical cup for a flat headed punch on the right hand side.

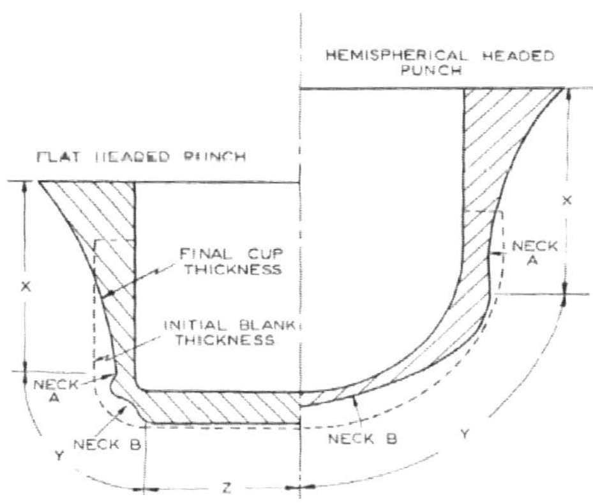

Figure 2. Sections through deep drawn flat ended and hemi spherical cups with flat and hemispherical-headed punches

In hemi-spherical punches, making allowance for bending over the punch profile radius is not essential, whereas in the more general case of drawing with a flat-headed punch, making allowance for bending over the die and punch profiles, has not yet been solved. The punch load at any phase of drawing is determined by the forming region. If the blank is held rigidly at the die to prevent radial drawing the process becomes one of pure stretch-forming. Extensive and detailed experimental and theoretical investigations of cup-drawing have been carried out by different researchers, aiming at reducing the different defects in the process and improving the quality of the produced parts, [2-35].

2. Materials, equipment and experimental procedures

2.1 Materials

The specimens were circular discs of 180 mm diameter and 0.42 mm thickness made from carbon steel with the following w.t. percentages: 0.22% C and 0.5%Mn and the remainder is Fe. They were annealed before being used. Their mechanical behavior in the annealed condition is shown in Figure 3.

2.2 Equipment and experimental procedures

The deep drawing tests were carried out using the die shown in Figure 4 which was designed and manufactured for this purpose. It consists of the following main parts: the upper and lower platens in line. They were made of galvanized steel, sleeves, blank and die holders which were all made from galvanized carbon steels. Compression springs and the punch and die which were made of X12M die steel of the chemical composition shown in Table 1.

Table 1. Chemical composition of X12M die steel

Element	C%	Mn%	Si%	Cr%	V%	Fe%
Wt. %	1.7%	0.35	0.4	0.12	0.3	Bal.

The punch and die were heat treated in accordance with the heat treatment recommended by the suppliers and the obtained hardness, as measured by Rockwell Hardness, is RC 67. All the punches and dies which were used for investigating the different parameters in this paper i.e. radial clearance percentages defined as, the radial clearance between punch and die / the blank thickness = C / to, where C is the clearance and to is the original thickness of the blank, punch and die profile radii were all made of the same material and heat treated to RC 67 and their diameters were measured using the Too Makers travelling microscope and their profile radii using shadow graph at magnification X20. The values of the used radial clearance percentages are shown in Table 2. Five punches and five dies with different profile radii were machined and ground under the same cutting conditions. Their dimensions are shown in Tables 3 and 4 respectively.

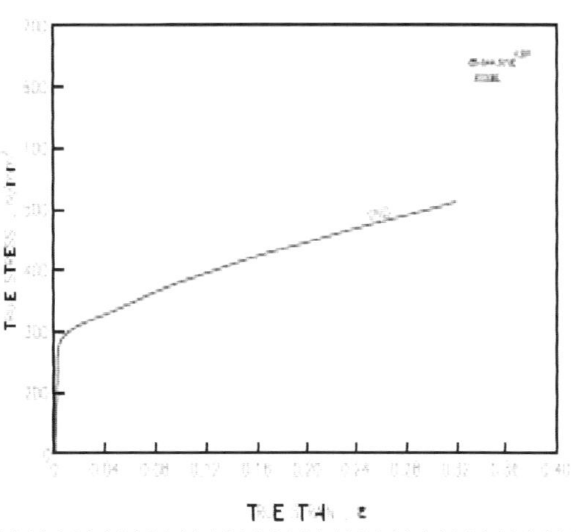

Figure 3. Mechanical behavior of blanks material

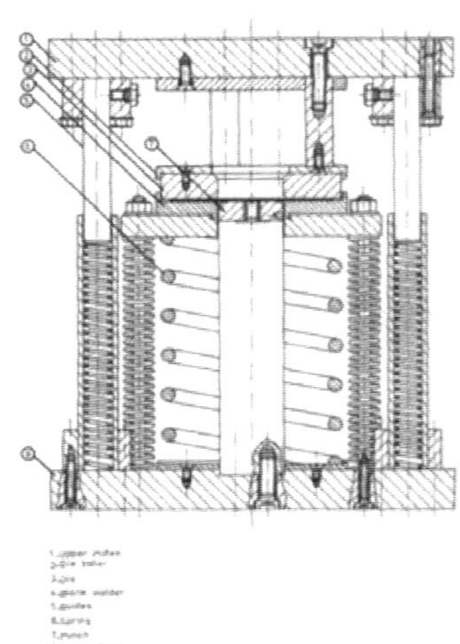

Figure 4. Th deep drawing die

Table 2. The values of radial clearance and radial clearance percentages

Symbol	**C1**	**C2**	**C3**	**C4**	**C5**	**C6**
Radial clearance mm	0.3976	0.4473	0.4943	0.5450	0.5695	0.7947
Radial Clearance %	94.8	106.6	117.8	129.9	135.7	189.4

Table 3. The values of punch profile radius, (mm)

Symbol	**rp1**	**rp2**	**rp3**	**rp4**	**rp5**
Punch profile radius (mm)	2	5	10	15	20
Rpn/pr	3.34	8.34	16.67	20.0	33.34

Table 4. The values of die profile radius, (mm)

Symbol	**rd1**	**rd2**	**rd3**	**rd4**	**rd5**
Die profile radius, (mm)	2	4	6	10	15
rdn / to	4.77	9.53	14.3	23.84	35.75

All the drawing tests were carried out on the Instron Universal Testing machine of 250 KN at constant cross head speed of 10mm / minute and the autographic record was obtained for each test.

3. Theoretical considerations

3.1 Determination of the radial and hoop stresses

As the punch starts to press the blank downwards through the die opening, the outer radius of the blank decreases continuously throughout the drawing pass under the action of an induced compressive radial stress. At any instant, assuming frictionless drawing condition, for simplicity, this radial stress 6rx at any radius x is determined from the following equation:

$$6rx / 1.1\ Y = \ln b / x \tag{1}$$

and the tangential stress, 6Θx is determined from :

$$6\Theta x / 1.1\ Y = \ln 1 - \ln b / x \tag{2}$$

3.2 determination of the maximum height after the first draw

Consider a circular blank of diameter 2a and of original thickness to is formed into a cylindrical cup of base diameter 2b and height H. where a and b are the radii of the blank and the base of the cylindrical cup.

For simplicity, assuming there is no change in thickness which is very justified from the experimental results of the work in this paper and the work of other researchers, applying the constant volume principle in plastic deformation to get:

$$a^2 \times to = b^2 \times to + 2\ b \times H \times to \tag{3}$$

simplifying to get

$$H = b/2\ (R^2 - 1)\ , \tag{4}$$

where R is the drawing ratio $R = 2a / 2b$ (5)

If R is taken = 2.2 which is the limiting drawing ratio, LDR, by substituting this value in eqn. (4). Then the maximum obtainable value of H is equal to 2b, i.e. it is not possible to obtain a cup whose height is bigger than its base diameter. This is very useful engineers and technicians in industry

IOP Conf. Series: Materials Science and Engineering **146** (2016) 012009 doi:10.1088/1757-899X/146/1/012009

because by looking at the drawing of the required part can decide if he can produce it from the first stage or he needs a redrawing process.

4. Results and discussion

In this section, the main parameters affecting the deep drawing process are given and discussed which include: the drawing ratio, holding down pressure, radial clearance percentage, punch and die profile radii and their effect on the maximum drawing force and the defects encountered in the process are presented and discuses.

4.1 Blank holding down pressure

Normally, two types of blank-holding down pressure are commonly used: clearance blank-holding and pressure blank-holding; the object in each case is to prevent wrinkling of the blank during radial drawing, but with the minimum of interference with free drawing. In the early work of reference, [10] on mild steel blanks it was shown that with clearance blank-holding, an initial clearance of 5 per cent was sufficient for this purpose. With pressure blank-holding the medium pressure necessary to prevent wrinkling was 400 psi of blank contact area and a clearance of 0.002 in when clearance blank–holding were used .The same was adopted in this research work. It was also found that increasing the force beyond this amount had little effect on the maximum punch load or on the final thickness in the base or on the profile radius of the produced cups, though the walls were thinner with the higher loads.

4.2 Drawing ratio

Drawing ratio is defined as the ratio of blank diameter to the throat diameter of the die. It was found that for any given drawing conditions the punch load increases with blank diameter in an approximately linear manner, over the whole of the useful range with slight tendency to drop near the limiting drawing ratio. It is worth mentioning in this respect that one should differentiate between the drawing ratio which is a geometrical parameter and the limiting drawing ratio which is a material property.

4.3 Radial clearance between punch and die

Radial clearance between punch and die throat may affect the drawing process directly by controlling the freedom of the walls either to thicken or to taper and pucker. It can be seen from Figure 4b that the maximum drawing force is greatly influenced by the radial clearance particularly when its value is less than the blank thickness i.e. the case known as ironing condition where it increased more than three folds. As the clearance increases above the blank thickness it becomes less affective until it reaches a constant value e.g. it can be seen that. The best radial clearance percentage for the steel used material was the difference in the maximum drawing force all C / to % above 100 % within the tested range does not exceed 10 %, however at 130 % the produced cups have least wrinkling and ears height. This is in agreement with the results reported in reference [10] for mild steel. Hence a net radial clearance of about 30 per cent is suitable for general purposes, with free drawing and a reduction of 50 per cent, and this has the sanction of practical experience. However, increasing it beyond this value may allow a bell-mouth to persist near the rim of the cup, which would be practically objectionable if the deep drawn products are required to undergo a re-drawing operation. A more important feature than the drawing force is the local strain in the blank, which may lead to local necking and finally to cracks and fracture. Radial stresses tends to thicken the blank at its rim, while bending and sliding over the die profile and the punch head tends to thin it. The most serious thinning arises from the stretching over the punch head and particularly between the punch head and the die to reduce the thinning as much as possible it is required to maintain high frictional condition on the punch while maintaining low friction everywhere else similar to mandrill drawing.

The effect of radial clearance percentage on the thickness strain distribution on the used steel is shown in Figure 5. It indicates that the maximum thinning is 6 % which occurred at 106 % radial clearance percentage. In general, there is not a great difference between the values of thinning among the investigated values of C / to %. It is worth noting from Figure 5 that all the thinning occurs at punch profile zone of the blank which is subjected to bending and tension.

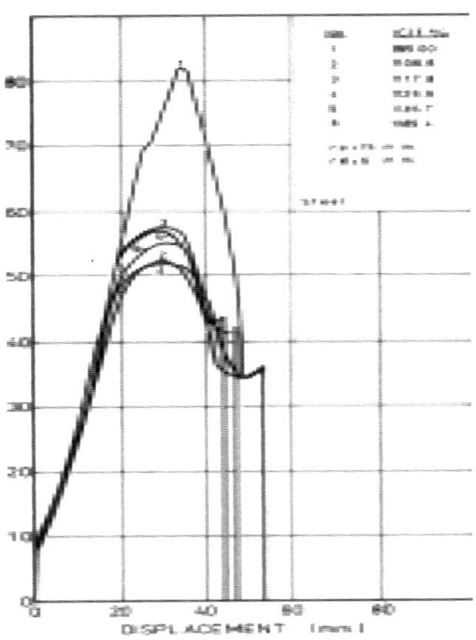

Figure 4b. Autographic records of different C/ t %

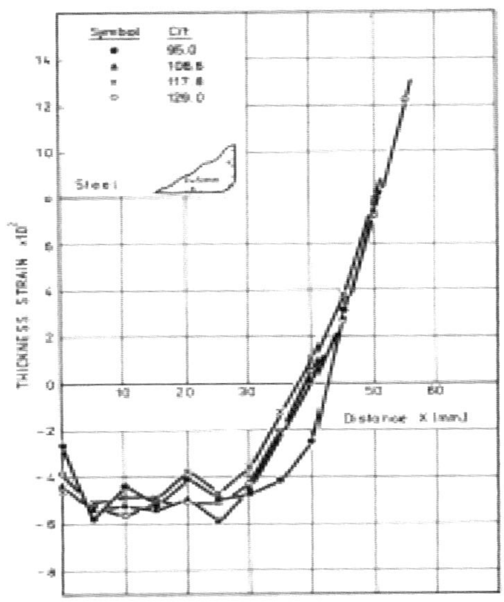

Figure 5. Thickness strain distribution

4.4 Die profile radius

In this section the die profile radius was varied between 2 and15 mm, as shown previously in Table 4, section 2, while keeping the punch profile radius constant. This was carried out on five punch profile radii namely: 2, 5, 10, 15 and 20 mm. as shown previously in tables. The autographic records for all the combined values of dies and punches were obtained from which the maximum drawing force was determined for each test. It was found that the maximum drawing force decreases with increase of the die profile radius, rd, whereas its liability for wrinkling increases. The optimum value for the used steel material was found at rd = 6 mm i.e. about 15 to. Also the sharper is the die profile radius, the greater is the maximum punch load, because of the increased process work due to plastic bending under tension. Consequently, decreasing the profile radius below 15 times the blank thickness lowers the limiting drawing ratio, (LDR). The effect of die profile on the essentially practical problem of success or failure of the drawing is shown in Figure 6. Each point near the critical range represents four tests. It was also found that the more generous is the punch radius, the more gradual is the rise of punch load and the longer is the punch travel and the maximum punch load is almost unaffected. This is an indicator of the limit of consistency in the cup-drawing.

4.5 Punch profile radius

The effect of the punch profile radius was investigated by using five different profiles: 2, 5, 10, 15 and 20 with constant die and constant radial clearance percentage, as was shown previously in table 3. The autographic record was obtained for each test. The effect of the punch profile radius on the thickness strain distribution is shown in Figure 9.

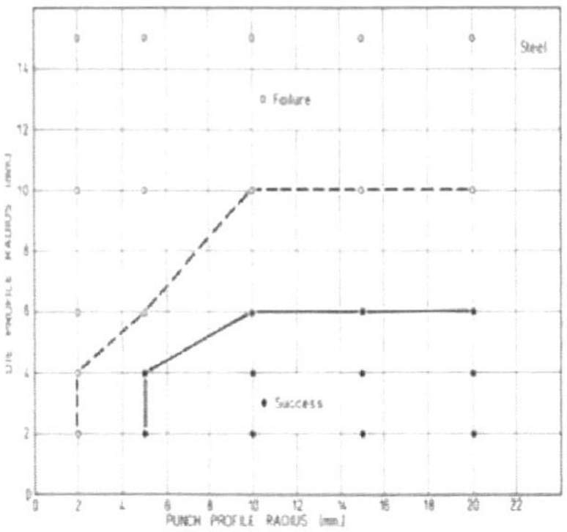

Figure 6. Effect of die profile on drawing capacity

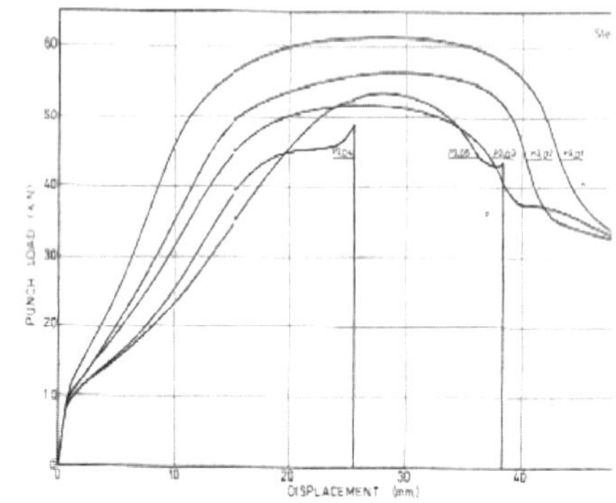

Figure 7. Effect of die profile radius on the autographic records at constant punch profile

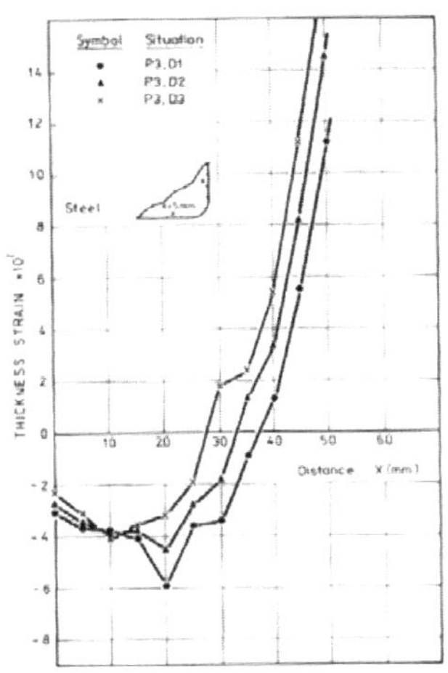

Figure 8. Effect of die profile radius

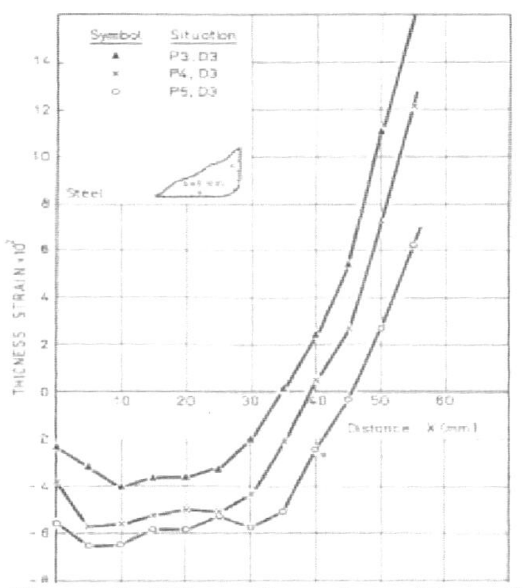

Figure 9. Effect of punch profile radius on the thickness strain distribution on the thickness strain distribution

4.6 Defects encountered in deep drawing

The following defects might take place in the deep drawing process

4.6.1 Ironing

Iron consists principally in reducing the wall thickness of the cup by restricting the clearance between the punch and the die to a value less than the blank thickness. The punch load is of primary importance in ironing because it determines the tension in the cup walls and hence the maximum reduction possible for a given punch load. A theoretical study of ironing has been reported in reference [21] and an experimental investigation was carried out using hemispherical heade punches for different condition of wall thinning, die profile and lubrication [22].

4.6.2 Galling

It is the transfer of some particles from the cup wall into the part of the die in contact with it. It causes damage to the die surface. It depends on the hardness o the blank and the lubricant effectiveness. It can be avoided by keeping the die surface in very good surface quality and using efficient lubricants to reduce friction between the cup and the die surface in contact with it.

4.6.3 Orange peeling
It occurs at the outer surface of the cup when the grain size of its material is large it can be avoided by reducing the grain size prior to drawing either by heat treatment or by adding grain refining the grains by the addition of the appropriate refiners.

4.6.4 Earing
It is caused by the planar anisotropy which is due to variation in the mechanical behavior of the sheet from its plane to any other direction inclined or perpendicular to it they are normally even in number 2 or 4 or 6. The worst number is 8 in case of brass blank. They appear on the upper part of drawn cup and is treated by trimming. The photograph of Figure 10 clearly shows the ears on the steel specimen.

4.6.5 Tearing
It might take place at the inner region of the anular part of the rim near the die profile if the holding down pressure is high which stops the blank from sliding and bending over the die profile radius, or it might take place in the maximum thinning region in the clearance region near the punch profile which is subjected to bending and stretching as clearly indicated in Figure 11.

4.6.6 Wrinkling
This defect starts in the rim region of the blank and travels towards the middle region, the clearance zone between the punch and die profiles, by the continuation of the deep drawing process, as indicated in Figure 11. Examination of the available literature reveals that experimental research work on this important defect is relatively little compared to the analytical and theoretical work, [23-27]. The critical dimensions of the rim at the onset of wrinkling was given in reference [23] where it was stated that wrinkling depends on properties of the blank thickness and its mechanical properties in addition to the shape and dimensions of the die set and the method by which the holding down pressure is applied. The experimental work of reference, [24] revealed that the resistance to wrinkling increases with the increase of strain percentage; which explains the use of the draw-beads in the deep drawing process to reduce the liability to wrinkling. The authors of reference [25] found the critical values of stress and thpunch displacement at which wrinkling starts to occur. They also agree with authors of reference [26] that both the critical stress and displacement increase with the increase of strain in the process.

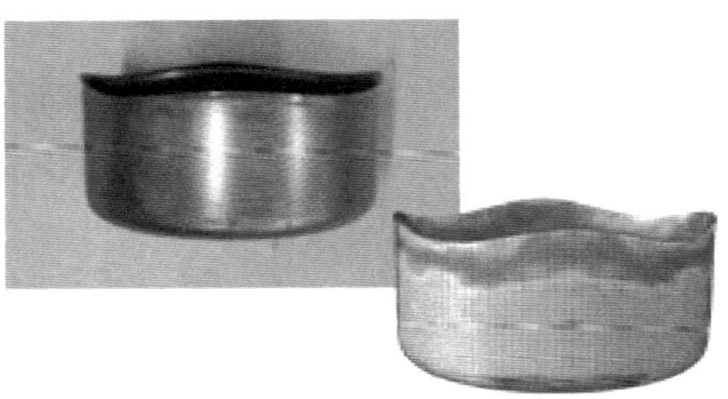

Figure 10. Photograph showing ears

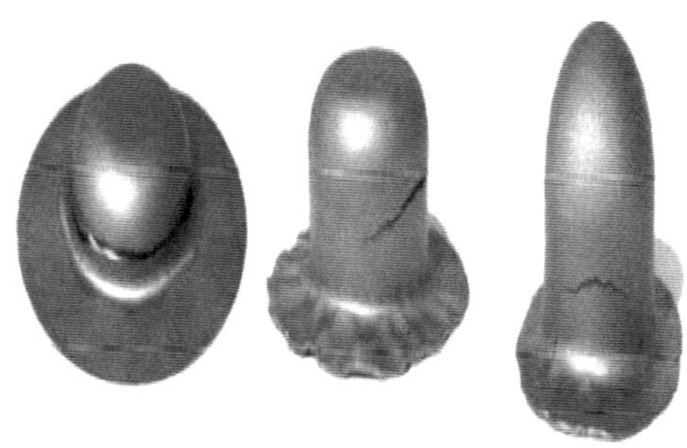

Figure 11. Tearing and wrinkling

5. Conclusions
Within the experimental limitations, regarding the investigated geometrical errors the following points may be concluded:

i). The maximum drawing force is greatly influenced by the radial clearance between the punch and the die, particularly when it is less than the blank thickness i.e. the case known as ironing condition, it increased more than three folds the clearance increases above blank thickness it becomes less affective until it reaches a constant value. The best radial clearance percentage for the steel used material was found at 130 % which produced least wrinkling and ears height. However increasing it a beyond this value causes bell shaped cup.

ii). The maximum drawing force decreases with increase of the die profile radius, rd, whereas its liability for wrinkling increases. The optimum value for the used steel material was found at rd = 6 mm which equals about 15 to.

iii). The maximum drawing force increases with increase of the punch profile radius, rp , up to rp = 15 mm then it starts to decrease. However, it was less affected by rp compared to rd.

iv). Although the research on the deep drawing process has been going on for more than nine decades, it is far from being complete and further work is required to get rid of the defects encountered in the process and renders it cost effective.

6. Acknowledgment

The author is grateful to the Applied Science Private University, Amman, Jordan for partial financial support granted to this research (Grant No.DRGS-2015).

7. References

[1] Johnson W, Mellor P B 1980 Engineering Plasticity, The Camelot Press Ltd., Southampton, U.K
[2] Sommer M H 1925, Machinenbou, 4, 1171
[3] Eksergian GL 1926, Trans. ASME,48, 609
[4] Eksergian GL 1927, Metal Industry, 23, 483
[5] Siebel E and Pomp A 1929, inst. Eisenfforsch, 11, 139
[6] Sachs G 1934-1935, Proc. Inst. Of Automobile Eng., 29, 588
[7] Oehler G 1939, Metallwistschaft, 16, 1059
[8] Swift H W 1939-1940, Proc. Inst. Of Automobile Eng., 34, 361
[9] Jevons J D 1945 The Metallurgy of Deep Drawing and Pressing, 2nd Ed, Chapman & Hall.
[10] Swift H W 1948, Engineering, 166, 333 and 357
[11] Chung S Y, Swift H W 1951 Cup-drawing from a Flat Blank, Proc. Inst. Mech. Engrs.165. 199.
[12] Hesenberg W C F 1954 A simple Account of some of Professor Swift Work on Deep Drawing, B.I. S.R.A. Report MW/1954
[13] Alexander, J M 1960, Metal Review5, No 19
[14] Chiang D C, Kobayashi 1966, J. Engng. Industry, 88, 443
[15] El-Sebaie M G, P B Mellor 1972, Int. J. Mech. Sci.,14, 535
[16] El-Sebaie M G, P B Mellor 1973, Int. J. Mech. Sci., 15, 485
[17] Panknin W 1976, Sheet Met. Ind. 53
[18] Gransow W G 1979, Sheet Met. Ind., 56, 561
[19] Gotoh M 1980, Int. J. Mech. Sci., 1980, 22, 367
[20] Yu X T, Johnson W 1982, Int. J. Mech. Sci., 24, 175. Lee S H, Kobayashi S 1974, Proc. 15th Int. M. T.D. R. Conf., Birmingham, UK
[21] Fukui S, Hansson, Analytical Study of Wal Il Ironing, Considering Work-hardening. Annls of C.I.R.P., 18, 593
[22] Freeman P, Leeming H, Ironing of Thin-walled Metal Cups-the Distribution of the Punch Load, B.I.S.R.A. Report No.MW/E/46/53
[23] Lange K, Dalhaimer R, 1967, Proc. 18th. M.T.D.R. Conf., 221
[24] Geckeler J.W, Angew, 1928, Math. Mech., 8, 341
[25] Senior B. W 1956, J. Mech. Phys. Solids, 4, 235
[26] Naziri H, Pearce R 1969, La Metallurgia Italiana, 40, 727
[27] Triantafyllidis N, Needleman A, 1980, J.Engng.Mat.Tech.102, 241
[28] Chen F K, Huang T B 2003 Formability of stamping magnesium-alloy AZ31 sheets J. Mater. Process. Technol. 142643–647

[29] Zhang H, Huang G, Kong D, Sang G, Song B 2011 Influence of initial texture on formability of AZ31B magnesium alloy sheets at different temperatures J. Mater. Process.Technol. 211 1575–1580

[30] Zhang H, Huang G, Fan J, Roven H J, Pan F, Xu B 2014 Deep draw ability and deformation behavior of AZ31 magnesium alloy sheets at 473 K, Mater. Sci.Eng.A608 234–241

[31] Liu C Y, Jing R, Wang Q, Zhang B, Jia Y Z, Ma M Z, Liu R P 2012 Fabrication of Al/Al3Mg2composite by vacuum annealing and accumulative roll-bonding process Mater. Sci. Eng. A 558 510–516

[32] Liu C Y, Wang Q, Jia Y Z, Jing R, Zhang B, Ma M Z, Liu R P, 2012 Microstructures and mechanical properties of Mg/Mg and Mg/Al/Mg laminated composites prepared via warm rollbonding Mater. Sci. Eng. A 556 1–8.

[33] Jafarian M, Khodabandeh A, Manafi S 2015 Evaluation of diffusion welding of 6061aluminumand AZ31 magnesium alloys without using an interlayer Mater. Des. 65160–164.

[34] Zhao Y, Lu Z, Yan K, Huang L 2015 Microstructural characterizations and mechanical properties in underwater friction stir welding of aluminum and magnesium dissimilar alloys Mater. Des.65675–681

[35] Yan J, Xu Z, Li Z, Li L, Yang S 2005 Microstructure characteristics and performance of dissimilar welds between magnesium alloy and aluminum formed by friction stirring Scripta Mater. 53585–589

IOP Conf. Series: Materials Science and Engineering **146** (2016) 012010 doi:10.1088/1757-899X/146/1/012010

Fatigue crack growth in an aluminum alloy-fractographic study

I. Salam, W. Muhammad and N. Ejaz
Institute of Industrial Control Systems, P. O. Box 1398, Rawalpindi, Pakistan

E-mail: [1]iftikhar.salam@gmail.com, [2] valimuhammad@yahoo.com, [3]noveedejaz@yahoo.com

Abstract. A two-fold approach was adopted to understand the fatigue crack growth process in an Aluminum alloy; fatigue crack growth test of samples and analysis of fractured surfaces. Fatigue crack growth tests were conducted on middle tension M(T) samples prepared from an Aluminum alloy cylinder. The tests were conducted under constant amplitude loading at R ratio 0.1. The stress applied was from 20,30 and 40 per cent of the yield stress of the material. The fatigue crack growth data was recorded. After fatigue testing, the samples were subjected to detailed scanning electron microscopic (SEM) analysis. The resulting fracture surfaces were subjected to qualitative and quantitative fractographic examinations. Quantitative fracture analysis included an estimation of crack growth rate (CGR) in different regions. The effect of the microstructural features on fatigue crack growth was examined. It was observed that in stage II (crack growth region), the failure mode changes from intergranular to transgranular as the stress level increases. In the region of intergranular failure the localized brittle failure was observed and fatigue striations are difficult to reveal. However, in the region of transgranular failure the crack path is independent of the microstructural features. In this region, localized ductile failure mode was observed and well defined fatigue striations were present in the wake of fatigue crack. The effect of interaction of growing fatigue crack with microstructural features was not substantial. The final fracture (stage III) was ductile in all the cases.

1. Introduction

Aluminum alloys are of vital importance for many structural applications in automotive, aircraft and nuclear industries. The principal design features of these alloy systems are their high strength with low density along with high corrosion resistance. The 6xxx series aluminum alloys are frequently used due to relatively high strength, good corrosion resistance and high toughness in addition to their good formability and weldability [1].

Thick-walled cylinders are an essential part of many industries e.g. food, petroleum, armament, nuclear and chemical processing. A major use of these cylinders is as storage for gas products. The fluctuating gas pressure inside the cylinders produces the conditions of cyclic loading on the inner wall creating axial and hoop stresses.

In many applications the cylinders are prone to cyclic stresses during their normal operation and large internal pressures produce high tension hoop stresses along the inner surface of the cylinder. The latter may result in the nucleation of the internal surface cracks which propagate due to cyclic action of high-pressure pulses. If the primary crack growth mechanism is slow, the cracks will be detected during routine maintenance by non-destructive testing (NDT) so that corrective measures can be taken

Content from this work may be used under the terms of the Creative Commons Attribution 3.0 licence. Any further distribution of this work must maintain attribution to the author(s) and the title of the work, journal citation and DOI.
Published under licence by IOP Publishing Ltd

before crack growth moves into a high risk regime. After the crack reaches a critical size, the ultimate failure may be catastrophic and result at pressures which are even lower than the design capacity of the cylinder. For this reason, it is quite necessary to analyze the crack propagation behavior in the cylinders under cyclic hoop stress to ensure their integrity against the fatigue failure. Many researchers have done valuable work to analyze the cylinders under fatigue loading [3-7].

In general, the fatigue process is characterized by three distinct regions [8].

Region I is associated with the growth of cracks with low ΔK's, and is commonly believed to account for a significant proportion of the fatigue life of a structure.

Region II has received the greatest attention as it is in this region the ''Paris'' crack growth law [3] can be applied, viz:

$$da/dN = C\Delta K^{m}$$

where C and m are experimentally obtained constants.

Region III Rapid crack growth occurs in region III and this region is typically thought to account for a small fraction of the total life.

During fatigue process the material undergoes physical changes making imprints of the process on the fracture surfaces. After the material fails, these imprints provide important information about the process going on especially at the crack tip. The study of the fracture surfaces to expose the fatigue behavior of the material on the basis of 'on-site observation' is called the fractographic analysis.

Fractographic techniques using scanning electron microscopes (SEM) are extensively used in analysis of the fracture surfaces. The correct determination of fatigue crack initiation and propagation has obvious and important implications for several steps such as materials selection, component design, manufacturing processes [9].

In present study, a detailed post-fracture fractographic analysis of the fatigue samples was conducted with the help of SEM. The samples under investigation were taken from an extruded aluminum alloy in two orientations and tested under fatigue conditions at different stress levels. The details of the study are given in reference [10]. During the fatigue test, the crack length was measured with the help of travelling microscope. Some tested samples were subjected to detailed fractographic analysis using SEM. The crack growth region was divided into three sub regions and the width of striations were examined in detail in these three regions. The results obtained from the two techniques (experimental and SEM analysis) were compared.

2. Experimental

1.1 Material characterization

The material investigated was AA 6061. The chemical composition and the mechanical properties are reported in reference [10].

1.2 Fatigue crack growth test

Fatigue crack growth tests were performed on middle tension M(T) specimens. Samples were cut from the cylinder in LR orientations, Figure 1. This orientation refers to the standard ASTM E 399 which provides the crack plane orientation code for bar and hollow cylinder. In this orientation the samples were perpendicular to the longitudinal axis of the cylinder 'L' while the notch direction was such that the crack propagation was along the radial direction 'R' as shown in Figure 1. Samples were machined according to the standard ASTM E 647. The machined samples were subjected to mechanical grinding and subsequently fine polishing.

Fatigue crack growth experiments were performed on a servo-hydraulic testing machine. Tests were conducted in the tension-tension mode under constant amplitude loading with *R* ratio 0.1. A sinusoidal waveform was applied at a loading frequency of 10 Hz. Tests were conducted at stress levels of 20, 30& 40 per cent of the yield strength of the material. Crack length was measured with the help of a traveling microscope, at a magnification of 100xand the data was recorded against the number of cycles. Tests were conducted in air at a temperature 22°C and approximately 30 per cent relative humidity.

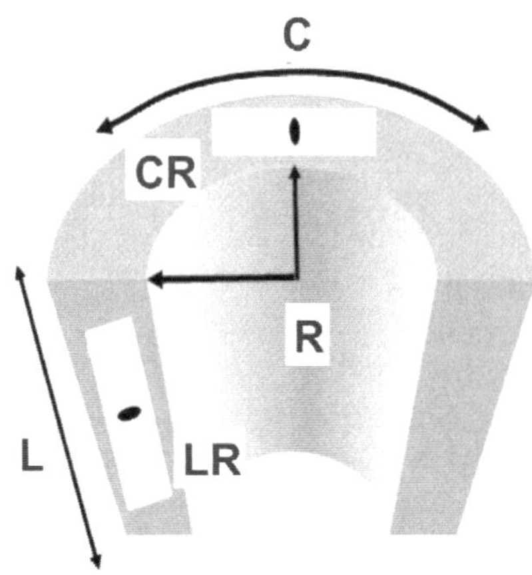

Figure 1. Orientation of the LR and CR samples representing tangential

3. Results and discussion

1.3 Fatigue crack propagation

The material was subjected to experiments to reveal the fatigue crack growth behavior of the cylinder in CR and LR directions. Figure 2 shows the plots for the crack length versus the number of cycles. It was observed that crack growth is faster in CR orientation as compared to LR orientation. This may be attributed to prior working of the material.

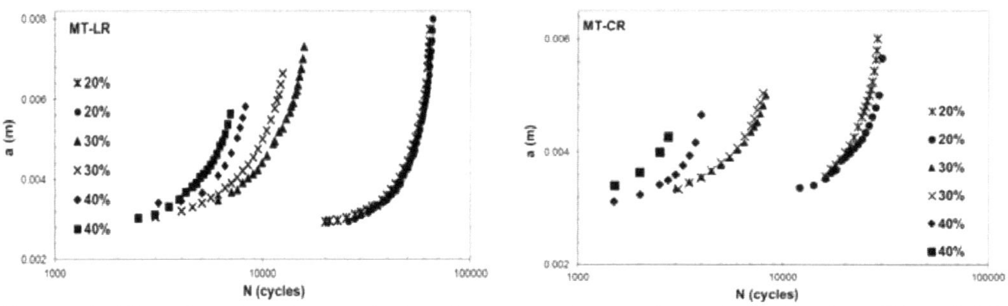

Figure 2. The plots for the crack length versus the number of cycles for CR and LR orientations

1.4 Stereography

The stereo-micrographs of representative fracture surfaces of both directions are shown in figure 3. The stereo-micrographs in both orientations showed typical fatigue crack surfaces, consisting of an initiation region marked as R1, followed by a smooth region representing the crack growth region marked as R2 and finally the rapid fracture region due to overload named as R3. It was noted that crack grows faster in CR orientation as compared to LR orientation. It was also observed that crack grows faster at higher stress level and vice versa in both orientations.

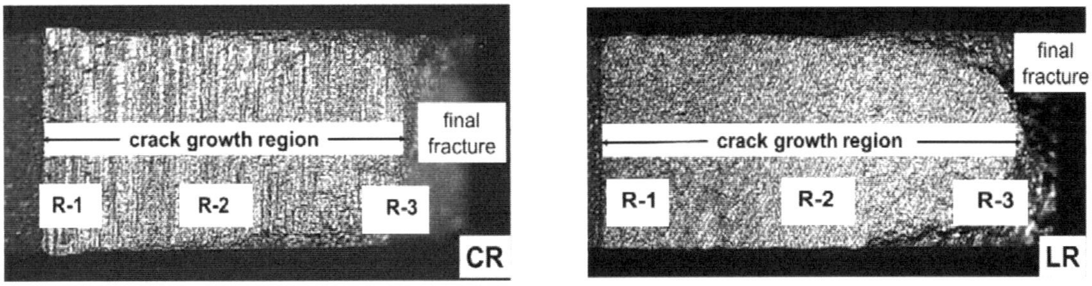

Figure 3. Stereo micrographs of fractured surfaces in both orientations

1.5 SEM analyses

The fracture surfaces of types of samples (CR, LR) were then subjected to quantitative SEM analysis. The lengths of fatigue striations were measured in three regions marked as R1, R2 and R3. Crack growth rate (CGR) was measured through fatigue striation spacing. The CGR was also calculated from already measured crack lengths during testing. This data was then plotted as shown in Figure 4 for both orientations. The Comparison of fatigue test and SEM results shows close agreement at low stress levels and differ significantly at higher stress levels. This difference was more prominent in LR samples than CR samples, being about three fold of fatigue test results. This may be due to the fact that in both orientations the crack front is almost parallel in R1 and R2 regions. However in R3 region of LR samples the crack front is projected out whereas in CR samples it is almost parallel, Figure 3. Thus when crack is measured during test, we are not at the actual crack tip in LR samples. This causes difference in fatigue and SEM data at high stress levels i.e. in region R3.

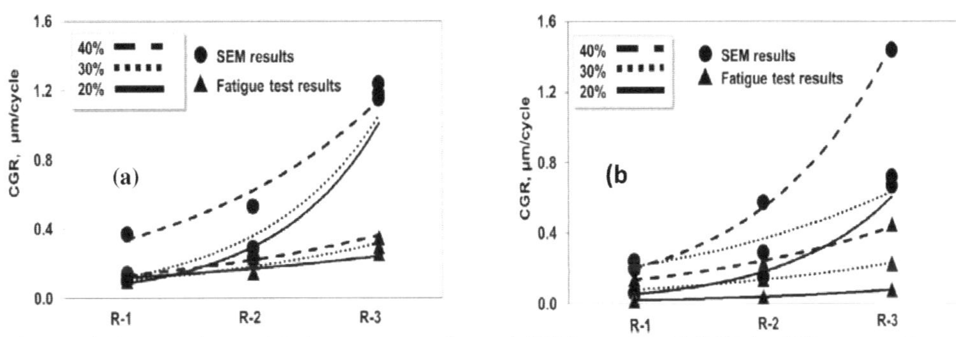

Figure 4. Comparison of fatigue test results and SEM results of CGR in different regions of a) CR & b) LR samples

During SEM analyses it was also observed that in R2 region (crack growth region), the failure mode changes from intergranular to transgranular as the stress level increases in both LR and CR orientations. In the region of intergranular failure the localized brittle failure was observed and fatigue striations were difficult to reveal as shown in Figure 5. However, in the region of transgranular failure the crack path is independent of the microstructural features. In this region, localized ductile failure mode was observed and well defined fatigue striations were present in the wake of fatigue crack, see Figure 6. The effect of interaction of growing fatigue crack with microstructural features was not substantial. The final fracture (R3) was ductile in all the cases.

IOP Conf. Series: Materials Science and Engineering **146** (2016) 012010 doi:10.1088/1757-899X/146/1/012010

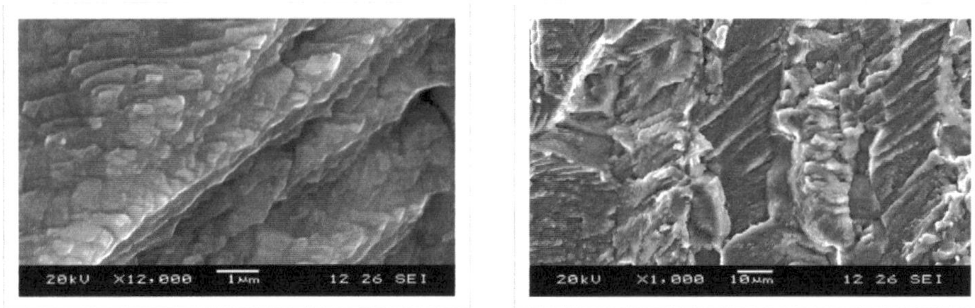

Figure 5. Localized intergranular fracture and unresolved striations at low stress level (R-1)

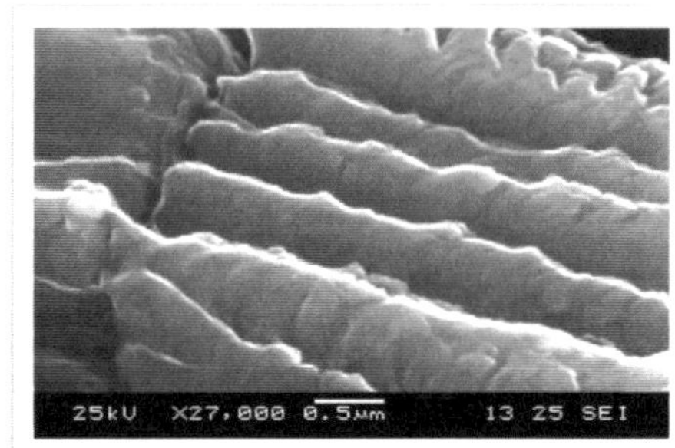

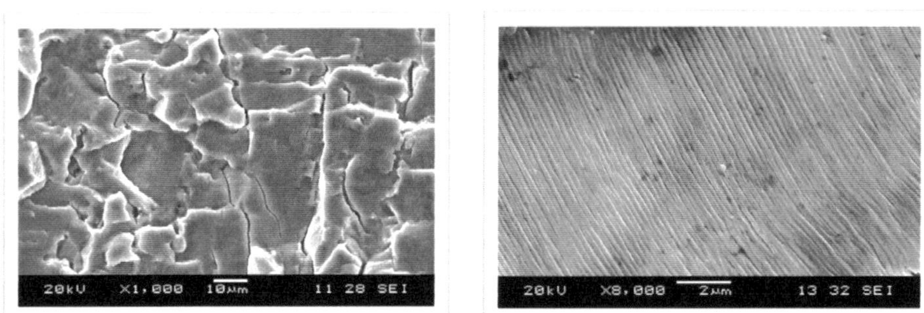

Figure 6. Localized ductile fracture and well defined striations at high stress level (R-2)

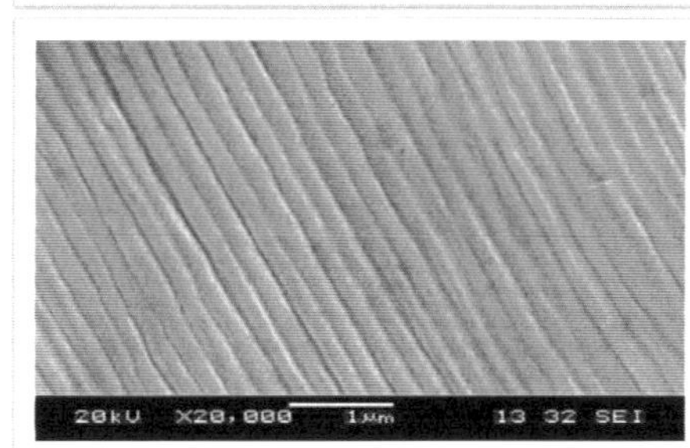

4. Conclusion

I. The fatigue crack grows faster in CR orientation as compared to LR orientation.

II. The fracture appearance was very different in CR and LR directions owing to prior deformation.

III. In stage II (crack growth region, R2), the failure mode changes from intergranular to trans-granular as the stress level increases

IV. Fracture surface morphology was changed from brittle to ductile with increasing stress level. The qualitative and quantitative CGR data was in agreement with each other at low stress level i.e. in intergranular region. However appreciable difference was observed at high stress level i.e. in the transgranular region.

IOP Conf. Series: Materials Science and Engineering **146** (2016) 012010 doi:10.1088/1757-899X/146/1/012010

5. References

[1] Malik M A, Salam I, Muhammad W and Ejaz N 2003 *J. Mech. Sc. and Tech.* 23 1261-1269
[2] Shlyannikov V N 2000 *Int. J. Pressure Vessels Piping* 77 227–34
[3] Rees D W A 1989 *Eng. Fract. Mech.*33(6)927–40
[4] Tomkins B 1973 *Int. J. Pressure Vessels Piping* 1(1)37–59
[5] Su B and Bhuyan G S 1998 *Int. J. Pressure Vessels Piping* 75(12) 879–86
[6] Al-Rubaie K S, Barroso E K L and Godefroid L B 2006 *Int. J. Fatigue*28 934–42
[7] Stanzl-Tschegg S E and Mayer H 2001 Int. J. Fatigue 23 S231–S237
[8] Alegre J M, Bravo P and Preciado M 2007 *Eng. Fail. Anal.* 14 396–407
[9] Tarpani J R, Ruckert C O F T, Milan M T, Silva R V, Rosato Jr A, Pereira R N, Bose W W, Spinelli D, 2004 *Eng. Fail Anal*11 547–559
[10] Muhammad W, Ejaz N and Salam I 2014 *IOP conf. series: Mats. Sc. and Eng.* 60 (012064) 1-9

Thermal, mechanical and electrical properties of polyanaline based ceramic nano-composites

M. Sohail, M. S. Khan and N. S. Khattak[1]
Shaheed Benazir Bhutto University, Sheringal Dir (Upper), KP Pakistan
[1]National Center of Excellence in Physical chemistry, University of Peshawar, (25120) Peshawar, KPK, Pakistan

E-mail: msohail2000@gmail.com

Abstract. Micro/nanohybrid materials have vast applications due to their great potentialities in the field of nanoscience and nanotechnology. Herein we report an investigation on the fabrication and physicochemical characterization of ceramic ($Fe_{0.01}La_{0.01}Al_{0.5}Zn_{0.98}O$) and hybrid ceramic-polyaniline nano-composits. Ceramic nano-particles were prepared by sol-gel technique while optimizing the molar ratios of the constituent's metal nitrates. The prepared inorganic particles were then embedded in the polymer matrix via one-pot blending method. The prepared ceramic particles and their composites with polyaniline were analysed under FT-IR, SEM and TGA. The presence of some chemical species was observed at the interface of the compositing materials. TGA analysis showed the thermal stability of the composite material. Frequency dependent dielectric properties were analysed and it was found that conducting polyaniline has an additional effect on the electrical behaviour of the composite. Rheology study showed enhanced mechanical properties of composite material as compared to their constituting counterparts.

1. Introduction

Nanotechnology is the study of the science of nano-structured materials having dimensions in the range from 1 to 100 nm. Various physical properties of these nano-structured materials vary from those of the corresponding bulk materials due to their small dimensions. In this context, extensive research has been done on organic-inorganic nano-composites due to their promised potential application in the last two decades [1-3]. Nano-composites is a class of advanced materials due to their distinctive properties that cannot be found in macrocomposites [4]. Material scientists, chemists, physicist and biologists are trying to fully exploit functional nano-composites so as to produce smart materials. Functional (organic-inorganic) nano-composites are generally composed of inorganic nano-particles embedded in the organic polymer matrix. These hybrid composites combine the benefits of both organic polymer (i.e. process ability, flexibility, ductility, dielectric and mechanical) and inorganic counterpart (i.e. thermal stability, rigidity) [5]. An important characteristic of organic-inorganic nano-composites is that inorganic fillers with nano-dimensions enhance the interfacial area in these composites as compared to traditional composites [6, 7]. Not only for academics but in most of the applied areas such as ionics, electronics, optics, energy, environment, medicine and biology, nano-composites represent new materials. Applications comprise functional smart coatings, smart membranes, fuel cells, sensors, photocatalysts, photovoltaic, nanophotonics, therapeutics, innovative cosmetics, microelectronics, imaging, drug release, packing industries and automobiles etc. [8-10]. In

Content from this work may be used under the terms of the Creative Commons Attribution 3.0 licence. Any further distribution of this work must maintain attribution to the author(s) and the title of the work, journal citation and DOI.
Published under licence by IOP Publishing Ltd

recent years, significant efforts have been made to devise and control the synthesis of hybrid composites with tailored morphologies.

Numerous organic-inorganic nano-composites have been prepared while adding a wide variety of nano fillers e.g. magnetic nano-particles, semiconductors and carbon nanotubes. A number of polymer's significant properties such as light weight, anti-corrosiveness, mechanical strength and dielectric behavior are used to fabricate these nano-composites [11]. Polyaniline (PANI) is an intrinsic conducting polymer widely used due to its facile synthesis, good environmental stability and process ability [12]. PANI based hybrid nano-composites have adopted a revolutionary progress. PANI has exceptional electrical properties that can be controlled significantly by protonation and charge-transfer doping. However, processing of such conducting polymers generally utilizes a large amount of organic solvents which are noxious to the environment.

Our synthesized ceramic-PANI composite is one of the types of hybrid nano-composites that combine the advantages of conducting polymer along with doping transition metal ions (ceramic). PANI in this composite also acts as a sensitizer that improves light absorption. Instead of strong organic solvents, aqueous media was used throughout the preparation process so as to make it a green pathway. Different physicochemical characteristics of the prepared nano-composits have been analyzed for its possible applications.

2. Experimental

2.1. Materials

All the chemicals applied in the preparation of ceramic nano-particles and composite were of analytical grade. Zinc nitrate, ferric nitrate, sulphuric acid (Scharlu, Spain), lanthanum nitrate (Merck, PAK), aniline (Acros Organics, USA), citric acid (Daejung, Korea) and potassium permanganate, potassium nitrate (BDH, England) were used as received.

2.2. Preparation of ceramic nano-particles

Sol-gel technique was used for the synthesis $Fe_{0.01}La_{0.01}Al_{0.5}Zn_{0.98}O$ ceramic particles. In a typical experiment, $Fe_2(NO_3)_3 \cdot 9H_2O$ (0.073g), $Al(NO_3) \cdot 3H_2O$ (1.242g), $La(NO_3)_3 \cdot 9H_2O$ (0.162g) and $Zn(NO_3)_3 \cdot 6H_2O$ (11.552g) were dissolved together in 100 ml double distilled (DD) under stirring with the slow addition of citric acid (5g) to this mixture. Liquid ammonia was used for keeping the environment of the solution neutral. The solution was then kept in an oven for drying at 80°C. A light bluish gel was obtained which was then grinded and sintered for 3 hours at 600°C. The diffused material was ground properly and particles in micro to nano dimensions were obtained.

2.3. Nano-composite synthesis

The *in-situ* free radical polymerization technique was used for composite formation. In 100 ml DD water, 250mg ceramic particles were dispersed. Then 25 ml of aniline solution (10% in 1 M HCl) and 25 ml of ammonium persulphate (0.1 M in I M HCl) were added to this suspension drop wise. The reaction mixture was then stirred for 3 hours at 5°C with subsequent overnight digestion. For removing soluble impurities and side products, the sample was centrifuged for 10 minutes twice at 4000 rpm and was dried in an oven at 70°C.

2.4. Characterization

Physicochemical analysis was done via FT-IR spectroscopy (IR Prestige-21 Japan), SEM (JEOL JSM-6700F Japan) and TG/DTA (Diamond, Perkin Elmer USA). Mechanical characteristics were studied using Anton Paar Rheometer (Physica MCR 301 Germany). Dielectric properties were analysed using RF Impedance/Material Analyzer (Agilent E4 997 A USA).

3. Results and discussion

3.1. Functional group determination

The formation of ceramic particles (C) and their compositing with PANI (C-PANI) was determined by FT-IR spectroscopy. Peak at about 3370 cm^{-1} denotes O-H stretching vibration in the particles. The same characteristic peak is shown in the composite at 3506 cm^{-1}. Zn-O stretching in ceramic particles is represented by the starting narrow peak at 430 cm^{-1}[13]. All other peaks in the composite curve show the corresponding IR modes for C-C, C-H, C-N and C=O groups. The narrowing and absence of certain peaks in the composite curve represent the successful dispersion of ceramic particles in PANI matrix. FT-IR results confirm the synthesis of particles and their insertion in the polymer matrix.

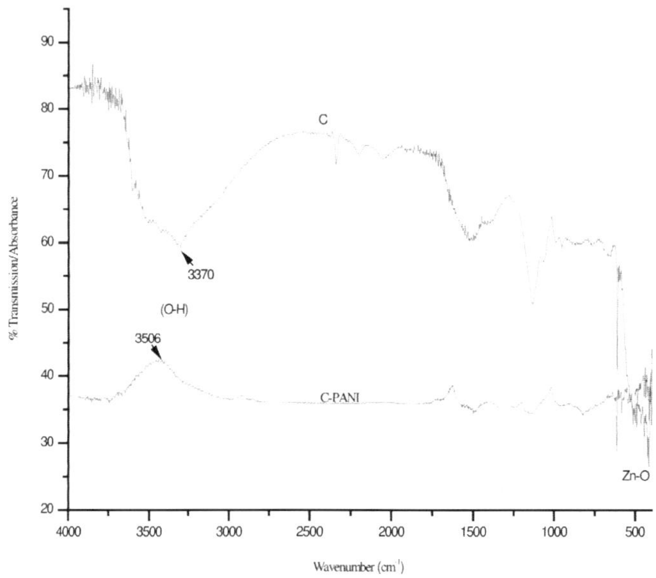

Figure 1. FT-IR spectra of ceramic particles (C) and composite (C-PANI)

3.2. Morphology

Spherical, cylindrical and triangular crystallites of the ceramic nano-particles are shown in Figure 2 (a). The particle size is in the range from nm to µm. A slight agglomeration and small pores are also observed that may be due to sintering process. The micrograph of the C-PANI composite indicates a smooth morphology which is due the immobilization of ceramic particles in the polymer matrix. PANI matrix keep the particles disperse and thus prevent them from aggregation. This property of the composite makes it suitable for its application in nanotechnologies.

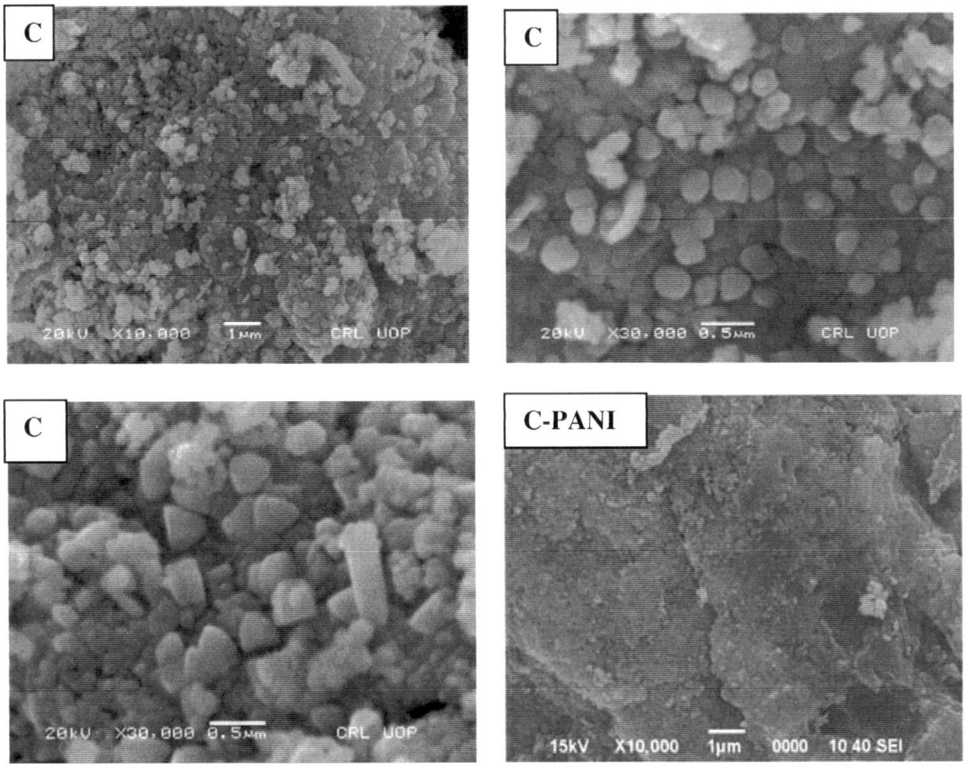

Figure 2. SEM micrographs of the ceramic particles and their composite with PANI

3.3. Thermal properties

Thermal properties were analysed under N_2 atmosphere in the temperature range from 30-700 °C. As shown in Figure 3, ceramic particles are thermally stable up to 600°C and do not shows any apparent weight loss. At about 115°C, water molecules evaporate from the particles. The second drop at 600°C (4% weight loss) reveals the creation of metal oxide phase in the ceramic material. TGA curve of C-PANI composite shows three main degradation steps. In the first step at 85°C, water and other volatile gases evaporate from the sample. In the 2nd step at 300°C, PANI chains degradation begins and completely decomposes at 520°C. In the third step at 600°C, destabilization of ceramic particles occurs and thus the composite loss its stability at this stage completely. As a whole, ceramic particles increase the thermal stability of the PANI in the composite due their tough nature [13].

3.4. Dielectric properties

Frequency dependent dielectric properties were analysed in the range from 1 MHz to 3 GHz at ambient temperature. For both inorganic particles and their composite with PANI, dielectric constant (ε') and loss tangent (tan δ) are strongly dependent on frequency as shown in Figures 4 and 5. With increasing frequency ε' and tan δ indicate a sharp decrease initially up to 1.70 GHz. This shows the unusual dielectric dispersion. Resonance type behaviour occurs at high frequency (2 GHz) which may be due to the lack of matching between the polarization of applied electric frequency and induced electric moments.

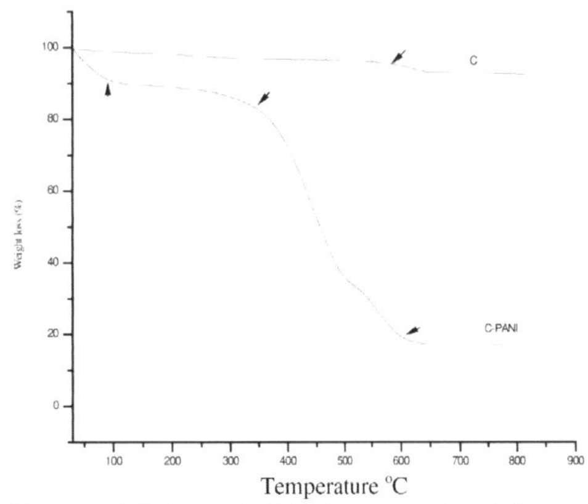

Figure 3. Thermal (TGA) analysis of the prepared materials C and C-PANI

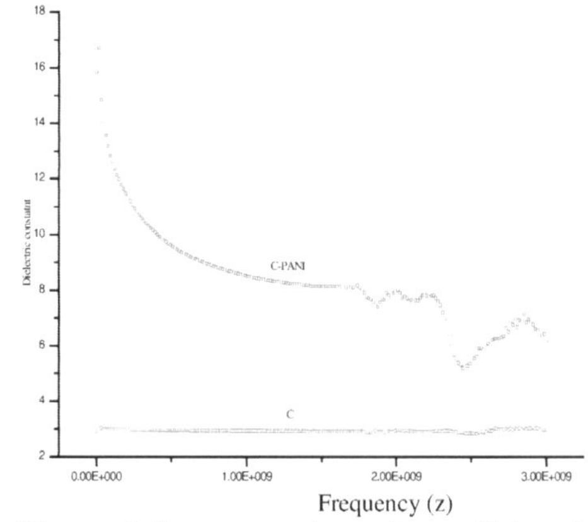

Figure 4. Frequency dependence dielectric constant of C and C-PANI composite

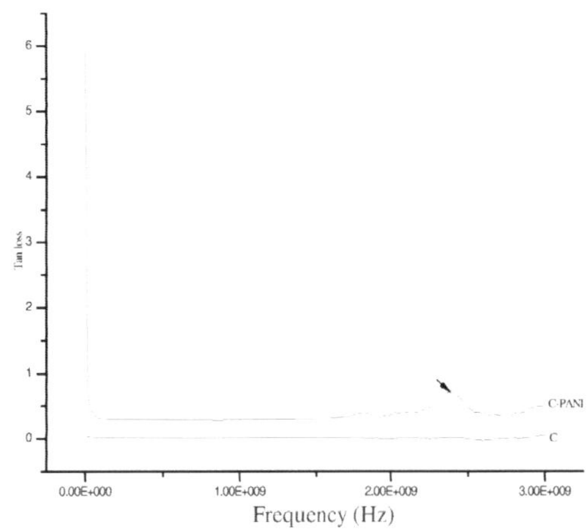

Figure 5. Frequency dependence Tan loss at 300K for C and C-PANI

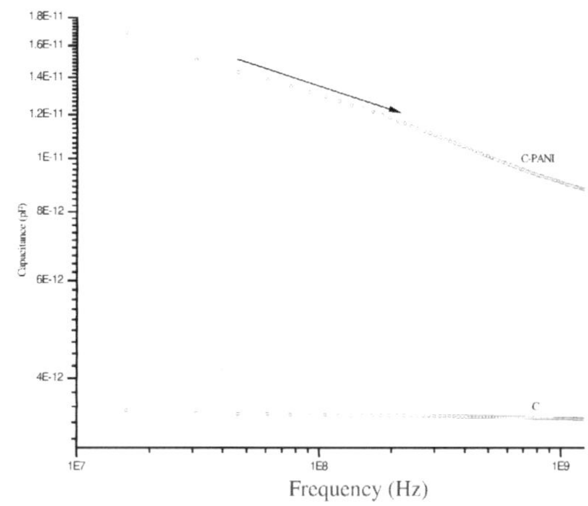

Figure 7. Frequency versus capacitance plot for C and C-PANI composite

Capacitance (Figure 7) of both the materials also shows higher value at low frequency region due to decrease in space charge region at the electrode surfaces. The decrease in capacitance with increase in frequency might be due to moderately blocked charge carriers adjacent to the electrodes [14].
Frequency dependent *AC* conductivity was found using relation (1);

$$\sigma_{ac} = \omega\varepsilon'\,\varepsilon'\tan\delta \qquad (1)$$

Where ω is angular frequency, ε′ is the permittivity of free space, ε′ is dielectric constant and tan δ is the loss tangent. ($\omega = 2\pi\nu$)
From the Figure 8, it is clear that the *AC* conductivity of the composite increases with the addition of PANI due to its intrinsic conducting property. This also signifies that hopping of electrons conducting mechanism is dominant for which a friction of energy is needed.

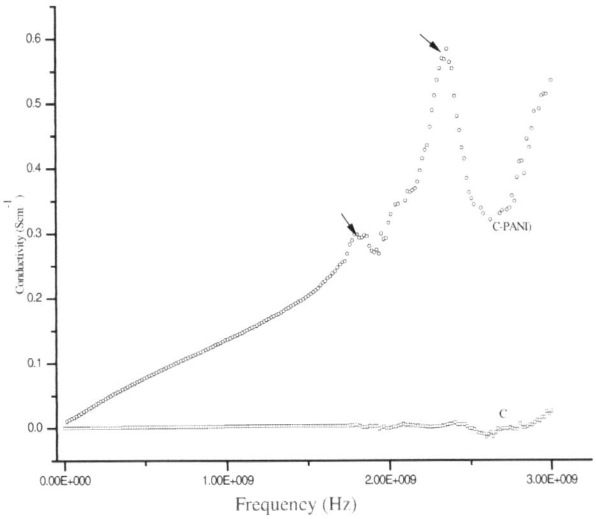

Figure 8. AC conductivity of C and C-PANI as a function of frequency at 313K

3.5. Mechanical behaviour

Rheological properties were measured in order to investigate the mechanical effects of ceramic particles and polymer on the properties of the composite. Shear stress and viscosity of the ceramic particles and their composite with PANI were determined as a function of shear rate at 30, 40 and 50 °C. The shear stress of ceramic particles decreases with increasing shear rate at all temperatures and exhibit equilibrium at high shear rate values [15]. However, this property for the composite first decreases and then increase at high shear rate values as shown in Figure 8 a and b. Particles attraction leads to their aggregation. Shear stress dismisses such aggregates via rising shear rate which is known as shear thinning. The variation of viscosity with shear rate is shown in Figure 9 (a, b). It is clear from these plots that at low shear rate, the viscosity is high and then it drops at high shear rate values at all the three temperatures for both the particles and composite. It shows that in low shear rate region, each system exhibits a strong tendency of shear thinning and Newtonian behaviour. It is accredited to the fact that low shear rate is unable to disturb the metal ions in the ceramic particles and the complex composite system. As the shear rate increases, the structures break down occurs and thus the viscosity of the system decreases. The noticed increase in shear stress and decrease in viscosity of the composite shows the pseudo- plastic nature of ceramic-polymer system.

Figure 10 (a, b) shows that the behaviour C-PANI nano-composite confirms well to the Bingham model [15]. In this model, the shear stress is represented by $\tau = \tau_0 + \eta_0\gamma$, where τ is the shear stress, τ_0 denotes yield stress; η_0 is the zero field viscosity while γ is the shear rate. Both the particles and the composite display Bingham fluid behaviour in the low shear rate region at various temperatures.

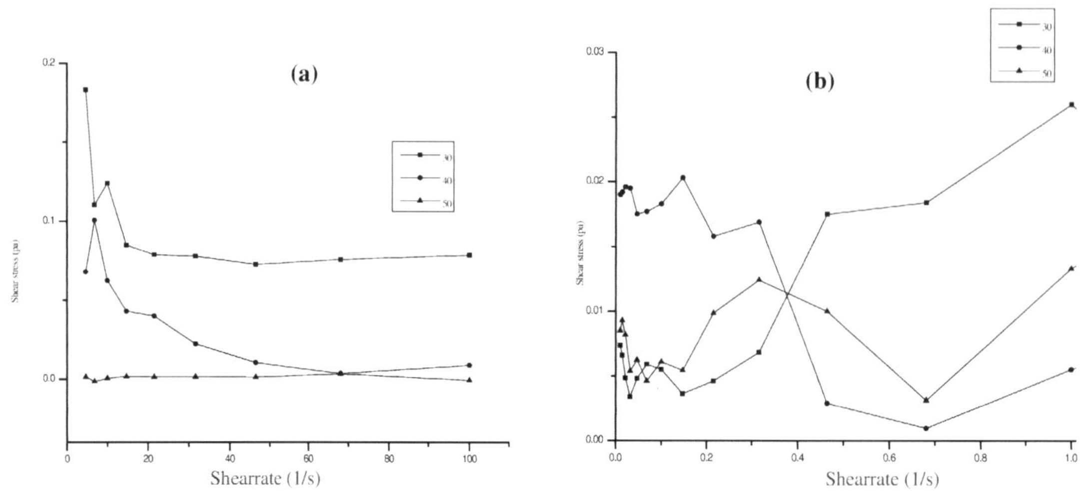

Figure 9. Shear rate versus shear stress plot for (a) ceramic particles and (b) composite with PANI

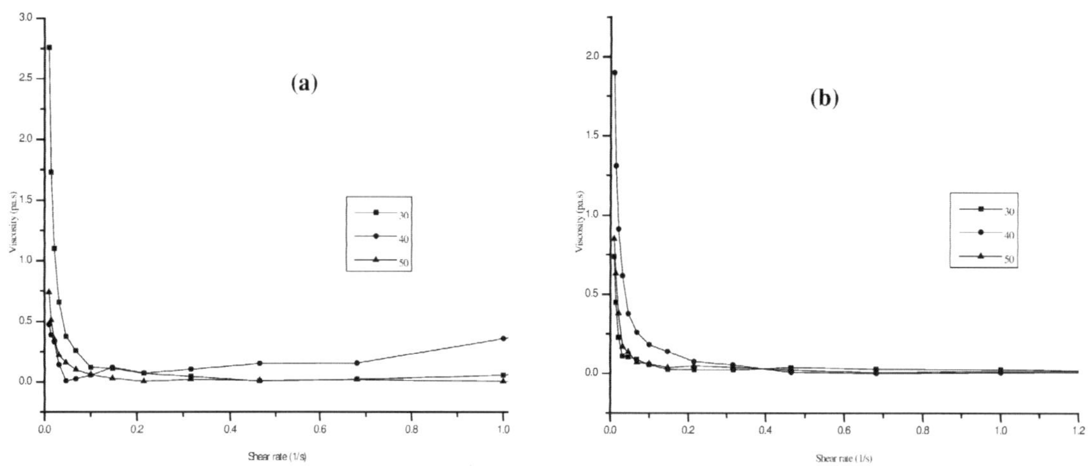

Figure 10. Shear rate versus viscosity plots (a) C and (b) C-PANI composite

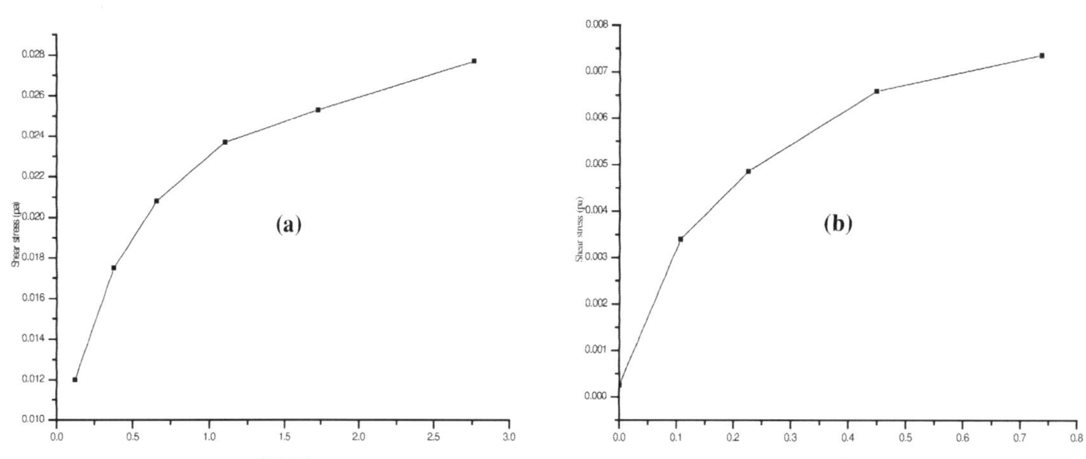

Figure 11. Bingham fluid behaviour of (a) C and (b) C-PANI composite

4. Conclusions

Inorganic ceramic particles were synthesized by sol-gel technique. The prepared particles were successfully incorporated in PANI matrix via *in-situ* free radical polymerization pathway. The composites were characterized by different physicochemical techniques. FT-IR confirmed the synthesis of the particles and composites. SEM showed that the particle are in range nm-μm and have dispersed well in the polymer matrix. Dielectric properties of the ceramic have been enhanced by the PANI. Particles and composites also showed good rheological properties. It is proposed that the prepared material can be used in embedded capacitors, switching windows and electromagnetic shutters.

5. Acknowledgements

The National Centre of Excellence in Physical chemistry (University of Peshawar) is highly acknowledged for providing financial support. I also acknowledge the consistent help and support of my loving spouse during the preparation of this manuscript.

6. References

[1] Yu Y Y, Chien W C, Wu T H and Yu H H 2011 *Thin Solid Films* 520 1495-1502
[2] Yu Y Y, Chien W C, Tsai T W and Yu H H 2011 *Mate. Chem. Phys.* 126 962-972
[3] Yu Y Y, Chien W C, Lin J M and Yu H H 2011 *Thin Solid Films* 519 4731-4736
[4] Liu B T, Tang S J, Yu Y Y and Lin S H 2011 *Colloid. Surf.* A. *Physicochem. Eng. Aspect.* 377 138-143
[5] Park C H, Lee H S, Lee K H, Kim D H, Kim H R, Lee G H, Kim J H and Im S 2011 *Org. Electron.*12 1533-1538
[6] Bichler S, Feldbacher S, Woods R, Satzinger V, Schmidt V, Jakopic G, Langer G and Kern W 2012 *Opt. Mate.* 34 772-780
[7] Zhao F, Zhang Z, Liu Y, Dai Y, Chen J and Ma D 2012 *Org. Electron.* 13 1049-1055
[8] Nakayama N and Hayashi T 2008 *Prog. Org. Coat.* 62 274-284
[9] Wua C H, Li H, Fong H H, Pozdin V A, Estroff L A and Malliaras G G 2011 *Org. Electron.* 12 1073-1089
[10] Khan M S, Gul R, Wahid M S 2013 *J. Polym. Eng.* 33(7) 633-638
[11] Yuvaraj H, Jae-Jin S and Kwon T L 2012 *J. Supercritic. Fluid.* 71 45-63
[12] Kargirwar S R, Thakare S R, Choudhary M D, Kondawar S B and Dhakate S R 2011 *Adv.Mat. Lett.* 2 (6) 397
[13] Shashi K, Susheel K and Amit K 2013 *J. Allo. Comp.* 578 249-256
[14] Jung-Rag Y, Jeong-Woo H and Kyung-Min L 2009 *Trans. Elect. Electron. Mate.*10 116-120
[15] Moeen S J, Vaezi M R, Yousefi A A and Ghasemi E 2011 *J. App. Polym. Sci.* 2534-2539

Forming of a super plastic sheet metal made of MgAZ31 alloy

Adnan I. O. Zaid, Mustafa A. A. Al-Matari[1], M. A. H. Nazzal[2]
Applied Science Private University, Mechanical and Industrial Engineering Department, Amman 11931 Jordan
[1]Industrial Engineering Department, The University of Jordan, Amman 11942, Jordan
[2]Mechatronic Engineering Department, German-Jordan University, Madaba, Jordan

E-mail:adnan_kilani@yahoo.com

Abstract. Metal forming industries are constantly looking for advanced innovation, economical and energy efficient techniques. Superplastic forming has a great potential to be one of those advanced forming methods. It is a near net shape forming process which uses a unique type of materials where elongation exceeds 200% during a controlled forming conditions, e.g. temperature, pressure, and strain rate. Most of superplastic materials are formed by gas technique at elevated temperature. The main objectives of the research work in this paper were: to study the effects of the forming schemes on the forming time and thickness distribution of the formed and device a method to improve the forming part thickness and its uniformity distribution and the forming time. In this paper, a hydraulic and heating system were designed and manufactured to facilitate the experimental investigation. The superplastic magnesium alloy AZ31, Mg AZ31, was formed at 350°C with different strain rates to investigate the effect of the forming pressure profiles on the thickness uniformity of the superplastic formed part. The pressure profiles were generated based on Dutta and Mukherjee analytical approach. Finally, a variable strain rate method is modified to improve the uniformity of the thickness distribution of the formed part and reduce the forming time; which is a major limitation of superplastic forming.

1. Introduction

1.1 Super plasticity and superplastic Materials

Many superplastic materials follow this constitutive equation:

$$\sigma = k\boldsymbol{\varepsilon}^{n}\,\dot{\epsilon}^{m} \quad (1)$$

Where σ is the flow stress, ε is the true strain, $\dot{\epsilon}$ is the true strain rate, n is strain hardening index, m is the strain rate sensitivity index, and k is the material constant. In this equation the value of n is very small, i.e. strain independent and because the increase in strain rate will cause decrease in the strain hardening then the equation becomes

$$\sigma = k\,\dot{\epsilon}^{m} \quad (2)$$

The value of m depends on the material; it can be calculated from the slope of logarithm of stress versus logarithm of the strain rate. The sigmoidal shape can be divided into three regions where different microstructural mechanisms are believed to dominate the deformation behavior, [4]. It is noteworthy to mention that as the grain size increases, the maximum value of m decreases; which

Content from this work may be used under the terms of the Creative Commons Attribution 3.0 licence. Any further distribution of this work must maintain attribution to the author(s) and the title of the work, journal citation and DOI.
Published under licence by IOP Publishing Ltd

IOP Conf. Series: Materials Science and Engineering **146** (2016) 012012 doi:10.1088/1757-899X/146/1/012012

yields to premature failure during deformation. This is why it is recommended to form fine-grained alloys.
The strain rate can be defined as the change in strain with respect to time.

$$\dot{\epsilon}=\frac{d\varepsilon}{dt}=\frac{d[ln(\frac{l}{lo})]}{dt}=\frac{1}{l}\frac{dl}{dt}=\frac{v}{l} \quad (3)$$

Where *t* is the time, and *l* is the instantaneous length.

2. Material, equipment, and experimental set-up

2.1 Materials

The used specimens were square sheets of the dimension 50 cm x 50 cm x 1.6 mm thickness and made from magnesium alloy, MgAZ31,having the chemical composition shown in Table 1.

Table 1. Composition of the MgAZ31 alloy in wt. %

Element	Al	Zn	Mn	Fe	Cu	Ni	Si	Ca	Be	Sr	Ce	Mg
Wt.%	3.1	1.0	0.42	0.006	0.003	<0.003	<0.1	<0.01	<0.005	<0.005	<0.01	Bal.

The mechanical behavior of this alloy represented as true stress versus true strain curves at different strain rates and at constant temperature of 350 °C is shown in Figure 1, and the logarithmic stress strain rate diagram at the same temperature is shown in Figure 2.

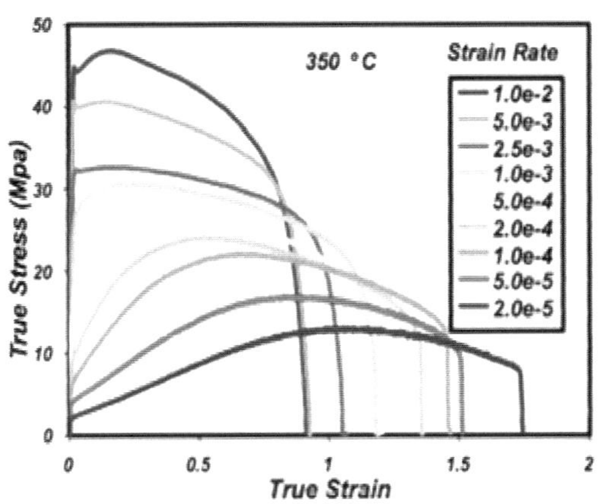

Figure 1. The stress strain curves for MgAZ31 at different strain rates and 350 °C

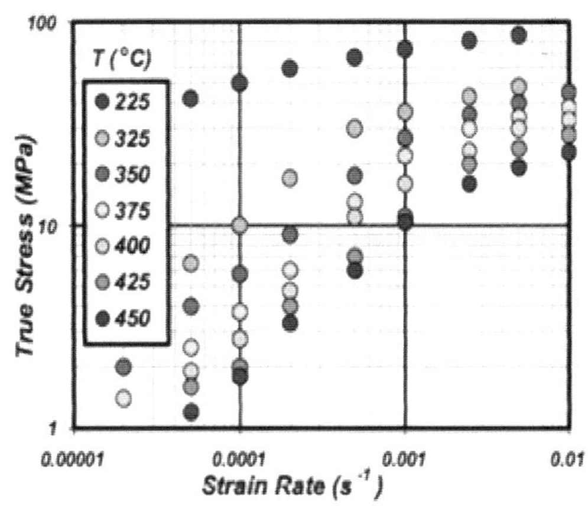

Figure 2. Logarithmic stress strain rate curve for MgAZ31 alloy

2.2.1. Hydraulic press

A four-column100 tons hydraulic press 1 m X 1 m X 2 m dimensions is designed, manufactured and assembled to carry out the experimental work. Figure 3 shows a 3 D drawing of the assembled press. The reason for using a four-column press is to fix and stabilize the working table (component 3) when it slides in the vertical direction. The upper, lower and moving tables are built from hollow steel tubes with 10 mm wall thickness. Details of its design, manufacturing and the different components associated with it are given in, [5].

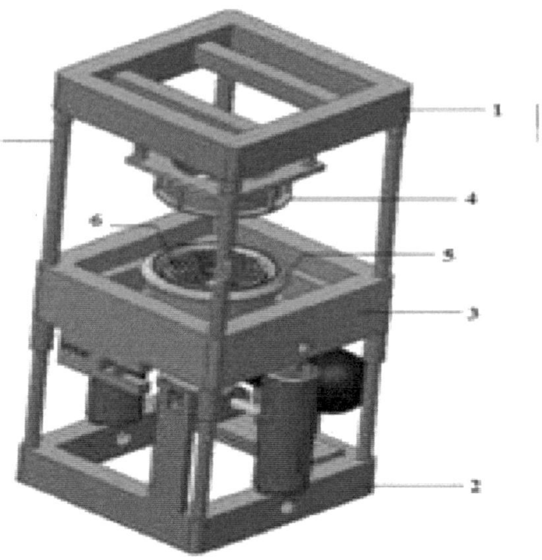

Part No.	1	2	3	4	5	6	7
Part name	Reservor	Electric motor	Coupling system	Pump	Rod	Piping line	Column

Figure 3. Three dimensional drawing of the hydraulic press and its components

Part No.	1	2	3	4	5	6	7
Part name	Reservor	Electric motor	Coupling system	Pump	Rod	Piping line	Column

Figure 4. A photograph showing the hydraulic system

The hydraulic system, shown in Figure 4, consists of 10 HP electrical motor, two hydraulic cylinders, a pump, a reservoir, and control valve. It allows the compressed oil to travel through the piping system to the rods at a pressure up to 70 bars.

3. Theoretical considerations

3.1 Forming pressure time history

Selection of the forming pressure time history for superplastic forming is very critical as it plays a major role in the thickness uniformity of the formed sheet. In this work, the pressure profile is generated based on the analytical approach given [6]. They derived an equation that relates the required gas pressure to the material parameters such as flow stress, strain rate, and also the sheet and die geometries to form the sheet at a constant strain rate. The equation is as follows:

$$P = 4\frac{Si}{a}\sigma \exp(-\dot{\epsilon}t)\,[\exp(-\dot{\epsilon}t)\,(1\text{-}\exp(-\dot{\epsilon}t))]^{1/2} \qquad (4)$$

Where a is the radius of the die and S_i is the initial sheet thickness.

3.2. Forming pressure profiles at different strain rates

In this work, pressure profiles are generated for four constant strain rates, namely: $5x10^{-3}$ s^{-1}, $2.5x0^{-3}s^{-1}$, $1x10^{-3}$ s^{-1}, $5x10^{-4}$ s^{-1}. Figures 5-8 inclusive show the pressure profiles separately for each strain rate at a magnified scale for clarity. It is worth mentioning that at each strain rate, forming is stopped when the dome height reached 180 mm; which is equivalent to an effective strain of 0.7 at the pole of the dome.

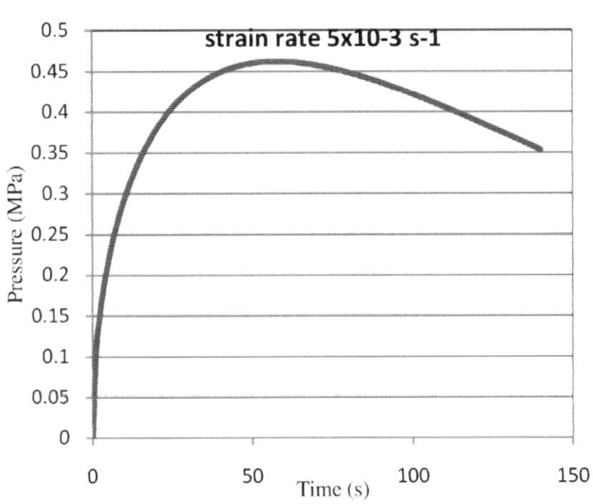

Figure 5. Pressure profile for a strain rate of $5x10^{-3}$ s^{-1}

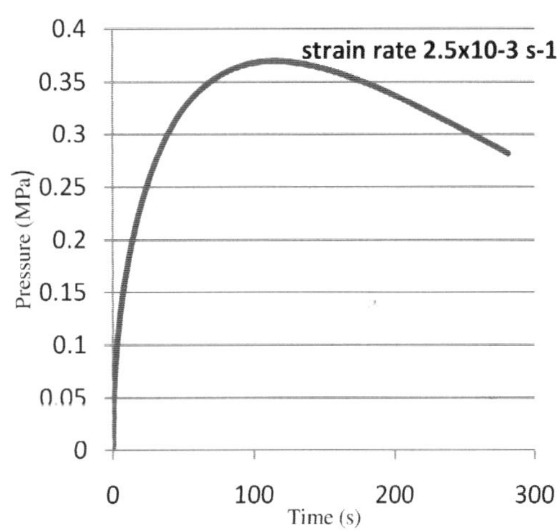

Figure 6. Pressure profile at a strain rate of $2.5x10^{-3}$ s^{-1}

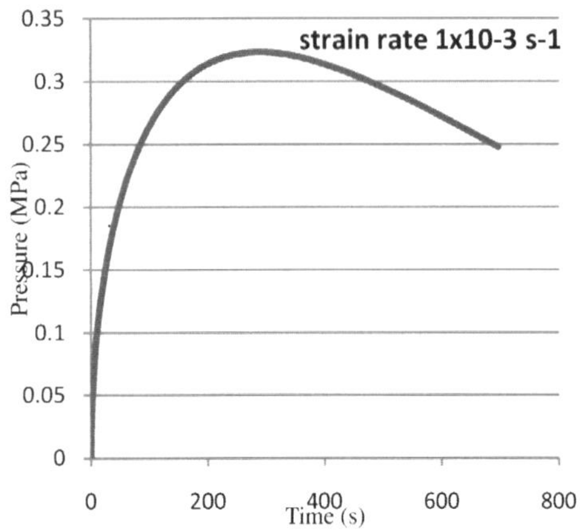

Figure 7. Pressure profile at a strain rate of $1x10^{-3}\ s^{-1}$

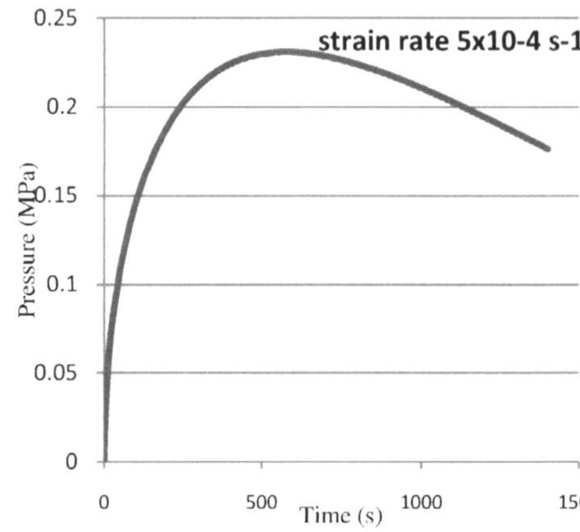

Figure 8. Pressure profile for a strain rate of $5x10^{-4}\ s^{-1}$

It can be seen from this figure that the pressure increases gradually at the beginning of deformation of the sheet until it reaches its maximum value. This is attributed to the large radius of curvature of the sheet at the initial stages of deformation. Therefore, a small amount of pressure is required to form the sheet at the beginning of deformation. As the radius of curvature decreases, more pressure is required to form the sheet. When more deformation takes place, the thickness decreases at the polar region of the dome. As a result, the forming pressure decreases so that a constant strain rate is maintained in the polar region of the dome.

4. Results and discussion

4.1. Variation of dome height with time for different strain rates

According to reference [7], the dome height (H) can be calculated from the following equation:

$$H = a\ [(\frac{S}{Si})-1]^{\ 0.5} \tag{5}$$

Where a is the radius of the die, S_i is the initial thickness and S the instantaneous thickness.
The instantaneous thickness at the pole can be found from the following equation, [6]. (Dutta and Mukherjee, 1992).

$$S = S_i \exp\ (-\ \dot{\epsilon} t) \tag{6}$$

The theoretical dome heights as a function of time for strain rates of $5x10^{-3}\ s^{-1}$ and $5x10^{-4}\ s^{-1}$ are shown in Figures 9 and 10 respectively.

Figures 11 and 12 shows the variation of dome height with time analytically and experimentally at strain rates of $2.5\ x\ 10^{-3}s^{-1}$ and $1\ x\ 10^{-3}s^{-1}$ respectively. It can be demonstrated from these two figures that the amount of plastic deformation at the same time is higher at the high strain rate for both theoretical and experimental values. Also shown on the curve, the difference between the experimental and theoretical values obtained from reference [7]. This is due to elastic recovery, which is not considered in the analytical equation. This is why in designing and manufacturing of dies, to obtain the required dimensions, elastic recovery is also considered in the die design process. Furthermore, these two figures indicate that the theoretical equation suggested by Pilling and Ridley overestimates the calculations of e height.

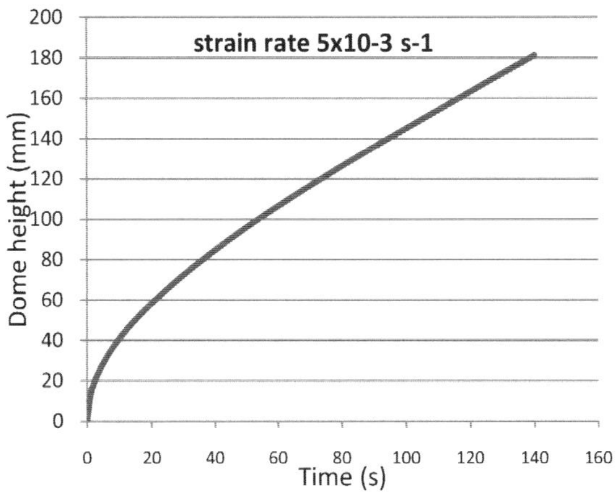

Figure 9. Dome height versus time at a strain rate of $5x10^{-3}$ s^{-1}

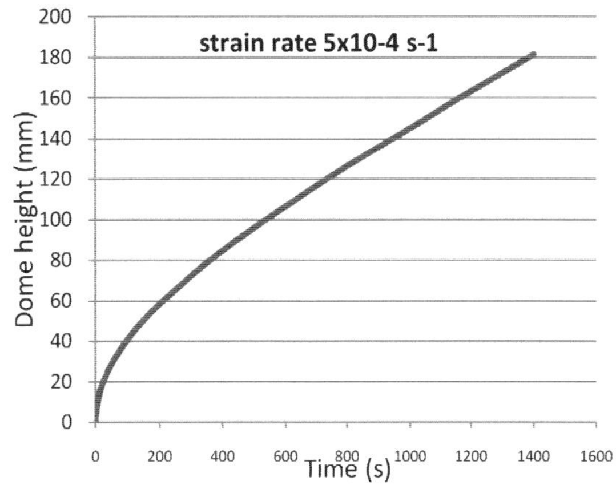

Figure 10. Dome height versus time at a strain rate $5x10^{-4}s^{-1}$

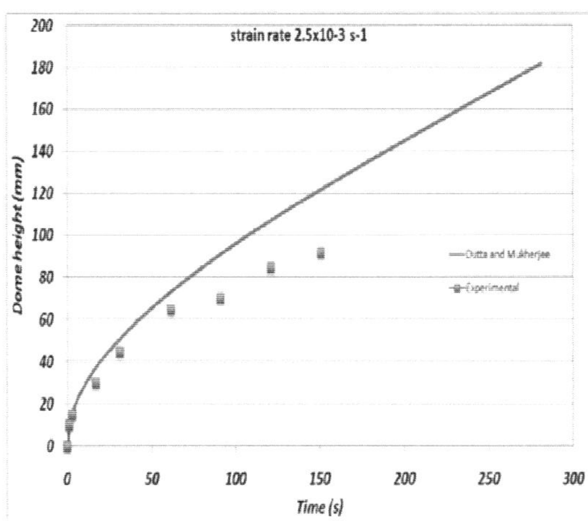

Figure 11. Theoretical and experimental presentation of dome height versus time at $2.5x10^{-3}$ s^{-1}

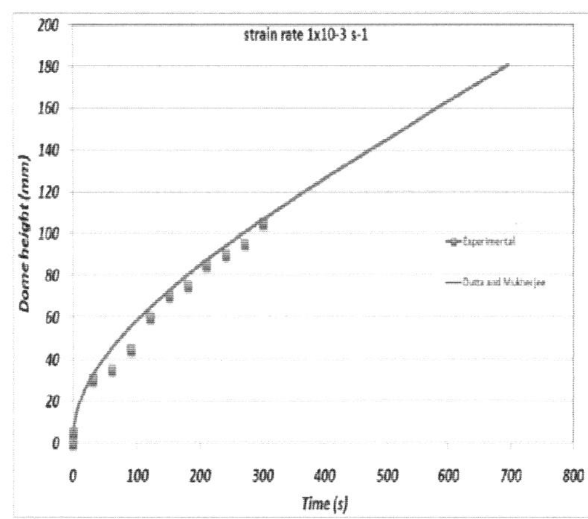

Figure 12. Theoretical and experimental presentation of dome height versus time at $1x10^{-3}$ s^{-1}

4.2 Variation of dome height with the distance from the pole

Figures 13 and 14 show the variation of the dome height with the distance from the dome pole at strain rates of 1 x $10^{-3}s^{-1}$ and 2.5 x $10^{-3}s^{-1}$ respectively.

It can be seen from figures 13and 14 that the height of the dome at the pole at $1x10^{-3}s^{-1}$ is higher than that at $2.5x10^{-3}s^{-1}$ being 108 mm at $1x10^{-3}s^{-1}$ and 92 mm at $2.5x10^{-3}s^{-1}$ strain rate. This indicates that the AZ31Mg alloy possesses higher degree of super-plasticity at the lower values of strain rate within the experimental limitations.

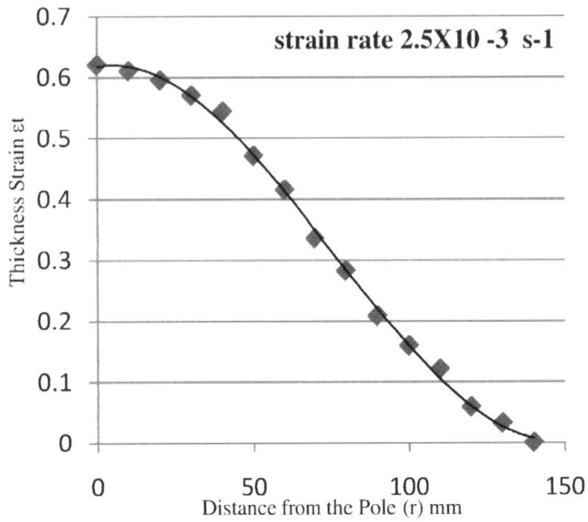

Figure 13. Dome Height versus Distance from the Pole at Strain Rate of $2.5x10^{-3}s^{-1}$

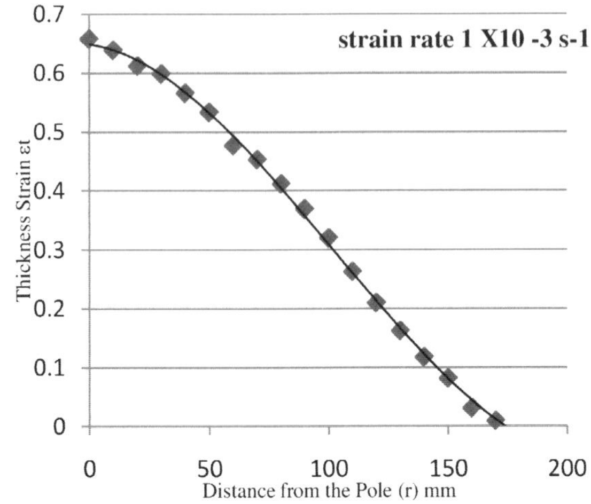

Figure 14. Dome Height versus Distance from the Pole at Strain Rate of Rate $1x10^{-3}s^{-1}$

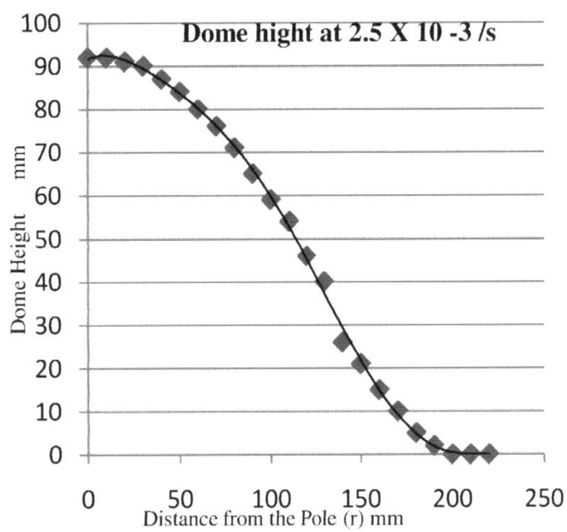

Figure 15. Thickness strain versus distance from the pole (r) at strain rate of 2.5 x $10^{-3}s^{-1}$

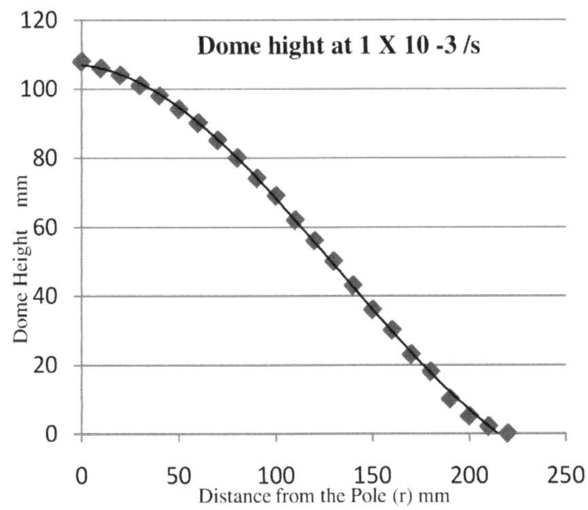

Figure 16. Thickness strain versus distance from the pole (r) at strain Rate 1 x $10^{-3}s^{-1}$

4.2 Sheet thickness distribution at different strain rates

To determine the final thickness strain in the formed sheets, the deformed sheets into dome shapes were sectioned along their diameters, ground and polished, and the thickness was measured at the pole of the dome and at locations of 10 mm from each other were measured using a digital micrometer, from which the thickness strain εt distribution along the sheet diameter was obtained using the following equation:

$$\varepsilon t = \ln\left(\frac{S}{Si}\right) \qquad (9)$$

Where S is the final measured thickness and S_i is the original sheet thickness which is 1.78 mm in average. This was carried out for the deformed sheet at each strain rate and the results are then presented graphically as thickness strain versus distance from the center. These are shown in Figures 15 and 16 at strain rates of $1x10^{-3}s^{-1}$ and $2.5x10^{-3}s^{-1}$ respectively.

It can be seen from Figures 15 and 16 that at the lower strain rate, 1 $X10^{-3}s^{-1}$, the thickness strain is slightly higher than at 2.5 $X10^{-3}s^{-1}$ strain rate, being 65.8% thinning compared to 62.1% i.e. about 3.7% higher. This is explained by the degree of the super-plasticity of the AZ31Mg alloy possessing higher degree of super-plasticity at the lower strain rate. The photographs of Figure17 show the deformed sheets into domes, and Figure18 shows a photograph of the dome with crack initiation at the pole. This is expected as the maximum plastic deformation occurred at the pole, where maximum thinning has occurred.

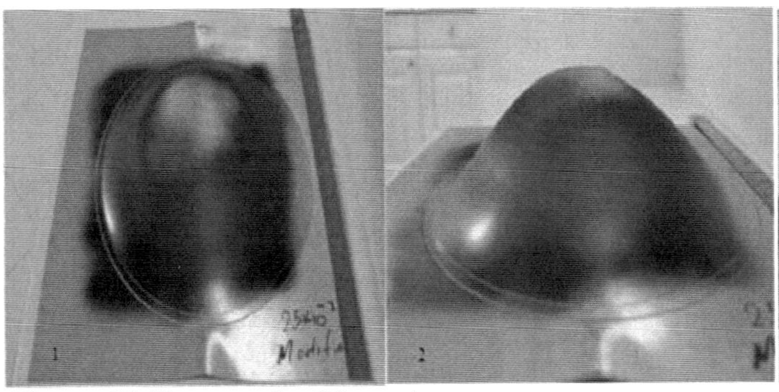

Figure 17. Photograph of formed parts

Figure 18. Photograph with crack initiation at the pole

4. Conclusions
The following points are concluded:

I. The superplastic, (SP) forming of the AZ31Mg alloy at 400 °C on industrial and conventional level is proved to be carried out successfully, as this alloy cannot be formed successfully due to its low ductility.

II. The maximum possible successful deformation of this alloy, represented by the maximum polar height and the amount of thinning at the pole, is found to vary with the strain rate in the forming process, being higher at the lower strain rate 1 $X10^{-3}s^{-1}$, the thickness strain is slightly higher than at 2.5 $X10^{-3}s^{-1}$ strain rate, being 65.8% thinning compared to 62.1% i.e. about 3.7% higher.. This indicates that the alloy possesses higher degree of superplastic behavior at the lower strain rate values.

III. The height of the dome at the pole at 1 $X10^{-3}s^{-1}$ is higher than that at 2.5 $X10^{-3}s^{-1}$ being 108 mm at 1 $X10^{-3}s^{-1}$ and 92 mm at 2.5 $X10^{-3}s^{-1}$strain rate.4. The superplastic forming time of this alloy is high compared to the forming time of the ordinary engineering material. This agrees with the previous findings of other researchers, [8-9].

5. Acknowledgment
The first author (Adnan I. O. Zaid) is grateful to the Applied Science Private University, Amman, Jordan for partial financial support granted to this research (Grant No.DRGS-2015).

6. References
[1] Chandra N 2002 Constitutive Behavior of Superplastic Materials, International Journal of Non-Linear Mechanics, 37, 461-484

[2] Abu-FarhaFadi K 2007 Integrated approach to the superplastic forming of magnesium alloys, PhD thesis, University of Kentucky Doctoral Dissertations, 493

[3] Kalpakjian S, Schmid S R 2002*Manufacturing Processes for Engineering Materials* 4th ed., New Jersey: Pearson Education Inc

[4] Zaid A I O 2004Superplasticity and Superplastic Materials: Applications and Recent Developments *Proc of 2nd International Conference on Frontiers of Advanced Engineering Material*, 37-44

[5] Al-Matari M A A 2014 Sheet Metal Forming of a Light weight Superplastic Alloy The University of Jordan, Amman, Jordan

[6] Dutta A, Mukherjee A K1992 Superplastic Forming: an Analytical Approach, *Materials Science and Engineering,* A157, 9-13

[7] Pilling J, Ridley N 1989 *Super-plasticity in Crystalline solids* The Institute of Metals, London, UK

[8] Khraisheh M K, Zbib H M 1999 Optimum Forming Loading Paths for Pb-Sn Superplastic Sheet Materials, *ASME Journal of Engineering Materials and Technology*121, 341-345

[9] Khraisheh M K , Abu-Farha F K, Weinmann K J 2007Investigation of Post-Superplastic Forming Properties of AZ31 Magnesium Alloy,CIRP Annals - Manufacturing Technology, 56, 289-292

Optical and structural properties of Cr and Ag thin films deposited on glass substrate

A. Rauf, K. Ahmed, F. Nasim, A. N. Khan and A. Gul
Institute of Industrial Control System, P. O. Box 1398, Rawalpindi, Pakistan

E-mail: abdulrauf3134@yahoo.com

Abstract. Most of the rotating or noting patterns are being developed by using silver plating through chemical coating. Silver layers deteriorate with the passage of time and become less reflective while undergo through cleaning process due to its softness and the results become unpredictable. In this paper an alternate method for development of above mentioned pattern has been demonstrated. Chromium (Cr) and Silver (Ag) thin films of 200nm and 160nm thick respectively have been realized using electron beam evaporation (PVD technique) on quartz substrate. Structural analysis has been carried out by XRD and SEM while optical transmission/reflection has been studied using spectrophotometer. XRD analysis shows that Ag coated thin films exhibit FCC structure while Cr coated thin films reveals a BCC structure. SEM analysis shows almost smooth and uniform surfaces in both cases. After passing through high and low temperature cycles it was found that the results of pattern structures developed by chromium coating were more reliable than obtained through silver platting process.

1. Introduction

Thin film chromium (Cr) and silver (Ag) coatings are used in photo masks, developing the desired reticules and pinhole on glass substrate integrated circuits, optical beam splitters etc. There are some other applications like thin film resisters, capacitor plates and antimagnetic shielding materials [1-2]. The chromium oxide (Cr_2O_3) thin films exhibit high hardness, high wear and high corrosion resistance which are important properties for protective coating applications [3-4].Hardness and wear resistance are important properties of chromium oxide for protective coating applications. In this regard by controlling the coating deposition parameters high hardness and good wear resistance can be obtained. Chromium thin films are characterized either entirely on the electrical properties [5-10] or entirely on structural properties [11-14].

There are different techniques to deposit Cr/Cr_2O_3 coatings e.g., sputtering, electron beam evaporation, CVD process, electrochemical deposition, reactive pulsed laser ablation technique and chemical spray pyrolysis depending on different applications.

Optical applications of Cr_2O_3 / Cr thin films such as solar energy conversion and solar energy shielding films for windows have been made [3, 15-17] indicating an appreciable order of transmittance in the visible and near IR region where as the reflectance spectrum indicated a moderate value over the UV, visible and NIR wavelength region.

In this paper optical and structural properties of chromium (Cr) and silver (Ag) thin films deposited on glass substrate are explored for further optical applications.

Content from this work may be used under the terms of the Creative Commons Attribution 3.0 licence. Any further distribution of this work must maintain attribution to the author(s) and the title of the work, journal citation and DOI.
Published under licence by IOP Publishing Ltd

2. Experimental

Chromium (Cr) and silver (Ag) thin films of thickness 200nm and 160nm respectively have been deposited through PVD technique on quartz substrates 25mmdiameter and 2mm thickness. The samples (05 each) were properly cleaned with a mixture of either and alcohol (80% x 20% by volume) in clean environment i.e., class 100 supper clean table before placing them in an ultra-high vacuum system. Thin film coating technique was employed by using optical design software. Balzers vacuum coating machine was used for film deposition at the base pressure of 2.0 x 10^{-7} Torr. 99.99% pure chromium and silver grains of 50μm were used as source materials. Figure 1 shows the schematic diagram of thin film vacuum coating system. The detailed experimental technique is explained elsewhere [18].

During vacuum coating process, 50 ~ 55nm thick layer for both the cases was achieved in one routine process. The samples were again cleaned with the same mixture and the process is repeated for another 50 ~ 55nm thick layer in order to remove the pin holes from the coated surface. The process is repeated till the desired layer thickness and smooth surfaces for both cases were achieved.

The microstructure of coated samples were examined using X-ray diffractometer. The surface morphology of coated surfaces was characterized by SEM. The optical properties (Transmission/ Reflection) of coated surfaces were measured using SHIMADZU UV-3100 photo-spectrometer in the spectral range 300 to 3200nm.

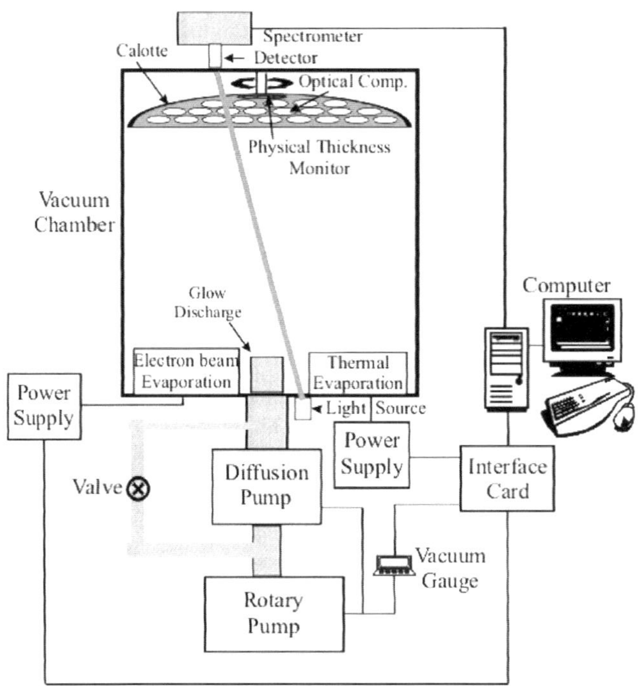

Figure 1. Schematic diagram of thin film coating system

3. Results and discussions

Cr and Ag thin films developed through e-beams evaporation on glass substrate were analyzed. Two methods i.e., XRD and SEM were used to study the structural behavior while the optical studies were carried out using the photo-spectrometer.

3.1.Structural analysis

Figures 2 & 3 show the X-ray diffraction (XRD) patterns and optical microscopy of e-beam deposited 200nm Cr thin films on quartz substrates. It can be seen from Figure 2 that only one peak with 2θ about 43.8° is indicative with bcc (110) orientation structure as indicated elsewhere [1]. Due to efficient vacuum system, base pressure in the range of 2.0 x 10^{-7} Torr, almost no impurities are present at the surface of chrome coated thin films suggesting the preferred orientation of e-beam deposited Cr films on glass substrate is bcc (110). Figure 3 represents the optical micrograph of chromium coated thin films on quartz. It is clear from the figure that a smooth pattern is achieved which also supports the high purity chromium thin films.

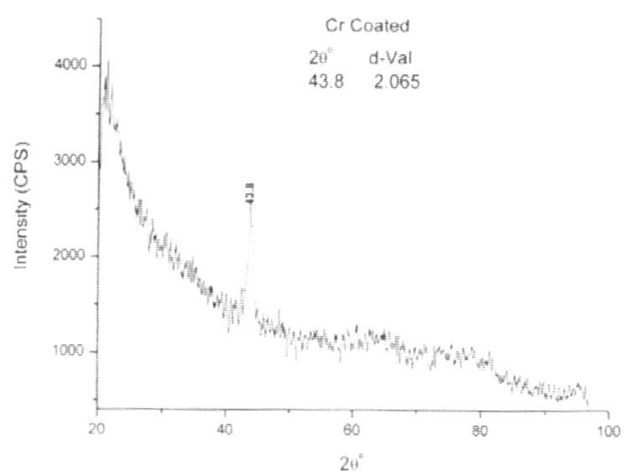

Figure 2. XRD pattern of Cr coated thin films on quartz

Figure 3. Optical Microscopy of Cr coated films on quartz

The analysis of the elemental compositions for chromium thin films deposited on glass substrate has been carried out through SEM-EDS spectrum analysis. Figure 4 shows the EDS spectrum of Cr films. It is clear that Cr peaks are prominent on the spectrum. Only one Si peak is observed which may be due the quartz substrate material. This also indicates that a smooth and homogenous coating is achieved.

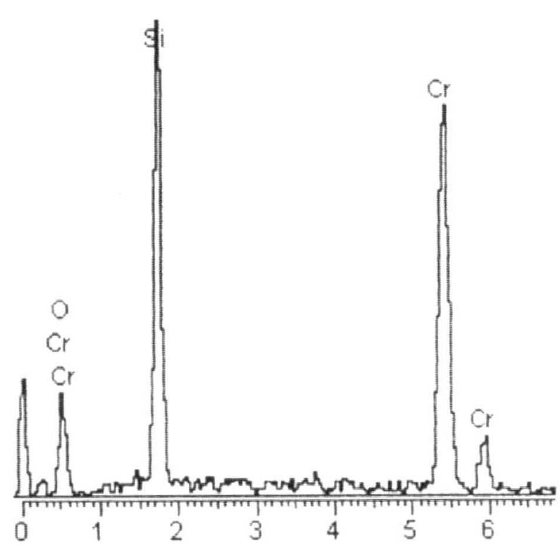

Figure 4. SEM-EDS spectrum of Cr coated thin films on quartz

Figures 5 & 6 show the X-ray diffraction (XRD) patterns and optical microscopy of e-beam deposited 160nm Ag thin films on quartz substrates. From figure5 it is clear that four peaks with 2θ at 37.35°, 43.5°, 63.75° and 76.65° of silver coated films at different orientation with fcc structure have been found. The corresponding (h k l) values are (111), (200), (220) and (311). It is also clear from the Figure 5 that only silver peaks are present throughout the coated surface indicating the purity of the coated films. Figure 6 shows the optical micrograph of silver coated thin films showing a smooth, highly pure and homogenous structure of the silver plated films.

IOP Conf. Series: Materials Science and Engineering **146** (2016) 012013 doi:10.1088/1757-899X/146/1/012013

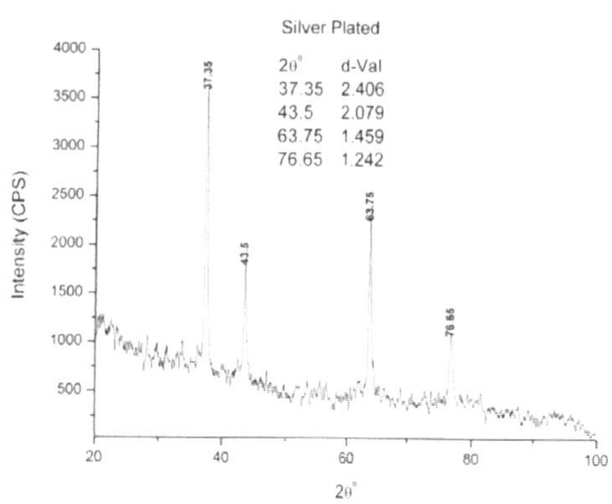

Figure 5. XRD pattern of Ag coated thin films on quartz

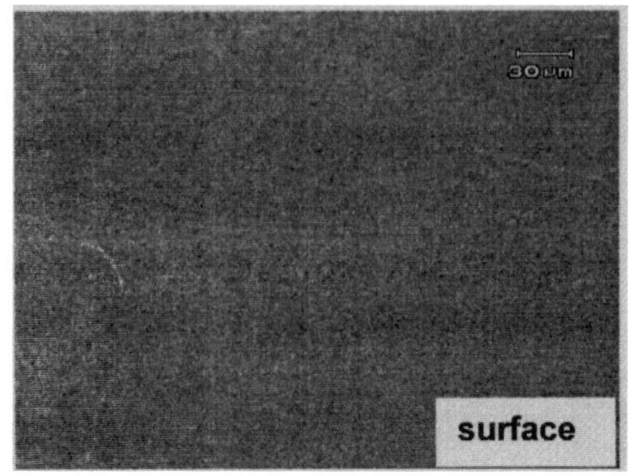

Figure 6. Optical Microscopy of Ag coated thin films on quartz

Figure 7 shows the SEM-EDS spectrographs of silver coated thin films deposited on quartz. It can be seen from the figure that only one element i.e., silver is present throughout the coated surface. This indicates the sustainability of vacuum coating system.

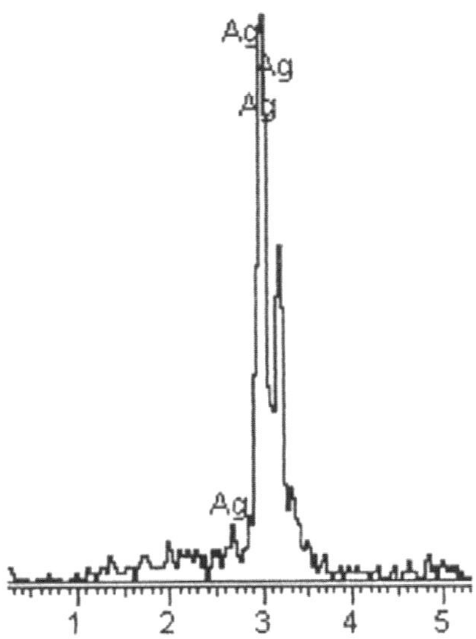

Figure 7. SEM-EDS spectrum of Ag coated thin films on quartz

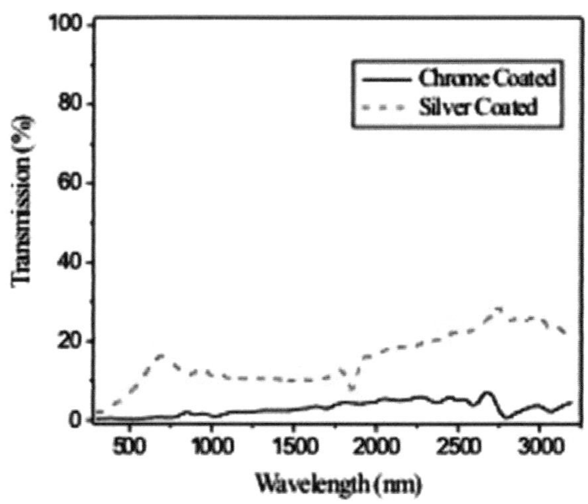

Figure 8. Transmission Vs. wavelength for Cr and Ag coated samples

3.2.Optical analysis

Spectral transmittance/reflectance T/R of Cr and Ag thin films has been measured at wavelength $0.3< \lambda <3.2$ μmusing UV-Visible spectrophotometer. Figure 8 shows the transmission vs. wavelength spectra for 200nm and 160nm thick Cr and Ag coated thin films respectively.

The optical studies indicate that for Cr coated samples, there is no transmission in the visible range i.e., from 300 ~ 700 nm wavelength. The transmission increases from 1000 nm and reaches maximum value of about 4.26% at 2100 nm and then decreases drastically to minimum value (0%) at 2750 nm. The reason of such a low transmission is due to the high film thickness (200 nm). As the coated film thickness increases the components become mirror like and hence transmission decreases.

In case of Ag coated samples the transmission increases upto 18% in the visible range (from 300 ~ 700 nm wavelength). The transmission decreases to about 10% in the IR region (800 ~ 1800 nm wavelength) and then starts increasing beyond this region. It should be noted that the film thickness in this case is less (160 nm) as compared to Cr film thickness so the transmission in the visible region appears. In the IR region, the behavior of the spectrum is almost same instead that the magnitude of transmission is somewhat higher due to less film thickness.

4. Conclusions

Cr and Ag thin films of thickness 200 and 160 nm respectively have been prepared on glass substrate by e-beam evaporation at a pressure of 4.66 x 10-5 Torr. XRD study shows that the thin films are amorphous in nature, while SEM study shows that the growth of thin films is uniform and smooth throughout the surface. The elemental composition has been studied through SEM-EDS analysis indicating the pure Cr and Ag elements are present throughout the coated surfaces. Optical studies in Cr coated samples indicate a very small transmission in IR region where as no transmission in visible spectrum showing the mirror like behavior. In case of Ag coated samples there is an appreciable transmission in visible region while the same behavior is observed in IR region due to the less film thickness as compared to Cr coated samples. After passing through the environmental tests it was found that the results of pattern structures developed by chromium coating were more reliable than obtained through silver platting process. It is due to the fact that the bonding of Cr with the glass is strong as compared to Ag and hence Cr coating with stands in sever environment conditions.

5. Acknowledgement

The authors acknowledge the structural analysis group in obtaining the XRD and SEM data and Mr. Sajid Mehmood (optical analysis group) for developing thin films and valuable contributions for providing the optical analysis data.

6. References

[1] Kulkerni A K and Chang L C1997 Thin Solid Films 301 17
[2] Holloway P H1980 Solid State Technology 23 109
[3] Julkarnain M D, Hossain J, Sharif K S and Khan K A 2011 J. of Optoelectronics and Advanced Materials,13(5) 485
[4] Pang X, Gao K, Lou F, Yang H, Qiao L, Wang Y and Volinsky A A 2008 Thin Solid Films 516 4685
[5] Gould P A 1965 Br. J. Appl. Phys. 16 1481
[6] Hara H and Sakata M 1977 J. Phys. Soc. Jpn. 43 468
[7] Udachan L A, Shivaprasad S M, Ashrit P V and Angadi M A 1980 Phys. Status Solidi 60 K191
[8] Mehanna E A, Arajs S, Helbig H F, Aidun R and Kattan N A 1987 J. Appl. Phys. 61(8) 4273
[9] El-Hiti M A, Ahmed M A and El-Shabasy M 1989 J. Mater. Sci. 24 329
[10] Angadi M A and Udachan L A 1981 J. Mater. Sci. 16 1412
[11] Ravipati D P, Haines W G and Dockendorf J L 1987 J. Vac. Sci. Technol. A5 (4) 1968
[12] Duan S L, Artman J O, Wong B and Laughlin D E 1990 J. Appl. Phys. 67(9) 4913
[13] Wang D D and Oki T 1990 J. Vac. Sci. Technol. A8 (4) 3163
[14] Norenberg H and Neumann H G 1991 Thin Solid Films 198 241
[15] Hutchins M G 1983 Surf. Tech.20 301
[16] Geotti-Bianchini F, Guglielmi M, Polato P and Soraru G D 1984 J. Non-Crystal. Solids 63 251
[17] Xiaolu Pang, Kewei Gao, Fei Luo, Yusuf Emirov, Alexandr Levin A and Alex Volinsky A 2009 Thin Solid Films 517 1922
[18] Ahmed K, Khan A N, Rauf A and Gul A 2014 IOP Conf. Series: Materials Science and Engineering 60 012016

Prediction of the fatigue life distribution for aluminum through its mechanical characteristics

S. Z. Ramadan and Adnan I. O. Zaid
Mechanical and Industrial Engineering Department,
Applied Science Private University, Amman 11931 Jordan

E Mail: s_ramadan@asu.edu.jo; adnan_kilani@yahoo.com

Abstract. A novel and reliable theoretical model based on the Birnbaum-Saunders (BISA) distribution is presented from which the fatigue life can be determined. Experimental verification of the model is in progress and will be published in due course.

1. Introduction

According to a poll conducted in 1985 by the American Society for Quality Control, reliability was the second most important attribute among the ten most important products attributes [1]. This result is expected as the impact of product failure can range from minor injuries and/or loss to sever injuries/death and/orloss. Therefore, the precise prediction of failures can save lives and money. Usually, the reliability of the products is determined through one of several types of life testing. The primary objective of these tests is to quantify the reliability of the product, which can be used to determine whether a set of goals for the product are met or not. Typically the result of the reliability test is a set of failure times that is analyzed statistically to predict the reliability distribution through curve fitting. One of the disadvantages of this method is the high cost and time in some circumstances especially when regular life tests are used. Another disadvantage is the low accuracy of the test when accelerated life tests are used due to extrapolation. Moreover, this approach pays no attention to the strong relationship between the mechanical properties and the reliability distribution of the product as it uses parametric or nonparametric statistical methods to predict the reliability from the failure data and not from the mechanical properties of the product.

Little work has been done to predict the life of products through methods other than life testing. Good references for fatigue life prediction methods can be found in [2]. Ref [3] predicted the fatigue life of the tibial component of a polymeric PMMA space rutilizingmultiaxial fatigue coupled with computational simulations and material properties. Ref [4] used the S-N curve and the crack growth rate curvet predict the fatigue life. Ref [5] proposed a method to predict the fatigue-life based on linear elastic analysis and used it to predict the fatigue for different materials subjected to constant amplitude multiaxial proportional loading. Ref [6] Used ΔK–N curve utilizing finite element analysis to predict the fatigue life of spot welds starting from coarse finite element meshes and ending at one unique ΔK–N curve.

Among the little available non life-testing prediction methods, prediction through fitting the mechanical properties to the life distribution parameters' was absent. The primary objective of this paper is to propose a method for fatigue failure mode life prediction of carbon steel shafts through their mechanical properties. The primary advantage of this method is in saving time and money. Once the relationship between the mechanical properties and the reliability distribution parameters for the

Content from this work may be used under the terms of the Creative Commons Attribution 3.0 licence. Any further distribution of this work must maintain attribution to the author(s) and the title of the work, journal citation and DOI.
Published under licence by IOP Publishing Ltd

carbon steel shaftsis known, the reliability distribution for any carbon steel shaft can be induced from the mechanical properties of that shaft without conducting reliability tests.

The Birnbaum-Saunders (BISA) distribution was used as the life distribution for the fatigue mode failure for these shafts. Different carbon steel specimens were prepared through heat treatment to generate different micro-structures. The mechanical properties of each type of micro-structure were determined through standard compression test. A number of standard specimens for each micro-structure were prepared and subjected to a simple stress accelerated life test based on standard fatigue test to generate failure times. Bayesian inference was conducted on these failure times to estimate the BISA parameters. The parameters of the BISA were related to the mechanical properties of the specimens such that the shape parameter and the scale parameter of the BISA were expressed as a function of the mechanical properties of the carbon steel.

2. Fatigue mechanism

Fatigue is the progressive and localized structure damage for the part when it is subjected to cyclic loading. The fatigue failure stress values are well below the ultimate tensile strength or even the yield strength of the material. Fatigue failure process can be divided into periods: the crack initiation period and crack growth period. The crack initiation period starts by initiation of micro-cracks due to local accumulation of dislocations, high stresses at local points, or plastic deformation around inhomogeneous inclusions or other imperfections in or under the contact surface [7]. The crack initiation period usually covers a large portion of the fatigue life under stress amplitudes just above the fatigue limit. The microstructure of the specimen, along with the applied stress and the geometry of the specimen, play a vital role in determining the position and mode of the fatigue crack initiation [7]. Consequently, at the same applied stress and geometry, the microstructure of the specimen determines the mode and position of the fatigue crack. The crack growth period follows the crack initiation period in which the cracks propagate until they cause a permanent damage to the part. The fatigue life of the member is the time (cycles) corresponding to those two periods.

The fatigue life of the specimen under fatigue failure mode can be modeled using variety of distributions like Weibull, Gamma, Lognormal and BISA (known as fatigue life distribution). The strength of BISA distribution is that it is based on the renewal theory and a set of physical properties of the fatigue process [9]. The main theory behind the BISA distribution is that:

1- The fatigue failureis generated by the cycles of the stress that causes degradation or crack growth until a critical crack size is reached, which causes a point fatigue failure.
2- In each cycle, the crack will extend and these extensions are statistically independent. The central limit theorem guarantees that the total crack extension is approximately normally distributed.

The BISA distribution defines a random variable T as the life time in cycles with the following probability density function (pdf) and cumulative density function (CDF):

$$f_T(t;\,\alpha,\beta) = \frac{t-\beta}{\alpha\sqrt{2\pi\beta t^3}}exp\left(\frac{\beta-t}{\alpha\sqrt{\beta t}}\right) \tag{1}$$

$$F_T(t;\,\alpha,\beta) = \frac{1}{\sqrt{2\pi}}exp\left(\frac{\beta-t}{\alpha\sqrt{\beta t}}\right) \tag{2}$$

Where α and β are the shape and scale parameters respectively such that$t, \alpha, \beta > 0$. According to Ref. [10] the scale parameter β is the ratio of the critical crack extension to the average incremental crack extension per cycle. Therefore, β represents the median number of cycles to failure as the time to the pth quantile of the BISA distribution is:

$$t_p = \frac{\beta}{4}\left(\alpha z_p + \sqrt{\alpha^2 z_p^2 + 4}\right)^2, \tag{3}$$

where z_p is the pth quantile of the standard normal distribution [11].

3. The effect of the mechanical characteristics on the fatigue life

In general, the metallurgical and /or the mechanical characteristics of an engineering material can be changed either by heat treatment if the material is liable for treatment such as the carbon steels, or more recently refining its structure by the addition of some rare elements. For example, the addition of Ti or Ti+B to Al and Al alloys to their melts prior to solidification to grain refine their metallurgical structure as they solidify in columnar structure with large grain size which tends to reduce their mechanical strength and surface quality [13]. It is worth mentioning in this respect that the addition of other alloying elements in the presence of the main grain refining element may affect its refining efficiency. For example, V, Mo, and C improves grain refining efficiency, whereas other elements like Cr, Zr, and Ta poison the grain refinement, i.e make the grains larger in size.

Figure 1 shows the effect of adding Ti-B at a rate of 0.1% which corresponds to the peritectic point at the Al-Ti-B phase diagram which is the ration normally used in industry.

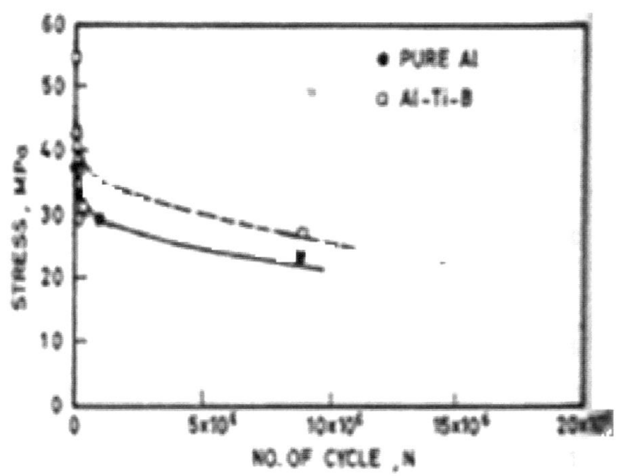

Figure 1.Effect of adding Ti-B at a rate of 0.1% Al-Ti-B

The equation that governs the strength of most of the materials is given by:

$$\sigma = k \in^n \tag{4}$$

where k is the strength factor of the material, $\in$ is the true strain, and n is the work or strain hardening index. It is easy to notice that as k increases, the strength of the material increases as well. This matter increases the material's ability to resist fatigue, hence increases the fatigue life. On the other hand, increases n decreases the strength of the material as $n$and $\in$ both are less than 1. Thus, increasing n, reduces the material's resistant to fatigue, consequently, decreases its fatigue life.

Changing the microstructure of the material with refiners changes its k and n values and consequently, changes its fatigue life. Changing the fatigue life changes the shape and scale parameters of the fatigue life distribution of the material.

As mentioned earlier, the aim of this paper is to find the equations that relate the mechanical characteristics i.e., k and n with the fatigue life distribution parameters i.e., α and β of Al-Ti-B alloys. For this purpose, different Al-Ti-B alloys were prepared using different weight percentages of Ti-B and their $k$and η values were found using compression test. For each alloy, number of specimens were prepared and a Type I simple step accelerated fatigue life test was conducted on them. The resulting censored data was fitted to BISA distribution using Bayesian inference to determineα and β for each alloy. Non-linear exponential-based regression model was used to find the equations relating α and β to k and $n$for Al-Ti-B alloys such that

$$\alpha = \exp(a_0 + a_1 k + a_2 n) \quad (5)$$

$$\beta = \exp(b_0 + b_1 k + b_2 n) \quad (6)$$

4. Life testing

The purpose of life testing is to assess the probability that the product can perform its intended tasks adequately in its useful life under normal operating conditions. Life tests are usually conducted under accelerated conditions to accelerate the failures and hence reduce the testing time. Unfortunately, the data collected under the abnormal conditions is not reliable as the data collected under normal conditions as the accelerated conditions impose abnormal operating circumstances on the specimens; therefore, extrapolation between the abnormal and the normal circumstances must be done to predict the life of the specimen under normal operating conditions.

To relate the parameters of the life distribution under abnormal conditions to the parameters under normal conditions, stress-acceleration models are usually used. The log-linear stress-acceleration model is widely used in literature to relate the mean time to failure of the distribution to the stress level. Under this acceleration model, usually the scale parameter is the only parameter affected.

In this paper a Type-I right censoring simple step accelerated fatigue life test (SSAFLT) is conducted to generate failure times. Figure 2 shows a schematic diagram for the SSALFT under Type-I right censoring.

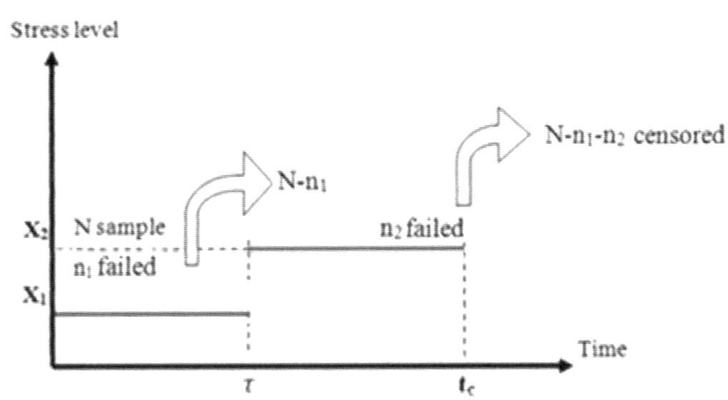

Figure 2. Schematic diagram for Type-I right censoring for SSAFLT

The SSAFLT under Type I censoring can be described as follows: n samples are stressed under x_1 stress level for τ time (cycles) during which n_1 samples fail. A higher stress, x_2, is applied on the survived ones until the end of the test time t_c (cycles) such that $x_2 > x_1$. In this paper, the SSAFLTs were carried out on Wholer Testing Machine at the laboratories of the Mechanical and Industrial Engineering Department in the Applied Science University.

Cumulative exposure model is assumed to account for the cumulative damage under x_1 stress. The cumulative exposure model is inspired from palmgren-miner rule which states that the fatigue failure is expected when such life fractions sum to unity, that is, when 100% of the life is exhausted such that $\sum_{j=1}^{n} \frac{N_j}{N_{fj}} = 1$, where$N_j$ is the number of cycles under stress j and N_{fj} is the number of cycles to failure for stress j from the S-N curve.

Figure 3 shows a conceptual graph for palmgren-miner ruleunder SSAFLT. The figure shows that the specimen will start at the first curve with β_1 for τcycles and then will continue on the second curve with β_2 until the end of the test cycles t_c. During τ cycles, damage (fatigue) will be accumulated in the specimen due to the stress x_1 and when the specimen is subjected to the second stress x_2, the specimen will not befresh; hence the accumulated fatigue caused by the stress x_1 must be accounted for.

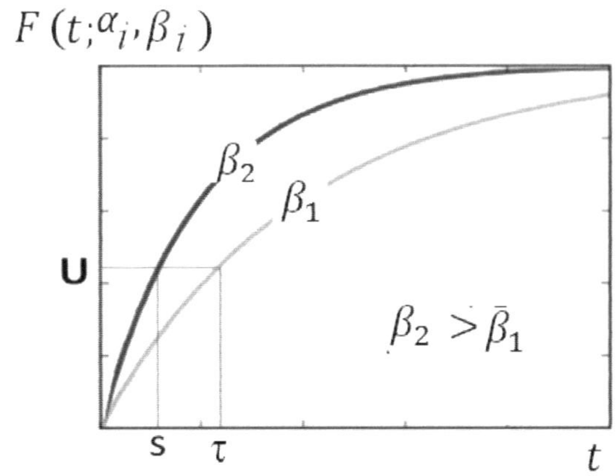

Figure 3. A conceptual graph for palmgren-miner rule in case of SSAFLT

The accumulated damage is manifested as a shift of the value s in the time scale for the second step in the test. The value of s as a function of the BISA parameters andτcan be derived as:

$$s = \frac{\beta_2\left[\zeta^2\left(\frac{\tau}{\beta_1}\right) + 2\right] + \sqrt{\left[-\beta_2\left[\zeta^2\left(\frac{\tau}{\beta_1}\right) + 2\right]\right]^2 - 4\beta_2{}^2}}{2} \qquad (7)$$

5. Bayesian Inference

Consider a vector of failure data points $\boldsymbol{t}$, and a life distribution with parameters $\boldsymbol{\theta}$. The posterior distribution of the parameters given the data is:

$$f(\theta; t) = \frac{f(\theta)L(t;\theta)}{f(t)} \qquad (8)$$

Where $L(t;\theta)$ is the likelihood of the data given the model parameters θ, $f(\theta)$ is the joint prior distribution of the parameters, and$f(t) = \int_\theta f(\theta)L(t;\theta)$ is the preposterior marginal distribution of t. Since the preposterior marginal distribution of t is constant, then the unnormalized posterior distribution of the parameters given the data can be expressed as:

$$f(\theta; t) \propto f(\theta)L(t;\theta)$$

Gibbs sampling can be used to sample from $f(\theta; t)$. Gibbs sampling is built in WINBUGS software which is a stand-alone program to allow practical MCMC methods available to applied statisticians.

The cumulative distribution function considering the cumulative exposure model for the BISAis expressed as:

$$F(t;\alpha,\beta_1)=\begin{cases}\frac{1}{\sqrt{2\pi}}exp\left(-\frac{\zeta^2\left(\frac{t}{\beta_1}\right)}{\alpha_1^2}\right) & 0<t<\tau\\ \frac{1}{\sqrt{2\pi}}exp\left(-\frac{\zeta^2\left(\frac{t-(\tau-s)}{\beta_2}\right)}{\alpha_2^2}\right) & \tau\leq t\leq t_c\end{cases}, \qquad (9)$$

and the probability density function is:

$$f(t;\alpha,\beta_1,\beta_2)=\begin{cases}\frac{t-\beta_1}{\alpha\sqrt{2\pi\beta_1 t^3}}exp\left(\frac{\beta_1-t}{\alpha\sqrt{\beta_1 t}}\right) & 0<t<\tau\\ \frac{\left(t-(\tau-s)\right)-\beta_2}{\alpha\sqrt{2\pi\beta_2\left(t-(\tau-s)\right)^3}}exp\left(\frac{\beta_2-\left(t-(\tau-s)\right)}{\alpha\sqrt{\beta_2\left(t-(\tau-s)\right)}}\right) & \tau\leq t\leq t_c\end{cases}, \qquad (10)$$

Moreover, the reliability function for the survived units is:

$$R(t_c;\alpha,\beta_1,\beta_2)=1-\frac{1}{\sqrt{2\pi}}exp\left(-\frac{\zeta^2\left(\frac{t_c-(\tau-s)}{\beta_2}\right)}{\alpha_2^2}\right) \quad t\geq t_c \qquad (11)$$

Under Type-I right censoring, the likelihood function has three parts: one for the censored data and the other two corresponding to the two steps. In this model, the likelihood function will be written such that the log-linear relation between β_i and the stress level x_i is assumed as in equation (1):

$$ln(\beta_i)=a+bx_i \qquad (12)$$

Where a and b are the model parameters such that their values depend on the product under test and the test method used.

The likelihood function for n samples utilizing the cumulative exposure model and the log-linear model is:

$$
\begin{aligned}
L(t; a, b, \alpha) \\
= \prod_{i=1}^{n_1} \left(\frac{t_i - \exp(a + bx_1)}{\alpha \sqrt{2\pi \exp(a + bx_1) t_i^3}} exp \left(\frac{\exp(a + bx_1) - t_i}{\alpha \sqrt{\exp(a + bx_1) t_i}} \right) \right) \\
\times \prod_{i=1}^{n_2} \left(\frac{(t_i - (\tau - s)) - \exp(a + bx_2)}{\alpha \sqrt{2\pi \exp(a + bx_2)(t_i - (\tau - s))^3}} exp \left(\frac{\exp(a + bx_2) - (t_i - (\tau - s))}{\alpha \sqrt{\exp(a + bx_2)(t_i - (\tau - s))}} \right) \right) \\
\times (n - n_1 - n_2) \left[1 - \frac{1}{\sqrt{2\pi}} exp \left(- \frac{\zeta^2 \left(\frac{t_c - (\tau - s)}{\exp(a + bx_2)} \right)}{\alpha^2} \right) \right]
\end{aligned}
\quad (13)
$$

6. Specimens preparation

For our purposes, Al-Ti-B was used as a refiner for pure Al where 20 micro-alloys of Al-Ti-B were prepared.

6.1. *Base metal*

99.8% purity commercial aluminum was used in preparing the specimens for this study. The chemical composition of the 99.8% purity commercial aluminum is shown in Table 1.

Table 1. The chemical composition (Wt%) of the 99.8% purity commercial aluminum

Element	Fe	Si	Cu	Mg	Ti	V	Zn	Mn	Na	Al
Wt. %	0.09	0.05	0.005	0.004	0.004	0.008	0.005	0.001	0.005	bal.

The chemical composition of the commercially available ternary Al-Ti-B master alloy is shown in Table 2. The ternary Al-Ti-B master alloy was used to manufacture the Al- %Ti- %B and its micro-alloys.

Table 2. Chemical composition, wt% of the ternary Al-Ti-B alloy

Element	Ti	B	Fe	Si	V	Al
Wt%	4.6	0.92	0.12	0.09	0.12	Balance

High purity aluminum and titanium powder were used in manufacturing the Al-Ti binary alloy. Graphite crucible and stirring rods were used in all of experiments.

6.2. *Al base metal preparation*

The high purity aluminum wires were pickled in HNO_3 to remove any contamination, and then they melted in a graphite crucible inside an electric furnace at about 800°C. The molten was then poured in a hollow cylindrical brass rods of 10 mm inside diameter and 55 mm external diameter and left to solidify. After the solidification, the rods were rolled into sheets of 2 mm thickness, 10 mm width and 240 mm length.

Ternary Al-Ti-B grain refiner is commercially available and where provided by ARAL company in Jordan

Fatigue tests specimens (Al-(Ti-B)$_{x\%}$) of the dimensions shown in Figure 4 were machined into the standard fatigue specimen as per ASTM.

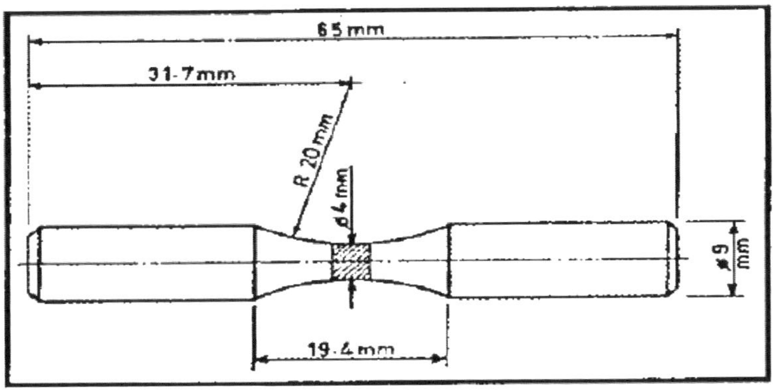

Figure 4. The standard fatigue specimen

The specimens were machined using a Colchester CNC-1000 lathe machine at a cutting spindle speed of 1030 rpm (30m/min), a depth of cut 0.5 mm and a feed rate of 50 mm/min. All specimens were machined at the same cutting conditions, and thus, have the same surface roughness, which is essential to avoid any discrepancies in the fatigue result which may be caused by different cutting conditions.

7. Experimentation

Twenty different micro-structures will be used in the experimentation. For each micro-structure the mechanical properties need to be determined and 50 samples need to be prepared from it. These 50 samples will be subjected to SSAFLT and the BISA parameters will be estimated using WINBUGS. This will provide 20 sets of parameters corresponding to the 20 different sets of mechanical properties. Since model parameters are independent, we can perform regression analysis between the mechanical properties and the scale and shape parameters of the distribution individually to come up with two functions that relate the mechanical properties to the BISA parameters and thus to the reliability of the specimens.

Due to the limited allowable space for the paper to fulfill the requirements and the lengthy experimentation the verification of this theoretical model will be the subject of another paper.

8. Conclusion

A novel and reliable theoretical model based on the Birnbaum-Saunders (BISA) distribution is presented from which the fatigue life can be determined. Experimental verification of the model is in progress and will be published in due course.

9. Acknowledgment

The authors are grateful to the Applied Science Private University, Amman, Jordan for partial financial support granted to this research (Grant No.DRGS-2015).

10. References

[1] Quality progress 1985 18 (Nov.), 12-17

[2] Weicheng Cui 2002 A state-of-the-art review on fatigue life prediction methods for metal structures *J Mar SciTechnol*7, 43–56

[3] Carnelli D1, Villa T, Gastaldi D, Pennati G 2011 Predicting fatigue life of a PMMA based knee spacer using a multiaxial fatigue criterion *J ApplBiomaterBiomech*.9(3), 185-92. doi: 10.5301/JABB.2011.8917.

[4] Wang XS et al (1999). "Prediction of fatigue life of carbon steel using only the tensile strength". In: Wu XR, Wang ZG (eds) Proceedings of the 7th International Fatigue Congress (Fatigue'99). Higher Education Press, Beijing, pp 845–850

[5] Guechichi H, Benkabouche S, Amrouche A, Benkhettab M 2011 A high fatigue life prediction methodology under constant amplitude multiaxial proportional loadings *Materials Science And Engineering:* A528, 4789-4798

[6] Henrysson, H F 2000 Fatigue Life Predictions of Spot Welds Using Coarse FE Meshes *Fatigue and Fracture of Engineering Materials and Structures*, 23, 737-746.

[7] Fadiga G, Sraml M 2009 Fatigue crack initiation and propagation under contact loading*Engineering Fracture Mechanics*79, 1320-1335

[8] Beden S M, Abdullah S, Arffin A K 2009 Review of fatigue crack initiation and Propagation models of metallic components *European Journal of Scientific Research*28, 364-397

[9] Birnbaum Z W, Saunders S C 1969 A New Family of Life Distributions *J.A ppl. Prob.* 6,319-327

[10] Engelhardt M, Bain L J, Wright F T 1981 Inference on the parameters of the Birnbaum–Saunders fatigue life distribution based on maximum likelihood estimation *Technometrics* 23(3):251–256

[11] Chang D S, Tang L C 1994 Percentile bounds and tolerance limits for theBirnbaum–Saunders distribution. Communications in Statistics, Part A Theory and Methods 23, 2853–2863

[12] Fatemi A, Yang L 1998 Cumulative fatigue damage and life prediction theories: a survey of the state of the art for homogeneous materials *Int J Fatigue* 20, 9–34

[13] Zaid A I O, Al-Qawabah S M A 2012 Effect of Zirconium Addition on the Grain Size and Mechanical Behavior of Aluminum Grain Refined by Titanium Plus Boron (Ti+B) in the as Cast and Cold Extruded Conditions *Key Engineering Materials*

IOP Conf. Series: Materials Science and Engineering **146** (2016) 012015 doi:10.1088/1757-899X/146/1/012015

Study of TiO_2 nano-tubes using electrochemical anodization method for applications in dye-sensitized solar cells

T. Ghani[1,2], M. Mujahid[1], M. Mehmood[2] and Q. Shahzad[2]
[1]School of Chemical and Materials Engineering,
National University of Sciences and Technology, H-12 Islamabad, Pakistan
[2]National Center for Nanotechnology, Department of Metallurgy and Materials Engineering, Pakistan Institute of Engineering and Applied Sciences,
P. O. Nilore, Islamabad, Pakistan

E-mail: tayyaba.ghani@scme.nust.edu.pk

Abstract. TiO_2 nano-tubes are getting strong attraction in many fields due to their unique properties. They are important in biomedical application, Dye sensitized solar cells, sensor and photocatalys is applications, etc. Our prime interest is to grow these tubes for dye-sensitized solar cells with high conversion efficiency and low production cost. In this research, we have synthesized TiO_2 naonotubes by anodizing 25 μm thick titanium foil at 40V using two-step anodization method. The electrolyte used is the ethylene glycol with varying concentration of NH_4F and fixed concentration of deionized water. Effect of different concentrations of the electrolyte on tube crystal structure has been studied. It is observed that crystallinity increases with increased concentration of fluoride ions. It is found that two-stepanodization method results in more crystalline and open structures. Scanning electron microscopy is utilized to study the surface morphology and tubes growth, whereas observation of the crystal structure of nano-tubes is made by X-ray diffraction

1. Introduction

Titanium oxide (TiO_2) is a famous and versatile oxide being used for many applications such as bio-sensors, gas sensors, as a photo-catalyst, in biomedical application, for photo degradation and photovoltaic devices [1-3]. Over the last few years, many researchers have focused their research on the processing and production of nanostructured titanium oxide such as nanopowders, nanowires, nanorods, nano-tubes utilizing different methods, e.g., anodization, sol-gel, vapour deposition, hydrothermal, etc., [4-7].TiO_2 has three different crystalline forms as rutile, anatase and brookite. Anatase is more active phase for photovoltaic applications having better electronic and surface chemistry properties as compare to other forms, however, it is rare in abundance.

Our main concern is to grow titania nano-tubes for Dye Sensitized Solar Cells (DSSCs). DSSCs are based on nano-crystalline semiconductor oxide layer on conducting surface, dye sensitizer and electrolyte [8]. The first DSSC was reported by Miecheal Gratzel and O' Regan in 1991 [9]. It was made by nanocrystalline titania layer produced by nano particles. Since, electrons, passing through the TiO_2nanocrystalline film produced by nanoparticles, trap at grain boundaries and irregular pores structure result in low electron diffusion. One-dimensional nano-structures of TiO_2, e.g., nanorods,

Content from this work may be used under the terms of the Creative Commons Attribution 3.0 licence. Any further distribution of this work must maintain attribution to the author(s) and the title of the work, journal citation and DOI.
Published under licence by IOP Publishing Ltd

nano-tubes, nanowires provide efficient charge transport. Among these one dimensional structure, nano-tubes based on TiO_2 proves to be very effective to reduce the possibility of charge recombination in DSSCs. Zwilling [10] and Gong [11] were the early reporters of TiO_2 tubes in aggressive electrolytes by anodization method. Later on many researchers synthesize these tubes with same method but different electrolytes [6-7, 12-14]. Now a days, ethylene glycol and glycerol with the addition of ammonium fluoride are frequently used to fabricate smooth and open ended tubes.

For the fabrications of TiO_2 nano-tubes, we have preferred anodization method as it is a relatively low cost process and easily controllable. By this method two types of oxides can be grown one is barrier type by using solutions of sulfuric, phosphoric and acetic acid, second is nanoporous and nanotubular type oxide utilizing chlorine base solutions, aqueous and organic solutions containing fluoride ions. This method allows the variation of different parameters, e.g., voltage, electrolyte concentration, temperature, etc., and accordingly to control the tubes morphology and structure. We have fabricated nano-tubes using double anodization method. It is observed that crystallinity of tubes increases with two step anodization. It is found that annealing is the important step to convert amorphous TiO_2into crystalline anatase form desired for DSSCs.

2. Experimental steps

A 25 μm thick Ti foil was anodized for at 50 V at room temperature. The electrolyte used is ethylene glycol containing 0.5 wt % Ammonium fluoride (NH_4F) and 3 wt % deionized water. Before anodization, samples were cleaned by sonicating in different solvents such as ethanol, acetone and deionized water. For double anodization experiments, oxides produced by first anodization were removed by sonicating in deionized water. The second step anodization was performed in same electrolyte for same experimental conditions. For characterization of samples Scanning Electron Microscopy (SEM) and X-Ray Diffraction (XRD) were utilized.

3. Results and discussions

Valve metals such as Ta, W, Ti, etc., can produce their oxides in oxygen containing environment. These naturally produced and ultrathin oxides remain present on their surface and affect experimental growth of anodically produced oxides. Therefore it is necessary to clean the surface before anodization. Sonicating in ethanol and acetone and finally deinonized water removed dust and debris from the surface. Anodization experiments were performed at 40 V keeping temperature, time and electrolyte concentration as constant. These experiments were usually performed at room temperature for one to two hours. Figure 1 shows surface of TiO_2 nano-tubes prepared after first anodization at 40 V at room temperature. The surface of tubes is not very clear instead debris are present. Tubes seem to entangled in the form of branches and groups, also most of the tubes are closed from their ends.

To get the regular pattern, we have utilized the two step anodization method as people usually do with aluminium oxide. It is clear from the Figure 2 that after removal of TiO_2 produced by first anodization, surface of TiO_2 has regular hexagonal pattern.

Figure 1. Top surface of TiO_2 nano-tubes after single anodiztion at 40 V in ethylene glycol containing 0.5 wt% NH_4F and 3 wt% deionized water at room temperature

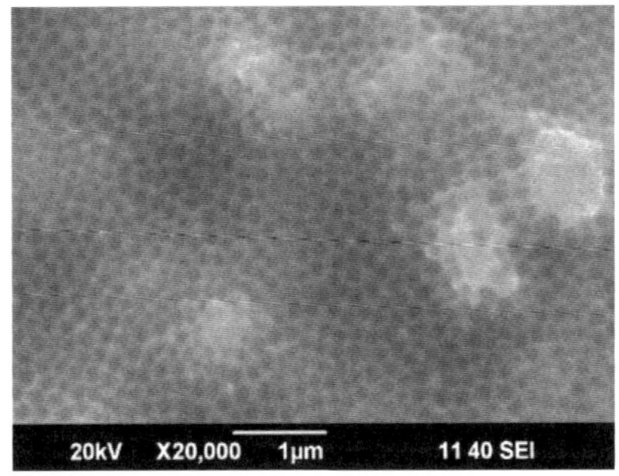

Figure 2. Ti foil surface showing hexagonal pattern after removal of TiO_2 tubes produced by anodization at 40 V

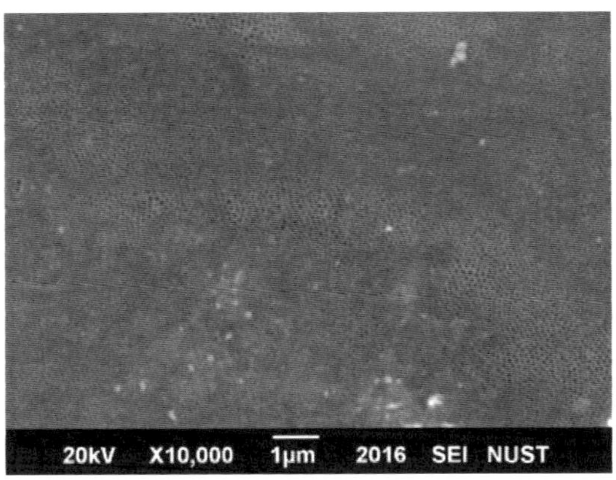

Figure 3. Top surface of TiO_2 nano-tubes after two step anodization at 40 V in ethylene glycol containing 0.5wt% NH_4F and 3 wt% deionized water at room temperature

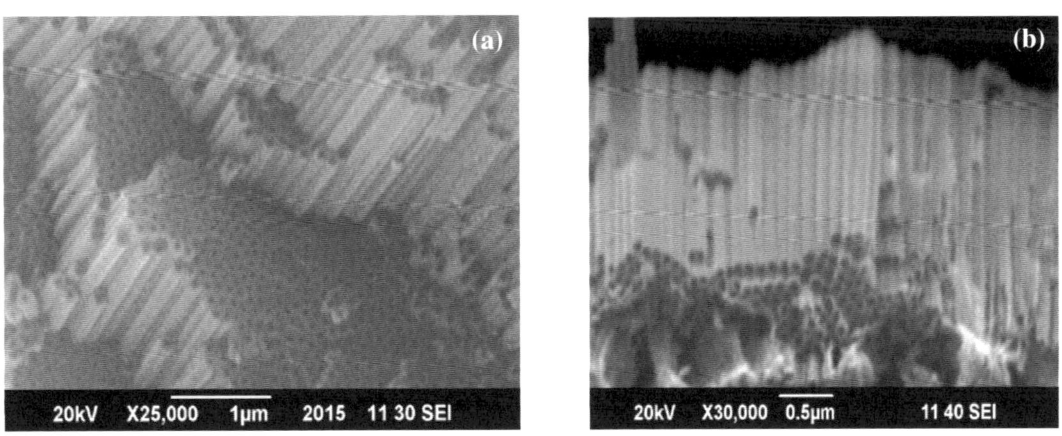

Figure 4. Cross-section of TiO2 tubes fabricated by two step anodization experiment at4 0V

Upon second anodization with same experimental conditions regular ordered pattern is formed as presented in Figure 3. Still at some part of surface, tubes are covered. It is expected that doing such experiment at higher voltages results in open ended surface. Cross-sectional images in Figure 4 show that tubes are very smooth. The SEM image b in the same Figure is taken by breaking the produced oxide intentionally. The different steps in Figure shows that inner surface of tube grows in conical shape.

It is found that oxide produced is very fragile, amorphous and brittle and, therefore, during sample handling, some tubes get broken. The surface of oxide produced by single anodization is rough and have irregular pore pattern with foreign particles or debris present on it. Most of the tubes are closed at the ends. Special etching processes by mechanical and chemical treatments to remove these debris are required for clean and open ended surfaces. The resulting samples get reduced in thickness after such etching processes. However utilizing the two step anodization method, such steps could be avoided.

It is observed in our experiments that oxides got peeled off or rolled during drying and annealing processes. After first anodization, TiO_2 broke into pieces after being detached from the substrate. However, after second anodization, oxide is detached in the form of thin (few micrometer) layer with area equal to the substrate. It is expected that it can be used as free standing membrane in energy applications. However, more work is required to understand such thin sheets structure, stability, their mechanical and optical properties.

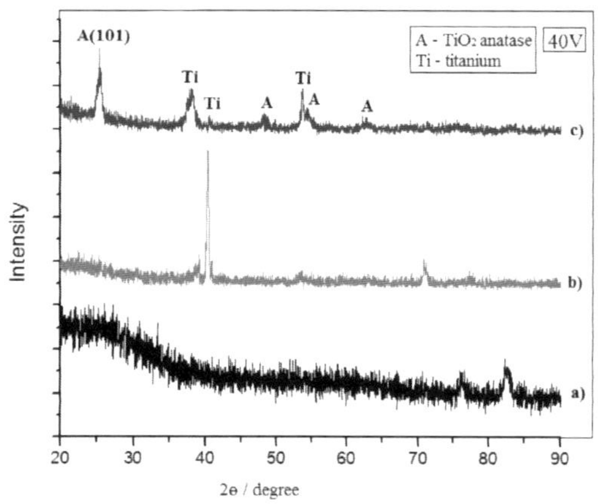

Figure 5. XRD plot of TiO_2 nano-tubes after annealing at 550 °C, as anodized (a), after first anodization (b), after second anodization (c)

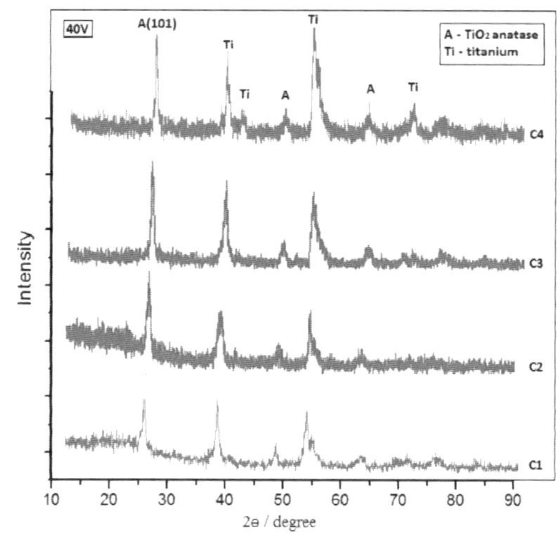

Figure 6. XRD plot of TiO_2 nano-tubes after two step anodization and annealing at 550 °C for different concentrations of NH_4F(C1= 0.3wt %, C2= 0.4 wt%, C3 = 0.5 wt%, C4 = 0.6wt%)

XRD plots (taken at 5° grazing angle) of Figure 5 show that as anodized sample is amorphous. Whereas after annealing at 550 °C, anatase polymorph's small peak (101) is obtained for the samples singly anodized at room temperature in ethylene glycol electrolyte. The crystallinity of tubes increases for doubly anodized samples. It may be possible that the increase in crystallinity after second anodization is due to the ordering nature of the oxide. The peaks of Ti are present as a substrate effect.

Figure 6 shows the effect of increased concentration of NH_4F. It is observed that crystallinity of tubes increases with increasing the concentration of F^- ions.

4. Conclusion

In this paper we have presented a study of TiO_2 nano-tubes by performing anodization experiment at 40 V. Templates are prepared by two step anodization experiments. It is observed that after second anodization tubes with open pores and very regular pattern could be obtained. Such results could be enhanced at higher voltages. The two stepanodization experiments have improved the crystallinity of tubes. Crystallinity also increases with increase in concentration of NH_4F. This is due to the increased concentration of fluoride ions. It is expected that such nanostructured templates, when used as photoanode in DSSCs, would result in efficiency greater than the found values with nano-tubes so far.

5. References

[1] Roy P, Berger S and Schmuki P 2011*Nanoscience*50 2904
[2] Kowalski, Kim D and Schmuki P 2013 Nano today 8 235
[3] Lee H C, Zhang L F, Lin J L, Chin Y L and Sun T P 2013 *Sensors*13 14161
[4] Hamadanian M, Jabbari V, Gravand A and Asad M 2012 *Surface and Coating Technology* 206 4531
[5] Chauhan R, Kushwaha R and Bahadur L 2013 *Mater. Chem. Phys.*139 525
[6] Paulose M, Prakasam H E, Varghese O K, Peng L, Popat K C and Mor G K 2007 *J. Phy. Chem. C* 111 14992
[7] Macak J M, Tsuchiya H, Ghicov A, Yasuda K, Hahn R, Bauer S, Schmuki P 2011 *Solid State and Material Science*11 3
[8] Gratzel M 2003 *J. photochem. photobiol.*4 145
[9] Oregan B and Gratzel M 1991 *Nature*353 737
[10] Zwilling V, Aucouturier M and Darque-Ceretti E 1999 *Electrochim. Acta*44 921
[11] Gong D, Grimes A C and Varghese A O 2011 *J. Mater. Res.*16 3331
[12] Kuo Y Y and Chien C H 2013 *Electrochim. Acta*91 337
[13] Albu S P, Ghicov A, Macak J M, Hahn R and Schmuki P 2007 *Nano let.* 7 1286
[14] Zhang J, Li S, Ding H, Li Q, Wang B, Wang S and Wang H 2014*J. Power Sources*247 807

IOP Conf. Series: Materials Science and Engineering **146** (2016) 012016 doi:10.1088/1757-899X/146/1/012016

Effect of temperature and loading on output voltage of lead zirconate titanate (PZT-5A) piezoelectric energy harvester

Z. Butt and R. A. Pasha
Department of Mechanical Engineering, University of Engineering and Technology, Taxila 47050, Pakistan

E-mail: zubairbutt64@yahoo.com, asim.pasha@uettaxila.edu.pk

Abstract. Energy harvesting is the process of acquiring energy from the external sources and then further used to drive any system. Piezoelectric material was operated at various temperature but the characterization of the material mostly performed at room temperature. The depolarization in piezoelectric material occurs when the material is heated to its curie temperature and when mechanical stresses are high to disturb the properties of the material. The aim of this paper is to study the performance of lead zirconate titanate (PZT-5A) piezoelectric material under various temperatures and loading conditions. The output voltage of piezoelectric material decreases with increase of temperature. It was found that output voltage from the harvester increases when loading increases while its temperature decreases.

1. Introduction

Piezoelectric materials are an inventive step used for energy harvesting. Energy harvesting system uses piezoelectric materials where distortions produced by different resources are directly transformed to electrical voltage through piezoelectric effect. With the ever increasing energy needs, extracting and manipulating more and more energy sources has become a need of the day. There are two types of energy harvesting schemes macro scale and micro scale energy harvesting. Piezoelectric energy harvesting schemes comes under the category of micro scale energy harvesting scheme [1]. The first application of piezoelectricity appeared in sonar in which quartz crystals was used to produce ultrasonic waves during World War I [2]. Piezoelectric materials generate electrical voltage when mechanical stress is applied known as direct effect of piezoelectric. The inverse piezoelectric effect is the ability of a material to produce mechanical distortion when electric field is applied. The energy harvesting by using piezoelectric material uses direct piezoelectric effect. The direct piezoelectric effect was discovered in 1880 by Pierre and Jacques Curie in various naturally occurring substances including Quartz and Rochelle salt [3]. If piezoelectric materials are heated to a threshold temperature, the crystal pulsations may be so strong that domains become chaotic and element become completely depolarized. This critical temperature is called Curie temperature [4]. Salem Saadon and Othman Sidek in 2011 presented a review about vibration based piezoelectric energy harvesting technology which is carried out by researcher during the last three years [5]. Seon-Bae Kim et al. studied the behavior of piezoelectric materials at various thermal conditions. The output power at different temperature has been predicted by the change of piezoelectric coupling coefficient. He showed that output power decrease with increase of temperature [6]. Poorna Mane et al. studied the effect of temperature, pressure, frequency, and load resistance on the energy harvesting prospective of the device. The output voltage and power were calculated as 108V and 11641μW at 20°C, 275.8kPa,

Content from this work may be used under the terms of the Creative Commons Attribution 3.0 licence. Any further distribution of this work must maintain attribution to the author(s) and the title of the work, journal citation and DOI.
Published under licence by IOP Publishing Ltd

2.5Hz and 2MΩ. [7]. In 2012 Kyungrim Kim et al. designed and fabricated a shear-mode piezoelectric accelerometer using $YCa_4O\ (BO_3)_3$ single crystals (YBCO). The accelerometer was tested at high temperatures ranging from room temperature to 1000°C and at frequencies ranging from 50 Hz to 350 Hz [8]. In 2011 Sang-Hyo Kima, Jin-Hee Ahn etal, examine the piezoelectric effects on various loading conditions for the possibility of harvesting energy from the bridges by converting potential energy into electrical energy using piezoelectric material [9]. A comparative study of single crystals for high temperature applications of piezoelectric devices was conducted in 2009. Result suggest that only $YCa_4O\ (BO_3)_3$ (YCBO) and $La_3Ga_5SiO_4$ (LGS) was operated at temperatures higher than 500°C and PZT performs best at room temperature [10].

Our aim in this paper was to determine the effect of temperature and loading on output voltage of lead zirconate titanate (PZT-5A) piezoelectric material at 5 Hz frequency.

2. Material

In this research work a rectangular lead zirconate titanate (PZT-5A) specimen was selected for required experimentation. Lead zirconate titanate (PZT) is a perovskite ceramic material that shows marked piezoelectric effect. It can be prepared by using different methods e.g. sputtering and jet molding etc. It is a pyro-electric material that generates voltage when temperature changes so that's why it is also used as a temperature sensor. It can be classified in to two groups hard and soft PZT depend upon the mechanical and electrical characteristic. It can be used to make ultrasonic transducer and ceramic capacitor. It can be used in sonar technology, surgical instruments and ultrasonic bending and welding because of its high permittivity, large coupling factor, high Curie temperature and high charge coefficient [11]. The modeling dimensions of specimen are given in Table 1 and properties of the specimen as given by supplier are given in Table 2.

Table 1. Dimension of specimen

Composition	Trade	Part No.	Dimension (mm)
PZT-5A	Piezo system Inc.	T107-A4E	L=20,W=10 T=5

Table 2.Specimen properties as provided by supplier (PIEZO SYSTEM INC.USA)

Sr. No.	Description	Notation	Value	Units
1	Density	P	7750	Kg/m^3
2	Mechanical Quality factor	Q_m	80	
3	Piezoelectric voltage coefficient	g_{33}	24×10^{-3}	Vm/N
4	Piezoelectric strain coefficient	d_{33}	374×10^{-12}	m/volt
		d_{31}	-17×10^{-12}	
5	Dielectric constant	ε_{33}^{T}	1.504×10^{-8}	F/m
6	Elastic compliance constant	S_{11}^{E}	16.4×10^{-12}	m^2/N
7	Initial depolarization field	E_c	5×10^{5}	V/m
8	Polarization field	E_p	2×10^{6}	V/m
9	Curie temperature		350	°C
10	Thermal expansion coefficient		$\sim 4 \times 10^{-6}$	Meters/meter °C

3. Experimental setup

For the sake of experimentation, a setup was designed in such a way that load cell is fixed on a base of mild steel iron sheet. A 12V dc motor with a cam shaft mounted by a screw mechanism was used to provide alternating loading conditions. The rectangular specimen was fixed with nut and bolts on a load cell in such a way that it's lower and upper both sides faces copper electrodes. These electrodes acted as cathode and anode. The schematic diagram of specimen is as shown in Figure 1. The mechanical load was applied on the top surface via screw mechanism and function generator was used to electrically shock the specimen at 5Hz frequency. The arbitrary function generator was used to generate a series of sinusoidal signals. An energy harvesting circuit or ac to dc converter was used to convert ac voltage into dc voltage is as shown in Figure 2. For electrical and thermal insulation mica sheet was used which is resistant to both electrical and thermal conductivity. For on spot heating we used heating filament in a circuit and temperature gun was used to monitor surface temperature. The voltage response is investigated on digital oscilloscope at different temperatures and loading conditions. The experimental setup used for energy harvesting is as shown in Figure 3.

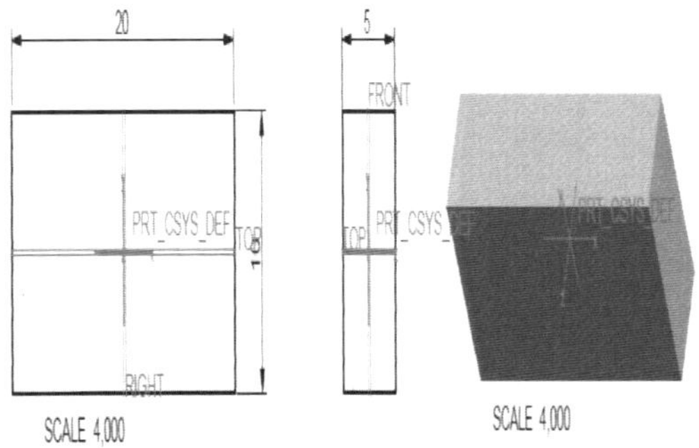

Figure 1. Schemtic diagram of specimen

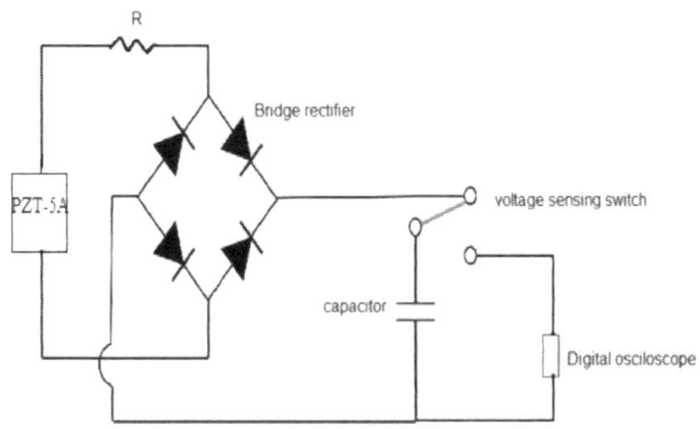

Figure 2.AC to DC converter

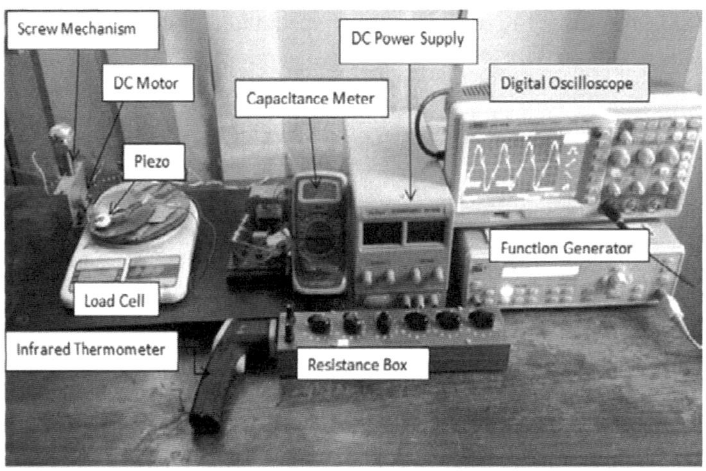

Figure 3. Experimental setup for energy harvesting

4. Results and discussion

Figure 4 shows that on increasing the temperature at different loading conditions the output voltage decreases accordingly. Vary the temperature up to 350°C to checkout its response on oscilloscope at different loading conditions. It is observed that temperature at 20°C does not influence the output voltage at different loading conditions which shows that the every transducer made from this type of material may be successfully used up to this temperature. When the temperature is between 50°C and 300°C the properties of Piezoelectric ring changes due to depolarization effect and the output voltage decreases. At temperature 350°C they degrade very rapidly and tending to zero. The maximum dc voltage of 5.93V was obtained at 30N, 5Hz frequency and 20°C temperature. So it is highly applicable to use piezoelectric material at low temperature and maximum loading conditions.

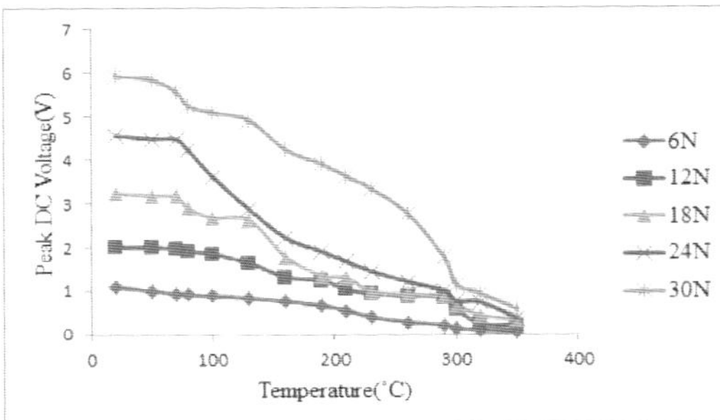

Figure 4. Effect of temperature on Peak dc voltage at 5Hz frequency and different loading conditions

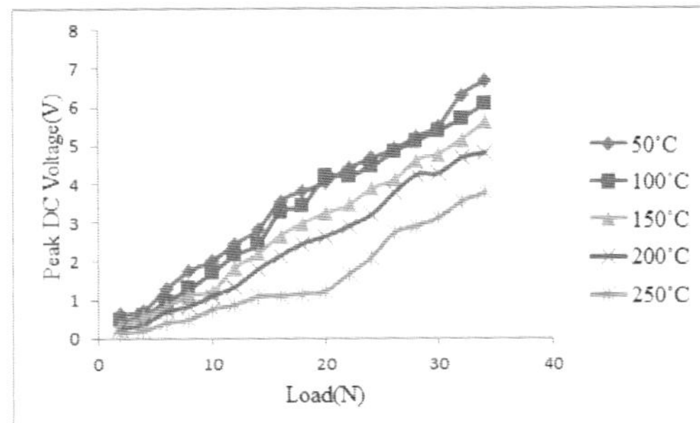

Figure 5. Effect of loading on Peak dc voltage at 5Hz frequency and different temperature conditions

Figure 5 shows the linear relationship between load and output voltage at different temperature conditions. The data indicate that temperature has negative effect on the output voltage. It is observed that temperature at 50°C, 100°C, 150°C, 200°C and 250°C present a significant decrease in voltage due to depolarization effect at variable loading conditions. The maximum voltage of 6.67V was obtained at 36N, 50°C and 5Hz frequency. It was seen that output voltage increases when the load increases and temperature decreases.

5. Conclusions

The behavior and performance of lead zirconate titanate (PZT-5A) piezoelectric material was investigated as a function of environment temperature and different loading condition. The output voltage of the material decreases with the increase of temperature. Following conclusions are obtained from this research work:

I. It was found that temperature up to 20°C does not influence the output voltage. When temperature increases at different loading conditions the output voltage decreases.
II. The output voltage of piezoelectric material increases when loading increases and temperature decreases.
III. It is determined that lead zirconate titanate (PZT-5A) piezoelectric material perform well below Curie temperature. So it is highly applicable to use this type of material at low temperature and for greater loading conditions.

6. Acknowledgments

This project was funded by University of Engineering & Technology Taxila, Pakistan. My deep gratitude goes to my M.Sc. supervisor, Professor Dr. Riffat Asim Pasha for his supervision, continuous support, encouragement, research approach and expertise in this research work.

7. References

[1] Tanvi Dikshit, Dhawal Shrivastava., et al. Energy harvesting via piezoelectricity. International Journal of Information Technology July – December, 2010; Vol. 2 No. 2; ISSN 0973 – 5658
[2] Dr. Sabar D. Hutagalung. School of Materials & Mineral Resources Engineering, University Sains Malaysia EBB 443
[3] Piefort, Finite element modeling of piezoelectric active structures, Doctoral Dissertation (2000-2001), University Libre De Bruxelles, Active Structures Laboratory, Department of Mechanical Engineering and Robotics
[4] Piezoelectric Ceramic Characterization by T. L Jordan & Ziounaies VA23681
[5] Salem Saadon, Othman Sidek, A review of vibration-based MEMS piezoelectric energy harvesters. Energy Conversion and Management, 52(2011): pp. 500–504

[6] Seon-Bae Kim., et al., Temperature effects on output power of piezoelectric vibration energy harvesters. Microelectronics Journal, 42(2011): pp. 988–991

[7] Poorna Mane., et al., Studying the effects of temperature on energy harvesting using pre-stressed piezoelectric diaphragms, Department Of Mechanical Engineering, Virginia Commonwealth University, 601 West Main Street, Richmond, VA, USA 23284

[8] Kyungrim Kim., et al., Design, fabrication and characterization of high temperature piezoelectric vibration sensor using YCOB crystals. Sensors and Actuators, A178 (2012): pp. 40–48

[9] Sang-Hyo Kima, Jin-Hee Ahna, et al., Analysis of piezoelectric effects on various loading conditions for energy harvesting in a bridge system. Sensors and Actuators, 167 (2011): pp. 468–483

[10] V.Bedekar, J.Oliver, et al., Comparative study of energy harvesting from high temperature piezoelectric single crystals, Journal of Applied Physics.48 (2009) 091406

[11] D. Berlincourt and H.H.A. Krueger, Technical Publication TP-226 Properties of piezoelectricity Ceramics. www.morgan-electroceramics.com

Determination of plasma temperature and electron density of iron in iron slag samples using laser induced breakdown spectroscopy

T. Hussain[1], M. A. Gondal[2] and M. Shamraiz[3]
[1,3]National Institute of Vacuum Science and Technology,
P. O. Box 3125, Islamabad, Pakistan
[2]Laser Research Laboratory, Physics Department, King Fahd University of Petroleum & Minerals (KFUPM), Box 5047, Dhahran 31261, Saudi Arabia

E-mail: thminhas7@yahoo.com

Abstract. Plasma temperature and electron density of iron in iron slag samples taken from a local plant is studied. Optimal experimental conditions were evaluated using Nd: YAG laser at 1064 nm. Some toxic elements were identified and quantitative measurements were also made. Plasma temperature and electron density were estimated using standard equations and well resolved iron spectral lines in the 229.06-358.11 nm region at 10, 20, 30 and 40 mJ laser pulse energy with 4.5 μs delay time. These parameters were found to increase with increase in laser pulse energy. The Boltzmann distribution and experimentally measured line intensities support the assumption that the laser-induced plasma was in local thermal equilibrium. It is worth mentioning that iron and steel sector generates tons of solid waste and residues annually containing variety of contaminants which can be harmful to the environment and therefore knowledge, proper analysis and investigation of such iron slag is important.

1. Introduction

Knowledge of the plasma temperature is vital to understand the dissociation, atomization, ionization and excitation processes occurring in the plasma and helpful in utilizing the plasma to maximize analytical potential of LIBS. The method most frequently used for determination of excitation temperature of the plasma is the Boltzmann plot method. In this method, the relative intensity of the thermometric species is measured and used for the calculation of the excitation temperature of the plasma by assuming the local thermal equilibrium (LTE) condition. The LTE as sumptionis justified by the fulfilment of the Mc Whirter criterion [1, 2].For LTE to exist in the plasma, the electron density has to be high enough so that the collision rates exceed the radiative rates. Plasma temperature can be determined from the ratio of the intensities of:

I. Neutral to neutral lines and
II. Ion to neutral lines, usually for the same element

Laser-induced plasmas have been used with increasing frequency as spectroscopic sources for the measurement of Stark widths [3].The plasma temperature of iron slag sample was estimated using Boltzmann plot of LIBS intensity for neutral iron lines and Stark broadening.

This study is made due to the reason that classification, identification and knowledge about electron density and plasma temperature of iron in iron slag has been the burning issue in various research laboratories of the world and a variety of techniques are being used for this purpose[4].The

Content from this work may be used under the terms of the Creative Commons Attribution 3.0 licence. Any further distribution of this work must maintain attribution to the author(s) and the title of the work, journal citation and DOI.
Published under licence by IOP Publishing Ltd

interaction of high power laser beam with material has been investigated by various researchers [5]. Due to huge demands of iron and steel to fulfil the needs of the construction industry, automobile industry and various other industrial sectors, environmental issues related to iron slag are gaining importance. The LIBS system was designed and developed at KPUPM. Prior to application of LIBS for quantitative analysis of slag samples, the plasma parameters such as temperature and electron density were measured as these parameters play vital role in the sensitivity of LIBS system. When high power laser pulse falls on a solid target, the irradiation in the focal spot can lead to rapid local heating and intense evaporation. The ablated material compresses the surrounding gas and leads to the formation of shock wave. In this process a number of phenomenons including heating, melting, evaporation, atomization, excitation and ionization are involved. The emission spectrum produced from the plasma plume can be collected, spectrally resolved and used for the identification and quantification of the emitting species present in the ablated material [6].Pulsed laser induced plasma has a very short temporal existence and is transient in nature with a fast evolution of the characteristic parameters that are dependent on irradiation conditions such as incident laser intensity and irradiation spot size. The characteristics of the plasma plume are dependent on laser irradiance, target composition and atmospheric condition [7].

The hot vapor plasma can interact with the surrounding atmosphere in two ways:

I. The expansion of the high-pressure vapor drives a shock wave into atmosphere, and
II. Energy is transferred to the atmosphere by a combination of thermal conduction, radiative transfer and heating by the shock wave.

The subsequent plasma evolution depends on irradiance, size of vapor plasma bubbles, target vapor composition, ambient gas composition, pressure and laser wavelength. The important quantities such as radiative transfer, surface pressure, plasma velocities and plasma temperature are strongly influenced by the nature of the plasma. It is therefore, the emission lines of the plasma can be used to determine the temperature and the density of electrons [8].The plasma is initiated and sustained by inverse bremsstrahlung absorption during collisions among atoms and ions, electrons and the gas species. This is the reverse of the well-known bremsstrahlung process in which high-energy electrons, upon traversing a gas or solid, emit radiation as they slow down. The theoretical considerations on plasma production and heating by means of laser beams have been proposed by several workers [9].

2. Experimental setup

2.1 LIBS system

Nd: YAG Laser (Spectra Physics, Model GCR100) in Q-switched mode with ability to deliver maximum pulse energy of one joule with a pulse width of 8 ns and operating at a 10 Hz was employed for production of plasma spark at the sample surface. A convex lens having focal length 30 mm was used to focus Nd-YAG laser pulse onto the surface of sample at fundamental wavelength 1064 nm. The range of pulse energy was 20-120 mJ. The laser energy was measured with a calibrated energy meter (Ophir Model 300) for the study of dependence of LIBS signal on incident laser energy. The spectrometer used was Ocean Optics LIBS 2000+ system. The light from the plasma spark is collected by a collimating lens using UV graded fused silica 1 meter, multimode sampling fiber with SMA connector and is transferred to LIB2000 + Spectrometer (Ocean Optics). Our LIBS 2000+ has four spectrometers modules to provide high resolution (FWHM 0.1 nm) in the 200 nm to 620 nm wavelength region. The detector has a gated CCD camera having 14,336 pixels. This makes it possible to measure a LIBS spectrum over broad spectral range (200–620 nm) simultaneously with high spectral resolution (0.1 nm). The emission is observed at a 45°angle to the laser pulse.

2.2 Sample preparation
Iron slag samples were collected from a major steel manufacturing plant situated in Jubail, Saudi Arabia. These samples were grinded in spectro mill ball pestle impact grinder (Model 1100-11, Chemplex Industries, Inc. 10707, USA) to suitable mesh size for preparation of pellets. Generally, the Stark widths are measured using a single sample with a small content of the element of interest, in order to reduce self-absorption [10-12].Two slag samples under investigation were pressed to pellets shape by applying a load of 8.6× 10^7 Nm^{-2} for half an hour by placing the homogenous mixture in a dic. These pellets have diameter of 20 mm and thickness of 10 mm. While for testing on ICP, the samples were digested with aqua regia, ashed, dissolved in nitric acid and analyzed using inductively coupled plasma emissions spectroscopy. As Mgis present in abundance in iron slag samples, its pellets were also prepared by mixing it in KBr.

3. Results and discussion

3.1 Optimization of various parameters of LIBS
Prior to the application of LIBS system to iron slag samples, the LIBS spectrometer was carefully optimized for the improvement of LIBS efficiency. The major parameters which were optimized are as under:

I. Laser pulse energy
II. Dependence of LIBS signal on pressure for pellets formation
III. Sample rotation speed
IV. Influence of laser energy on binding materials

3.2. Dependence of LIBS on laser energy.
The dependence of LIBS signal intensity on incident laser energy was carried out by recording the spectra of standard samples at different laser energies. For this purpose sample of Mg in known concentration were prepared. Laser produced plasma emission spectra from pellet samples prepared in KBr matrix, was recorded in the 400- 620 nm region at 15, 20, 25, 30, 35 and 40 mJ energy per laser pulse. A time delay of 4.5 μs was applied at which maximum signal was recorded. The laser pulse energy dependence of LIBS signal intensity for Mg is depicted in Figure 1. It is clear from Figure 1 that LIBS signal increases with increase in laser energy. LIBS signal shows strong linear dependence on the laser energy which is clear from the least square fit of 0.999.

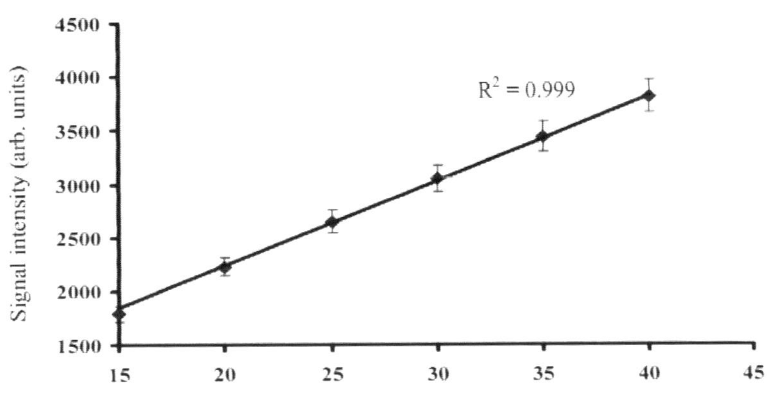

Figure 1. Plot of dependence of the intensity of Mg 518.36 nm emission line on laser energy

The higher laser energy generates optically thick plasma and strong back ground continuum, so, saturation region can be expected at higher laser pulse energies. It was noted during the experimentation that a pulse of 25 mJ was enough in case of these elements under investigation to obtain sufficient intensity and good precision.

3.3. Dependence of LIBS signal intensity on pressure for pellet formation

The trend of increase in intensity of emission lines with the pressure applied for the formation of pellets is depicted in Figure 2. The emission line intensity increases linearly with applied pressure. This is due to the fact that with increase in pressure pellets become more dense and rigid which help in reduction of the dust produced by ablation from the powder samples. The dust generated by powder samples hinders the incident laser energy to the target sample. It also generates more complexity in the excitation process and increases the value of Relative Standard Deviation (RSD) in the acquired data. In addition to above mentioned facts, the density of trace elements in pellet also increased due to high pressure for pressing the pellets, which could be attributed to high density of compressed material inside pellet.

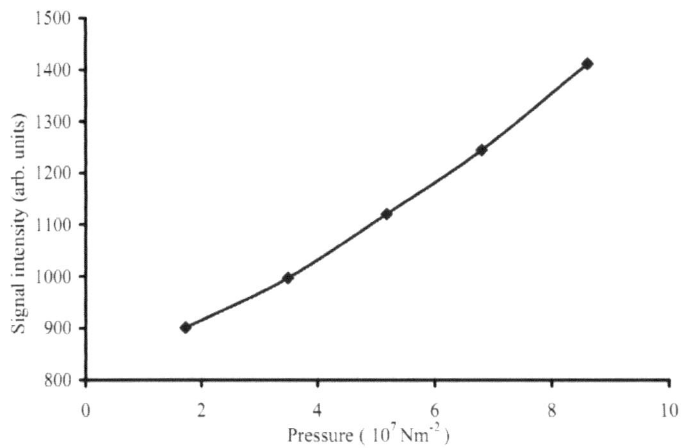

Figure 2. Dependence of the emission line intensity of Mg 518.36 nm line on the pressure applied to form the pellets

3.4. *Effect of sample rotation speed on LIBS signal intensity*

A typical plot of LIBS signal intensity versus the sample rotation speed is presented in Figure 3. It can be seen that the LIBS signal intensity increases linearly with sample rotation speed. For the stationary sample, the emission intensity decreases with time and is almost zero after approximately 100 seconds (1000 laser shots for repetition rate 10 Hz) from the start of the acquisition cycle. The decrease in the LIBS intensity is due to the crater produced by the laser pulses in a stationary sample. The size of the crater increases with time. The crater not only affects the sample ablation rate but also the plasma images produced onto the optical fiber. The sample rotation speed of 12 rotations per minute (rpm) was applied to acquire data for most of the test samples where maximum LIBS signal intensity was recorded.

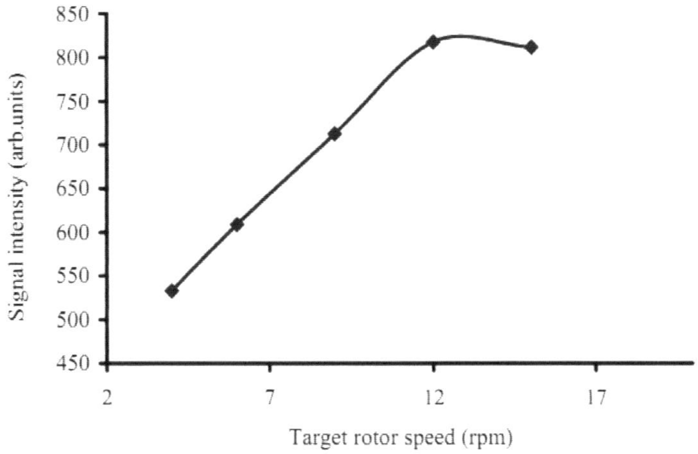

Figure 3. Variation of the intensity of the LIBS signals Mg 518.36 nm emission line with the target rotation speed

3.5. *Influence of Laser Energy on binding materials*
The dependence of Mg signal intensity on the incident laser energy and the used binding material was also studied in this experimental work. Among different binding materials like potassium bromide, poly vinyl alcohl, starch, silver and aluminium, KBr found to be better [13]. The dependence of Mg signal on laser pulse in the matrix of KBr is presented in Figure 1. The LIBS signal for Mg 518.36 nm line intensity is plotted as a function of incident laser energy. It is clear from Figure 1 that LIBS signal is linearly dependent on the incident laser energy which is indicated by the least square fit with R2= 0.999. With higher irradiation laser energy, more material is ablated from the pellet and thus the LIBS signal intensity increases.

3.6. Plasma temperature of iron slag (sample)
The plasma temperature of iron slag sample was calculated using equation 1 and 2 along with Fe lines in the 249.06 to 358.11 nm spectral region, instead of two lines method. The intensity of a spectral line is given by equation 1 [14].

$$I = \frac{hcgAN}{4\pi\lambda U} e^{-E/KT} \tag{1}$$

Where in this equation, h is Planck's constant, c is the velocity of light, λ is the wavelength of the optical transition between two levels, g is the statistical weight of upper state, E is the energy of the upper state, U is the partition function, K is the Boltzmann constant and T is the electrons' temperature. By taking the natural logarithm of both sides above equation can be written as:

$$\ln\left(\frac{I\lambda}{Ag}\right) = -\frac{E}{KT} + \ln\left(\frac{hcN}{4\pi U}\right) \tag{2}$$

In the above equation, for a given species, the last term is constant, therefore, plotting ln (Iλ/Ag) against E for number of spectral emission lines, will give a straight line of slope$\frac{-1}{KT}$ and intercept $ln\left(\frac{hcN}{4\pi U}\right)$. Thus by measuring slope, plasma temperature can be determined. In order to be more accurate in temperature determination using LIBS technique, those lines were chosen which were well resolved and also the transition probabilities were known. The spectra were recorded with a 4.5 μs delay time and 20 mJ laser pulse energy. Table 1 gives the values obtained from the measurements to calculate the plasma temperature. The plasma temperature was calculated at 10, 20, 30 and 40 mJ laser pulse energy and was found to be 3626 K, 3876 K, 4149 K and 5554 K respectively. The NIST database was used for the calculations of plasma temperature. The Boltzmann distribution and experimentally measured line intensities support the assumption that the laser-induced plasma was in local thermal equilibrium.

Table 1. Spectroscopic constants of iron lines used in temperature determination for slag sample-1

Wavelength	$A*10^8 s^{-1}$	g_i	g_k	Ei(eV)	E_K(eV)
249.06	3.8	5	7	0.087	5.06
252.28	2.9	9	9	0	4.9
271.90	1.4	9	7	0	4.5
344.09	0.084	7	5	0.05	3.65
358.119	1.02	11	3	0.85	4.3

The variation of temperature of Fe laser produced plasma in slag sample no.1 at delay time of 4.5 μs and laser pulse energy 10, 20, 30 and 40 mJ is shown in Figure 4 below. This figure clearly indicates that with the increase in the laser pulse energy the plasma temperature is also increased.

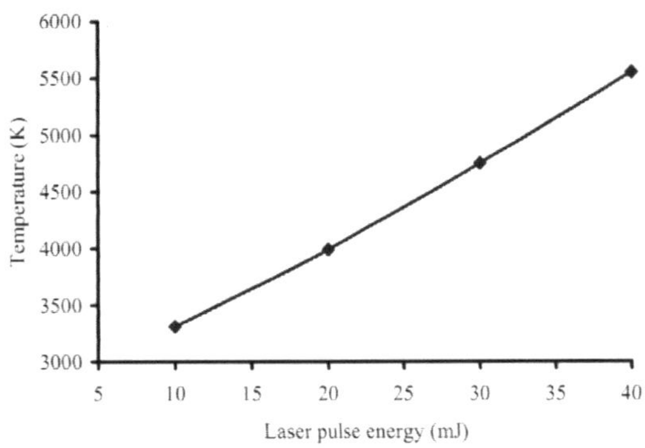

Figure 4. Variation of temperature of Fe laser produced plasma in slag sample no.1 versus laser pulse energy recorded at delay time of 4.5μs

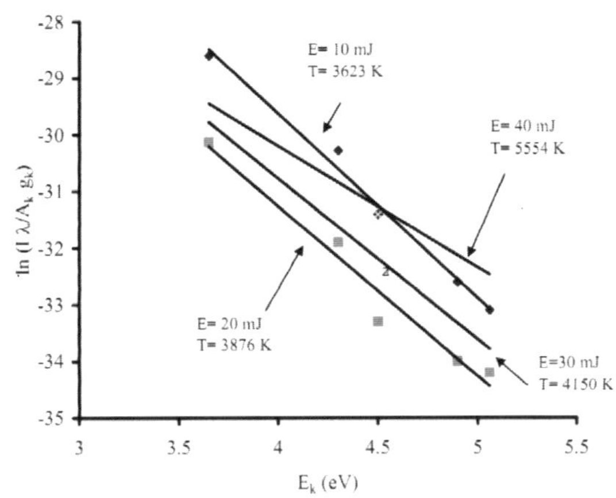

Figure 5. Boltzmann plot for temperature determination of Fe laser produced plasma slag sample no.1 at delay time of 4.5 μs and laser pulse energy 10, 20, 30 and 40 mJ, respectively

Boltzmann plot for temperature determination of Fe laser produced plasma slag sample no.1 is shown in Figure 5.

3.7. Electron density

In this experimental work the electron densities for Fe lines for iron slag sample were calculated using equation no. 3:

$$Ne \geq 1.6 \times 10^{16} \Delta E^3 T_e^{1/2} \qquad (3)$$

Where in the equation no. 3, ΔE is the largest observed transition energy for which the local thermal condition holds and T_e is the excitation temperature (K). The electron density and plasma temperature at laser pulse energy of 10, 20, 30 and 40 mJ are given in Table 2 at 4.0 μs delay time.

Table 2. Plasma temperature and electron densities at different laser pulse energies of sample no.1

Laser pulse energy (mJ)	Temperature(K)	Electron density (cm^{-3})
10	3623	$1.18x10^{16}$
20	3875.9	$1.22x10^{16}$
30	4149.3	$1.27x10^{16}$
40	5554	$1.46x10^{16}$

From Table 2 it is clear that with increase in laser pulse energy the plasma temperature and electron densities both increase.

3.8. Analysis of iron plant slag

Slag outputs in iron and steel production depend on the feed material chemistry and the type of furnace. Typically for an ore feed of 60 - 66% iron blast furnace, slag production ranges from about 220 to 370 kilograms per metric ton of pig iron produced. Lower grade ores produce much higher slag fraction. Iron scraps material collected from various locations, contain variety of contaminants. The contaminants present in scrap material can be harmful to the environment and therefore proper analysis of such iron slag is important. The primary hazardous components of iron slag are zinc, lead, cadmium, nickel and chromium [15]. As demand of iron is increasing day by day rapid elemental identification is required for waste management of slag. There are numerous health risk involved to the steel/iron factory workers at workplace and surrounding environment due to solid waste generated from the iron and steel mill industry. The major problem is due to a large volume of the waste generated and of hazardous metals present therein. Quantitative elemental analysis of slag samples was also conducted besides plasma temperature and electron density measurements. The elements like magnesium, lead, copper, calcium, chromium, maganese, sulphur, zinc, titanium, strontium, vanadium, and zirconium in slag samples were identified and quantitative measurements were made. Results obtained on iron slag samples are briefly presented here.

3.9. LIBS spectra of different slag samples

Figures 6 and 7 depict the emission spectra of iron slag samples recorded at 4.5 μs delay time in 200-620 nm regions. Here the laser pulse energy was 20 mJ. The distance between the optical fiber and the plasma was 40 mm. The selected lines of iron were used to study the effect of different experimental parameters and to compare the signal intensities with the calibration curves of the elements under investigation. Elements detected in slag samples and comparison of concentration detected with LIBS and ICP is presented in Table 3.

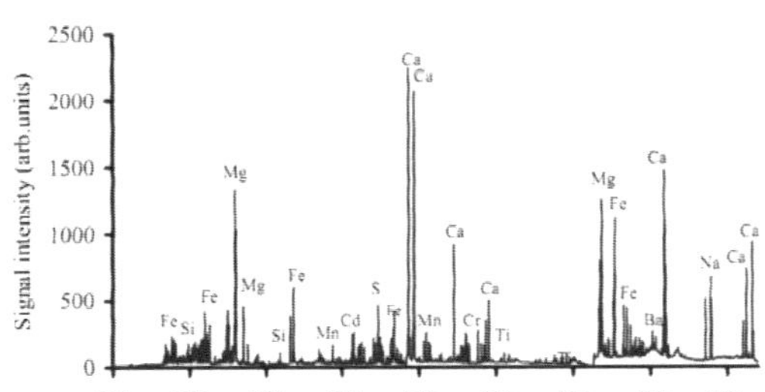

Figure 6. Typical spectrum of slag sample-1 in 200 to 620 nm region recorded at delay time of 4.5 μs and laser pulse energy of 20 mJ

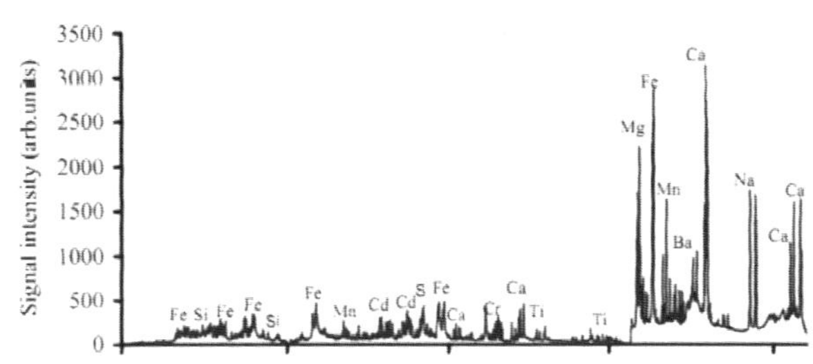

Figure 7. Typical spectrum of slag sample-2 in 200 to 620 nm region recorded at delay time of 4.5 μs and laser pulse energy of 20 mJ

Table 3. Elements detected in different slag samples and comparison of LIBS with ICP

Element	Wavelength (nm) (Striganov and Sventitski, 1968)	Sample1		Sample2	
		LIBS (ppm)	ICP (ppm)	LIBS (ppm)	ICP (ppm)
Ba	553.5	564.06	553	542.28	514
Ca	393.3	188700	185000	188720	185000
Cd	361.2	54.4	52	59.97	58.8
Cr	427.3	1866.6	1830	1438.2	1410
Fe	526.9	281520	276000	314100	308000
Mg	518.2	80194	79400	63529	62900
Mn	403.4	16564	16400	16463	16300
Na	588.9	1131.2	1120	751.44	744
P	438.5	2747.2	2720	2272.5	2250
S	373.8	1397.4	1370	1458.6	1430
Si	288.1	79285	78500	64741	64100
Ti	498.2	5635.8	5580	4231.9	4190

From Table 3 it is clear that Ba, Cr, and S are present in very high concentrations. For example in sample no. 1, the concentration of Ba = 564 ppm, Cr = 1866 ppm and S = 1397.4 ppm while the maximum permissible limits set by EPA and other regulatory authorities for these elements are: Ba = 2, Cr = 1 and S = 800 ppm respectively. Barium is harmful for skin if it is inhaled during breathing or water containing it is taken. Similarly chromium concentrations is detected to be higher than the maximum permissible limit and breathing a high level of chromium can cause irritation to the nose. However, a limited amount of chromium is an essential nutrient that helps the body to use sugar, protein and fat. During several industrial processes (such as iron slag formation) sulphur compounds are added to the environment that is hazardous to animals as well as humans. These sulphur compounds are toxic and have unpleasant smell: may have other affects on human health [16-18]. For slag samples the precision of LIBS measurements depend on the complexity of the sample, homogeneity, and the reproducibility of the laser shots. Typical RSD values for LIBS are in the range 1–10% and for slag samples, the precision was 2%. The relative accuracy (RA) is calculated using equation no. 4

$$RA = \frac{|d| + SD \times \frac{t_0.975}{\sqrt{n}}}{M} \qquad (4)$$

In above equation, d is the difference between the LIBS measurement and the ICP (standard method). SD is the standard deviation of LIBS measurement, M is the measurement from standard method, n is the number of measurements and t_0 .975 is the value at 2.5% error confidence. It can be seen from Table 4, the relative accuracy is in the range of 0.001 – 0.049, which is quite acceptable for any good instrument.

Table 4. Comparison of LIBS with ICP, relative accuracy, limit of detection and delay times for different elements detected in slag sample.

Element	Wavelength (nm) (Striganov and Sventitski, 1968)	Sample 1		Relative accuracy (RA)	Standard deviation (S.D) LIBS	LOD LIBS (ppm)	Delay time (μs)
		LIBS (ppm)	ICP (ppm)				
Ba	553.5	564.06	553.0	0.020	0.098	14.0	5.5
Ca	393.3	188700	188500	0.001	0.122	11.0	5.0
Cd	361.2	54.4	52.0	0.049	0.129	10.0	4.0
Cr	427.3	1866.6	1830	0.020	0.091	7.0	5.5
Fe	526.9	281520	276000	0.028	0.132	6.0	4.5
Mg	518.2	80194	79400	0.010	0.115	9.0	5.0
Mn	403.4	16564	16400	0.012	0.095	6.0	4.5
Na	588.9	1131.2	1120	0.015	0.119	5.0	4.0
P	438.5	2747.2	2720	0.018	0.125	13.0	4.5
S	373.8	1397.4	1370	0.020	0.128	16.0	4.5
Si	288.1	79285	78500	0.017	0.130	10.0	4.0
Ti	498.17	5635.8	5580	0.013	0.135	9.0	5.0

3.10. Calibration curves for elements under investigation

In order to test the validity of LIBS spectrometer for quantitative analysis on actual iron slag samples, the spectrometer was calibrated for each trace metal of our interest. For the construction of the calibrations curves, two stoichiometric samples of iron slag were prepared. Pure metals in powder form were mixed with the matrix material KBr. Then, pellets were pressed in a 10 bar pellet press. In order to test the homogeneity of our samples, several LIBS measurements were performed at different locations at the surface of the pellets. Four samples of known concentration such as 100,000, 10,000, 1000 and 100 ppm of elements under investigation were prepared in the matrix of KBr and LIBS spectra were recorded for these four concentrations of each element. All these spectra were recorded with an average of 20 laser shots at three different locations on the sample surface. The calibration curves for Cd and Cr are depicted in Figures 8-9.

Thus in this study, two slag samples collected from the local iron plant were characterized qualitatively as well as quantitatively for elemental detection. Ba, Ca, Cd, Cr, Fe, Mg, Mn, Na, P, S, and Si were detected in reasonable abundance in these samples. The results showed excellent correlation with the results obtained from a standard method such as inductively coupled plasma analysis. The calibration curves were effective in quantifying the trace metal concentration in slag samples and for estimation of spectrometer limits. The detected concentration of some of the elements such as Ba, Cr and S was higher than the permissible safe limits.

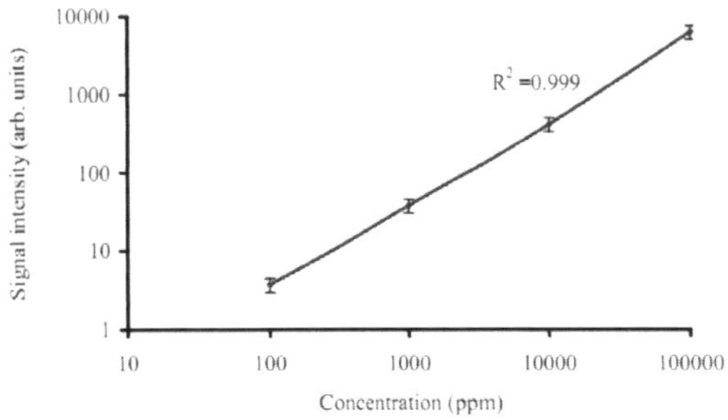

Figure 8. Signal intensity versus Cd concentration at 4.5 μs time delay and laser pulse energy 20 mJ. Error bars denote 95% confidence limits and are based on triplicate measurements.

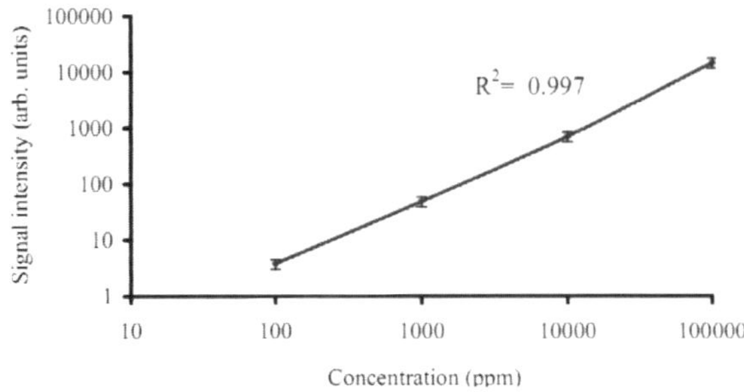

Figure 9. Signal intensity versus Cr concentration at 4.5 μs time delay and laser pulse energy 20 mJ. Error bars denote 95% confidence limits and are based on triplicate measurements.

4. Conclusion

Laser Induced Breakdown Spectroscopy (LIBS) is applied for the determination of plasma temperature and electron density of iron in iron slag samples. The LIBS spectrometer was carefully optimized for the improvement of LIBS efficiency. The plasma temperature was calculated at 10, 20, 30 and 40 mJ laser pulse energy using standard equations and well resolved spectral lines in the 229.06-358.11 nm region and was found to be 3626 K, 3876 K, 4149 K and 5554 K respectively. The NIST database was used for the calculations of plasma temperature. The Boltzmann distribution and experimentally measured line intensities support the assumption that the laser-induced plasma was in local thermal equilibrium. It was observed that with the increase in the laser pulse energy the plasma temperature is also increased. Electron densities for Fe lines for iron slag sample were calculated which is also increased with increase in pulse energy. Quantitative elemental analysis of slag samples from a local plant was conducted besides plasma temperature and electron density measurements. Concentration of Ba, Cr, and S was found to be exceeding maximum permissible limits. Concentration of detected elements by LIBS was compared with ICP. The results showed excellent correlation with the results obtained from inductively coupled plasma analysis.

5. Acknowledgements

The support by the Physics Department of King Fahd University of Petroleum and Minerals is gratefully acknowledged. Authors are highly thankful for those who helped during research work.

6. References

[1] McWhirter R W P 1965 Plasma Diagnostic Techniques *Pure and Applied Physics vol 21*eds Huddlestone R H and Leonard S L (New York: Academic Press) 201-264
[2] Aguilera J A and Aragón C 2004 *Spectrochim. Acta Part B*59 1861
[3] Lesage A 2009 *New Astron. Rev.* 52 471
[4] Gondal M A, Hussain T, Yamani Z H and Barkry A H 2007*J of Envir. Sciene and Health Part A*42 767-775
[5] Aguilera JA and Aragon C 2004 Characterization of laser-induced plasma by spatially resolved spectroscopy of neutral atoms and ion emission. Comparison of local and spatially integrated measurements. *SpectrochimicaActapart B*.591861-1876
[6] Cabalin LM and Laserna J J1998 Experimental determination of laser-induced breakdown thresholds of metal sunder nanosecond Q-switched laser operation *Spectrochimica Acta PartB* 53723-730
[7] Abdellatif GandImam H2002 A study of laser plasma parameters at different laser wave length *Spectro chimica Acta Part B*.571155-1165. Muller K and Stege H 2003 Evaluation of the analytical potential of the Laser- induced breakdown spectroscopy for the analysis of the historical glasses*Archaeometry*45421-433

[8] Aragon et al., 2001 Aguilera and Aragon2004 Bengoechea et al., 2006 Tognoni et al., 2007
[9] Muller K and Stege H 2003Evaluationof the analytical potential of the Laser- induced breakdown spectroscopy for the analysis of the historical glasses *Archaeometry* 45421-433
[10] Mayo R and Ortiz M and Plaza M 2008 *J. Phys. BAt. Mol. Opt. Phys.*41 095702
[11] Mayo R and Ortiz M 2008 *J. Phys. B: At. Mol. Opt. Phys.*41 225702
[12] Alonso-Medina A and Colón C 2006 *Astrophys. J.*672 1286
[13] Gondal M A, Hussain T, Yamain Z H and Baig M A 2007 Role of various binding materials for trace elemental analysis of powder samples using laser induced breakdown spectroscopy *Talanta*72 642-649
[14] Bekefi,G.(1976). Principles of laser plasma, John Wiley & Sons, New York.
[15] World Bank Handbook 1998 Pollution Prevention and Abatement, Iron and Steel manufacturing; Draft technical background document, environment department World Bank (Washington: DC) July Issue pp 327-331
[16] Lee PN and Corning DM1993 *Experimental toxicology.* The basic issues Cambridge: The Royal Society of Chemistry), pp405-440
[17] (Saudi-Arabian: Royal Commission/Jubail Industrial City) *Environmental Regulation*
[18] *Standards* September1999. *Standard Methods for the examination of water and wastewater*18th ed. 1992 (Washington DC: American Public Health Association)

Investigation into the effect of some additives on the mechanical strength, quality and thermal conductivity of clay bricks

Adnan I. O. Zaid, A. Qandil[1], M. A. A. Qattous[2]
Mechanical and Industrial Engineering Department,
Applied Science Private University, Amman 11931 Jordan
[1]Civil Engineering Department, Applied Science Private University
Amman 11931 Jordan
[2]Industrial Engineering Department, University of Jordan, Amman, Jordan

E-mail: adnan_kilani@yahoo.com

Abstract. It was repeatedly reported that the clay bricks industry in Jordan is facing both weak mechanical strength and poor quality which caused marketing problems where it is expected to serve the increasing demand of housing in the country especially after the political crises in the neighboring countries Iraq and Syria. It is therefore anticipated that improvement of the mechanical strength and quality of the produced clay evaluation of the brick industry in Jordan is worth investigating. In this paper, theoretical and experimental investigation obtained from field visits to the factories producing clay bricks were carried out. Furthermore, the effect of using some additives from locally available materials namely: Battn El-Ghoul Clay, Suweileh sand and Olive extracts on the mechanical strength, thermal conductivity and surface quality of the produced bricks is investigated and discussed. The experimental results indicated that thermal conductivity, color and durability were all enhanced and the ultimate compressive strength was reduced but remained higher than the acceptable value for brickwork.

1. Introduction and literature review

Brickmaking industry is probably the second earliest industry of mankind after agriculture. The archaeological excavations have shown that the oldest sun-dried bricks have been found beneath the foundations of the old city of Jericho in the Jordan River valley, just a little north of the Dead Sea. This was about 9000 to 10000 years where no molds were used at that time. These excavations indicate that clay has existed in Jordan and have been in use since the old ages, [1-2]. Ever since, the art of brickmaking had advanced to the process of firing. As early as 1955 MILLS [3] published a book about clay bricks from the stage of manufacture to the stage of use. This book seems to be the first book in the field. Later, the mineralogical aspects of clay using differential thermal analysis is given in [4]. Simple manuals for brickmaking to be used with villagers to enable them to make good and cheap bricks by themselves were introduced in [5-6]. The use of bricklaying as a craft is given in references [7-10]. However, the first introduction of the heavy clay technology was forwarded in 1969 in reference [11]. A useful detailed review of brickwork suitable for building based on engineering

Content from this work may be used under the terms of the Creative Commons Attribution 3.0 licence. Any further distribution of this work must maintain attribution to the author(s) and the title of the work, journal citation and DOI.
Published under licence by IOP Publishing Ltd

design and scientific research which provided the recent developments in brickwork technology from which prefabricated and pre-stressed bricks were manufactured are given in [12]. A new trend in the research of brick work was reported in where a good manual which describes a method of stabilizing sun-dried adobe bricks which are water proof and exceptionally strong and have a noticeable quality of durability, where little maintenance and keep-up are necessary were reported in 1972, [13,14]. After 1972, voluminus research papers appeared in the literature on clay bricks all over the world both in indusrial and developing countries in attempts to improve its mechanical strength, durability and surface quality, [15-44].

2. Materials, Equipment and experimental proceadures

2.1Materials

The raw materials used in this paper were taken from three factories producing building clay brick and denoted by A, B and C. Two specimens were taken from each factory and designated as A1, A2 for factory a, B1 and B2 for factory B and C1,C2 for factory C.

2.2 Experimental Procedures

Chemical analysis was carried on each specimen before and after grinding in two different Government Establishments, namely: the Natural Resources Authority, NRA, and the Jordan Phosphate Mines Company as illustrated in Tables 1 and 2 respectively.

Table 1. Chemical Analysis carried out at NRA after grinding raw material (X-ray Fluorescence Analysis)

Sample	MgO%	Al_2O_3%	SiO_2%	%CaO	%Fe_2O_3	%L.O.I
A1	1.01	14.60	44.32	14.40	7.08	17.60
A2	0.24	12.55	74.66	1.70	4.07	5.78
B3	0.97	14.37	39.02	18.20	7.23	19.28
B2	0.19	20.33	70.20	0.60	3.92	7.79
C1	1.11	13.70	38.14	18.10	7.54	19.83
C2	0.43	11.81	51.32	16.90	2.56	16.88
C3	0.90	13.20	49.18	15.30	5.94	7.25

Table 2. Chemical Analysis carried out at Jordan Phosphate Mines Company after grinding Raw

Sample No.	% Na_2O	% MgO	% Al_2O_3	% SiO_2	% P_2O_5	% K_2O	% CaO	% TiO_2	% MnO	% Fe_2O_3
A1	0.0	2.23	12.94	41.30	0.13	1.37	12.37	1.20	0.06	7.21
A2	0.0	0.21	9.92	71.29	0.15	0.81	3.00	1.24	0.03	4.36
B3	0.02	2.38	12.87	33.12	0.05	1.66	16.76	1.13	0.07	7.20
B2	0.03	0.27	16.84	64.34	0.19	0.56	0.11	1.91	0.03	4.85
C1	0.03	2.47	13.10	33.15	0.03	1.66	16.67	1.12	0.07	7.33
C2	0.14	0.87	10.03	51.86	0.18	0.77	12.75	0.96	0.04	2.73
C3	0.70	2.01	14.08	46.09	0.16	1.24	13.88	1.22	0.07	6.51

3. Results and discussions

The results of testing the fired briquettes are shown in Figures 1 to 5 inclusive. The tests which were carried out on the fired briquettes are the compression test, water absorption tests, and the thermal conductivity test. It is worth mentioning that the sample in the first two tests consisted of 5 specimens at least, while in the third test it was ranging from 2 to 3 specimens and in some rare cases it consisted of 4 to 5 specimens. Throughout the measurements taken to establish the data presented in this chapter, care was taken to minimize the error, and as there were many sources of error throughout this work, in the weighing of raw materials and additives, in the mixing and forming stage, in the drying stage, in the firing stage,and finally in testing the fired briquettes. Although, care was taken in manufacturing and testing of all the briquettes of all samples under the same conditions, it was decided to consider a coefficient of variation of 10% as a criterion to accept or reject the specimen from the sample. Normally this percentage will be as low as 2% in metals. In Figure 1, which shows the ultimate compressive strength versus the percentage of each additive, all the additives showed various amounts of decrease in the ultimate compressive strength. For the additive Batn El—Ghoul Clay, the decrease in the compressive strength is minimum and the fired briquettes are still acceptable, for load bearing brickworks even for high percentages of Batn El-Ghoul. Clay. in the case of adding Suweileh Sand the decrease in compressive strength is more pronounced than the previous case. Despite this pronounced decrease, Suweileh Sand is still acceptable for non-load bearing brickworks. In the case of Olive Extracts, still more decrease in the compressive strength was noted; therefore only an additive less than 2% could be used to consider Olive Extracts acceptable for load—bearing bricks and 3% could be considered suitable for non-load bearing bricks.

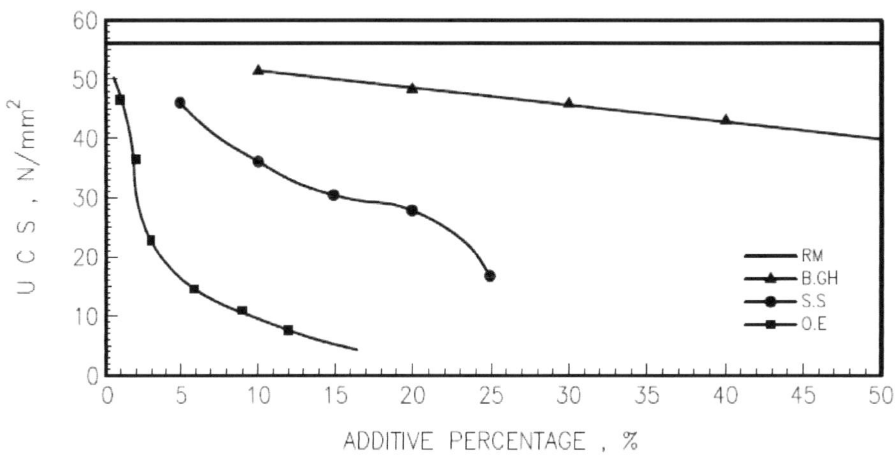

Figure1. Ultimate compressive strength versus additive percentage

Figure 2 shows the percentage of water absorbed (24—hours cold submersion) versus additive percentage for each of the additives. In this Figure it is clear that Batn El-Ghoul Clay shows a small increase over the raw materials, although for additions of 10% and 20% of Batn El—Ghoul Clay the percentage absorbed water is less than that for the raw materials. For Suweileh Sand nearly no great change in the percentage absorbed water occurs when increasing the additive percentage, and all the data points fall a little above those of the raw absorbed is noted for all the percentages, this great increase is clearly noted when the percentages exceeded the 3%. It is. worth mentioning that there is some inflection or turning points at different percentages in the three additives which are due to surface bores or cracks during forming of briquettes, neglecting these inflections will make the trends in the above discussion more clear. Figure 2 shows percentage water absorbed (5-hours boiling) versus additive percentage for each of the additives. In this figure, Batn El-Ghoul Clay starts nearly from the same percentage absorbed by the raw materials and increases until an additive percentage of 30% is reached where it starts to decrease, but still remains above that absorbed by the raw materials, until an additive percentage of 40% is reached. Then it starts to increase again. Suweileh Sand starts above Batn El-Ghoul Clay, then starts decreasing until 10% of additive, then it starts to increase until 20% of additive is reached where it starts to decrease again. Olive Extracts shows a clear sharp increase in the percentage water absorbed. It is worth mentioning that the inflection points of 40% Batn El-Ghoul and the 5% Suweileh Sand could be due to improper mixing of the raw materials and the additives at the aforementioned additive percentages.

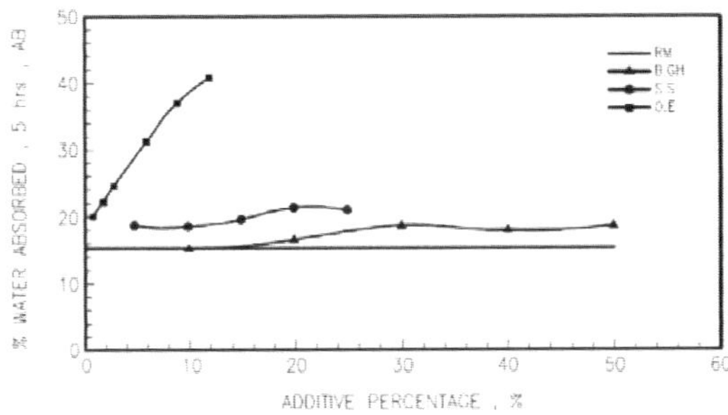

Figure 2. Percentage water absorbed (5-hoursboiling) versus additive percentage

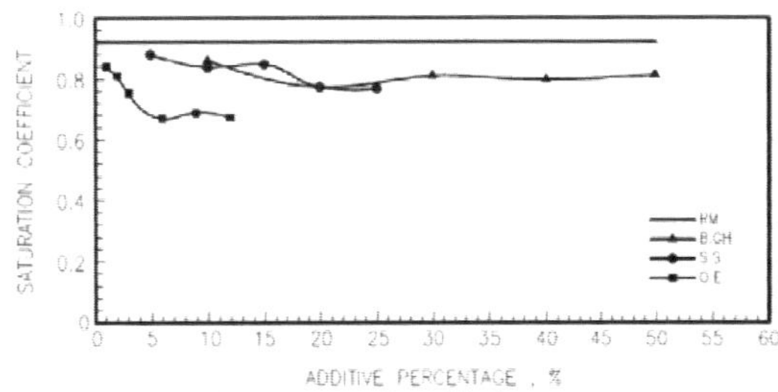

Figure3. Saturation Coefficient versus additive percentage

Figure 3 shows the saturation coefficient versus the additive percentage for each of the additives. In this figure all the three additives start below the raw materials which have a saturation coefficient in the range of negligible weathering (NW). However Batn El—Ghoul Clay shows the best results to be used in sever weathering (SW) nearly for all additive percentages, but Suweileh Sand shows the ability to be used in moderate weathering (MW) if it is added at percentages less than 15%. However, it could be used in severe weathering (SW) if it is added at percentages of 20% or 25%. On the other hand, Olive Extracts could be used in moderate to severe weathering if added at 1,2 and 3% otherwise it will be used in negligible weathering. The above discussion is based on ASTMc62-87(64).

Figure 4, which shows the percentage apparent porosity versus additive percentage for each additive, could be discussed in the same manner as the aforementioned discussion of Figure 3 since the percentage apparent porosity is a function of the absorbed water percentage, after boiling for 5-hours, where the apparent porosity is the product of the specific gravity (GS) by the absorbed boiled water percentage.

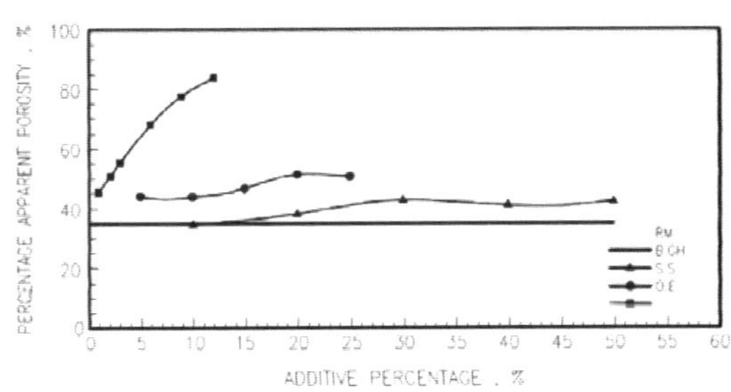

Figure 4. Percentage apparent porosity versus additive percentage

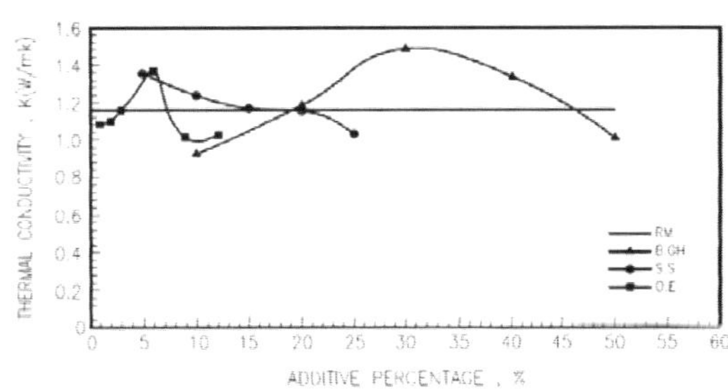

Figure 5. Thermal conductivity versus additive percentage

Figure 5 shows the thermal conductivity versus the percentage for each additive. It can be seen from this figure that increasing the Batn. El-Ghoul clay percentage resulted in increase of the thermal conductivity until 30% addition, then a decrease in the thermal conductivity initiates, the best thermal conductivities were at 10% and 50% of Batn El—Ghoul Clay increase, followed by a decrease in the thermal conductivity. This could be attributed to the presence of mice which behaved as a conductor at this low firing temperature.

When the percentage of apparent porosity was increased the thermal conductivity started to decrease. In case of Suweileh Sand, the thermal conductivity decreases continuously with the increase of Suweileh Sand percentage and a sharp decrease in the thermal conductivity occurred at 20% addition of the sand. This could be due to the increase in porosity upon increase of the percentage of Suweileh Sand. However, the high thermal conductivity at the beginning is due to the large grain size of the

Suweileh Sand. The most interesting additive was Olive Extracts, Figure 5. It is clear that the Olive Extracts starts at lower thermal conductivity than that of the raw materials and starts increasing slowly until 3% of Olive Extracts is reached where an increase in the thermal conductivity was pronounced; after which a sharp decrease starts. The low thermal conductivity at the beginning is due to the high percentage apparent porosity existed even at 1% of Olive Extracts, but increase in the thermal conductivity is attributed to the increase of boreealthough the porosity is still increasing. However, the effect of increase in the bore diameter on the thermal conductivity could not outweigh the effect caused by the high porosity existed, therefore the thermal conductivity started to decrease again. On the whole, it may be concluded and suggested to add Suweileh Sand in a percentage ranging from 15-25, and Olive extracts in percentages between 1-3 or 9-12. One of the main properties tested in this work was the color of the briquettes, where an acceptable red color was obtained when the raw materials were used alone. This red color is the result of using materials under the acceptable Cao: Fe_2O_3 ratio and using fine grinding with proper mixing. The homogeneity in color and mixing is clearly noted when the samples were crushed in the compression test. The effect of adding Batn El-Ghoul Clay on the color produced. The following addition percentages specimens were used:10, 20, 30, 40 and 50% respectively and the produced briquette colors were compared with that produced from the original raw materials. It was found that at 10 and 20% addition of Batn El—Ghoul Clay the color was changed to dark red while at 30% the white scum started to appear and increased more at higher additive percentages, 40 and 50%. The reason for this white scum or efflorescence is the presence of potassium oxide and/or magnesium oxide. Similarly, the effect of adding Suweileh Sand on the produced color was also investigated using the addition percentages of 5,10,15,20 and 25% respectively. The results indicated no great change had occurred, only a little enhancement of the red color was obtained by increasing the sand percentage, in which the percentage of Fe_2O_3 in Suweileh Sand is appreciable, so the overall CaO: Fe_2O_3 ratio is slightly decreased . It is worth mentioning that in these two additives it was noted that the surfaces of the obtained briquettes were fine and smooth. Furthermore, in the case of Suweileh Sand addition early no change in volume was noted. The effect of adding Olive Extracts on the produced color was also investigated but with different percentages namely: 1,2,3,6,9 and 12%. It was noticed that the color was changing between red to pink at the first two percentages, (1 and 2%), whereas at 3% it started to change to pink, this was very clear at 6% addition, and at the 9% a pale color started to appear, which was more pronounced at 12%. This pale color appeared pink in some specimens and yellow in others. The surfaces of the obtained specimens were good and acceptable at the first two percentages, while the appeared bores started to destroy the surface at the 3%. This distortion was more pronounced at higher percentages, (9% and 12%). The large bore diameters and the pale colors are attributed to the burning of the carbon aqueous (organic) matter of the Olive Extracts which includes wood.

3. Conclusions

The following points are concluded:

i) To obtain clay bricks free from line blowing fine grinding of the raw materials is essential to a particle size about 500 microns.

ii) The homogeneous color could be obtained only if proper mixing and homogeneous firing are ensured together with a high temperature enough to allow the iron compounds, mainly the ferric oxide, (Fe_2O_3) to react with the clay ingredients.

iii) Using the three investigated additives which are locally available namely: Batn El-Ghoul clay, Suweileh sand and Olive Extracts, resulted in improvement of the saturation coefficient. This enables the clay brick to be used under severe weather conditions. Batn El-Ghoul may be added with any percentage for severe weather conditions, Suweileh sand is recommended for moderate weather conditions and Olive Extracts if added at percentages less than 3% can be used for severe weather conditions.

iv) The three used additives resulted in decrease of the ultimate compressive strength. However, if they added in the appropriate percentages:10% Batn El-Ghoul, 15% Suweileh sand and less than

3% Olive Extracts they can produce good results i.e. low thermal conductivity, good compressive strength, good saturation coefficient and good red color.

4. Acknowledgment

The first and second authors are grateful to the Applied Science Private University, Amman, Jordan for partial financial support granted to this research (Grant No.DRGS-2015).

5. References

[1] Brownell, W E 1974 Structural Clay Products, Springer Verlag, Wien
[2] Sarraf, F 1982 Clay and Clay Bricks in Jordan, Previous Studies and Existing Situation, Royal Scientific Society, Building Research Center, Amman
[3] Mills A P 1955 Materials of Construction; Their Manufacture and Properties, 6th edition, John Wiley and Sons Inc
[4] Mineralogical Society 1 1957 The Differential Thermal Investigation of Clays, London
[5] Mathur G C 1960 Bricks Making, National Building Organization, Ministry of Work, Housing and Rehabilitation, India
[6] The Brick and Tile Institute of Ontario 1965 Clay Masonry Manual, Ontario, Canada.
[7] McKay J K Brickwork, 2nd edition, Longman, London, 1968
[8] Nash WG 1966 Brickwork One, 1st edition, Hutchinson, Technical Education, London
[9] Nash, W G 1968 Brickwork Two, 1st edition, Hutchinson, Technical Education, London
[10] Hodge J C 1971 Brickwork for Apprentices, 3rd edition, Edward Arnold Publishers Ltd, London
[11] Clews, F H 1969 Heavy Clay Technology, 2nd edition, Academic Press Inc., London
[12] Lancer D 1792 Elements of Load Bearing Brickwork, 1st edition, Pergamon PRESS, London
[13] International Institute of Housing Technology1972 California State University, Fresno Foundation, The Manufacture of Asphalt-Emulsion-Stabilized Soil Bricks and Bricks Makers Manual, California
[14] Appall I S 1972 Low-Cost Durable Shelters, Build International, March/April,100-103.
[15] Swindle- Dressler Company 1972 Feasibility Study for Ceramic Industry, Jordan; Drafts Phase 1 Report, Parts 1 & 2, U.S.A
[16] knizek I 1978 Brickmaking Plant; Industry Profile, UNIDO, Vienna
[17] Krienger M 1978 Home Owners Encyclopedia of House Construction, McGraw-Hill Inc
[18] Kumar A 1979 Building Materials and Components, 2nd edition, MIR Publishing, Moscow
[19] Gupta R B and Sharma, B.R. 1981Insulating Materials, In: Electrical Engineering Materials, 2nd edition, smithSumatraHonda, New Delhi, 126-200
[20] Research on lime Stabilized Soil Building Blocks 1982 Report of Mission, Royal Scientific Society, Amman, Jordan
[21] Country Case Study on Jordan 1982 Draft Report, Royal Scientific Society, Amman
[22] Building Research Establishment (BRE) BREPAK BLOCK MAKING MACHINE , London, 1982
[23] Al-Kass R M, Ha M, Khalil N, AL-Takarli S 1984 Effect of Fine Calcite Grains Present in the Soil on the Properties of Clay Bricks, Journal of Building Research, Building Research Center, Baghdad, 3, 1-31
[24] Al-Saudi N,Mageed I, Ahmad K 1986 Free S well Properties of Bentonite / Kaolinite Mixture, Journal of Building Research, Building Research Center, Baghdad, **5**, , 29-63
[25] Habitat United Nations Centre for Human Settlements (UNCHS), Manual on Basic Principles of Each Application, Nairobi, 1986
[26] Habitat United Nations Centre for Human Settlements (UNCHS) 1986 Manual on Production of Rammed Earth, Adobe and Compressed Soil Blocks, Nairobi
[27] Al-Taie M, Al-Ramadan K 1987 Improving the Refractoriness Under Load of the Fire – Clay Bricks, Journal of Building Research, Building Research Center, Baghdad, 6,60-77

[28] Al-Kvass R M, Al-Ali B T1988 Lightweight Clay Bricks with Expanded Polystyrene additions, Journal of Building Research, Baghdad, 7,1-10
[29] Jurban B A, Habbali S M, Hamdan M A, ZaidA I O 1988 Some Mechanical and Thermal Properties of Clay Bricks for the Jordan Valley Region, Journal of Materials and Structures, 21, 364-369
[30] Awawdeh M 1991Assessment of BATN EL-GHOUL CLAY, JORDAN, M.Sc. Thesis, Leicester University, U.K
[31] Sultan SA1993 Foam Clay Blocks Industry From Local Material in Jordan, Proceedings of the First International Conference on Implementing Local Material in Industrial Applications, 1, Amman, Jordan, 133-141
[32] Khreis R I 1993 Evaluation of Jordanian Silica Sand for Molding in the Foundry Industry – Shell Process, Proceedings of the First International Conference on Implementing Local Materials in Industrial Applications,1 , Amman, Jordan, 254-311
[33] Dwairi I M, Sallam M 1994 Evaluation of Bentonitic Clays of Azraq for Foundry Sand Uses, The First Jordanian Mining Conference, Amman Jordan, April, 23-25
[34] Raouf ZA , Al-Mousawi MJ 1993The Use of Local Materials in Plastic Concrete Wall for Seepage Control, Proceeding of the First International Conference on Implementing Local Materials in Industrial Applications,1. Amman, Jordan, Sep. 6-9, 72-85
[35] Pierzchlewicz J 1993 New Structural Trends Regarding Clay Hollow-Blocks for External Walls, Proceedings of the First International Conference on Implementing Local Materials in Industrial Applications, 1. Amman, Jordan, 93-102
[36] Ali G A, Pierzchlewicz J, Sivakugan N 1999, Soils and Aggregates in Construction Industry, Conference on Implementing Local Materials in Industrial Applications, 1. Amman, Jordan, 191-201
[37] British Standards Institution, BS 3921: 1958, U.K.
[38] American Society for Testing Materials (ASTM), C62-87, U.S.A
[39] United Nations Industrial Development Organization (UNIDO), 1984 Small-Scale Brick-making, Vienna
[40] Bain J A 1971 A Plasticity chart as an aid to the identification and assessment of industrial clays, Clay Minerals, 9, 1-17
[41] Bain J A, Highly D E 1978 Regional Appraisal of Clay Resources – Challenge to the Clay Mineralogist, Proceedings of the International Clay Conference, 437-446
[42] Bowles J,Liquid and Measurements, 4th edition, McGraw Hill Inc., New York, 1992, 21-32
[43] Cusson Company, usermanual thermal conductivity apparatus, U.K
[44] Grimshaw R W 1971 Miscellaneous Properties of Ceramic Materials, In: The Chemistry and Physics of Clays and Allied Ceramic Materials, 4th edition. John Wiley and Sons, New York, 905-966

IOP Conf. Series: Materials Science and Engineering **146** (2016) 012019 doi:10.1088/1757-899X/146/1/012019

Multiple shock compressions in tri-amino-tri-nitro-benzene based energetic materials

T. Hussain[1,2], **L. Yan**[1], **F. Huang**[1]
[1]Department of Physics and Applied Mathematics,
Pakistan Institute of Engineering and Applied Sciences, Islamabad, Pakistan
[2]State Key Laboratory of Explosion Science and Technology,
Beijing Institute of Technology, Beijing 100081, P. R. China

E-mail: hushan@bit.edu.cn

Abstract. Various hazard and vulnerability scenarios for energetic materials involve multiple shock compression. Triaminotrinitrobenzene (TATB) based explosives are comparatively insensitive to shocks of low impacts and therefore considered suitable materials for the devices vulnerable to low shock impacts. Multiple shock compression may further desensitize these materials. Although the desensitization criterion for HMX based plastic bonded explosive PBX9404 has long been determined, no such criterion for TATB based explosives has yet been determined. In the present paper the desensitization criterion for TATB based explosive is proposed, based on recently performed experiments. By using a model which employs the proposed criterion to account for the desensitization effects in TATB based explosives, numerical simulations of the multiple shock experiments have been performed. The calculated results agree closely with the experimental results.

1. Introduction

TATB based explosives belong to the class of insensitive explosives. These explosives have been of great interest because of their remarkable insensitivity and relative invulnerability to accidental shocks. The explosives LX-17 (92.5% TATB/7.5% kel-F 800) and PBX9502 (95% TATB/5% kel-F 800) have been extensively studied and show non ideal behaviour. Their failure diameters are comparatively large [1-2] and detonation waves are more severely affected by abrupt geometry changes than other explosives [3]. Their detonation and shock initiation properties have stronger dependence on the initial temperature and particle size distribution [2, 4]. Unlike other explosives, the detonation waves in unconfined explosives fail even at 97% of their Chapman-Jouguet (CJ) velocities [1]. The shock initiation threshold criterion $P^2\tau$ has been observed to hold for many explosives [5]. However, Honodel et al. [6] noticed the departure of TATB explosives from $P^2\tau$ = *constant* criterion at low pressures. His analysis suggested that at low shock pressures TATB based explosives may behave like homogeneous explosives. Souers [7] discussed the initiation thresholds for TATB based explosives. Haskins et al. [8] proposed a modified criterion for the prediction of shock initiation threshold for superfine TATB. It is now an established fact that passage of a shock which is weak enough to detonate the explosive, but strong enough to compress the explosive above it Hugoniot elastic limit (HEL), desensitizes the explosive for the succeeding shock which could otherwise detonate the explosive if the explosive were not preshocked. Desensitization criteria for PBX-9404 has been experimentally determined by Campbell [9] by subjecting PBX-9404 to pre-shocks in the

Content from this work may be used under the terms of the Creative Commons Attribution 3.0 licence. Any further distribution of this work must maintain attribution to the author(s) and the title of the work, journal citation and DOI.
Published under licence by IOP Publishing Ltd

pressure range 1.0 to 2.4 GPa. Using the experimental data, a relation was proposed between the preshock pressure (*P*) and the time required (τ) for desensitization of PBX-9404 as $P^{2.2}\tau = 1150$ (kbar$^{2.2}$ μs). Unfortunately no such experiments have been performed for LX-17 explosive and consequently no such relation for the explosive is available to date, although the need has been felt by many authors [10, 11]. Recently Vandersall et al. [12] performed shock desensitization and detonation quenching experiments on LX-17 using a Teflon flyer fired from a gas gun to impact at one end of a cylindrical sample of LX-17 and generate a single low amplitude shock wave to pre-shock the explosive sample just prior to detonating it from the other end. However some more experiments are needed to exactly determine the limits of the pressure range and the time window for desensitization.
We have proposed a desensitization model for PBX-9404 and applied it for simulation of various preshock desensitization experiments for PBX-9404 in a recent study. In the present study we tried to adapt the desensitization model for LX-17 using the experimental results of Vandersall [12], and simulated the desensitization experiments on LX-17 reported in literature [10, 12, 13]. A brief overview of the modelling of desensitization effects is given in the next section. The available preshock desensitization experiments are simulated in section 3 for validation of the model.

2. Modelling multiple shocking effects

Most of the phenomenological reactive flow models in use today do not have capability to account for multiple shocking effects such as desensitization induced by preshocking. Some special treatments is required in these models to deal with such problems. For example, JTF model [14] accounts for preshock desensitization effects by restricting creation of hot spots to the first shock only, while the second shock is allowed to provide the adiabatic heating only. Wescott-Stewart-Davis (WSD) model [15] has later been improved to a temperature dependent form [16] and is made capable to predict the reduction of the reactivity with multi-shock loading. Although the reactive burn model CREST [17] that utilizes entropy-dependent reaction rates, has been successful updated [18] for predicting the persistent dead zones in corner turning experiments [13] on PBX 9502 explosive at different initial temperatures without any shock desensitization treatment, however, the detail of the improved CREST model for PBX9502 has not been published yet.

We have tried to annex the desensitization model for LX-17 in Lee Tarver ignition and growth model, as it is one of the popular reactive burn models in use today and has been embedded in various commercial hydrocodes. It has been parameterized for a number of explosives and has been used to solve a variety of problems [4, 19-28]. Lee-Tarver model [24, 28] employs separate Johns-Wilkins-Lee (JWL) equations of state (EOS) for both the solid unreacted explosive and the reaction products, and a reaction rate equation. The three term reaction rate equation for the standard model is of the form

$$\frac{d\lambda}{dt} = R_I + R_{G_1} + R_{G_2}, \qquad (1)$$

where

$$R_I = I(1-\lambda)^b(\eta - 1 - C_{\text{crit}})^x,\ R_{G_1} = G_1(1-\lambda)^c\lambda^d p^y \text{ and } R_{G_2} = G_2(1-\lambda)^e\lambda^g p^z,$$

while λ is the fraction reacted, *t* denotes the time, ρ_0 is the initial density, ρ is the current density, C_{crit} is the critical compression for ignition and *P* is pressure. *I, b, x,* G_1, G_2, *c, d, e* and *g* are constants. R_I is the ignition term which represents the creation of hot spots; while R_{G1} and R_{G2} are the growth and completion terms, respectively. The limit for R_I is from $0 < \lambda < \lambda_{\text{igmax}}$, for the R_{G1} is from $\lambda_{\text{G1min}} < \lambda < \lambda_{\text{G1max}}$ while R_{G2} is calculated from $\lambda_{\text{G2min}} < \lambda < 1$.

The standard equation (1) has no mechanism to deal with desensitization and thus results rapid reaction by second shock. Several attempts have been made to modify the reaction rate equation (1) to simulate the desensitization experiments. For example, to account for desensitization of LX-17 by reflected shocks, Tarver [29] added a condition in the reaction rate equation that if a critical range of pressures was applied to the explosive, the reaction rate was forced to zero. Whitworth [30] modified the reaction rate equation by making it dependent on the preshock pressure P_{shk}, in such a way that the pressure=*min* (P, P_{shk}) was used to find the reaction rate, where P is local pressure. A qualitative agreement was reported for the calculations of Mulford's precursor shock experiment [31]. However, contrary to the experimental results for detonation quenching experiments [9], the model gives increased rate of quenching with decreasing precursor shock pressure. The modification proposed by Oliveira [25] introduces the desensitization rate depending upon the degree of desensitization(φ), which varies from 0 to 1. The critical compression α in the ignition term R_I is thus modified accordingly. The switching limit of the growth term from 0 to $\lambda_{G1max}(\varphi)$ was also changed, defined by $\lambda_{G1max}(\varphi)= \lambda_c\varphi$, where $\lambda_c > 0$. This modification has some drawbacks mainly due to empirical nature of the parameters and the form of the shape function.

We have proposed a desensitization model which could successfully reproduce the preshock desensitization experiments for an HMX based explosive EDC-37 [32, 33], and could qualitatively reproduce the detonation quenching and reflected shock experiments for LX-17 explosive. This model has been used to simulate the desensitization effects during shaped charge jet penetration in covered explosives [34]. Here we briefly describe the model, the details can be found elsewhere. Our model makes use of the assumption [35] that the preshock shock desensitization takes place for the preshock pressure P in certain range ($P_1 < P < P_2$) specific to the explosive. Based on the experiments, it is also believed that the shock desensitization is a time dependent process. Our model utilizes $P^n\tau$ desensitization criterion suggested by Campbell [9] to measure the degree of desensitization, where n=2.2 for PBX 9404. Thus the degree of desensitization φ is defined as

$$\varphi = \frac{1}{P^n \tau} \int_{t_1}^{t_2} P^n(t)\, dt \qquad (2)$$

where the time t_1 corresponds to the pressure P_1 below which it is assumed no desensitization occurring to the explosive and after which the desensitization begins and the time t_2 corresponds to the pressure P_2 where the desensitization is maximum, i.e. at $P=P_1$, $\varphi = 0$ and at $P=P_2$, $\varphi=1$. The critical compression C_{crit} has to be increased to account for the loss of the potential hot spot generation sites in the desensitized explosive, and is defined as

$$C_{crit}(\phi) = C_{crit} + \phi\, C_I\,, \qquad (4)$$

The growth term coefficients are adjusted as follows

$$G_1(\phi) = (1-\phi)G_1\,,\ \text{and}\ G_2(\phi) = (1-\phi)G_2 \qquad (5)$$

where C_I is an empirical constant adjusted for an explosive by comparing the calculation results with the experimental results, to inhibit the ignition term in reaction rate equation (hence inhibiting the detonation in the preshocked explosive).

2. Desensitization criteria for LX-17 and simulation of desensitization experiments

As discussed earlier, Campbell [9] fitted a curve for the pressures and time durations to completely desensitize the PBX9404 explosive samples and found that the points lying on $P^n\tau$ curve correspond to complete desensitization of the sample. It can be noted that n=2.2 for desensitization while n=2 for shock initiation criterion. The TATB based explosive have been observed to significantly deviate [6] from the $P^2\tau$ shock initiation criteria which many other high explosives follow [5]. Therefore it is quite possible that the desensitization criterion would be different for TATB based explosives. A series of experiments such as those of Campbell [9] is indeed required to find n in $P^n\tau$ desensitization criterion. Recently, Vandersall et al. [12] performed a first set of experiments using a gas gun and manganin pressure gauges to study the desensitization effects and calibrated the empirical constants in the Oliveira's desensitization model. In order to quantitatively measure the interaction of weak shock with the LX-17 detonation waves, Vandersall et al. [12] used a 101 mm gas gun to launch Teflon flyers at low (277 to 491 m/s) velocities to impact LX-17 targets containing embedded pressure gauges to record the pressure histories. The target explosive LX-17 was in the form of thin disks with pressure gauge packages inserted in between, with the total explosive thickness approximately 28 mm. The target explosive was detonated at the opposite ends at correct times using the detonators and LX-10 boosters. The LX-10 detonation waves promptly initiated LX-17 detonation waves that then travelled through the charges and interacted with the weak nonreactive shock waves caused by flyer impact on the opposite end. Both of the waves interact within the LX-17 targets. The simulation setup for the experiment is shown in Figure 1. The pressure history at different gauge points shows that the impact of Teflon flyer with velocity of 277m/s (shot # 4829) could not desensitize the target, while the impact of flyer with 357m/s (shot # 4817) was able to desensitize the target. These flyer velocities correspond to impact pressures of 0.52 and 0.68 GPa for the time durations of 3.0 and 3.3 μs respectively. There was a third shot also, with flyer velocity of 491 m/s impacting on one side of the target 10 μs earlier than the detonators were fired on the opposite end, however, the gauge record is not reliable probably due to stretching of pressure gauges by the preshock pressure.

By using the user equation of state option [36], we implemented the modified the Lee-Tarver ignition and growth model in LS-DYNA. If the input pressure (P) lies within the specific range for desensitization ($P_1 < P < P_2$), the degree of desensitization calculated and stored as a history variable for every computational element and is updated after every computational time step. Owing to the cylindrical symmetry of the problem, two-dimensional axi-symmetric Lagrange grids having 10 elements per mm in target explosive as well as in Teflon were used. The mesh density is enough [10, 37, 38] for Lee-Tarver model to sufficiently resolve the reaction zone. The thickness of the gauges was not mentioned in Vandersall et al., therefore we assume that it would be the same as used in recent experiments carried out in the same laboratory by the authors [39]. The gauges were placed at 0, 8, 14, 20, 25 and 28 mm from the impact surface of the target, and were modelled by Teflon with the Grüneisen equation of state [40], while the target LX-17 was modelled by Lee-Tarver equation of state [10, 37, 38]. The corresponding parameters are given in Table 1 and Table 2 respectively.

Table 1. Material parameters and Grüneisen equation of state for the Teflon flyer

Density	Shear modulus	Yield Stress	Sound Speed	Grüneisen equation of state parameters				
ρ_0 (g/cm^3)	G (Mbar)	σ (Mbar)	C(cm/μs)	S_1	S_2	S_3	γ_0	α
2.15	2.33	0.0050	0.134	1.93	0.0	0.0	0.9	0.0

Initial density ρ_0 =1.905 g/cm3

Table 2. The parameters for modified Lee-Tarver model for LX-17.

Reactants JWL EOS	A = 778.1 Mbar	B = -0.05031 Mbar	R_1 = 11.3	R_2 = 1.13	ω = 0.8938	C_v = 2.487x10^{-5} Mbar/K	T_0 = 298°K
Product JWL EOS	A = 14.8105 Mbar	B = 0.6379 Mbar	R_1 = 6.2	R_2 = 2.2	ω = 0.5	C_v = 1.0x10^{-5} Mbar/K	E_0 = 0.064 Mbar
Reaction Rates	I = 4.0x10^6 μs^{-1} λ_{igmax} = 0.02	a = 0.22 b = 0.667 x = 7.0	G_1 = 4500 Mbar^{-3}μs^{-1} λ_{G1max}= 0.8	c = 0.667 d = 0.866 y = 3.0	G_2 = 30 Mbar^{-1}μs^{-1} λ_{G2min}= 0.8	e = 0.667 g = 0.667 z = 1.0	
Desensitization Parameters	n = 1	$P^n\tau$ = 96 Kbar μs	C_1 = 0.1				

In the present simulation, as a first step, we were able to get the reported impact pressures and the time durations using a 3.5 mm Teflon flyer impact LX-17 target covered by 1.5 mm thick Teflon buffer plate. Later on, simulations were performed using this arrangement for application of the desensitization model for LX-17.

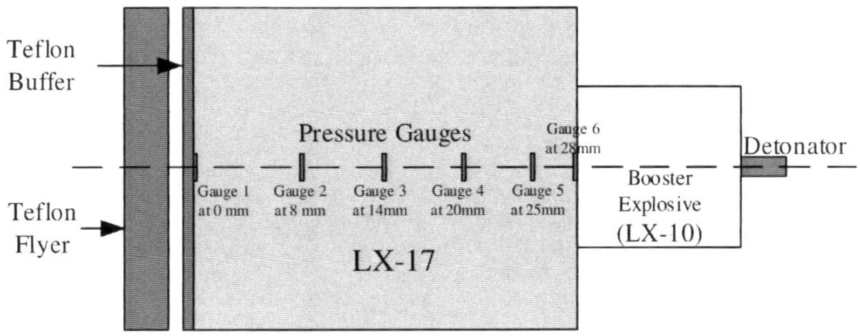

Figure 1. The simulation setup for the shock desensitization experiments

The requirement for our model is the desensitization criteria $P^n\tau$ for LX-17, which is not available at present. However we tried different values of n and $P^n\tau$ to have as good match to the experimental pressure history records as possible. We find that n=1 and $P^n\tau$=9.6 gives the best match, and these values are used in the present calculations. It implies that for LX-17, the desensitization rate linearly varies with pressure. This may be the reason why Vandersall et al. was able to reproduce the experimental data with Oliveira's model. However such results can also be obtained with other pairs of n and $P^n\tau$, which reflects the flexibility of the model as well as emphasises the need to perform further experiments to exactly determine the desensitization criterion.

The thickness of the LX-10 booster was adjusted so as to make the detonation wave arrive at gauge point at 28 mm depth at the time as given in the experiment (plot 3 in Vandersall et al.). In case of shot # 4817, the weak shock travels inside the explosive up to the depth of 22.6 mm at time 8.5 μs when it interacts with the detonation wave arriving from opposite direction. It can be observed from the experimental results as well as in calculation results that the pressure at the gauge points the material with increasing degree of desensitization. Within one microsecond it comes across the fully desensitized material with maximum degree of desensitization (φ=1), and the pressure starts decreasing at the next gauge points (at 14, 8 and 0 mm depths).

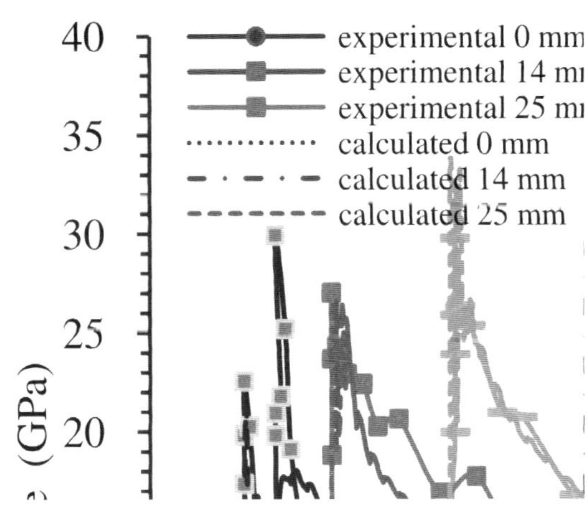

Figure 2. The experimental (shot # 4817) and calculated pressure histories at various gauge points inside the target using Lee-Tarver model

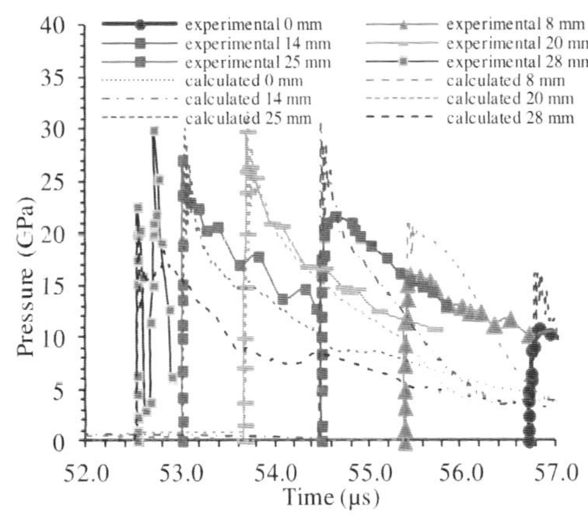

Figure 3. The experimental and calculated pressure histories at various gauge points inside the target using the modified Lee-Tarver model

In case of original Lee-Tarver model without any desensitization treatment, the calculation results do not reproduce the experimental results, as shown in Figure 2. However using the modified model, the match between the calculated and experimental pressure histories is excellent, as shown in Figure 4. The dotted lines in Figures 2-4 represent the calculated histories while the solid lines represent the experimental histories. In case of shot # 4829, no desensitization was observed, and the comparison between the experimental and calculated pressure histories is shown in Figure 4.

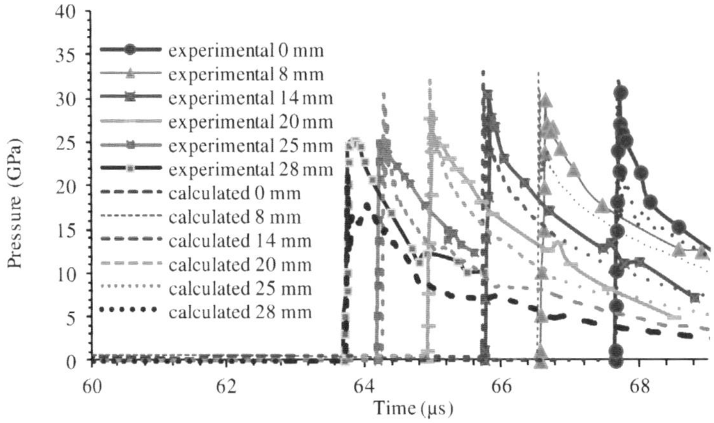

Figure 4. The experimental (shot # 4829) and calculated pressure histories at various gauge points inside the target using modified Lee-Tarver model. No desensitization was observed.

3. Conclusions

The present modification of Lee-Tarver model can be used to account for the desensitization effects. The extension does not change the functionality of the original Lee-Tarver model, unless the explosive is subjected to a weak preschock, which desensitizes the explosive. The present study demonstrates that the use of the desensitization criterion ($n=1$, $P^n\tau=9.6$) for LX-17 yields an excellent match with the experimental results. However, such pair is just estimation; the desensitization criterion (n, $P^n\tau$) is needed to be exactly determined for TATB based explosives by further desensitization experiments such as in references [9, 12] for simulation of multiple compression effects. The desensitization model used here is flexible enough to make use of other (n, $P^n\tau$) pairs by modifying the empirical parameters accordingly.

4. Acknowledgements

This paper is supported by National Natural Science Foundation of China (Grant No. 11272059, 11221202 and 11390362) and Program for New Century Excellent Talents in University Grant No. NCET-12-0037).

5. References

[1] Campbell A W and Engelke R. *Proc., int. Symp. Detonation, 6th.*, Coronado, CA, 1976, p 642 52
[2] Campbell A W, *Propellants, Explos., Pyrotech.*, 1984, 9(6), 183-7
[3] Mader C L, Zumbro J D, Ferm E N, Gittings M and Weaver R. *Proc., int. Symp. Detonation, 12th.*, San Diego, CA, 2002, p 55-60
[4] Urtiew P A and Tarver C M, *Combust., Explos. Shock Waves*, 2005, 41(6), 766-76
[5] Longueville Y D, Fauquignon C and Moulard H. *Proc., int. Symp. Detonation, 6th.*, 1976, p 105-14
[6] Honodel C, Humphrey J, Weingart R C, Lee R S and Kramer P E. *Proc., int. Symp. Detonation, 7th.*, Annapolis, MD, 1981, p 425-34
[7] Souers P C and Vitello P, *Propellants, Explos., Pyrotech.*, 2007, 32(4), 288-95
[8] Haskins P J and Cook M D. *Proc., int. Symp. Detonation, 14th.*, Coeur d'Alene, Idaho, 2010, p 1421-9

[9] Campbell A W and Travis J R. *Proc., int. Symp. Detonation, 8th.*, Albuquerque, New Mexico, 1985, p 1057-68
[10] Tarver C M, *J. Physical Chem. A*, 2010, 114(8), 2727-36
[11] Tarver C M, *J. Energ. Mater.*, 2012, 30(3), 220-51
[12] Vandersall K S, Garcia F, Tarver C M and Fried L E. *Proc., int. Symp. Detonation, 15th.*, San Francisco, CA, 2014, p
[13] Souers P C, Andreski H G, Cook C F, Garza R, Pastrone R, Phillips D, Roeske F, Vitello P and Molitoris J D, *Propellants, Explos., Pyrotech.*, 2004, 29(6), 359-67
[14] Johnson J N, Tang P K and Forest C A, *J. Appl. Phys.*, 1985, 57(9), 4323-34
[15] Wescott B L, Stewart D S and Davis W C, *J. Appl. Phys.*, 2005, 98(5), 053514
[16] Desbiens N, Matignon C and Sorin R, *J. Phys.: Conf. Ser.*, 2014, 500(15), 152004
[17] Handeley. *Proc., int. Symp. Detonation, 13th.*, Norfolk, VG, 2006, p 864-73
[18] Whitworth N J, *J. Phys.: Conf. Ser.*, 2014, 500(5), 052050
[19] Tarver C M, *J. Physical Chem. A*, 1997, 101(27), 4845-51
[20] Tarver C M and Chidester S K, *J. Press. Vessel Technology*, 2005, 127(1), 39-48
[21] Gustavsen R L, Sheffield S A, Alcon R R, Forbes J W, Tarver C M and Garcia F, *AIP Conf. Proc.*, 2002, 620(1), 1019-22
[22] Tarver C M and McGuire E M. 2002, p
[23] Lee E L and Tarver C M, *Phys. Fluids (1958-1988)*, 1980, 23(12), 2362-72
[24] Tarver C M, Hallquist J O and Erickson L M. *Proc., int. Symp. Detonation, 8th.*, Albuquerque, NM, 1985, p 951–61
[25] DeOliveira G, Kapila A, Schwendeman D, Bdzil J, Henshaw W and Tarver C. *Proc., int. Symp. Detonation, 13th.*, Norfolk, Virginia, 2006., p 13-23
[26] Starkenberg J. *Proc., int. Symp. Detonation, 12th.*, 2002, p 1001-9
[27] Ceder R, On the relation between the ignition and growth model and shock initiation criteria, in Proc., int. Symp. Ballistics, 28th 2014: Atlanta, GA. p. 150-60
[28] Tarver C M and Hallquist J O. *Proc., int. Symp. Detonation, 7th.*, Annapolis, MD, 1981, p 488-97
[29] Tarver C M, Cook T M, Urtiew P A and Tao W C. *Proc., int. Symp. Detonation, 10th.*, 1993, p 696-703
[30] Whitworth N J and Maw J R, *AIP Conf. Proc.*, 1996, 370(1), 425-8
[31] Mulford R N, Sheffield S A and Alcon R R, *AIP Conf. Proc.*, 1994, 309(1), 1405-8
[32] Gustavsen R L, Sheffield S A, Alcon R R, Winter R E, Taylor P and Salisbury D A, *AIP Conf. Proc.*, 2002, 620(1), 999 1002
[33] Winter R E, Sorber S S, Salisbury D A, Taylor P, Gustavsen R, Sheffield S and Alcon R, *Shock Waves*, 2006, 15(2), 89
[34] Hussain T, Tanver A, liu y and Huang F, *J. Energ. Mater.*, 2015
[35] Wescott B L, Stewart D S and Davis W C. *Proc., int. Symp. Detonation, 13th.*, Norfolk, Virginia, 2006, p 744-50
[36] *LS-DYNA®Keyword User's Manual*, Livermore Software Technology Corporation, Livermore 2014, pp. Appendix-B
[37] Urtiew P A, Vandersall K S, Tarver C M, Garcia F and Forbes J W, *Russ. J. Physical Chem. B.*, 2008, 2(2), 162-71
[38] Tarver C M and Chidester S K, *AIP Conf. Proc.*, 2007, 955(1), 429-32
[39] Frank G, Kevin S V and Craig M T, *J. Phys.: Conf. Ser.*, 2014, 500(5), 052048
[40] McQueen R G, Selected hugoniots, 1969

IOP Conf. Series: Materials Science and Engineering **146** (2016) 012020 doi:10.1088/1757-899X/146/1/012020

Application of advanced polymeric materials for controlled release pesticides

M. Rahim, M. R. Hakim and H. M. Haris
School of Chemical Sciences, Universiti Sains Malaysia,
11800 Pulau Pinang, Malaysia

E-mail: kpk566@gmail.com

Abstract. The objective of this work was to study the capability of advanced polymeric material constituted by chitosan and natural rubber matrices for controlled release of pesticides (1-hydroxynaphthalene and 2-hydroxynaphthalene) in aqueous solution. The released amount of pesticides was measured spectrophotometrically from the absorbance spectra applying a standardized curve. The release of the pesticides was studied into refreshing and non-refreshing neutral aqueous media. Interestingly, formulation successfully indicated a consistent, controlled and prolonged release of pesticides over a period of 35 days.

1. Introduction

Pesticides are increasingly used over the past 60 years for the yield improvement of agricultural production [1].However, each year 30,00,000 metric tons of pesticides have been applied, Worldwide [2] while 90 % of the applied conventional pesticides are lost in the environment leading to high cost and severe environmental pollution particularly surface and groundwater contamination [3]. According to a survey, three million pesticide poisoning cases occur every year with 220000 deaths, Worldwide [4]. Controlled release formulations (CRFs) can significantly protect the food crops damage and number of applications [5]. The superior advantages of CRFs are safe handling and less human toxicity [6].

Recently, the researchers have been focused to formulate a CRF system in order to produce the pesticide active ingredients over a long period. The application of CRF can minimize the spreading of pesticides residues into natural environment [7]. Therefore, the use of biodegradable and eco-friendly materials is of great interest for the development of CRF. In addition, several materials have been used as controlled-release carriers. However, natural, bio degradable, nontoxic and environmental friendly materials are strongly recommended [8]. Therefore, in the present study, concentration was focused on advanced materials comprising of chitosan encapsulated in natural rubber matrices.

Content from this work may be used under the terms of the Creative Commons Attribution 3.0 licence. Any further distribution of this work must maintain attribution to the author(s) and the title of the work, journal citation and DOI.
Published under licence by IOP Publishing Ltd

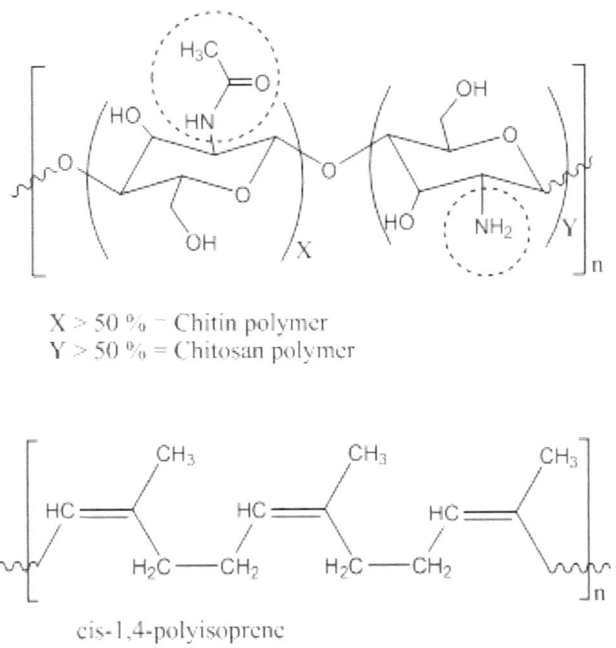

Figure 1. Chemical structure of chitin, chitosan and natural rubber

2. Preparation of advanced polymeric materials

Chitosan was dissolved in 2 % acetic acid containing selected pesticides such as 1-hydroxynaphthalene or 2-hydroxynaphthalene. The mixture was stirred for 1 hour using mechanical stirrer to prepare a gel. Natural rubber latex with known amount of curing ingredients as specified in Table 1 were mixed. The advanced polymeric material (APM) was prepared through encapsulation of the chitosan gel by natural rubber matrices. The resulting capsules were isolated and dried in oven at 60 °C for 48 hours.

Table 1. Chemical composition of curing ingredients

S.No.	Curing Ingredients	Amount (phr[a])
1	Natural rubber	100
2	Chitosan	28
3	Sulphur	1.5
4	Zinc-oxide active	1.8
5	ZDEC (Zinc diethyldithiocarbamate)	1.0
6	Antioxidant[2,2′-methylene-bis(6-tert-butyl-4-methylphenol)]	1.0

[a]phr: Parts per hundred of rubber

2.3. Characterization

The APMs were characterized by Fourier Transform Infrared Spectroscopy (FT-IR) and Thermogravimetric Analysis (TGA).

2.4. Release study

In order to study APMs for slow release of pesticides, the materials were studied under two different ways "without refresh" and "with refresh". First (without refresh), known amount of the APMs was immersed in 50 mL of distilled water and the bottles were closed with screw caps and protected from light in order to prevent photo degradation of active ingredients. After the selected time intervals, the pesticide concentration was monitored by UV-VIS spectrophotometer. The experiment was continued upto 35 days.

Second (with refresh), the APMs were also studied with refreshing the release medium after each 24 hours. The known amount of APMs was submerged in 50 mL of distilled water, after each 24 hours the concentration of released compounds was monitored and the medium was refreshed by 50 mL fresh distilled water. The "with refresh" study was also continued for 35days.

3. Results and discussion

APMs were prepared, characterized and successfully used for the controlled release of pesticides. The TGA and derivative thermo gravimetric (DTG) thermograms of partially cross linked natural rubber (NR), chitosan (CTS) and APM have been shown in Figure 3 and 4, respectively. Chitosan indicated two-step decomposition process, first step decomposition of 6.2 % indicates the liberation of volatile compounds and bound water molecules. The second stage rapid decomposition (weight loss of 58.1 %) shows thermal cleavage and chain scission of C—C and C—O bonds indicating main degradation temperature of 309.26 °C as shown in DTG thermogram [9-10].

Natural rubber matrices indicate single step degradation with rapid weight loss of 97.3 %, and the main decomposition temperature was observed 383.68 °C, as shown in the DTG curve. APM shows two step thermal decomposition, the first step decomposition of 8.4 % and the second decomposition of 83.4 % was due to bound water and thermal cleavage of cis-1,4-polyisoprenes, respectively.

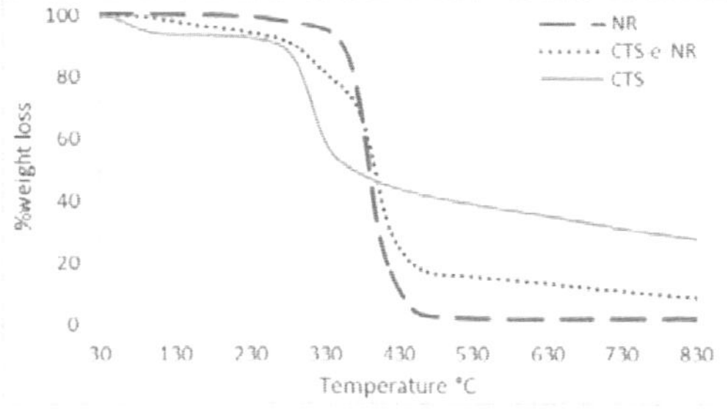

Figure 3. TG thermograms of CTS, NR and CTS-e-NR

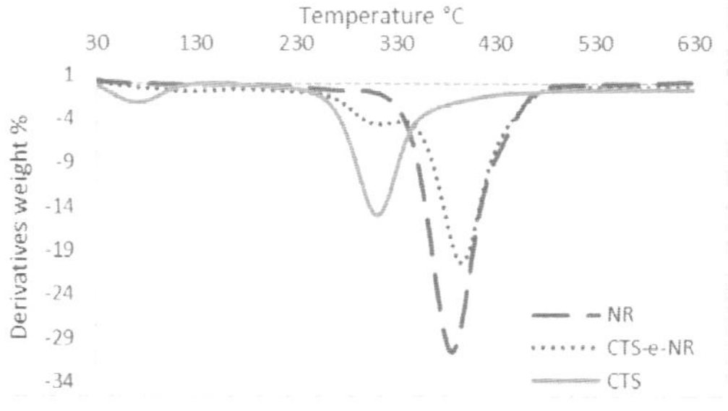

Figure 4. DTG curves of CTS, NR and CTS-e-NR

The FTIR spectra of chitosan, partially cross-lined natural rubber (CNR) and APMs have been shown in Figure 5. The FTIR spectrum of the APMs indicate that there is no alteration of the existing peaks and production of no new peaks. The spectra indicate that the interaction between chitosan and natural rubber matrices are totally physical.

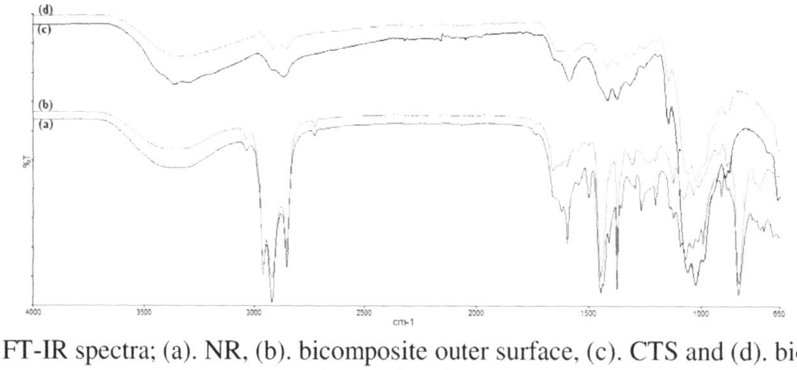

Figure 5. FT-IR spectra; (a). NR, (b). bicomposite outer surface, (c). CTS and (d). biocomposite internal surface

3.1. Pesticide release study

The objective of this work was to study the capability of APMs constituted by chitosan and natural rubber matrices for controlled release of pesticides in aqueous solution. Therefore, the APMs were studied for the release behaviour of 1- and 2-hydroxynaphthalene. The with-out-refresh results indicate that pesticides release gradually, and effective concentration of the pesticides was observed upto 20 days, as shown in Figure 6. With-refresh, the effective amount was released upto 16 days. In a conclusion, the release profile of the pesticides from APMs was found satisfactory: a lower pesticides release was observed at the early stage of the release experiments that lasted for 35 days, as presented in Figure 7.

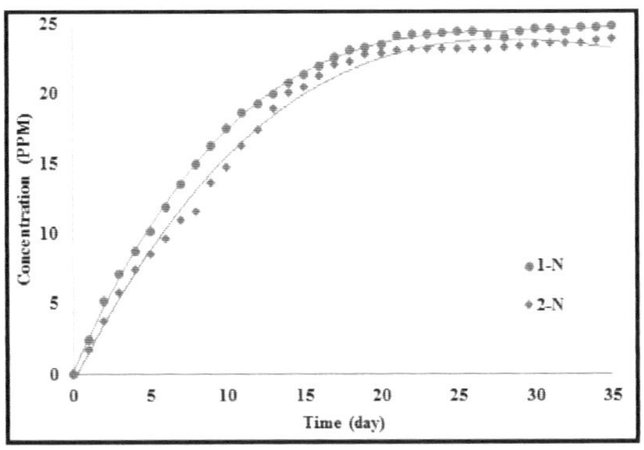

Figure 6. The release behaviour of naphthol compounds (without refresh)

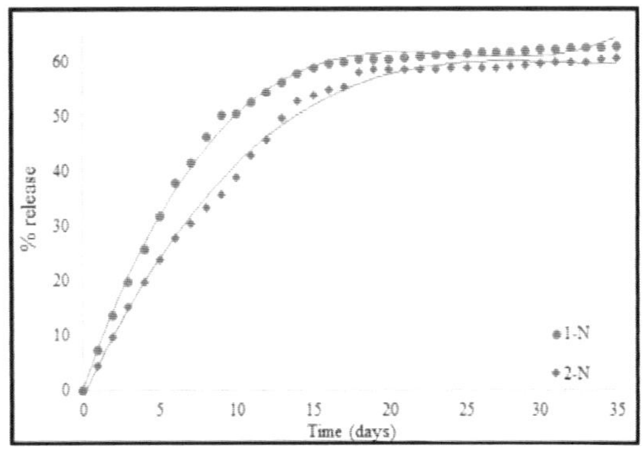

Figure 7. Accumulative % release of naphthol compounds (with refresh)

4. Conclusion

A polymeric advanced material consist of chitosan encapsulated in natural rubber matrix was prepared and characterized by FT-IR and TGA. The APMs were studied for the controlled release of pesticides such as 1-hydroxynaphthalene and 2-hydroxynaphthalene. The release of pesticides were found controlled, consistent and prolonged over a prolonged period of 35 days. Therefore, the formulation was strongly recommended for the controlled-release of pesticides on industrial scale.

5. Acknowledgement

The authors would like to appreciate Universiti Sains Malaysia (grant no. 1001/PKIMIA/814124) and TWAS (The World Academy of Sciences) and Universiti Sains Malaysia for providing TWAS-USM Fellowship.

6. References

[1] Damalas, C.A. and I.G. Eleftherohorinos, *Pesticide Exposure, Safety Issues, and Risk Assessment Indicators.* International Journal of Environmental Research and Public Health, 2011. 8(5): p. 1402-1419

[2] Pimentel, D., *Environmental and economic costs of the application of pesticides primarily in the United States.* Environment, development and sustainability, 2005. 7(2): p. 229-252

[3] Yi, Y., et al., *Gelation of photocrosslinkable carboxymethyl chitosan and its application in controlled release of pesticide.* Carbohydrate Polymers, 2011. 86(2): p. 1007-1013

[4] Eddleston, M., et al., *Pesticide poisoning in the developing world—a minimum pesticides list.* The Lancet, 2002. 360(9340): p. 1163-1167

[5] Muro-Suné, N., et al., *Predictive property models for use in design of controlled release of pesticides.* Fluid phase equilibria, 2005. 228: p. 127-133

[6] Ye, Z., et al., *Photo-responsive shell cross-linked micelles based on carboxymethyl chitosan and their application in controlled release of pesticide.* Carbohydrate Polymers, 2015. 132: p. 520-528

[7] Li, J., et al., *Controlled release and retarded leaching of pesticides by encapsulating in carboxymethyl chitosan /bentonite composite gel.* Journal of Environmental Science and Health, Part B, 2012. 47(8): p. 795-803

[8] Bruna, F., et al., *Organohydrotalcites as novel supports for the slow release of the herbicide terbuthylazine.* Applied Clay Science, 2008. 42(1): p. 194-200

[9] Alhwaige, A.A., et al., *Biobased chitosan/polybenzoxazine cross-linked films: preparation in aqueous media and synergistic improvements in thermal and mechanical properties.* Biomacromolecules, 2013. 14(6): p. 1806-1815

[10] Peniche-Covas, C., W. Argüelles-Monal, and J. San Román, *A kinetic study of the thermal degradation of chitosan and a mercaptan derivative of chitosan.* Polymer Degradation and Stability, 1993. 39(1): p. 21-28

IOP Conf. Series: Materials Science and Engineering **146** (2016) 012021 doi:10.1088/1757-899X/146/1/012021

Production of carbon nano-tubes via CCVD method and their corrosion protection performance in epoxy based coatings

M. A. Raza, F. A. Ghauri, M. S. Awan[1], A. Farooq and R. Ahmad
Department of Metallurgy and Materials Engineering, CEET, University of the Punjab, Lahore, Pakistan
[1]Ibn-e-Sina Institute of Technology H-11/4, Islamabad, Pakistan

E-mail: mohsin.ceet@pu.edu.pk

Abstract. Good yield of carbon products was obtained by catalytic chemical vapor deposition (CCVD) technique using 100-500mg of ferrocene catalyst at temperature of 900 °C and acetylene flow rate of 150-200cc/min. The effects of amount of ferrocene, temperature and hydrocarbons precursors on the yield of carbon nanomaterial's was calculated and characterized by X-ray diffraction (XRD), scanning electron microscope (SEM) andenergy-dispersive X-ray spectroscopy (EDS). Good yield of carbon nanomaterials primarily consisted of carbon nanotubes (CNTs) and carbon nanoparticles was obtained. CNTs obtained after purification were dispersed in epoxy resin to produce composite coatings which were coated on stainless steel 316L. The coated stainless steel samples' corrosion behavior was studied using open circuit potential (OCP), cyclic polarization and electrochemical impedance spectroscopy (EIS) techniques. Results showed that epoxy coating containing 4 wt. % of CNTs offered improved corrosion resistance to stainless steel.

1. Introduction

Carbon nanomaterials such as carbon nanotubes (CNTs)are employed in numerous applications due to their remarkable electronic, mechanical and chemical properties [1]. Production of CNTs with high yield and good quality at commercial scale needs extensive study. Catalytic chemical vapor deposition (CCVD) a commercially viable process is widely used for the production of CNTs [2-3]. Many parameters are considered during CCVD process including temperature, amount of catalyst, carbon source, etc. which require optimization [4-5].

Corrosion is a process which causes destruction of metals by their electrochemical interaction with environment. This needs to be reduced to increase the life time of metals which will save capital. Several methods are employed to reduce corrosion of metals but organic coatings in the form of paints or resins are widely used due to their ease of application [6]. To protect metals in marine conditions epoxy-based organic coatings are applied due to their high adhesion and superior corrosion resistance[7]. However, due to their hydrophilic nature defects in coatings formed which eventually result in corrosion of metal underneath [8-9].Therefore, epoxy coatings are needed to be modified to enhance their performance against corrosion. One of the easiest approaches is based on incorporating nanofillers into epoxy resin such as silica nanoparticles, carbon nanotubes, carbon black and carbon nanofibers [10-12]. Carbon nanotubes (CNTs) have hydrophobic nature and their addition in epoxy coatings have beneficial effect in reducing corrosion rates of metals.

Content from this work may be used under the terms of the Creative Commons Attribution 3.0 licence. Any further distribution of this work must maintain attribution to the author(s) and the title of the work, journal citation and DOI.
Published under licence by IOP Publishing Ltd

Few studies reported CNTs can increase corrosion resistance of steel, stainless steel or other metals appreciably[13]. Researchers reported that addition of carbon black nano particles or CNTs in the epoxy resin and studied corrosion resistance of such coatings[10]. The present study reports corrosion protection behavior of epoxy coatings which were prepared by mixing carbon nanomaterials (mainly CNTs) in epoxy resin. Carbon nanomaterials were produced by CCVD techniques and optimization of process parameters was carried out through manipulation of different parameters for enhancing the percentage yield of CNTs. The epoxy coatings containing CNTs were coated on 316L stainless steel and their effect on corrosion resistance of316L stainless steel was studied in simulated sea water.

2. Experimental

Ferrocene (a product of Merck), commercial grade xylene and acetylene were used for CCVD process. First of all, catalyst, ferrocene $(C_6H_5)_2Fe$, was weighed precisely and poured into a stainless steel chamber encapsulated with a heater which was sealed with high temperature RTV silicone gasket maker sealant. The stainless steel chamber had inlet for purgingacetylene gas and nitrogen gas into the stainless steel chamber to allow hydrocarbon source and carrier gas to be thoroughly mixed with the catalyst already placed in the chamber. The outlet of the chamber was connected to a quartz tube. The gases mixed with a catalyst came out from the outlet and entered into the quartz tube that was placed in a tube furnace and it was maintained at temperature of 800-1000 °C depending on the experimental conditions. One end of quartz tube was connected with the stainless steel chamber and other end was connected with a bubbler. The flow rate of gases was controlled so that a controlled amount of carrier gases and catalyst vapors enter into the reaction chamber (quartz tube). Before running experiments, only nitrogen gas was passed through the quartz tube to remove any air and then furnace was turned on to set temperature. When the desired temperature was attained in the quartz tube, heater was turned on to convert catalysts into vapors. As soon as catalyst vaporized, acetylene gas was introduced which was mixed with catalyst vapors and entered into the reaction zone. After 3-4 minutes of reaction, carbon product was seen in the quartz tube. The reaction continued for 30-60 min depending on experiment. After certain time, furnace was switched off and the product was allowed to cool to room temperature. Product of CCVD reaction was carefully retrieved from quartz tube and weighed to calculate % yield of the reaction. Stainless steel heating chamber was also disassembled to observe the amount of catalyst that was left behind. The schematic of setup is shown in Figure 1.

For purification of CCVD products, 3 g of the product was added in 100 ml concentratedHNO_3. The resulting mixture was stirred for 4 husing magnetic stirrer. After four hours, distilled water was added to make 500 ml solution which was stirred for 5 h. Filtration of solution was carried out and purified product was obtained after drying.

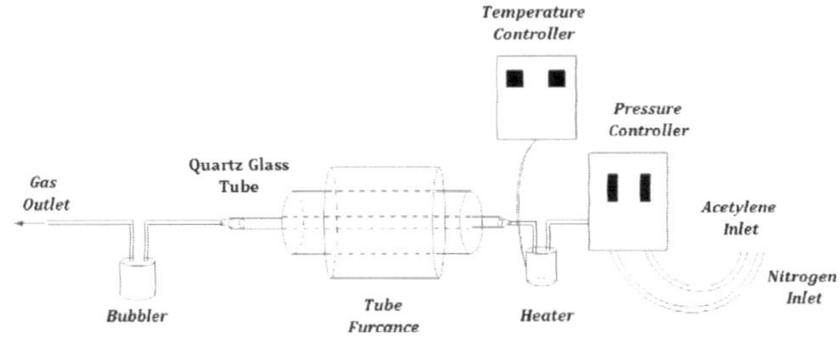

Figure 1. Schematic of CCVD process apparatus used in present study

The carbon products of CCVD reaction were characterized using field emission gun scanning electron microscope (FEG-SEM), energy dispersive x-ray spectroscopy (EDX) and X-ray diffraction (XRD).

IOP Conf. Series: Materials Science and Engineering **146** (2016) 012021 doi:10.1088/1757-899X/146/1/012021

CNTs were well dispersed in epoxy resin to produce CNT/epoxy coatings. These coatings were prepared by first dispersing a known amount of purified CNTs in the acetone by ultra-sonication (Sonicator, 35 Hz frequency) for 5 h. After 5 h, an appropriate amount of epoxy resin was poured in CNT/acetone dispersion. The resulting dispersion was again sonicated for 30 min so that CNTs get adequately disperse into epoxy resin. Later, acetone was evaporated from the dispersion. To the resulting dispersion of CNT and epoxy resin (a product of ICI, Pakistan), amide hardener was poured and was thoroughly mixed with a high speed mechanical stirrer. The ratio of epoxy and amide hardener was 80:20. The resulting dispersion of CNT/epoxy/hardener was coated on polished (2 µm diamond paste used for polishing) 316 L stainless steel samples with an applicator capable of applying coating thickness of 25 µm. Coatings were abbreviated as 316L+E-1% and 316L+E-4% coatings, having 1 wt. % CNTs and 4 wt. % CNTs in epoxy resin, respectively. Figure 2 shows a complete flow diagram of processes adopted in preparation of coatings.

Electrochemical measurement techniques, open circuit potential (OCP), cyclic polarization and electrochemical impedance spectroscopy (EIS) were performed to evaluate corrosion behavior of coatings using PC14/750 (Gamry Instruments, Inc., USA) and resultant spectra were analyzed by Echem analyst software version 5.68. Prepared samples were used as working electrode, graphite was used as auxiliary electrode and saturated calomel electrode (SCE) as reference electrode in freshly prepared 3.5% NaCl electrolyte.

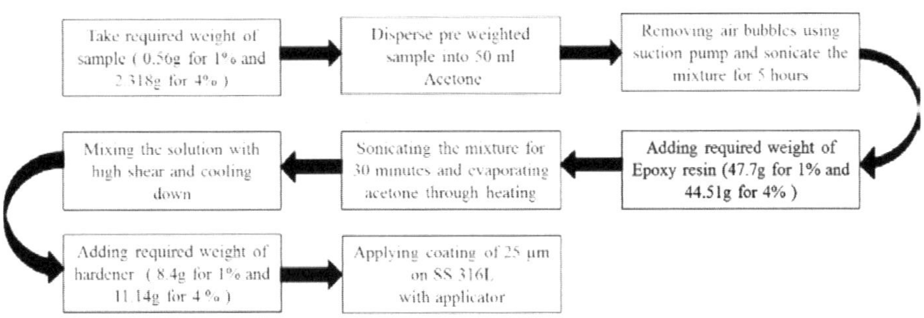

Figure 2. Flow chart of procedure adopted for preparation of coatings

3. Results and discussions

3.1. Effects of amount of catalyst, reaction temperature and carbon precursor type on carbon product

The yield of carbon product obtained after CCVD process was calculated by formula given in (1)[14],

$$\% \, yield = \frac{M_{total} - M_{cat}}{M_{cat}} \times 100 \qquad (1)$$

Where, M_{total} is the weight of carbon product and catalyst and M_{cat} is the weight of catalyst, respectively.

The effect of amount of catalyst on carbon product yield is presented in Figure 3(a). High yield of CCVD carbon nanomaterials was obtained when amount of catalyst was in the range of 100–500 mg and temperature, reaction time and acetylene flow rate were 900°C, 10 ± 5 min and 150–200 cc/min, respectively. Agglomerated metal catalysts may have formed due to large amount (1g) of catalyst. This agglomeration reduced the efficiency of carbon diffusion into the metal. In contrast, small amount of catalyst (100 and 500 mg) delivered higher amount of product.

Figure 3(b) shows the effect of reaction temperature on percentage yield of carbon product. When the amounts of catalyst, time and flow rate of acetylene were kept constant then temperature of the reaction zone also influenced the productivity of carbon product. High yield of carbon product was obtained in the temperature range of 800-900°C. As the temperature of the reaction zone increased further, the yield of carbon product decreased as high temperature decreased reactivity of metal catalyst with carbon and somehow combustion of carbon takes place [4].

Figure 3(c) shows the effect of type of hydrocarbon precursor on percentage yield of carbon product. Acetylene source provided higher yield than the combination of xylene and acetylene. The xylene promoted formation of nano sized carbon black particles in the reaction chamber as shown in Figure 4 (b). Since ferrocene was dissolved in xylene before CCVD process, it might have hindered evaporation of ferrocene or have reacted with ferrocene to inactivate the catalyst. It might be possible that xylene have diluted the amount of catalyst in the reaction chamber resulting in lower yield of carbon product.

Figure 3(d) shows XRD patterns obtained on the carbon products obtained from CCVD process using different parameters. XRD pattern (center) of a product obtained using 100–500mg of ferrocene catalyst at a temperature of 900 °C using acetylene flow rate of 150-200cc/min showed high intensity of (002) peak at 26° 2θ, indicating presence of graphitic plane having interplanar spacing of 0.33 nm. This confirms the formation of CNTs in our carbon product [15]. The peak at 47° is due to presence of iron particles. The XRD patterns obtained when excess of catalyst (Figure 3(d), top) and higher temperature (Figure 3(d), bottom) showed high intensity of iron peaks at 47° and low intensity of (002) graphite peak. It can be concluded from XRD analysis that good yield of CNTs can be obtained if ferrocene amount and temperature of reaction zone are controlled as mentioned above.

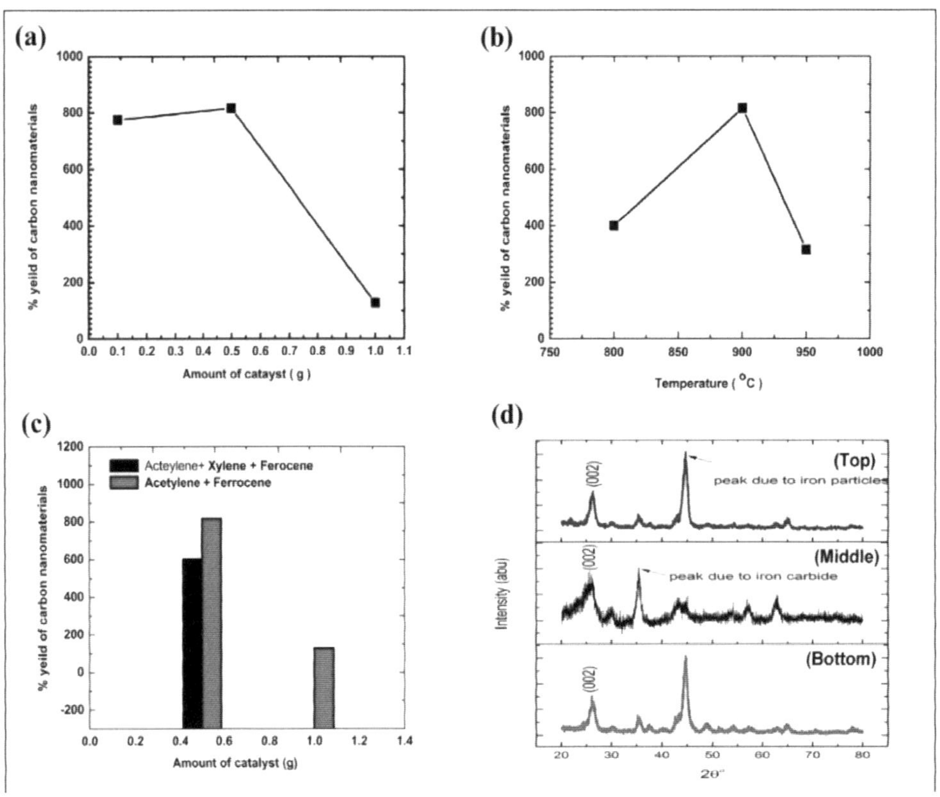

Figure 3. (a) Percent yield of carbon product with respect to catalyst (b) percent yield of carbon product with respect to temperature (c) percent yield of carbon product with respect to source of carbon precursor and (d) XRD patterns obtained at different temperatures respectively

Figure 4 shows scanning electron micrographs (SEM) micrgrphsof CNTs produced using different carbon precusors. Figure 4(a) represents carbon products obtained using acetylene gas and ferocene as catalyst. The product contains high yield of carbon nanotubes with average diameter of ~27 nm. In addition, carbon nanofibres (diameter=140 nm) (Figure 4(b)) and carbon black nanoparticles (diameter = 40 nm) (Figure 4(a)) were also observed in SEM analysis. On the other hand, SEM of the product obtained using ferrocene and combination of xylene and acetylene shows the product has mainly carbon particles having sizes in range of 0.6-1.0 μm (Figure 4(c)). This result clearly shows that accetylene in the presence of ferrocene can produce CNTs with high yield. The EDX analysis clearly confirms carbon and iron peak in CCVD products as shown in Figure 5.

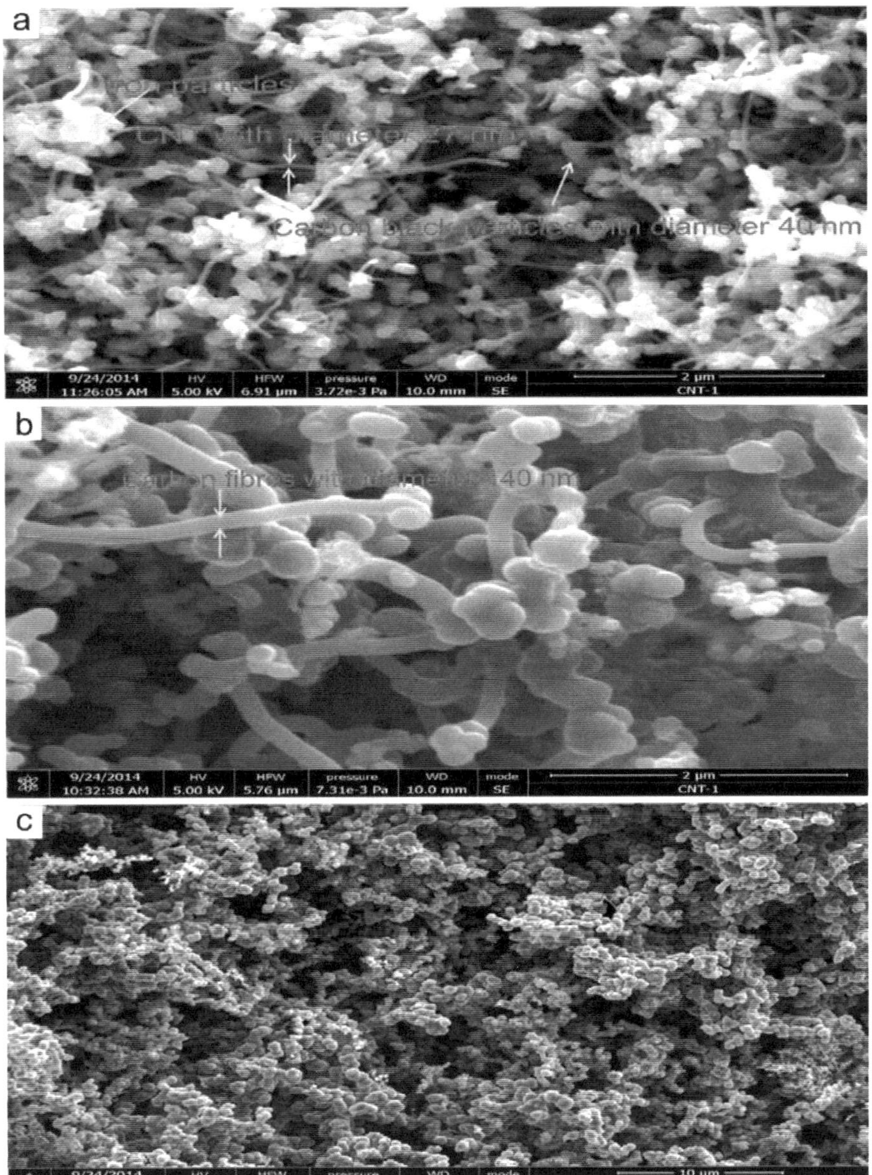

Figure 4. SEM of CCVD carbon product (a & b) ferrocene and acetylene only (c) ferrocene and combination of xylene and acetylene

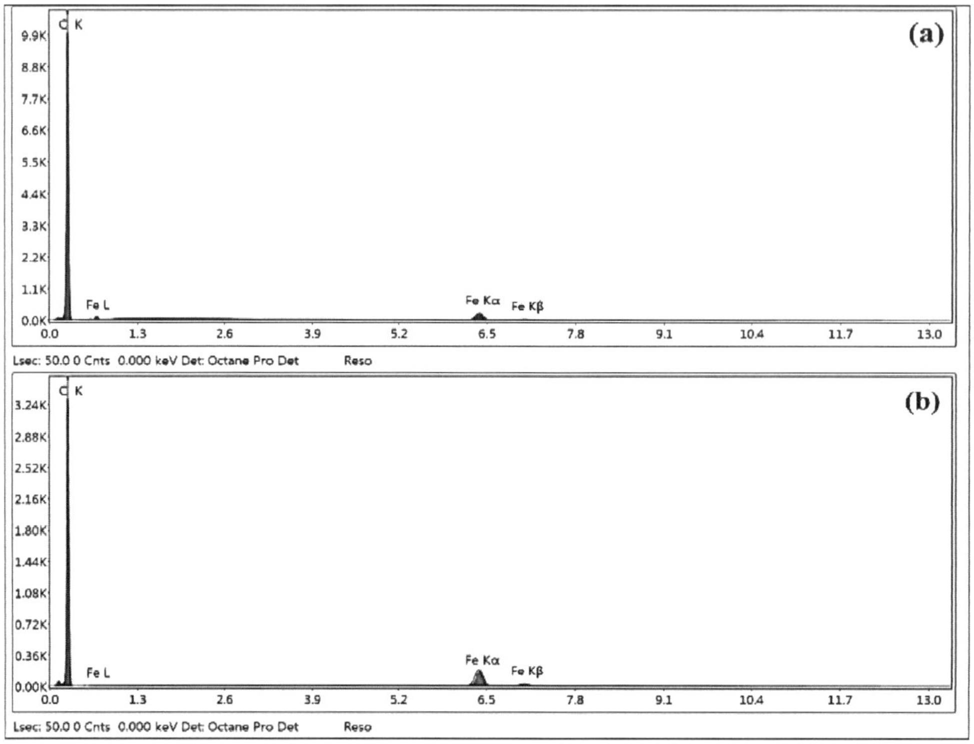

Figure 5. EDX analysis of CCVD product a) ferrocene and acetylene only (b) ferrocene and combination of xylene and acetylene

3.2 Electrochemical studies

Open circuit potential (OCP) of specimens were measured for 500s having stability of 0 mV/s. Cyclic polarization tests was carried out at forward scan rate of 5 mV/s and reverse scan rate of 2.5 mV/s with initial potential of -0.5 V to final potential of +0.3 V with respect to OCP to measure kinetic parameters. Figure 6 shows cyclic polarization curves of bare metal and epoxy containing CNTs in 3.5 % NaCl solution (simulated sea water).Epoxy coatings 316-E+1% and 316-E+4% showed higher breakdown potential (E_b = 475.5 mV) than bare stainless steel (E_b=111.9 mV) as shown in Figure 6 and Table 1. Higher breakdown potential indicates higher resistance to localized corrosion. Thus epoxy coatings containing CNTs showed higher resistance to localized corrosion. The positive hysteresis loops was observed for all coated samples. The area of hysteresis loop indicate amount of pitting on the samples. Although, epoxy coatings resisted initial localized corrosion on stainless steel surface but once the potential exceeded E_b limit the pitting phenomenon becomes more prominent which can be observed from the large area of hysteresis loop of epoxy coated samples as shown in Figure. 6. The increased pitting tendency of the coated samples was due to presence of pores or defects in the coatings.

Table 1 show kinetic parameters calculated from the curves in Figure 6 using Tafel relationship [32].Epoxy containing CNTs showed less corrosion rate as compared to bare metal. Corrosion current density of CNTs coatings 316-E+4% and 316-E+1% is 6.030 ($\mu A/cm^2$)and 67.50 ($\mu A/cm^2$), respectively, showing less corrosion current density as compared to bare metal 144 ($\mu A/cm^2$).

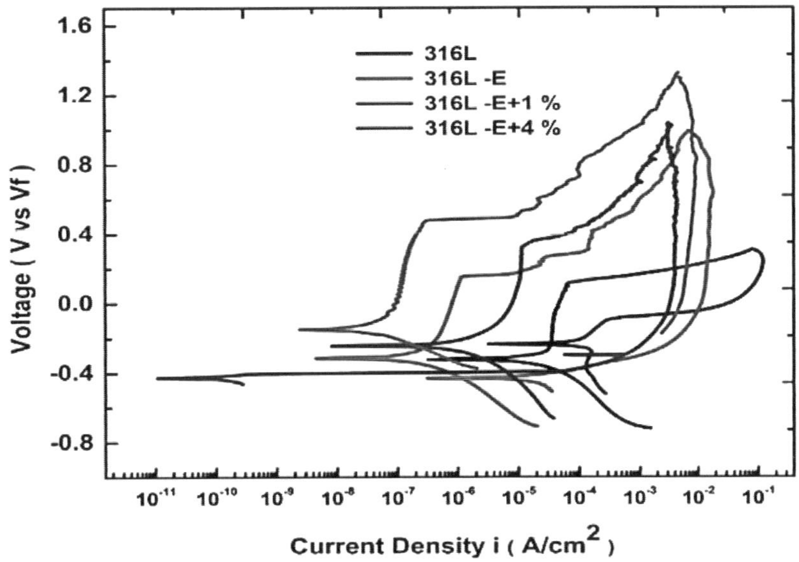

Figure 6. Cyclic polarization curves obtained in 3.5 % NaCl solution

Table 1. Kinetic parameters calculated from Polarization curves

Material	E_{corr} (mV)	Eb (mV)	I_{corr} ($\mu A/cm^2$)	Corrosion rate (mpy)	OCP (mV)
316L	-323	111.9	144.0	62.62	-221
316L-E	-314	150	110.0	48.03	-208
316-E+1%	-243	320	67.50	29.42	-163
316-E+4%	-148	475.5	6.030	2.627	127

In order to further evaluate corrosion behavior, EIS was conducted on coated samples in 3.5 % NaCl solution at AC voltage of ±10 $mV_{(rms)}$ amplitude with change in frequency from 10 mHz to 100 kHz. Figure 7(a) shows the Nyquist curves of coated samples obtained in 3.5 % NaCl solution. Nyquist curves tell about charge transfer mechanism. Higher the resistance to charge transfer, greater will be the area of Nyquist curve. Figure 7(b) showed bode plot indicated that 316L+ E-4% and 316L+E-1% coatings showed higher impedance |Z| than bare stainless steel. It was indicated that at low frequency CNTs-based coatings showed inductive behavior due to initiation of localized attacks caused by Cl^- ions. On the other hand, double layer capacitance behavior showed by bare stainless steel due to accumulation of charges eventually increases corrosion as indicated by Tafel values (Table 1).At higher frequencies 316L+E-1% and 316L+E-4%coatings showed higher |Z| values than bare metal which suggest that coatings were compact. The316L+E-1% coatings showed no significant reduction in corrosion due to more aqueous transport probably due to pores present in the coating and low percentage of CNTs. On the other hand, the coating316L+E-4% containing higher amount of CNT showed much better results due to higher wt. % of CNTs and hydrophobic nature of CNTs[16]which acted as barrier for aqueous transport through the coating.

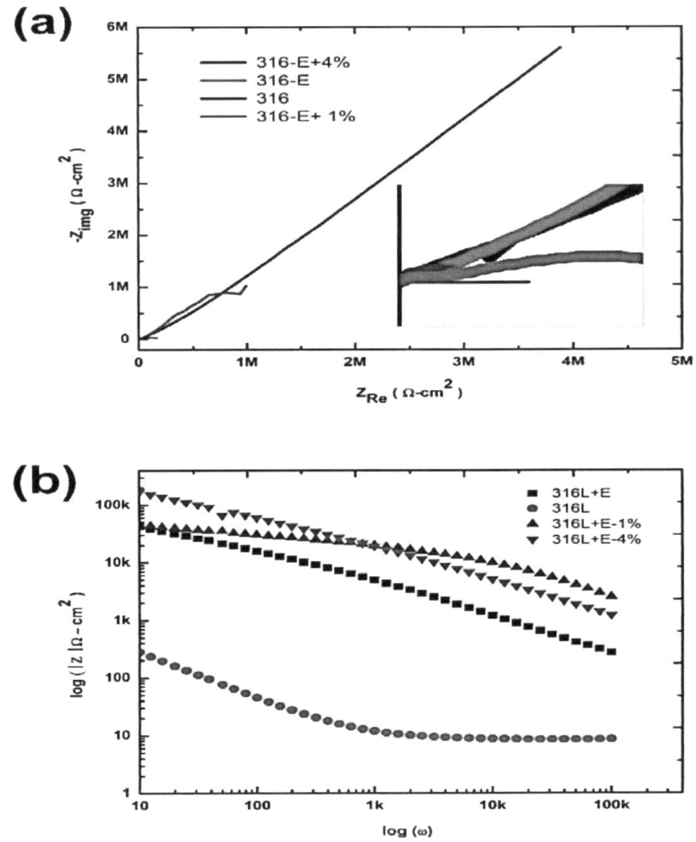

Figure 7. EIS curves obtained in 3.5 % NaCl solution (a) Nyquist curves
(b) Bode plot

4. Conclusions

I. Good yield from CCVD process can be obtained using 100-500mg of ferrocene at temperature of 900 °C using acetylene flow rate of 150-200cc/min.
II. SEM/EDX results showed higher yield of CNTs were obtained by using acetylene as hydrocarbons precursors than xylene plus acetylene.
III. Electrochemical results indicated that CNTs-based epoxy coatings increased corrosion resistance of bare metal to significant level and this increases with increase of wt. % of CNTs in epoxy resin which hindered the diffusion of aqueous solution in the coatings.

5. References

[1] M.S. Dresselhaus, G. Dresselhaus, P.C. Eklund, Science of Fullerenes and Carbon Nanotubes: Their Properties and Applications, Elsevier Science1996
[2] P.R. Birkett, A.J. Cheetham, B.R. Eggen, J.P. Hare, H.W. Kroto, D.R.M. Walton, Chemical Physics Letters, 281 (1997) 111-114
[3] J. Liu, H. Dai, J.H. Hafner, D.T. Colbert, R.E. Smalley, S.J. Tans, C. Dekker, Nature, 385 (1997) 780-781
[4] N.M. Rodriguez, Journal of Materials Research, 8 (1993) 3233-3250
[5] K.-E. Kim, K.-J. Kim, W.S. Jung, S.Y. Bae, J. Park, J. Choi, J. Choo, Chemical Physics Letters, 401 (2005) 459-464

[6] Z.W. Wicks, F.N. Jones, S.P. Pappas, D.A. Wicks, Organic Coatings: Science and Technology, Wiley2007
[7] G. Tesoro, Journal of Polymer Science Part C: Polymer Letters, 26 (1988) 539-539.
[8] M.T. Aronhime, X. Peng, J.K. Gillham, R.D. Small, Journal of Applied Polymer Science, 32 (1986) 3589-3626
[9] J. Zhou, J.P. Lucas, Polymer, 40 (1999) 5505-5512.
[10] X.H. Chen, C.S. Chen, H.N. Xiao, F.Q. Cheng, G. Zhang, G.J. Yi, Surface and Coatings Technology, 191 (2005) 351-356
[11] L.H. Yang, F.C. Liu, E.H. Han, Progress in Organic Coatings, 53 (2005) 91-98.
[12] B.M. Praveen, T.V. Venkatesha, Y. Arthoba Naik, K. Prashantha, Surface and Coatings Technology, 201 (2007) 5836-5842
[13] Journal of Composites for Construction, 5 (2001) 200-210
[14] C.-M. Chen, Y.-M. Dai, J.G. Huang, J.-M. Jehng, Carbon, 44 (2006) 1808-1820
[15] A. Cao, C. Xu, J. Liang, D. Wu, B. Wei, Chemical Physics Letters, 344 (2001) 13-17
[16] H. Jeon, J. Park, M. Shon, Journal of Industrial and Engineering Chemistry, 19 (2013) 849-853

A facile approach towards fabrication of super hydrophobic surface from functionalized silica particles

M. Awais, M. Jalil, U. Zulfiqar and S.W. Husain
Department of Materials Science and Engineering, Institute of Space Technology, Islamabad, Pakistan

E-mail: dr_swh@live.com

Abstract. A facile and cost effective method for fabrication of super hydrophobic surface on a glass substrate is studied. The surface is fabricated from surface functionalized silica particles, synthesized by sol-gel process with the average size of 200±10nm. These particles were functionalized with stearic acid to induce hydrophobicity followed by coating on a glass substrate. After coating, substrate was dried to remove excess solvent. The drying temperature was optimized and its effect on contact angle of hydrophobic surface was studied. It was observed that surface exhibits higher contact angle with increased drying temperature till the decomposition temperature of Stearic acid. Silica particles were characterized by using Scanning electron microscopy (SEM), thermal analysis was performed with Thermo gravimetric analysis (TGA) while the coated surface was studied using SEM and Contact Angle (CA) measurement.

1. Introduction

Super hydrophobic surfaces have gained importance due to their non-adhesive and non-wetting characteristics. These properties make them useful for many applications including self-cleaning, anti-corrosion and anti-icing [1]. Super hydrophobic surfaces exhibit contact angle greater than 150°[2] which make them non-wetting and non -adhesive. The increased water contact angle of a surface is due to its characteristic low surface energy and rough morphology. Surface roughness not only increases the contact angle but also decreases the solid liquid adhesion by creating air pockets between solid and the liquid interface [3-4].

Polymers [5], metal oxides [6-7] and nanoparticles [8-9] including titania and silica particles are being used for the fabrication of super hydrophobic surfaces. Silica based coatings offer multiple benefits which are; ease of fabrication, cost effectiveness, better adhesion and thermal stability[10]. Silica particles are generally synthesized by sol-gel process which allows regulating the particle size by changing reaction parameters. Previously, it has been studied that particle size can be tuned by changing the reaction temperature, concentration of precursor and co-solvent [11]. The synthesized silica particles are hydrophilic due to attached hydroxyl group. Fluorinated hydrocarbons are being used to alter the hydrophilic nature of silica particles to hydrophobic by replacing the hydroxyl group[12-13]. However, modification with fluorinated hydrocarbons is expensive and hazardous to health and environment. Stearic acid is another modifying agent which can be utilized for the modification purpose.

In this study, we have prepared and characterized silica particles to prepare hydrophobic surfaces.

Content from this work may be used under the terms of the Creative Commons Attribution 3.0 licence. Any further distribution of this work must maintain attribution to the author(s) and the title of the work, journal citation and DOI.
Published under licence by IOP Publishing Ltd

2. Materials and methods

2.1. Reagents

Tetraethoxysilane (TEOS, 98%) was purchased from Sigma Aldrich, Ethanol (99.8%) and Stearic acid was obtained from BDH Laboratory suppliers. Aqueous ammonia (33%) was supplied by Honeywell Burdick and Johnson and Methanol (HPLC) was purchased from LAB scan Analytical sciences.

2.2 Synthesis of silica particles

Silica particles were synthesized by taking inspiration from previously reported method [14]. 250 ml of methanol and 250 ml ammonium hydroxide solution were mixed for 10 min in a 1000 ml beaker using a magnetic stirrer. 5 ml of TEOS was added drop wise to the mixture and reaction was left to proceed for 30 min at 25°C. The product was centrifuged for 30 min at 4000 rpm followed by rinsing with methanol and drying.

2.3 Surface modification of silica particles

0.2g of the sesilica particle were added in 50ml ethanol in 100ml beaker. The mixture was sonicated using a probe ultra sonicator for 20 min. After sonication, 1g of stearic acid was added in the mixture in molten form followed by heating at 60°C for 2 hr.

2.4 Fabrication of super hydrophobic surfaces

The glass substrates were cleaned by rinsing with acetone followed by sonication to remove impurities. After that, the prepared solution of modified silica particles was dropped the substrate. The samples were subjected to different drying temperatures of 100°C, 180°C and 260°C to study the effect of temperature on contact angle.

2.5 Characterizations

Field Emission Scanning Electron Microscopy (SEM) (MIRA 3 TESCAN) was used to study the particle size, particles size distribution and the surface morphology of fabricated super hydrophobic surfaces. Thermal analysis of stearic acid, silica particles and modified silica particles was performed using TGA (METTLER TOLEDO).For contact angle measurement, an appropriate volume of water was dropped on the surfaces and captured using Nikon DSLR camera.

3. Results and discussion

Figure 1 shows the SEM micrographs of silica particles at two different magnifications. Non agglomerated and spherically shaped silica particles of size 200±10nm were synthesized using TEOS as precursor, ethanol as co-solvent and ammonia as catalyst. Previously it has been studied that changing the ratios of co-solvent and catalyst affects the particle size. Increased amount of ammonia increases the rate of hydrolysis and condensation and provides less time for nucleation which results in smaller particle size [11]. In the present study, spherically shape dsilica particles of size 200±10nm are produced by using the equal amount of ethanol and ammonia.

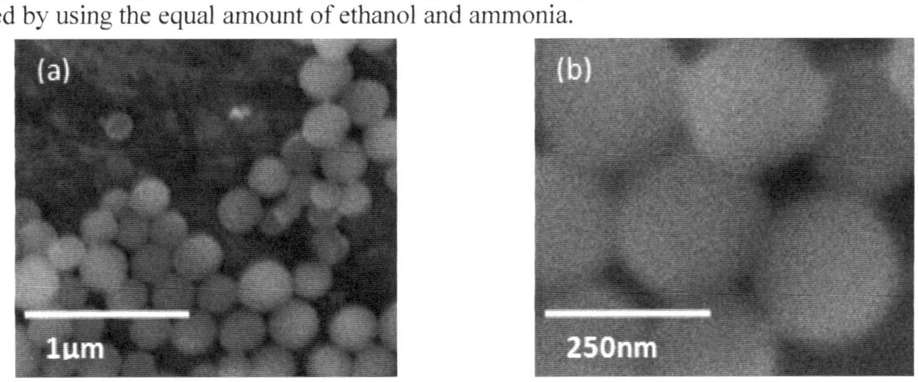

Figure 1. SEM images showing silica particles at two different magnifications

Figure 3 shows the thermo gravimetric analysis of stearic acid, silica particles and modified silica particles. The TGA results indicated that decomposition of stearic acid starts at about 200°C and completes at 310°C with 95% weight loss whereas the weight loss in pure silica particles is only due to evaporation of water molecules. However, the presence of stearic acid in modified silica particles was confirmed through the weight loss starting from 250° and completing at 370°. The initial weight loss which is observed in the as synthesized silica particles is not present in the modified particles due to evaporation of water as these particles were scratched from coatings subjected to drying temperature of 180°C. The water content would have been removed during drying. In a previous study, modification of anatase with stearic has been studied and weight loss in the range of 200°C to 450°C was recognized as the decomposition of stearic acid [15]. In another study, a significant weight loss of stearic acid modified tourmaline was observed in temperature range of 138°C - 380°C [16]. In the present work, thermal degradation in modified silica particles is observed in temperature range of 300°C-440°C. So it can be inferred from thermal analysis that coatings fabricated with these modified silica nanoparticles are stable up to 300°C.

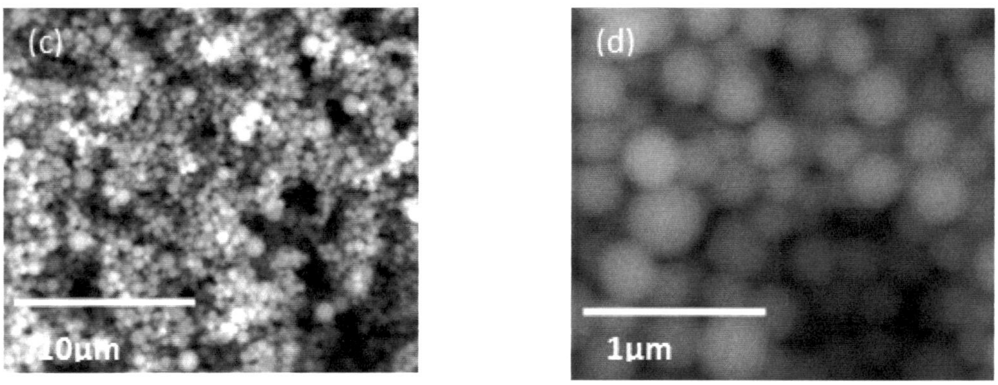

Figure 2. SEM images showing the surface morphology of substrate coated with modified silica particles at two different magnifications

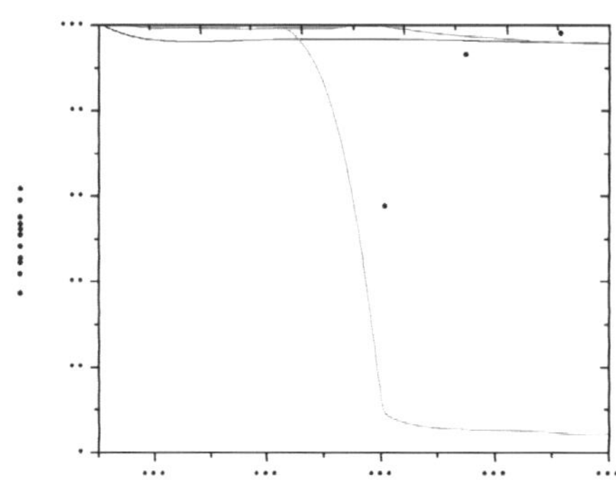

Figure 3. TGA profiles of (a) stearic acid, (b) silica particles, (c) modified silica particle

In Figure 2, the SEM micrographs show the morphologies of coated glass substrates. Coating of modified silica particles have produced the rough surface as indicated in the SEM images. Surface roughness plays very important role in hydrophobic characteristics of a surface [17]. The hydrophobicity in naturally occurring surfaces is thought to be because of rough hierarchical structure which creates repellence with water and makes the contact angle of water droplet greater than 150° [10].The role of surface roughness is explained by Wenzel and Cassie models, the first relates the wetting properties of a surface with roughness while the latter attributes it to the air pockets formed between surface grooves and liquid [10]. In the present study, it is observed that coating of silica particles has created roughness on glass substrate which contributes towards the hydrophobicity of surface as shown in Figure 5.

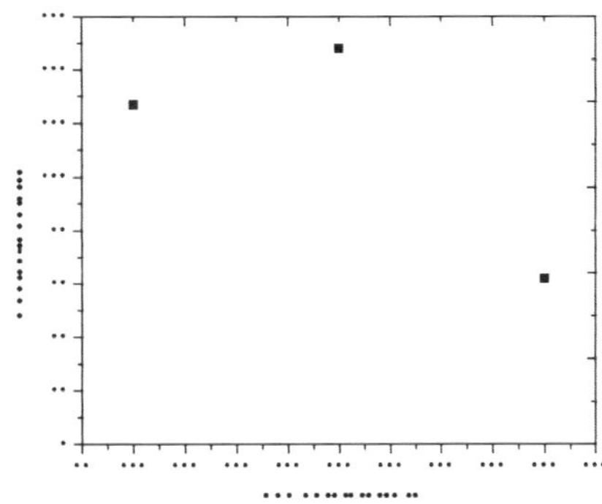

Figure 4. Variation in Contact angle with increasing drying temperature

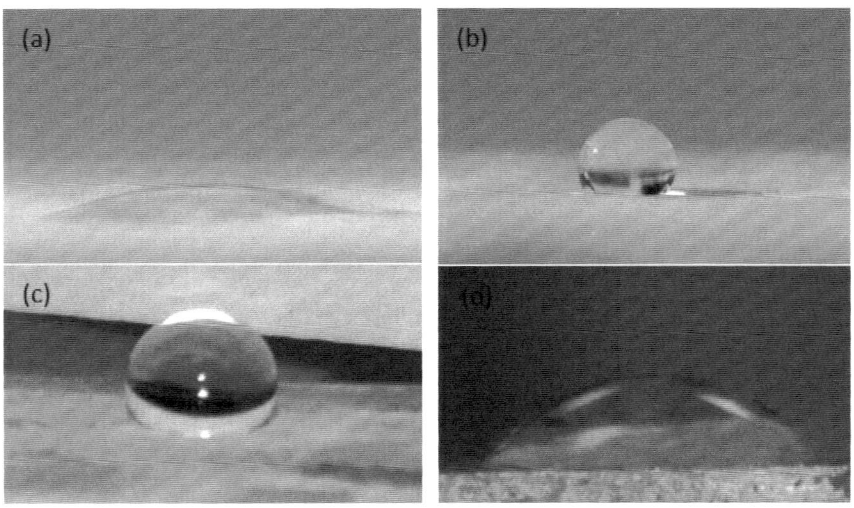

Figure 5. The contact angle of water droplet with (a) uncoated glass substrate, (b) coated glass substrate and dried at 100°C,(c) 180°C& (d) 260°C

Figure 4 graphically shows the trend of contact angle with increasing drying temperature. Figure 5 shows the contact angles of glass substrate before and after coating. Three samples were prepared and subjected to the drying temperatures of 100°C, 180°C and 260°C.It can be seen that the water contact angle is 130° at dying temperature of 100°C. The contact angle is increased to 150° when drying temperature was increased to 180°C. The contact angle dropped to 65° when drying temperature was increased to 260°C.The hydrophobic character of silica particles is owing to the long alkyl chain of stearic acid; the hydroxyl group attached to silica particle reacted with carboxyl group of stearic acid during modification [12].

4. Conclusion

Super hydrophobic surface has been prepared by coating modified silica particles on glass substrate. Silica particles were synthesized by sol gel process using TEOS as precursor. The prepared particles were modified with stearic acid followed by coating using dip coating process. Super hydrophobic surface with a maximum contact angle of 150° is produced in this study. Contact angle initially increases with increasing drying temperature then drops at 260°C. This drop is attributed to thermal destruction of alkyl chains attached to silica particles.

5. References

[1] M. Nosonovsky and B. Bhushan, "Superhydrophobic surfaces and emerging applications: Non-adhesion, energy, green engineering," *Curr. Opin. Colloid Interface Sci.*, vol. 14, no. 4, pp. 270–280, Aug. 2009

[2] Q. Ke, W. Fu, S. Wang, T. Tang, and J. Zhang, "Facile Preparation of Superhydrophobic Biomimetic Surface Based on Octadecyltrichlorosilane and Silica Nanoparticles," *ACS Appl. Mater. Interfaces*, vol. 2, no. 8, pp. 2393–2398, Aug. 2010.

[3] X. Sun, "Effects of the Based Membrane on the Hydrophobicity of Super-hydrophobic PES Membrane and its Structural Properties," *Mod. Appl. Sci.*, vol. 4, no. 2, p. p71, 2010

[4] X. Du, J. S. Li, L. X. Li, and P. A. Levkin, "Porous poly(2-octyl cyanoacrylate): a facile one-step preparation of superhydrophobic coatings on different substrates," *J Mater Chem A*, vol. 1, no. 4, pp. 1026–1029, 2013

[5] S. M. Kang, I. You, W. K. Cho, H. K. Shon, T. G. Lee, I. S. Choi, J. M. Karp, and H. Lee, "One-Step Modification of Superhydrophobic Surfaces by a Mussel-Inspired Polymer Coating," *Angew. Chem. Int. Ed.*, vol. 49, no. 49, pp. 9401–9404, Dec. 2010

[6] X. Liu, J. Zhang, T. Yang, L. Wang, Y. Kang, S. Wang, and S. Wu, "Self-assembled hierarchical flowerlike ZnO architectures and their gas-sensing properties," *Powder Technol.*, vol. 217, pp. 238–244, Feb. 2012

[7] C. Cheng and H. J. Fan, "Branched nanowires: synthesis and energy applications," *Nano Today*, vol. 7, no. 4, pp. 327–343, 2012

[8] X. L. Sun, Z. P. Fan, L. D. Zhang, L. Wang, Z. J. Wei, X. Q. Wang, and W. L. Liu, "Superhydrophobicity of silica nanoparticles modified with polystyrene," *Appl. Surf. Sci.*, vol. 257, no. 6, pp. 2308–2312, Jan. 2011

[9] A. L. Dubov, K. Perez-Toralla, A. Letailleur, E. Barthel, and J. Teisseire, "Superhydrophobic silica surfaces: fabrication and stability," *J. Micromechanics Microengineering*, vol. 23, no. 12, p. 125013, Dec. 2013

[10] A. B. Gurav, Q. Xu, S. S. Latthe, R. S. Vhatkar, S. Liu, H. Yoon, and S. S. Yoon, "Superhydrophobic coatings prepared from methyl-modified silica particles using simple dip-coating method," *Ceram. Int.*, vol. 41, no. 2, pp. 3017–3023, Mar. 2015.

[11] D. L. Green, J. S. Lin, Y.-F. Lam, M.-C. Hu, D. W. Schaefer, and M. T. Harris, "Size, volume fraction, and nucleation of Stober silica nanoparticles," *J. Colloid Interface Sci.*, vol. 266, no. 2, pp. 346–358, 2003

[12] L. Gao and J. He, "Surface hydrophobic co-modification of hollow silica nanoparticles toward large-area transparent superhydrophobic coatings," *J. Colloid Interface Sci.*, vol. 396, pp. 152–159, 2013

[13] C.-H. Xue, S.-T. Jia, J. Zhang, and J.-Z. Ma, "Large-area fabrication of superhydrophobic surfaces for practical applications: an overview," *Sci. Technol. Adv. Mater.*, vol. 11, no. 3, p. 033002, Jun. 2010

[14] A. G. Howard and N. H. Khdary, "Spray synthesis of monodisperse sub-micron spherical silica particles," *Mater. Lett.*, vol. 61, no. 8–9, pp. 1951–1954, Apr. 2007.

[15] L. Zhang, L. Chen, H. Wan, J. Chen, and H. Zhou, "Synthesis and Tribological Properties of Stearic Acid-Modified Anatase (TiO2) Nanoparticles," *Tribol. Lett.*, vol. 41, no. 2, pp. 409–416, Feb. 2011

[16] Y. Hu, "The Studies on the Surface Organic Modification of Tourmaline Powder with Stearic Acid and Its Composite," *Open J. Compos. Mater.*, vol. 04, no. 03, pp. 148–156, 2014

[17] Y. Liu, J. Liu, S. Li, Y. Wang, Z. Han, and L. Ren, "One-step method for fabrication of biomimetic superhydrophobic surface on aluminum alloy," *Colloids Surf. Physicochem. Eng. Asp.*, vol. 466, pp. 125–131, Feb. 2015

IOP Conf. Series: Materials Science and Engineering **146** (2016) 012023 doi:10.1088/1757-899X/146/1/012023

Effect of Zr addition on the mechanical characteristics and wear resistance of Al grain refined by Ti after extrusion

Adnan I. O. Zaid and S.M. A. Al-Qawabah[1]
Industrial Engineering Department, Applied Science Private University, Amman, Jordan
[1]Mechanical Engineering Department, Tafila Technical University, P. O. Box 13720, Amman 11942, Jordan

E-mail: adnan_kilani@yahoo.com; safwan1q@gmail.com

Abstract. Aluminum and its alloys are normally grain refined by Ti or Ti+B to transfer their columnar structure during solidification into equiaxed one which improves their mechanical behavior and surface quality. In this paper, the effect of addition of Zr on the metallurgical, and mechanical aspects, hardness, ductility and wear resistance of commercially pure aluminum grain refined by Ti after extrusion is investigated. Zr was added at a level of 0.1% which corresponds to the peretectic limit at the Al-Zr phase diagram. The experimental work was carried out on the specimens after direct extrusion. It was found that addition of Ti resulted in decrease of Al grain size, whereas addition of Zr alone or in the presence of Ti, resulted in reduction of Al grain size. This led to increase of Al hardness. The effect of the addition of Ti or Zr alone resulted almost in the same enhancement of Al mechanical characteristics. As for the strain hardening index,n, increase was obtained when Zr was added alone or in the presence of Ti. Hence pronounced improvement of its formability. Regarding the effect of Zr addition on the wear resistance of aluminum; it was found that at small loads and speeds addition of Ti or Zr or both together resulted in deterioration of its wear resistance whereas at higher loads and speeds resulted in pronounced improvement of its wear resistance. Finally, the available Archard model and the other available models which consider only the mass loss failed to describe the wear mechanism of Al and its micro-alloys because they do not consider the mushrooming effect at the worn end.

1. Introduction

Aluminum and its alloys are widely used materials in aircraft and automobile industries due to their attractive properties particularly their high strength-to-weight ratio, electrical and thermal conductivities in addition to their corrosion resistance, [1]. Unfortunately they solidify in columnar structure which tends to deteriorate their mechanical strength and surface quality. Therefore, they are normally alloyed or micro alloyed with other elements; e.g. aluminum–silicon (Al–Si) alloys find their applications in the automotive industry due to their high specific strength and wear resistance. The combination of high thermal conductivity and low coefficient of thermal expansion make them suitable for the internal combustion (IC) engine components such as cylinder block and piston. Sabatino and Arnberg [2] demonstrated that the addition of fluidity of the Al-Si alloys giving the good cast ability. Moreover, improvement in the tensile strength [3] and tribological performance [4] of Al–Si alloys with increase in the silicon content of the alloys was also observed. Morphology of the

Content from this work may be used under the terms of the Creative Commons Attribution 3.0 licence. Any further distribution of this work must maintain attribution to the author(s) and the title of the work, journal citation and DOI.
Published under licence by IOP Publishing Ltd

eutectic silicon in the hypoeutectic Al–Si alloy was effectively modified by chemical route typically with the impurity level addition of elements such as sodium 1.1[5], strontium [6], phosphorus [7] and scandium [8] by a process known as impurity induced twinning mechanism. However, in hyper-eutectic Al–Si alloys, with the presence of large primary silicon crystals, the modification and refinement by the aforementioned method is not effective. On the contrary, size and morphology of the silicon phase in Al–Si alloys was effectively refined by rapid solidification techniques such as spray forming and powder metallurgy. For instance, Cuietal [9] demonstrated that the primary silicon phase in Al–Si alloys can be effectively refined by spray forming. However, the presence of hard silicon phase which exists in the form of large grains, morethan100 microns inconventionally cast Al–Si alloyseutecticand primary silicon crystals in hypoeutectic and hypereutectic Al–Si alloys respectively is not recommended as it adverselyaffectstheir mechanical properties. Further information is given in references [10,11]. Another method of grain refining the structure of these alloys is by the addition of some grain refiners to their melt prior to solidification, mainly by addition of either titanium or titanium-boron. Several binary Al-Ti and Ternary Al-Ti-B master alloys are produced for this purpose and are now commercially available. Although the literature on the grain refinement in general and aluminum and its alloys in particular is voluminous and reviewed in reference [12]. The mechanism of grain refinement is still a controversial matter despite the different suggested reported models. Summary of these models is reported in [12-13]. Wear is an important phenomenon in machine elements and manufacturing processes, defined as the cumulative mass loss in the machine elements, cutting tools, punches and dies which leads to in change in their dimensions and causing their failure. The mechanism of wear is very complex due to the number of variables involved. Some of these variables is related to the material characteristics e.g. microstructure, mechanical strength, hardness and surface finish. Other variables include process parameters e.g. applied load, velocity, interface temperature, lubricant, corrosion of the surface and existence of foreign particles. The available literature on wear reveals that most of the published work is concerned with the effect of the above mentioned parameters on wear mechanism of aluminum and little work is published regarding the effect of addition of one or more grain refiners on its wear resistance both in the cast and after extrusion conditions.

In this paper, the effect of addition of Zr on the mechanical characteristics and wear resistance of commercially pure aluminum grain refined by Ti after direct extrusion is investigated.

1.1 Theoretical considerations
Most of the existing wear theories and mechanisms attempt to explain the wear process on a microscopic level, and relate the magnitude of wear to material properties and to the nature of the local involved parameters. The first mechanism relates the wear rate to the number of inert-atomic encounters between the opposing surfaces [14]. However, this mechanism has not been generally accepted. Later, attempts have been made to establish a new wear theory based the atomic level of wear, but this idea is still far from simple to be accepted, because the wear debris is generated simultaneously with the onset of motion, and the debris consists of aggregated particles of materials. The old reported theory is the most widely accepted theory which is used to predict the volume of the worn material from the following equation:

$$V=KPL/3H \quad (1)$$

Where V is the worn volume, P is the applied load, L is the sliding distance, and H is the material hardness and K is a constant, which was interpreted as the probability that a particular asperity contact could produce a wear particle which may be considered as a wear rate, [15]. Archad's theory although it gives indication in some cases of adhesive wear, it fails in other cases, particularly in determining the wear of aluminum and other soft materials like high conductivity copper, where in addition to the accumulated mass loss, there exists some plastic deformation at the worn end causing a mushroom like shape. This phenomenon was repeatedly reported in the literature,[12-13]. In the last two references the authors suggested a model which accounts for plastic deformation at the worn end and the shearing of the particles which accounts for the mass loss or the volume of the removed material.

2. Materials equipment and experimental procedures

2.1 Materials

Commercially pure aluminum of 99.8 % purity of the chemical composition shown in Table 1, was used throughout this work A binary Al-2.92 wt.% Ti and a binary Al-1.17 wt.% Zr master alloys were laboratory prepared and used to obtain 0.15 wt.% Ti and 0.1 wt.% Zr in the different alloys. Master and grain refined alloys were prepared in graphite crucibles and graphite rods were used for stirring the molten bath in all experiments.

Table 1. Chemical composition (wt. %) of commercially pure aluminium

Element	**Fe**	**Si**	**Cu**	**Mg**	**Ti**	**V**	**Zn**	**Mn**	**Na**	**Al**
Wt. %	0.09	0.05	0.005	0.004	0.004	0.008	0.005	0.001	0.005	Bal.

2.2 Equipment

The extrusion tests were carried out at room temperature, in cold condition, using the extrusion die shown in Figure 1, it consists essentially of a container made of steel D2 having an external diameter of 100 mm, internal diameter of 14 mm and 120 mm height. The punch is 13.96 mm diameter and 80 mm length made of D2 steel. The female die is circular of 10 mm internal diameter and 80 mm external diameter. Extrusion tests were carried out using the extrusion die shown in Figure 1.

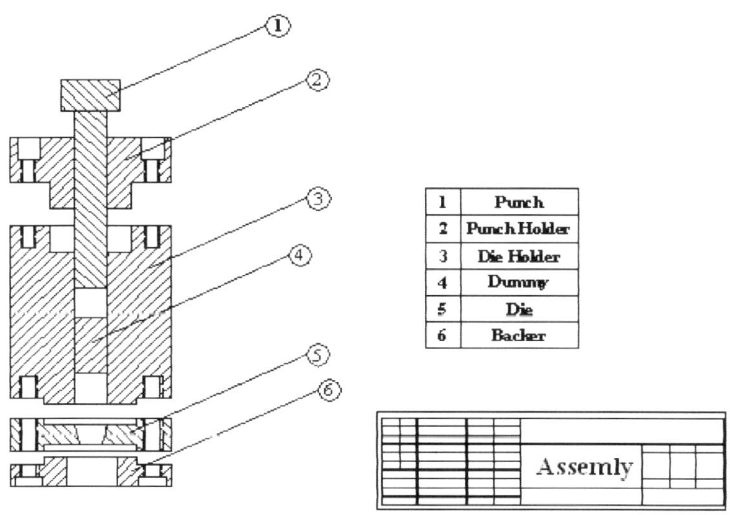

Figure 1. Extrusion die assembly

2.3 Experimental procedures

2.3.1 Preparing the binary master alloys and the micro-alloys

The experimental procedures started by preparing the two binary master alloys Al-Ti and Al-Zr. In each case titanium or zirconium was added into molten aluminum at 1100 C and stirred under a cry lite flux. The temperature was kept constant for half an hour before casting the alloy in a plate form of thickness less than 8 mm by spreading the melt over a thick cast iron plate. These master alloys were used in preparing the different Al micro-alloys, namely: Al-Ti, Al-Zr and Al-Ti-Zr. Standard tensile specimens were machined from Al and its prepared micro-alloys. Standard tensile tests were carried out on the Instron Universal Testing machine of 100 KN capacity at cross head speed of 5 mm/min. and the autographic record was obtained for each specimen from which themechanical behavior was determined.The wear tests were carried outusing the standard wear testing apparatus, (pin on rotating test) at different loads namely: 5, 10, 15, and 20 N and at different linear speeds, namely: 0.276, 0.801, and 3.01 m/sec. All the above mentioned tests were carried out on specimens both in the cast and after extrusion conditions.

3. Results and discussion

3.1 Effect of Ti and Zr on the microstructure of aluminum after extrusion

Figure 2 shows the effect of the extrusion process on the grain size of Al and its three micro-alloys which indicates that it resulted in pronounced decrease in grain size, for example it resulted in a decrease of 254.54%, 41%, and 51.3% in the case of Al-Ti, Al-Zr and Al-Ti-Zr micro-alloys respectively. The effect of Ti or Zr addition to Al either alone or together on its grain size after extrusion is explicitly shown in the photomicrographs of Figure 3.

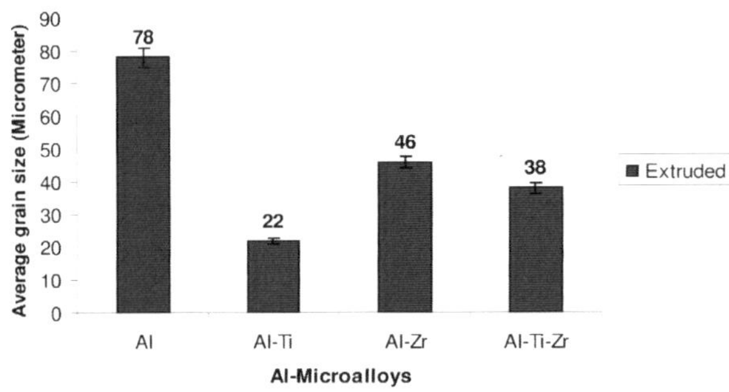

Figure 2. Effect of zirconium addition on the grain size of Aland Al-Ti after extrusion

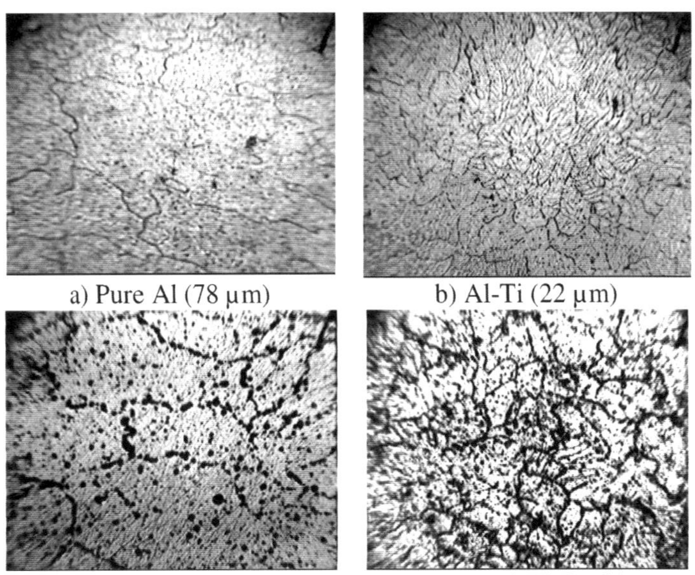

a) Pure Al (78 μm) b) Al-Ti (22 μm)
c) Al-Zr (46 μm) d) Al-Ti-Zr (38 μm)

Figure 3. Photomicrographs showing the general microstructure of Al and its micro alloys after extrusion at x250

3.2 Effect of Ti and Zr addition on the hardness of aluminum after extrusion

The histogram of Figure 4 shows the Vickers's micro-hardness of the Al and its micro-alloys Al-Ti, Al-Zr and Al-Ti-Zr after extrusion from which it can be seen that the extrusion process resulted in increase of the micro-alloy's hardness by the addition of Ti or Zr either alone or together by the following percentages: 75.68, 21.6 and 43.24 respectively. This is attributed to the grain refinement of the structure and the intermetallic compounds which exists in the main matrix.

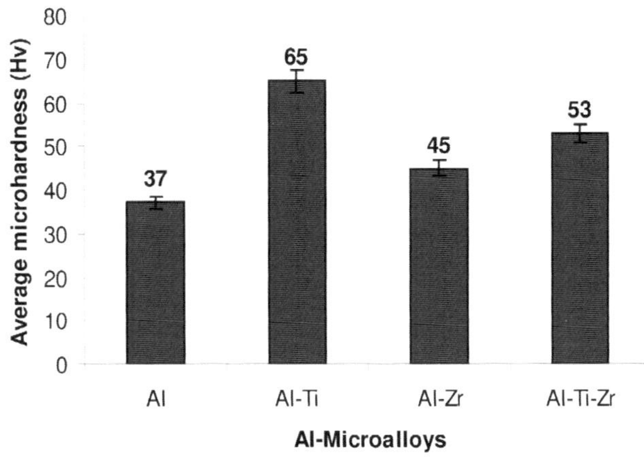

Figure 4. Effect of zirconium addition on the average micro-hardness of Al, Al grain refined by Ti after extrusion

3.3 Effect of Ti and Zr on the Mechanical Characteristics of Aluminum after Extrusion

Figure 5 shows the effect of Zr addition on the mechanical behavior of Al and Al grain refined by Ti. It can be seen from this figure that addition of Ti or Zr either alone or together to Al resulted in enhancement of its mechanical strength. The best improvement was achieved when they were added both together. Furthermore, it can be seen from Table 2 the effect on its mechanical characteristics, where an increase of 34.34 % was achieved in case of adding Ti or Zr alone and 7.37 % when added together in the flow stress at 20 % strain. Similarly, the strength coefficient was increased by 76.2 % when Ti or Zr was added alone and by 19.7% when added together. Regarding the effect on the strain hardening index ,n, it was highly increased by 154.5 % when Ti or Zr were added alone and 63.63 % when added together. This will result in enhancement of its formability; hence it reduces the number of stages required for forming at large strains.

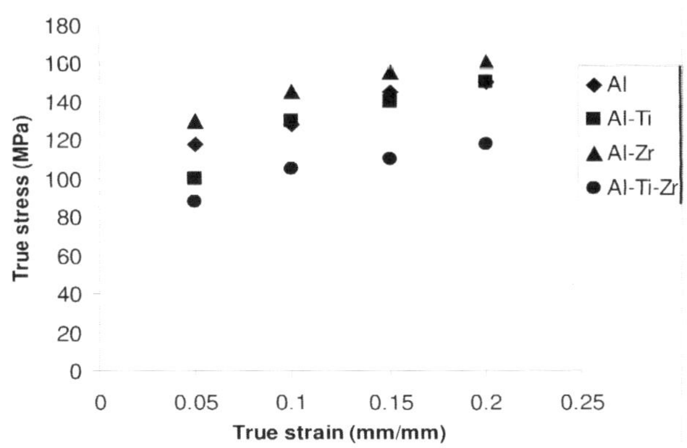

Figure 5. True stress- True strain curves of Al and its three micro-alloys after extrusion

Table 2. Mechanical characteristics of Al and its different micro-alloys after extrusion

Micro alloys	Flow stress (MPa)at strain=20%	Strain hardening index (n)	Strength coefficient (K) MPa	General equation of mechanical behavior
Al	99	0.11	118.6	$\sigma=118.6\varepsilon^{0.11}$
Al-Ti	133	0.28	209	$\sigma=209\varepsilon^{0.28}$
Al-Zr	133.4	0.28	209.3	$\sigma=209.3\varepsilon^{0.28}$
Al-Ti-Zr	106.3	0.18	142	$\sigma=142\ \varepsilon^{0.18}$

4.5 Effect of Zirconium addition on the wear resistance of aluminum and aluminum grain refined by titanium after extrusion

It can be seen from Figure 6 that the relationship between the accumulated mass loss and time for Al and its micro-alloys Al-Ti, Al-Zr, and Al-Ti-Zr at loads of 10 and 20 N and speed of 0.27m/s is almost linear.The commercially pure aluminum indicated less wear resistance than its micro-alloys, i.e. addition of Ti or Zr either alone or together to Al resulted in improvement of its wear resistance. Furthermore doubling the load to (20 N) at the same speed (0.27m/s) did not change the trend but the refiners became more effective, as illustrated by Figure 9 and Tables 3 and 4. It is worth noting that the wear rate, K, in the last column of Table 3 as calculated from Archad theoretical model is misleading because it indicates less value for Al than the K of its micro-alloys. However, the values of the reduction in height and the final area of the worn end which are the indicators of the mushrooming effect reveals that only addition of titanium resulted in reducing Al wear resistance. Therefore it is concluded that investigating the wear resistance to wear of soft materials like those in this paper should consider both the mass loss and the mushrooming effect at the worn end.

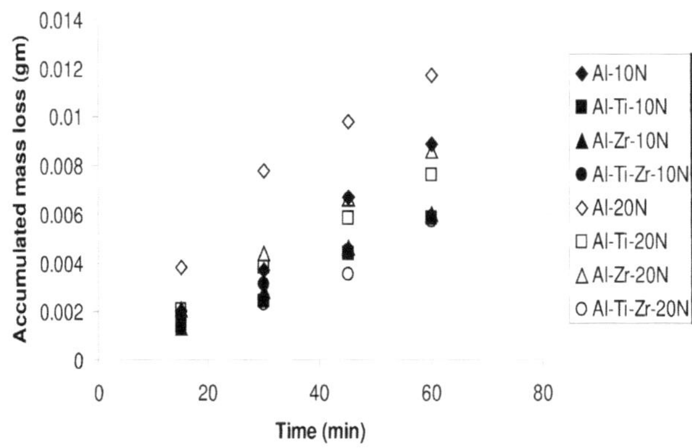

Figure 6. Effect of zirconium addition of Al and Al grain refined by Ti afterextrusion at (S = 0.27m/sec, loads = 10, 20 N)

Table 3. Mass loss and dimensional changes after one hr. wear for Zr addition to Al and Al grain refined by Ti after extrusion. (Load = 10 N, S = 0.27 m/sec, A_o = 38.48 mm^2)

Name	Mass loss (g)	D_f (mm)	Δ h (mm)	A_f (mm^2)	K Wear rate
Al	0.0089	7.80	0.18	47.76	0.03764
Al-Ti	0.0058	7.90	0.10	48.99	0.04309
Al-Zr	0.006	7.35	0.16	42.41	0.03086
Al-Ti-Zr	0.0059	7.35	0.19	42.41	0.03574

Table 4. Mass loss and dimensional changes after one hr. wear for Zr addition to Al and Al grain refined by Ti after extrusion.(load =20 N, S = 0.27 m/sec, A_o = 38.48 mm^2)

Name	**Mass loss (g)**	**D_f (mm)**	**Δ h (mm)**	**A_f (mm^2)**	**K Wear rate**
Al	0.049	7.90	0.15	48.99	0.1036
Al-Ti	0.0306	8.05	0.57	50.87	0.1136
Al-Zr	0.0364	7.50	0.58	44.16	0.0936
Al-Ti-Zr	0.0373	7.80	0.60	47.76	0.1130

Increasing the speed to 3m/s and at 10 and 20 N, the mass loss versus time for Al and its micro-alloys is shown in Figure 7, and the mushrooming parameters are explicitly shown in Tables 6 and 7, from which it is seen that the trend remained as before, except the addition of Ti has also resulted in enhancement of the wear resistance and all of them became more effective.

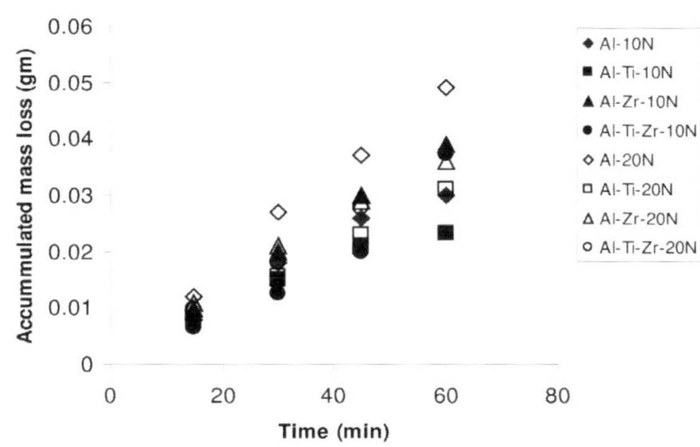

Figure 7. Effect of zirconium addition of Al and Al grain refined by Ti after extrusion at (S = 3m/sec, loads = 10, 20 N)

Table 6. Mass loss and dimensional changes after one hr. wear for Zr addition to Al and Al grain refined by Ti after extrusion. (Load = 10 N, S =3.01 m/sec, A_o = 38.48 mm^2)

Name	Mass loss (g)	D_f (mm)	Δ h (mm)	A_f (mm^2)	K Wear rate
Al	0.0244	8.35	0.55	57.39	0.0093
Al-Ti	0.0429	8.20	0.56	52.78	0.0286
Al-Zr	0.0194	8.35	0.50	57.39	0.6525
Al-Ti-Zr	0.0213	8.95	0.51	62.88	0.0116

Table 7. Mass loss and Dimensional Changes after one hr. Wear for Zr addition to Al and Al Grain Refined by Ti after Extrusion.(load =20 N, S = 3.01 m/sec, A_o = 38.48 mm^2)

Name	**Mass loss (g)**	**D_f (mm)**	**Δ h (mm)**	**A_f (mm^2)**	**K Wear rate**
Al	0.0117	11.5	0.22	103.82	0.0022
Al-Ti	0.0076	10.0	0.20	78.50	0.0025
Al-Zr	0.0086	9.55	0.20	71.59	0.0020
Al-Ti-Zr	0.0057	8.90	0.14	62.18	0.0015

4. Conclusions

The following points are concluded:

i. Addition of either Ti or Zr alone or together to Al resulted in grain refining of its structure. The best refinement is when Ti is added alone.

ii. Addition of Ti or Zr alone or together to Al caused improvement of its mechanical strength represented by increase of its flow stress, strength coefficient, ultimate tensile strength, its ductility and its strain hardening index. The last two items cause improvement of formability. This in turn will reduce the stages in the forming processes when large strains are involved. Also its hardness was increased. This is a great achievement because the increase in hardness is normally associated with decrease of ductility and formability.

iii. The Archad wear model cannot be used to explain the wear of Al and its micro-alloys due to the mushrooming effect at the worn end. This will lead to change of the part dimensions, which in some cases it can be more damaging than the mass loss.

5. Acknowledgment

The first author, (Adnan I. O. Zaid), is grateful to the Applied Science Private University, Amman, Jordan for (partial) financial support granted to this research (Grant No.DRGS-2015).

6. References

[1] Callister W D 2003 *Materials Science and Engineering, an Introduction*6th edition, John Wiley and Sons, Inc, New York

[2] Sabatino M D, Arnberg L 2009*Trans. Indian Inst. Met.* 62 321

[3] Zeren M 2005 *J. Mater. Process. Technol.* 169 292

[4] Torabian H, Patak J P, Tiwari S N 1995*J. Mater. Sci. Lett.* 14 1631

[5] Qiyang L, Qingchun L, Qifu L 1991*Acta Metall. Mater.* 39 2497

[6] Dahle A K,Nogita K,Zindel J W, McDonald S D , Hogan L M 2001*Metall. Mater. Trans. A*32949.

[7] Song X, Bian X, Zhang J, Liu X 2009*J. Mater. Sci.* 44 4856

[8] aghukiran N, Kumar R 2015Effect of scandium addition on the microstructure, mechanical and wear properties of the spray formed hypereutectic aluminum–silicon alloys, *Materials Science & Engineering A* 641 138-147

[9] Cui C, Schulz A, Schimanski K, Zoch H W 2009 J. Mater. Process. Technol. 209 522

[10] Pejaković V, Totolin V, Göcerler H, BrennerJ, Ripoll M R 2015Frictionandwear behavior Ofselectedtitaniumandzirconiumbased nitride coatingsinNa2SO4 aqueoussolutionunder low contact pressure, Tribology International 91267–273

[11] Baradeswaran A, Vettivel S C, Peruma A E, Selvakumar N, Issac R F 2014 Experimental investigation on mechanical behavior, modeling and optimization of wear parameters of B4C and graphite reinforced aluminum hybrid composites, *Materials and Design* 63 620–632

[12] ZaidA I O 2001Effect of some grain refiners on the mechanical properties of aluminum Proc. *7th International symposium on advanced materials*, 63-67, Islamabad, Pakistan.

[13] Zaid A I O, Al-Qawabah S M A 2010 Effect of Zr Addition on the Mechanical Behavior, Ductility and Wear Resistance of Aluminum Grain Refined by Titanium *Trans Tech publications*, Switzerland, *Key Engineering Materials* 442, 15

[14] Sarkar A D 1976 Wear of Metals", *International Series on Materials Science and Technology*, 33

[15] Archard J F 1953 Contact and rubbing of flat surfaces *J. of Applied Physics*24 981-988

Effect of molybdenum addition to ZA22 grain refined by titanium in the cast and after pressing by ECAP

Adnan I. O. Zaid, N. S. Al-Hunetti[1] and K. Y. S. Eyal-Awwad[2]
Applied Science Private University, Mechanical and Industrial Engineering Department, Amman 11931 Jordan
[1]University of Jordan, Mechanical Engineering Department, Amman 11936, Jordan
[2]Tafila Technical University, Mechanical Engineering Department, Amman, Jordan

E-mail: adnan_kilani@yahoo.com

Abstract. Zinc aluminum alloys are versatile materials which are widely used in manufacturing many industrial and engineering parts due to their attractive properties. The ZA22 has the extra advantage of possessing super plastic behavior within the temperature range from 350 to 375°C. The equal channel angular pressing, ECAP is a relatively recent manufacturing process by which heavy plastic deformation can be produced in materials resulting in grain refinement of its microstructure. It is, therefore, anticipated that if the ECAP process is applied to the ZA22 alloy after being grain refined by certain grain refiners may produce super plastic behavior in this alloy at room temperature, by this eliminating the heating process and its costs, hence widening its applications rendering it to be cost effective. In this paper, the effect of molybdenum addition at a rate of 0.1 % wt. to ZA22 grain refined by Ti on its metallurgical and mechanical characteristics in the cast condition and after applying the ECAP process is investigated. It was found that addition of Mo to ZA22 either in the non-refined or the refined by Ti resulted in refining its structure being more refined in the latter. The ECAP process resulted in further refinement of its structure of the ZA22-Ti, ZA22-Mo and the ZA22-Ti-Mo alloys. Regarding the mechanical behavior, it was found that addition of Mo to ZA22 resulted in pronounced reduction of its mechanical strength presented by the following values of the flow stress at 20% strain: from 451 MPa to 346 MPa, whereas pronounced increase in case of Ti addition i.e. by 22.22% and only increase of 1.1% when Mo is added in the presence of Ti. However the Vickers hardness HV was increased by 5% in case of Ti addition and 2.5% increase in case of Mo addition. Finally it was concluded that super plastic behavior was obtained at room temperature by the addition of Mo and the ECAP process.

1. Introduction

Zinc- aluminum alloys, in general are versatile materials which are widely used in a wide variety of engineering and industrial applications due to their attractive properties such as strength, toughness, rigidity, bearing load capacity, economical and clean cast ability. In many aspects they are superior to aluminum, magnesium, and copper alloys. Zinc aluminum die casting alloys are widely used in the automobile and air craft industries in manufacturing many mechanical parts such as carburetor bodies, fuel pumps bodies, wind-shield wiper parts, control panels, horns and parts of the hydraulic brakes. Furthermore, zinc-aluminum alloys are used in structural and decorative parts which include radiators, steering wheels, hubs and instrument panels, [1-2]. Other applications include electrical, electronic and

Content from this work may be used under the terms of the Creative Commons Attribution 3.0 licence. Any further distribution of this work must maintain attribution to the author(s) and the title of the work, journal citation and DOI.
Published under licence by IOP Publishing Ltd

appliance industries. Building hardware padlocks and toys are major areas of applications of these alloys,[3]. Recently, Ridge Tool Company has replaced its bearing gear covers from bronze into Zn-27% Al because they found that this alloy has many of the desirable characteristics of bronze like easy finishing, corrosion resistance and good wear resistance in addition that the Zn-27%Al cuts material cost by 50% and reduces weight by 43%beside the most important advantage that it has longer service life. A newly developed "nanometer-crystalline" in Japan, zinc-aluminum alloy with a molecular elongation of more than 100 percent is said to be so resilient as to make possible an earth-quake resistant damper that can protect buildings. Shutter mechanisms in cameras, and many other electrical and electronics consumer applications, [4]. For aeronautical industry where alloys are subject to multidirectional service requests, they must provide an optimal combination of mechanical strength, plasticity, toughness, fatigue resistance and good resistance to stress corrosion. Moreover, the mechanical properties of zinc-based alloys make them attractive substitutes for cast iron and copper alloys in many structural and pressure-tight applications. These zinc-based alloys have a distinct cost advantage over copper-based alloys [4]. The major advancement in the zinc industry over the past years has been the development of zinc alloys with higher aluminum contents. These new zinc–aluminum based alloys have high strength and hardness, improved creep and wear resistance and lower density, and, although developed originally for sand and gravity casting, they are now being used in growing amounts for pressure die casting. These alloys, which possess excellent casting and mechanical properties, have been increasingly used to replace traditional alloys such as aluminum, bronze, brass and cast iron in many industrial applications, [5]. Against these advantages of zinc-aluminum alloys they have the disadvantages of low creep resistance and solidification with dendritic structures of large grain size which tends to deteriorate their mechanical properties, specially the impact strength. Therefore, It is essential to grain refine their microstructure by some grain refining by some rare earth elements, e.g. Ti or Ti+B. This has engaged many researchers in the last six decades, when it was found that the presence of Ti in Al resulted in grain refinement of its structure from large columnar structure with large grain size into equi-axed refined structure with small grains which caused enhancement in its mechanical behavior and surface quality, [6-7]. The literature on grain refinement of Al and its alloys and zinc aluminum alloys is voluminous, [8-15]. Review of the grain refinement of Al and its alloys and zinc-aluminum alloys are given and discussed [8] and [9-10] respectively.

Another method of grain refinement beside the refinement by some rare earth elements is the severe plastic deformation, SPD, processes, which have engaged some researchers in the last two decades many studies have been directed towards material processing by applying severe plastic deformation, referred to later as SPD. The main function of SPD is imposing extremely large plastic strains on materials which in turn will result in achieving fine grain size structure in the material that is subjected to. The reduction of grain size to a micrometer level results in enhancement of the mechanical properties such as very high strength and hardness without loss of ductility, which may also be accompanied by low temperature super-plasticity. Significant grain refinement may be achieved in bulk polycrystalline metals through the application of SPD. In the early eighties Segal et al at Minsk in the former USSR originated the method of equal channel angular pressing referred to later as ECAP which involves subjecting massive billets to pure severe plastic shear strains without change in their cross sectional areas. Their objective, at that time, was to develop a metal forming process with a high strain rate, [16]. In the early nineties the method was further developed and applied as an SPD method for processing of structures with submicron and Nano metric grain sizes. ECAP can apply a high shear strain to materials through a specially designed die having two equally sized channels connected at a finite angle. The technique has been proven to be very useful in improving mechanical properties of metals and alloys, [17]. The technique is a viable forming procedure to extrude materials by use of specially designed channel dies without substantially changing the geometry by imposing severe shear plastic deformation. This technique has the potential for high strain rate super-plasticity by effective grain refinement to the level of the submicron-scale or even Nano scale, [16, 18]. Severe plastic deformation (SPD) is a new discipline of metal forming technology used to produce a very fine grained structure in metals and their alloys in order to improve their mechanical and physical properties(Fang, et al.,2006).In recent years, ultra-fine grained materials are currently of great scientific interest due to important mechanical and physical properties, increased fatigue life and high damage tolerances. Investigations of the mechanics of severe plastic deformation help engineers to come up with better designs for processes as well as better prediction of the results of processes. ECAP process is the most welcome process because of its nearly homogenous and effective mode of straining (simple shear). During deformation, the sample undergoes simple shear while there is no change in the cross-section geometry. In the ECAP process, a well lubricated billet of a square or round cross section of some length is pressed throughout a die of some angle, Φ, with a die relief angle, Ψ, as shown in Figure 1. ECAP is a useful tool for achieving exceptional grain refinement in bulk metallic alloys, and the grain sizes produced through ECAP are in the sub-micron range. Recently, several analytical studies reported that the magnitude of shear strain imposed to the sample is determined by the channel angle Φ and the angle associated with the arc of curvature ψ, [16-18].

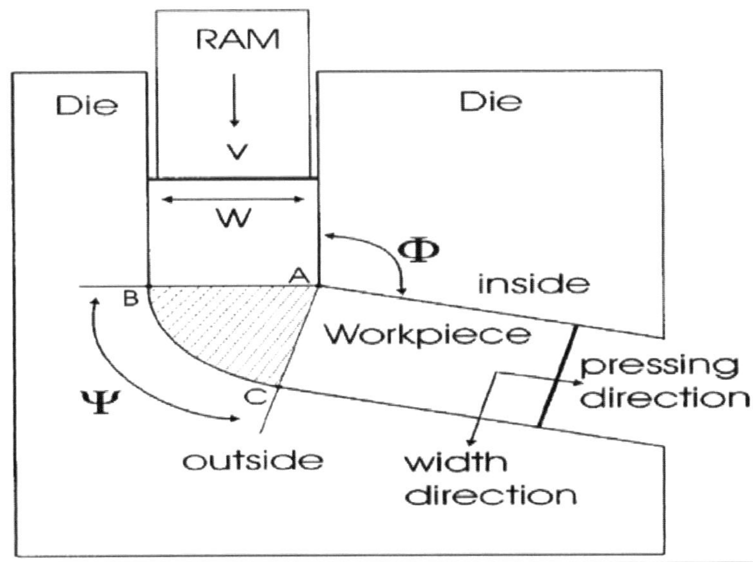

Figure 1. Equal chanel angular process

2. Materials, equipment and experimental procedures

2.1 Materials

Pure granular Zinc, High purity molybdenum, titanium and high purity aluminum were used in manufacturing main alloy ZA22 and the following binary master alloys: Al- Mo and Al-Ti, from which the different microalloys were made. Their chemical compositions were determined using Scanning Electron Microscope, SEM. Pure graphite crucibles were used in the melting process and graphite rods were used for stirring. The details of their manufacturing are given in [9].

2.2 Experimental procedure.

The experimental procedure started by designing manufacturing the ECAP die from H-13 and heat treated in accordance with the specified treatment cycle by the suppliers. Figure 2 shows a photograph of the designed and manufactured ECAP die.

Figure 2. ECAP die

3. Results and discussion

In this chapter, the results obtained through this investigation will be presented and discussed. As ZA22 alloy is industrially grain refined by either titanium, Ti, or titanium and boron, Ti-B, the obtained results will be dealt with under the following headings: Effect of molybdenum addition at a weight percentage of 0.1% on the grain size, microstructure, hardness, and mechanical behavior illustrated by the representative true stress, representative true strain, ZA22 grain refined by Ti and ZA22 grain refined by Ti+B. Finally, comparison is made between the effect of Mo addition at a rate of 0.1 % weight percentage to ZA22 alloy grain refined by Ti and ZA22 grain refined by Ti+B.

3.1 Effect of Addition of molybdenum on the Metallurgical and Mechanical Characteristics of ZA22 alloy Grain Refined by Ti

The effect of addition of Ti or Mo alone or together on the microstructure is explicitly shown in the photomicrographs of Figures 3 (a),(b),(c) and (d) for ZA22, ZA22-Ti, ZA22-Mo and ZA22l-Ti-Mo, respectively.

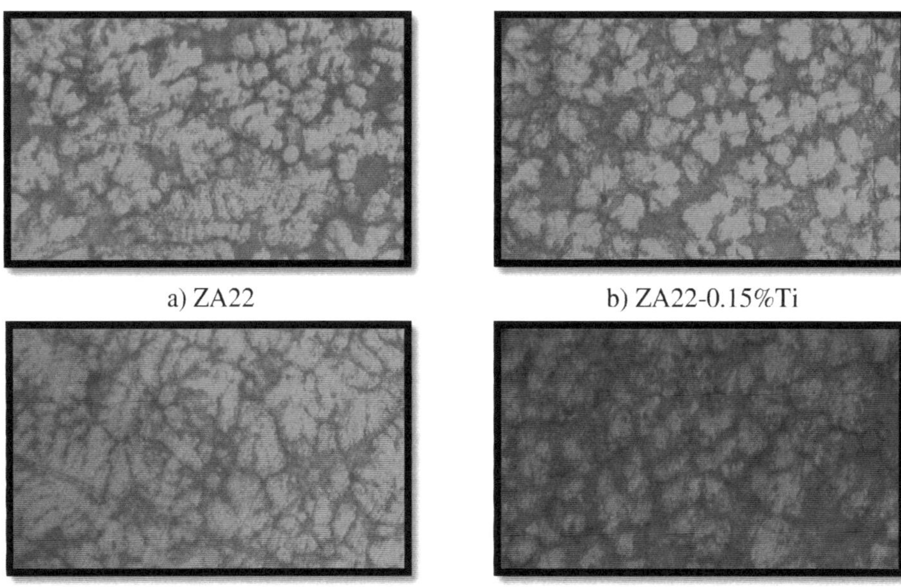

a) ZA22 b) ZA22-0.15%Ti

c) ZA22-0.1%Mo d) ZA22-0.15%Ti-Mo

Figure 3. Photomicrographs of ZA22 and its micro alloys: (a) ZA22, (b) ZA22-Ti, (c) ZA22-Mo, (d)ZA22l-Ti-Mo, in the as cast conditions, X 500

3.2 Effect of Mo Addition on the Hardness of ZA22 and ZA22 grain refined by Ti

It can be seen from the histogram of Figure 4 that addition of either Ti or Mo alone or both together resulted in enhancement of the hardness. The maximum increase is at Ti addition being 5.4 % followed by Mo addition (2.5 %) and the least little achievement is when both are added together, (0.7 %).

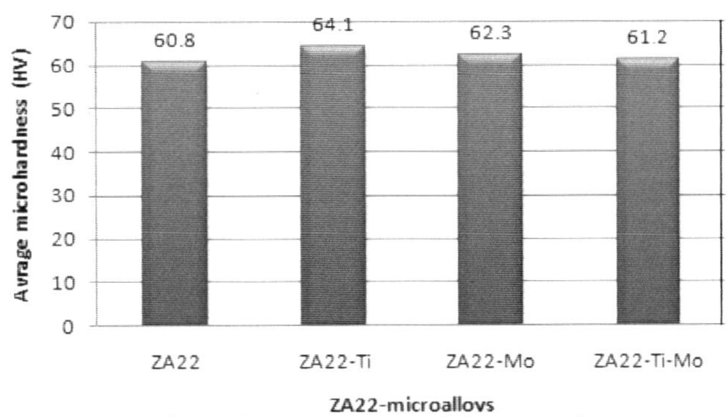

Figure 4. Effect of molybdenum addition on the average Vickers' smicro-hardness of ZA22 and ZA22 grain refined by Ti in the as cast condition

3.3 Effect of Mo Addition on the Mechanical Characteristics of ZA22 and ZA22 grain refined by Ti

Figure 5 shows the effect of Mo addition on the general mechanical behavior of ZA22 and the ZA22 grain refined by Mo and Ti micro-alloys represented by the true stress versus true strain curve of each of them. It can be seen from these curves that addition of Mo to ZA22 resulted in deterioration of its general mechanical behavior, e.g. a reduction in flow stress at 20% strain from 451 MPa to 349 MPa has occurred, whereas addition of Ti either alone or in the presence of Mo resulted in enhancement of its mechanical behavior. The enhancement is being more pronounced in the case of addition of Ti alone. These are explicitly illustrated from the results shown in Table 6 where the strength coefficient of the ZA22 has decreased from 836.4 MPa to 774.5 MPa in case of Mo addition but increased to 1150.8 MPa by adding Ti alone and to 933 by adding Ti+Mo. Regarding the work hardening index it can also be seen from Table 6: that it has increased from 0.364 of the ZA22 to 0.456, 0.495 and 0.458 in case of Ti, Mo, and Ti+Mo additions respectively. This may be attributed to the secondary interphases formed within the main matrix of the ZA22 alloy.

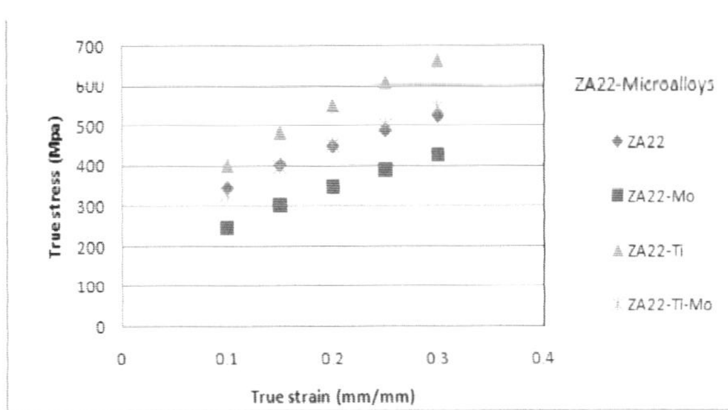

Figure 5. Effect of molybdenum addition on the True Stress true Strain of ZA22 and ZA22 grain refined by Ti in the as cast condition

Table1. Mechanical characteristics of ZA22 and its different micro-alloys in cast condition

Micro Alloys	Flow stress (MPa)at strain= 20 %	Strain hardening index (n)	Strength coefficient (K) MPa	General equation of mechanical behavior
ZA22	451	0.384	836.4	$\bar{\sigma} = 836.4\bar{\varepsilon}^{0.384}$
ZA22-Ti	552	0.456	1150.8	$\bar{\sigma} = 1150.8\bar{\varepsilon}^{0.456}$
ZA22- Mo	349	0.495	774.5	$\bar{\sigma} = 774.5\bar{\varepsilon}^{0.495}$
ZA22-Ti-Mo	456	0.458	953	$\bar{\sigma} = 953\bar{\varepsilon}^{0.458}$

Addition of Ti alone resulted in increase of its flow stress at 20 % strain as an increase of 22.4 % was achieved; also addition Ti+Mo resulted in a little achievement by (1%). However, addition of Mo alone resulted in decrease by (22.6 %). but addition of Mo alone or with Ti resulted in decrease of its flow stress at 20 % strain. as shown on Figure 6.

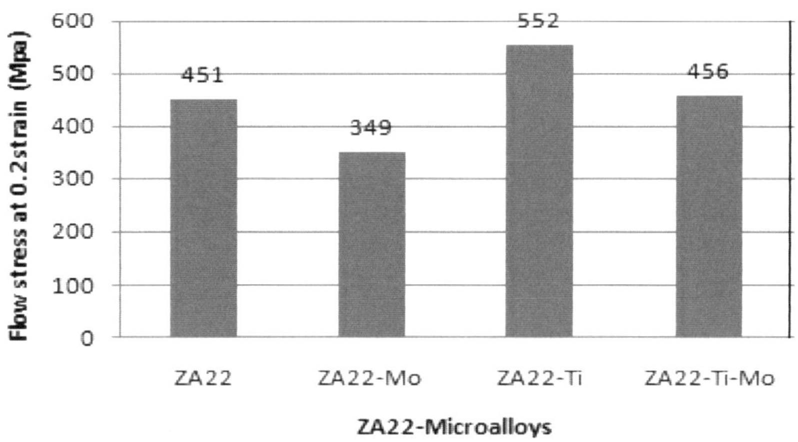

Figure 6. Histogram of the flow stress at 0.2 strain of ZA22 and ZA22

4. Conclusions

Addition of Mo to ZA22 resulted in the following:

a). In the as cast condition:

i). Modifying its metallurgical structure by reducing its grains size.

ii). Increase of its micro hardness, enhancing its mechanical behavior and increase of its strain hardening index, n, hence improves formability

b). After pressing by the ECAP process:

i). Further refinement of the grain size of ZA22 and its microalloys.

ii). Decrease in the microhardness and the flow stress but resulted in increase of the strength factor and increase of the strain hardening index, n, of ZA22 and its microalloys.

iii). Reduction in the extrusion force and work.

iv). The addition of Ti and Mo to ZA22 and its micro alloys indicated that they have gained some superplastic behavior at room temperature. This renders the process cost effective

5. Acknowledgment

The first author, (Adnan I. O. Zaid) is grateful to the Applied Science Private University, Amman, Jordan for partial financial support granted to this research (Grant No.DRGS-2015).

6. References

[1] Budinsk K G. 1992 Engineering Materials Selection, Fourth Edition, Prentice-Hall, Inc., NewJERSY

[2] Callister W D 1994 Materials Science and Engineering and Engineering: an Introduction, Third Edition, John Wiley and Sons inc., New York

[3] Nevison D C H 1998 Zinc and Zinc Alloys, American Society of Metals, ASM Handbook, Fourth Edition

[4] Şahin M, Cadirli E 2011Solidification Characteristics and Micro-structural Evolution of Zn-1.26 wt. % Al Alloy*Surface Review and Letters*18, 281-288

[5] Tu¨rk Durman, M, Kayali E S 2007 the effect of manganese on the microstructure and mechanical properties of zinc–aluminium based ZA-8 alloy, Journal Material of Science, 42, 8298–8305

[6] Cibula A1949-1950 *Journal Institute of Metals*76, 321-360

[7] Cibula A1951-1952 *Journal Institute of Metals*80, 1-15

[8] Zaid, A I O 2001 Grain Refinement of Aluminum and its *Alloys Proceedings of the Seventh International Symposium on Advanced Materials*, 2001, Islamabad, Pakistan,12-23

[9] Zaid A I O, AL-Dous Z M A 2006 Effect of Zirconium Addition on the Ductility and Toughness of Cast Zinc Aluminum Alloy-5 Grain Refined by Ti+B, *Proceeding of the Second International Conference on the Frontiers of Advanced Engineering Materials* Lahore, Pakistan

[10] Zaid A I O, Hussein M I J 2006 Effect of Zirconium Addition on the Mechanical Behavior and Wear Resistance of Zamak5 Alloy Grain Refined By Titanium.*Proceedings of the* 12th *International Conference on Machine Design and Production*Kusadasi-Turkey

[11] Pollard W A Pickwick K M, Jubb, J T, Packwood R I 1974 The Grain Refinement of Zinc-Aluminum Alloys by Titanium *Can. Metal. Soc.* 13

[12] Tsan C L, Nakata T Hayashi T 1976 Effect of Zirconium, Titanium, Manganese, and Nickel Additions on Dendrite Morphology of Zinc-Aluminum Alloys *IMONO*3,167-174

[13] Jones G P, Pearson J 1976 Factors Affecting Grain Refinement of Aluminum Using Titanium and Boron Additives Metallurgical Transactions 223-234

[14] Motohashi, Y, Kikuchi H, Shibata T 1985 Improvement in Mechanical Properties of Zinc-base Alloy for Metal Pattern. Effect of the Addition of Alloying Elements 33, 77-84

[15] Abdel Hamid A A 1992 Structure Modification of the α-Phase in Zn-Al Cast Alloys by Microalloying*Metalkd*83, 314-318

[16] Sanusi K OMakinde O D, Oliver G J 2012 Equal channel angular pressing technique for the formation of ultra-fine grained structures. *S Afr J Sci. Art,*108

[17] Fang D R,Zhang Z F, Wu S D, Huang C X, Zhang H, Zhao N Q, Li J J2006Effect of equal channel angular pressing on tensile properties and fracture modes of casting Al–Cu alloys, Materials Science and Engineering, 426, 305–313

[18] Rahul M R 2009 Development of A Continuous Equal Channel Angular Extrusion (ECAE) Process. Doctoral Dissertation, University of Mumbai, India

Analysis of degradation in UHMWPE a comparative study among the various commercial and laboratory grades UHMWPE

A. M. Azam, A. Ali[1], H. Khan[1], T. Yasin[1] and M. S. Mehmood[1,2]
Department of Physics, Umea University, SE-901 87 Umea, Sweden
[1]Department of Basic Sciences, University of Engineering and Technology, Taxila-47050, Pakistan
[2]Advance Polymer Laboratory, Pakistan Institute of Engineering and Applied Sciences, Islamabad-46050, Pakistan

E-mail: msajjad.82@gmail.com

Abstract. Oxidative degradation of the ultra-high molecular weight polyethylene (UHMWPE) limits the life of implants. This degradation can be monitored to estimate the service life of UHMWPE following the standard protocols as defined by American Standards for Testing Materials (ASTM). In this work, a comparative study has been carried on two commercially available UHMWPE grades i.e. GUR 1020 and GUR 1050 and one laboratory grade UHMWPE which was purchased from Sigma Aldrich. These powder samples were pressed while using hot press with controlled heating and cooling setup in open air under 200 bar of external pressure. These sheets were then subjected to accelerated aging in an oven at 80 °C for three weeks. The degradation of the UHMWPE was monitored by ATR-FTIR techniquefor three weeks. The oxidation index (OI) measurement showed that the commercial grade UHMWPE i.e. GUR-1020 and GUR-1050 degrade more as compared to laboratory grade UHMWPE. The values of OI after three weeks of accelerating aging were found 0.18, 0.14, and 0.09 for GUR-1020, GUR-1050, and Sigma Aldrich, respectively. In addition to this, it was found that commercial grades of UHMWPE suffer more structural alterations as compared to laboratory grade one. We hope that these results will be of particular and fundamental importance for the researchers and orthopaedic industry.

1. Introduction

Ultra-high molecular weight polyethylene (UHMWPE) belongs to polyolefin family which was introduced after the failure of polytetrafluoroethylene (PTFE) in 1962 by Sir John Charnley. UHMWPE has been used in a number of specific areas including bearing components, gears, sporting goods, guide rails and as a biomaterial in total joints replacement. More than 90% of the total joints replacement cases, hip cups and knee inserts are made of UHMWPE for immediate pain relief in aching patients [1-3]. Artificial joint substitution in elder people has been increasing in accordance with their desire for independent living and recovery of functional use. The Swedish Hip Arthoplasty Register (SHPR) which becomes active in 1979 has recorded types of prosthesis, factors concerning the operation and result in the form of complications. The Nordic Arthoplasty Register Association (NARA) has awarded a major grant from Nordic Council of Ministers in order to harmonize the generic descriptions of the used prosthetics in the region. According to SHPR Annual Report 2013,

Content from this work may be used under the terms of the Creative Commons Attribution 3.0 licence. Any further distribution of this work must maintain attribution to the author(s) and the title of the work, journal citation and DOI.
Published under licence by IOP Publishing Ltd

16,299 total hip replacement operations were operated in Sweden which corresponds to 324 per 100,000 inhabitants aged 40 years or older. International comparison of the countries reporting shows that Sweden has among the highest incidence means life expectancy is increasing and proportion of older people among the population increases [2]. In case of knee joint substitution, average life of UHMWPE joint is 15-20 years and life of an artificial joint depends on the degree of deterioration of UHMWPE. Bio-implant materials such as acetabular liners and tibial inserts are made from UHMWPE because of having biocompatibility, high wear resistance, lower coefficient of friction, suitable stiffness, abrasion resistance, toughness and relative low cost [4]. The extremely high viscosity of UHMWPE has made a challenging task in processing of reinforced UHMWPE composites. UHMWPE is almost impossible to process in the melt because of their high melt viscosity. Much effort has done in finding suitable processing methods [5]. In addition to this researchers are working to find the suitable method for stopping the degradation of UHMPWE by incorporating antioxidants, nano fillers, silane etc. [6]. However, some laboratories are using various available grades of UHMWPE which include commercial grades i.e. GUR-1020, GUR-1050, and some are using laboratory grade UHMWPE while purchasing from different suppliers of chemical industry. It is therefore, a comparative study among commercial and laboratory grades UHMWPE is necessary for having better understanding of degradation mechanism of UHMWPE.

In this work, a comparative study has been carried on two commercially available UHMWPE i.e. GUR 1020 and GUR 1050 and one laboratory grade UHMWPE purchased from Sigma Aldrich. These samples were subjected to accelerated aging for three weeks following the standard protocol as defined by ASTM standards [1]. The structural changes and oxidation degradation of these samples has been monitored by ATR-FTIR spectra taken after each week.

2. Materials and methods

2.1 Sample preparation

The UHMWPE (GUR 1020, GUR-1050, and Sigma Aldrich) was pressed into the sheets of 1mm thickness using hot press 190°C at 20MPa. A hold time of 10 minutes was given. After compression treatment samples were cooled down to room temperature at 20MPa. The samples were then divided into four groups, one was kept as non-aged sample and other three samples were put into oven for one, two and three weeks for accelerated aging, respectively at 80°C in air following ASTM standard guide (ASTM F2003-00) [1].The UHMWPE which was purchased from Sigma Aldrich was labeled as 'S' in the manuscript from here on.

2.2 Ftir measurement

The absorption of electromagnetic energy in the infrared (IR) region occurs due to vibrational modes of atoms in solids and liquids. Chemical species having different molecular formula are present in semi-crystalline materials such as polymers. FT-IR spectroscopy is extensively used experimental technique for the investigation of molecular structures of polymers It also investigates chemical bonds. Nicolet-6700 Fourier transform infrared spectrophotometer (Thermo Electron Corporation, Waltham, MA, USA) was used to record IR spectra in attenuated total reflectance mode and plotted in transmission mode. The spectra were obtained in the range of 4000 cm^{-1} to 500 cm^{-1} at a resolution of 6 cm^{-1} after acquiring 216 scans. The irradiated compressed sheets were served as protection for measurements after receiving from irradiation department. The spectra for all samples were recorded from three/four points and then averaged to decrease signal to noise ratio (SNR).

3. Result and discussion

The most effective way to accelerate the ageing of UHMWPE currently in practice is oven aging and researchers are using this method for estimating the service life of UHMWPE and its composites [3-4]. Un-aged and aged samples are tested using FTIR spectrometer on the recommended setting as mentioned above. The IR absorption bands of particular interest as far as degradation of these commercial and laboratory grade UHMWPE is concerned are listed in the Table 1.

Table 1. UHMWPE IR absorption bands of our interest

Description	IR absorption band (cm-1)
Alcohol & hydro-peroxide	3450-3350 [7-8]
Carboxylic acid, Aldehydes, ketones & Carbonyl species	1710-1740 [7-8]
Ethers and other -C-O-C groups	1100-1400 [7-8]
Trans-vinylene groups, C=C Unsaturated bonds	800-1000 [7-8]

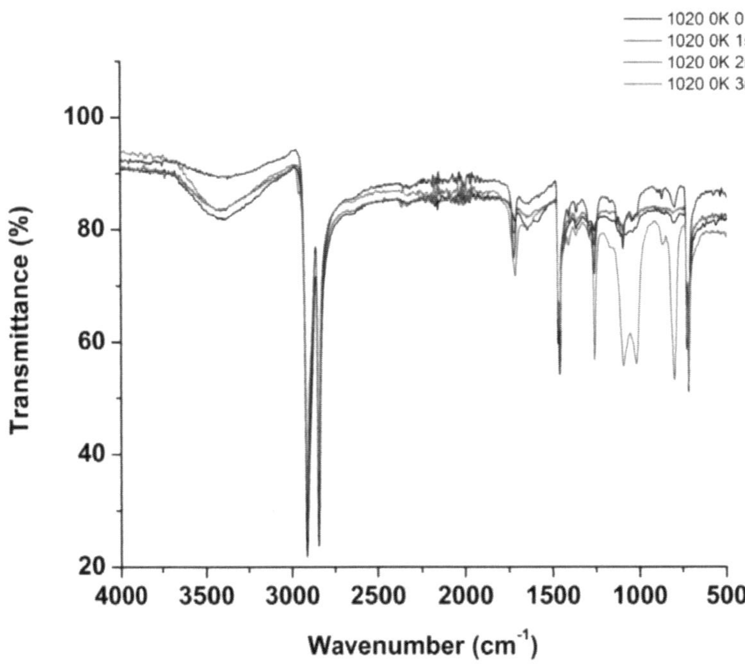

Figure 1. FTIR spectra of GUR-1020 grade of UHMWPE

Figure 1 show FT-IR spectra of sheets of UHMWPE prepared from GUR-1020. In the figure, it can be seen that each spectrum is labeled as 1020-0K-xW, where variable 'x' represent the time of oven aging i.e. it is 0 for control samples, 1 for one week aged, and so on up three weeks. It can be seen from the figure that spectra of GUR-1020 showan increase in absorption in the following bands after aging.

- Form 3000 cm^{-1} to 3750 cm^{-1} which corresponds to typical end products of polyethylene (PE) oxidation i.e. hydrogen bonded and non-bonded hydro-peroxide
- Form 1650 cm^{-1} to 1850 cm^{-1} which is due to absorption of carbonyl species i.e. C=O
- Form 1460 cm^{-1} to 1480 cm^{-1} and 2880 cm^{-1} to 3000 cm^{-1} which is associated with PE crosslinking
- Form 1100 cm^{-1} to 1400 cm^{-1} which is associated with –C-O-C– vibrations
- Form 800 cm^{-1} to 1100 cm^{-1} which is related with the absorption due to C=C un-saturation

It is evident from figure that in addition to increase in absorption due to material degradation, there is significant increase in absorption from 800 cm^{-1} to 1400 cm^{-1} which is mainly associated with –C-O-C– vibrations and C=C unsaturation. Furthermore, there is abrupt increase in the absorption of functional groups associated with these bands for three week aged samples. When all chemical groups are investigated, a similar trend has been observed that is hydroperoxides, carbonyl and vinylenes species have represented considerably higher content for longer aging time. There are no considerable contents are there for control and one week aged samples and found to increase remarkably as the aging time proceeds.

Shown in Figures 2 and 3 are the FTIR spectra of samples prepared from GUR-1050 and Sigma Aldrich grades of UHMWPE. The major changes are almost similar as mentioned above for GUR-1020. However, one can see from Figures 2 and 3 that except from the typical oxidation products and carbonyl species, absorption due to –C-O-C– vibrations and C=C unsaturation have considerable lower values. This fact suggests that GUR-1050 and Sigma Aldrich UHMWPE is more stable as compared to GUR-1020 when subjected to accelerated aging. The lower molecular weight and smaller chains of GUR-1020 might be responsible for this fact. However, further studies are in progress for detail investigation of this difference as far as stability of various UHMWPE grades is concerned.

Now coming to the quantitative measure of UHMWPE degradations due accelerating aging and for this purpose oxidation index (OI) of each sample was calculated following the standard protocol as defined in literature [7-8]. Basically, OI is the ratio of the peak height due to C=O stretching (^{o}C=0) components of ketone present at 1718 cm^{-1} to the absorption band at 2020 cm^{-1} which belongs to CH_2 bending vibrations (internal standard, $^{\delta}CH_2$, "twist" bend) i.e.

$$OxidationIndex(OI) = \frac{I_{1718}}{I_{2020}}$$

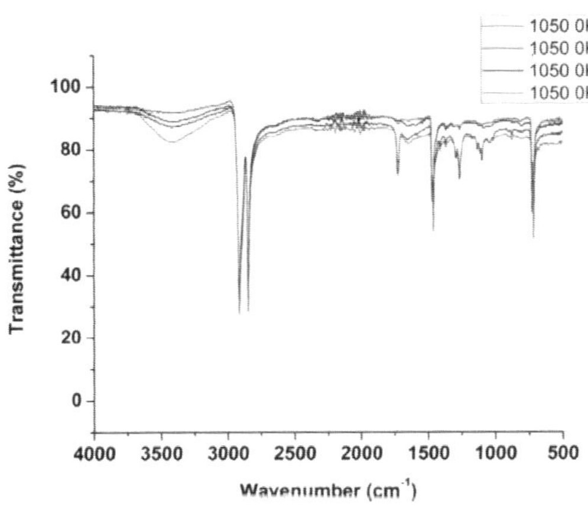

Figure 2: FTIR spectra of GUR-1050 grade of UHMWPE

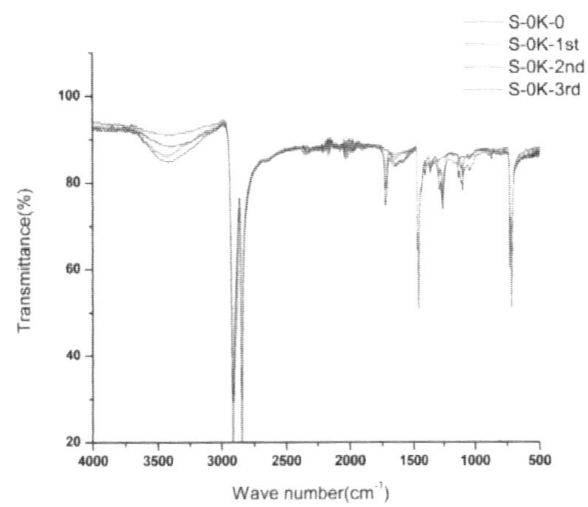

Figure 3. FTIR spectra of UHMWPE samples purchased from Sigma Aldrich

The values of OI are shown in Table 2 below:

Table 2. Oxidation Index (OI) of UHMWPE grades under investigation

UHMWPE Grade	Control	1-week aging	2-week aging	3-week aging
GUR-1020	0.02	0.10	0.13	0.18
GUR-1050	0	0.05	0.13	0.14
Sigma Aldrich	0.01	0.04	0.08	0.09

From figure it can be seen that, the vaule of OI index is almost zero for control samples, and found to increase with oven aging along with steep rise with increas in aging time. As expected all three grades i.e. GUR-1020, GUR-1050, and Sigma Aldrich have shown maximum values after 3rd week of aging. However, GUR-1020 is found to suffer more after three weeks of aging which 18 % and degradation in UHMWPE which is purchased from Sigman Aldrich is found less than 10 %. These qunatitative results further confirms that UHMWPE which is purchased from Sigma Aldrich (also used at laboratry scale for research work) is more stable when subjected to accelerated aging as compared to GUR-1020 and GUR-1050 which are the commercial products for medical implants. It is therefore, there might be slight difference in the results while using different grades of UHMWPE as far as radition induced sturctral changes in UHMWPE.

4. Conclusion

In order to analyze the degradation and thermal stablity three different grades (two commerical GUR-1020 and GUR-1050 and one labortary grade purchased from Sigma Aldrich) were subjeted to accelerated aging for three consective weeks and tested after each week by using ATR-FTIR spectrometer. The results obtained so far revealed that GUR-1020 suffer more as compared to other two grades i.e. GUR-1050 and Sigma Aldrich. The qunatitaive analysis showed that GUR-1020 degraded 18%, GUR-1050 degraded 14 %, and Sigma Aldrich UHMWPE degraded less than 10 %, respectively after three weeks of aging. These results showed that UHMWPE which is purchased from Sigma Aldrich and frequently used by researchers in labortory is more stable when subjected to thermal/accelerating aging.

5. Acknowledgment

The authors paticularly Dr. Malik Sajjad Mehmood would like to thank Dr. Muhmmmad Shah Jahan, University of Memphis, Memphis, TN, USA and Ticona Inc. for providing some samples/grades of UHMWPE for this research.

6. References

[1] Medhekar V, Thompson R W, Wang A andGrant McGimsey W 2003 Journal Applied polymer Science 87 814
[2] Taddei P, Affatato S, Fagnano C, and Toni, A 2006 Biomacromolecules 71912
[3] ISO 5834-4:2005 2005 Implants for surgery – UHMWPE - Part 4: Oxidation index measurement method
[4] Medel J, Garcia- Alvarez F, Gomez- Barrena E and Puertolas JA 2005 Polymer Degradation and Stabality 88 435
[5] Kasser M J 2009 The photochemistry of polyenyle radicals and its application to UHMWPE for use in replacement joints (University of Maryland: Maryland USA)
[6] Zhang H, Shi M, Zhang J and Wang S 2003 Journal of Applied Polymer Science892757
[7] Wang L H, Porter R S, Stidham H D and Hsu S L 1991 Macromolecules245535
[8] Kurtz S M 2009 UHMWPE: Biomaterials Handbook (BurlingtonUSA:Elsevier)

IOP Conf. Series: Materials Science and Engineering **146** (2016) 012026 doi:10.1088/1757-899X/146/1/012026

Investigation on the effect of titanium (Ti) addition to the Mg-AZ31 alloy in the as cast and after extrusion conditions on its metallurgical and mechanical characteristics

Adnan I. O. Zaid, Raghad and S. Hememat[1]
Applied Science Private University, Mechanical and Industrial Engineering Department, Amman 11931 Jordan
[1]University of Jordan, industrial Engineering Department Amman, Jordan

E-mail: Adnan_kilani@yahoo.com, [1]raghad_hemeimat@hotmail.com

Abstract. Magnesium-aluminum alloys are versatile materials which are used in manufacturing a number of engineering and industrial parts in the automobile and aircraft industries due to their strength – to –weight -ratios. Against these preferable characteristics, magnesium is difficult to deform at room temperature; therefore it is alloyed with other elements mainly aluminum and zinc to add some required properties particularly to achieve high strength -to- weight ratio. Grain refinement is an important technology to improve the mechanical propertiesand the microstructure uniformity of the alloys. Most of the published work on grain refinement was directed toward grain refining aluminum and zinc alloys; however, the effect of the addition of rare earth material on the grain size or the mechanical behavior of Mg alloys is rare. In this paper the effect of Ti addition on the grain size, mechanical behavior, ductility, extrusion force and energy, of Mg-AZ31 alloy both in the as cast condition and after direct extrusion is investigated.

1. Introduction

The increasing demand for light weight structures has forced the industry to look for new effective methods to produce light weight alloys. A major problem which hampers the wide spread such alloys like aluminum and magnesium alloys is their limited formability when the conventional manufacturing techniques are used. Magnesium and its alloys are the lightest constructional available materials on earth with a maximum density of 1.78 gm/cc and melting point of 650 °C, with good damping characteristics, but difficult to deform at room temperature, [1]. Magnesium and its alloys have a maximum density of 1.78 gm/cc. Mg and its alloys oxidize rapidly, accompanied by exothermic reaction, which may catch fire; therefore care must be taken during their machining. However products of Mg and its alloys are not fire hazard. Mg and its alloys are used in a wide variety of structural and nonstructural applications. Structural applications include automotive applications such as clutch and pedal support brackets and steering column lock housings. They are also used in industrial machinery such as textile and printing machines. Furthermore, they are used in commercial applications which include hand held tools, computer housings and ladders. Mg alloys are very useful in the parts which operate at high speed to minimize the inertial effect due to their light weights. They are also valuable for aircrafts due to their high strength- to- weight- ratio and good stiffness both at room and elevated temperatures. The ductility of magnesium alloys in the cold state is low; therefore it is recommended that these alloys are worked at hot temperatures. They are typically combined with

Content from this work may be used under the terms of the Creative Commons Attribution 3.0 licence. Any further distribution of this work must maintain attribution to the author(s) and the title of the work, journal citation and DOI.
Published under licence by IOP Publishing Ltd

manganese, aluminum, zinc, or zirconium for different applications and they can be used to produce components with superior mechanical properties. Magnesium and its alloys have recently received a great attention over the last decade, due to the increasing demand for the light weight structural materials replacing other metals in several industrial and engineering applications, [2]. Grain refinement is an important technology for improving the mechanical properties; surface quality and workability of both cast and wrought alloys.Grain refinement has an industrial history stretching back to 1949 when Cibula showed that the presence of Ti, particularly in combination of carbon or boron, produced a great refining effect in Al. Since then it became an industrial practice to grain refine Al and its alloys by adding Ti or Ti+B to the melt prior to solidification. Al- Ti binary and Al – Ti –B ternary master alloys are developed and commercially available for this purpose, [3-4]. Later, the grain refinement of zinc alloys was investigated in an attempt to enhance their potential to be used in many industrial applications, especially in the automotive and aircraft industries, [5]. Due to the great demand of Mg and its alloys, they have received great attention from the researchers to drive out results of their benefits focusing on achieving methods that might lead to refine the grains of the Mg alloys. Research works were conducted to test the formability of the Mg-AZ31 alloy and how it can be improved; it was found that as the Al content of the alloy increases, yield strength, ultimate tensile strength and the oxidation resistance all will increase; whereas the cast ability of the alloy will decrease, [6]. Later, a new equal channel angular extrusion (ECAE) processing was applied to control the microstructure and mechanical properties of AZ31 Mg alloy resulting in a very fine grains and a significant increase in both the ductility and yield stress of the alloy after the processing. Further results showed that the grain refinement in AZ31 magnesium alloy was successfully carried out using the ECAP (Equal Channel Angular Pressing) processing at different temperatures; reducing the grain size nearly four times after 2 passes of ECAP under 250 ^{o}C with a homogeneous grains. This also resulted in increasing the hardness, [7]. Research works were conducted to test the formability of the Mg-AZ31 alloy and how it can be improved; it was found that as the Al content of the alloy increases, yield strength, ultimate tensile strength and the oxidation resistance all will increase; whereas the castability of the alloy will decrease. The effect of Al on the alloy ductility, or maximum elongation during tensile testing, is less predictable since Al is capable of either increasing or decreasing mechanical strength depending upon the used heat-treatment.Initially, Mg-AZ31 alloy was also refined by adding alloying elements like Al-Ti-C master alloy with different Ti concentrations resulted in an efficient grain refinement of the alloy microstructure; this was explained by the forming of Al_4C_3 and TiC particles; it was also noted that the higher the percentage of Ti in the master alloy, the more efficient the grain refinement process, [8-9]. Researchers have also examined the superplastic behavior of the Mg-AZ31 alloy after extrusion and processing by the ECAP; it was demonstrated that an elongation of around 460% may be attained at a temperature of 150 ^{o}C, which reflects the potential for achieving superplastic behavior at low temperatures. Room temperature tensile properties of a superplastic Mg-AZ31 alloy were examined showing that the alloy behaved in a superplastic manner at low strain rates. The formability of the Mg-az31 alloy sheets at warm working conditions was investigated, [10,11].The effect of subjecting the Mg-AZ31alloy to severe plastic deformation after the addition of a grain refiner is rare. This formed the main objective of this work. In this paper, the effect of the addition of Ti to Mg-AZ31 alloy on its metallurgical and mechanical aspects in the cast and after direct extrusion is investigated.

2. Materials, equipment and experimental procedures

2.1 *Materials*

The following materials and alloys were used in this work:

Mg- AZ31 alloy, very high purity Ti and aluminum of 99.98 % purity and a flux made of 47% potassium chloride, 5% barium chloride and 48%Mg chloride to avoid oxidizing of the melt inside the electrical furnace. The Mg-AZ31alloy was imported from China and the , very high purity Ti and aluminum were supplied by ARAL, Arabic Co for manufacturing aluminum products.

2.2 Equipment and experimental procedure

The experimental procedure started with the design and manufacturing of the forward extrusion die followed by the preparation of the Mg-AZ31 and its microalloys before and after extrusion for the following tests: Vickers micro hardness test, standard tensile tests for mechanical behavior determination and specimens for metallurgical examination: grain size and photomicrographs of the general microstructure for each microalloy. All specimenswere machined on a CNC machine under the same cutting conditions of speed, depth of cut and feed rate.

3. Results and discussion

3.1 Effect of Ti addition on the grain size of Mg-AZ31 in the cast and after extrusion

The histogram shown in Figure 1 indicates that the addition of Ti to Mg-AZ31 alloy resulted in refining its grain size in the as cast condition; the grain size was reduced from 22 to 19 μm. Similarly, the histogram of Figure 1 also shows that the extrusion process resulted in further refinement of the grains of the Mg-AZ31 grain refined by Ti. The decrease in the grain size being 27% and 26% respectively. This agrees with previous findings in the literature. This is attributed to the heavy plastic deformation caused by the extrusion process. The grain refining effect is explicitly illustrated in the photomicrographs of Figure 2 (a), (b), in the as cast condition and Figure 3 (a), (b), after extrusion for Mg-AZ31, and Mg-AZ31-Ti respectively.

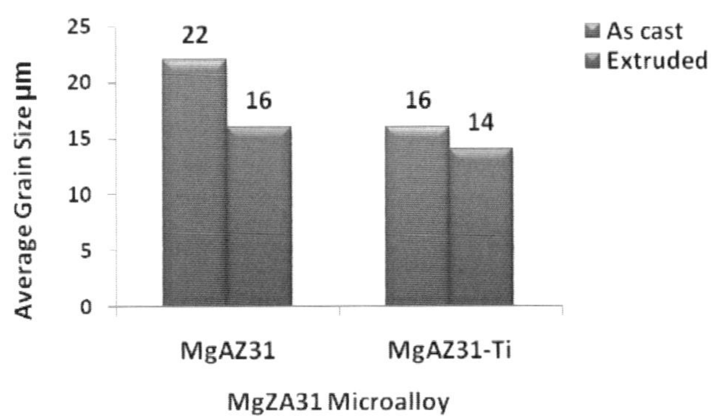

Figure 1. Effect of Ti addition on the grain size of Mg-AZ31 in the as cast and after extrusion

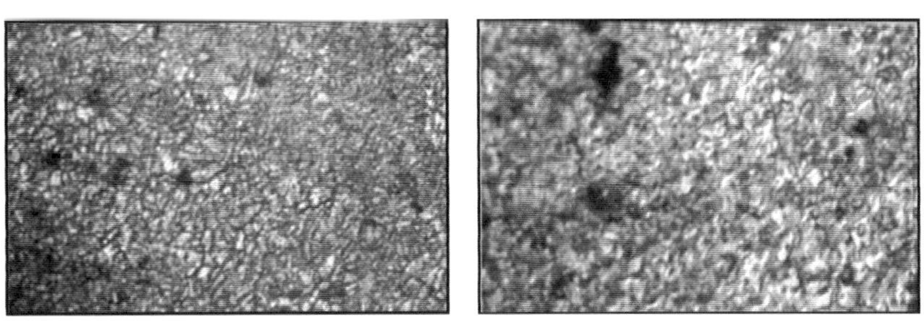

(a) Mg-AZ31 (22 μm) (b) Mg-AZ31 –Ti (19 μm)

Figure 2. Photomicrographs of Mg-AZ31 and Mg-AZ31-Ti in the as cast condition

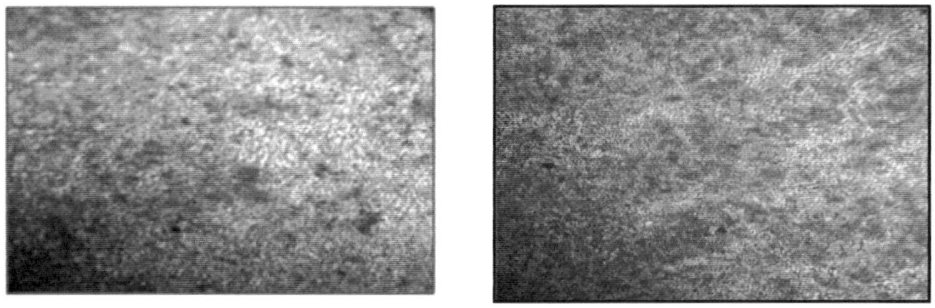

(a)Mg-AZ31 (16 μm) (b) Mg-AZ31 –Ti (14 μm)

Figure 3. Photomicrographs of Mg-AZ31 and Mg-AZ31-Ti after direct extrusion

3.2 Effect of Ti addition on the mechanical characteristics and formability of Mg-AZ31 both in the cast and after extrusion conditions

Figures 4 and 5 show the mechanical behavior of Mg-AZ31 and its microalloys both in the as cast condition and after extrusion respectively as determined from the compression tests. These figures indicate enhancement in its mechanical strength by Ti addition particularly at larger values of strain. This is in general agreement withthe results reported by in [12-14].

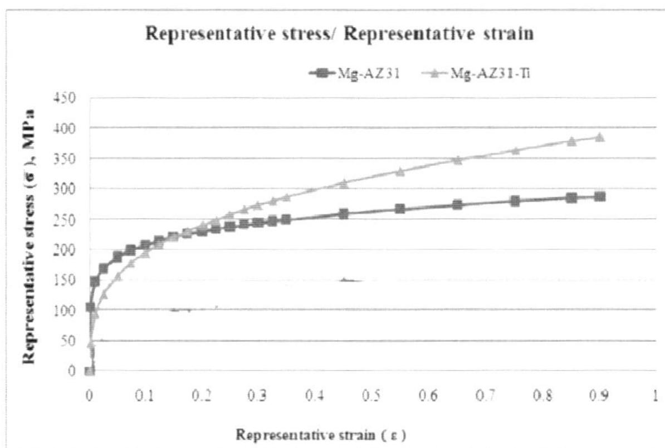

Figure 4. Representative stress (σ)- representative strain (ε) of Mg-AZ31 and its microalloy in the ascast condition

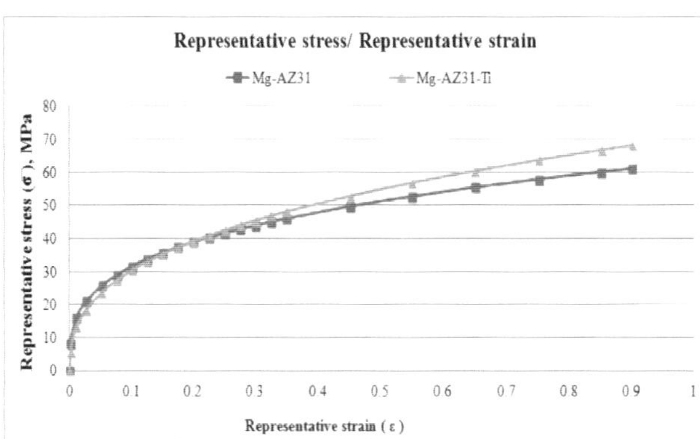

Figure 5. Rep. stress (σ), representative strain(ε) of Mg-AZ31 and its microalloy after extrusion

Table 1 gives a summary of the mechanical characteristics of Mg-AZ31 alloy and its microalloy Mg-AZ31-Ti in the as cast condition

Table 1. Mechanical characteristics of Mg-AZ31 and its microalloy in the cast condition

Mg-AZ31 and its microalloys	Flow stress strain=20 %	General equation of mechanical behavior	Strength coefficient (k) MPa	Strain hardening index (n)
Pure Mg-AZ31	230.1	$\sigma = 292\ \varepsilon^{0.148}$	292	0.148
Mg – AZ31 –Ti	241.3	$\sigma = 398\ \varepsilon^{0.311}$	398.1	0.311

It can be seen from Table 2 that the addition of Ti to Mg-AZ31 caused an increase in both the strength coefficient of the Mg-AZ31 alloy and the flow stress at 20% strain; this is attributed to the hard particles of the intermetallic compounds of alumides as discussed previously. Regarding the work hardening index, it can be seen that it has also increased by the addition of both Ti; this indicates better formability i.e. it can be formed to a larger strain before plastic instability occurs which in return will reduce the number of stages in the forming process when preforms are involved because the representative strain equals the strain hardening index ($\varepsilon = n$) at plastic instability. Table 2 shows the mechanical characteristics of Mg-AZ31 and its microalloys after the extrusion process. It clearly indicates that the extrusion process resulted in increase of the work hardening index from 0.148 to 0.299 for the Mg-AZ31 and from 0.311 to 0.366 for the Mg-AZ31-Ti.

Table 2. Mechanical characteristics of Mg-AZ31 and its microalloys after extrusion

Mg-AZ 31 and its microalloys	Flow stress (MPa) at strain=20%	General equation of mechanical behavior	Strength coefficient (k) MPa	Strain hardening index (n)
Pure Mg-AZ31	38.9	$\sigma = 63\ \varepsilon^{0.299}$	63	0.299
Mg – AZ31 –Ti	39.2	$\sigma = 70.7\ \varepsilon^{0.366}$	70.7	0.366

3.3Effect of Ti Addition on the Hardness of Mg-AZ31 both in the Cast and After Extrusion Condition

The effect of Ti addition on the hardness of Mg-AZ31 alloy both in the cast and after extrusion conditions is shown in the histogram of Figure 6. The histogram indicates that the addition of Ti to Mg-AZ31 alloy resulted in increase of its Vickers microhardness both in the as cast condition and after extrusion. This enhancement in case of Ti addition may be attributed to the formation of hard particles of the intermetallic compounds within the main matrix.

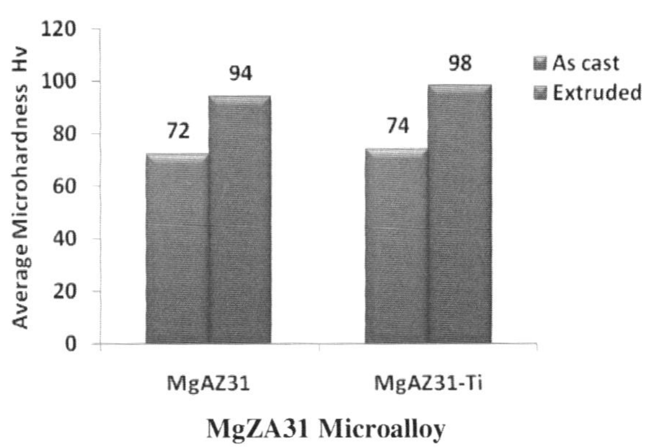

Figure 6. Effect of Ti addition on the hardness of Mg-AZ31 both in the as cast and after extrusion conditions

3.4 Effect of Ti addition to Mg-AZ31 on the extrusion force and energy

The experimental values of extrusion force and energy for the Mg-AZ31 alloy and its microalloy are determined from the autographic records of the extrusion process. Figures 7 and 8 shows the autographic records of the extrusion process of pure Mg-AZ31 and Mg-AZ31-Ti respectively. These curves were used to determine the extrusion force and energy as summarized in Table (3).It can be seen from this table that the extrusion force was reduced by 57% while energy of Mg-AZ31wAS reduced by 59% with the addition of Ti.

The Effect of extrusion conditions on microstructure and mechanical properties of micro alloyed Mg–Sn–Al–Zn alloys is given in [12] and the effects of Extrusion Speed on the Microstructure and MechanicalProperties of Mg-9Gd—3Y—1.5Zn—0.8Zr alloy is forwarded in [13].

Table 3. Experimental values of extrusion force and energy

Alloy	Extrusion Force (N)	Extrusion Energy (J)
Pure Mg-AZ31	15,365	207
Mg – AZ31 –T	6,598	85.3

3.5 Effect of Ti Addition to Mg-AZ31 on the ductility

The determination of ductility for the Mg-AZ31 and Mg-AZ31- Ti was obtained successfully in the as cast condition by the maximum elongation % and the maximum reduction in cross sectional area % as shown in the histograms of Figures 7 and 8 respectively, whereas after extrusion, it was not possible to obtain enough length of the extruded specimens of the Mg-AZ31 alloy and Mg-AZ31-Ti microalloy, although several tests were conducted, and the specimens from them were fractured after extruding 12 mm as shown in the photographs of Figures 9a and 9b for Mg-AZ31 and Mg-AZ31-Ti respectively Typical extrusion autographic records of the Mg-AZ31 and its Mg-AZ31-Ti microalloy are shown in Figures 10a and 10b respectively. The significant effect of adding trace amounts of Ti on the high temperaturedeformation behavior of fine-grained Mg–6Al–1Zn magnesium alloyis reported in [14].

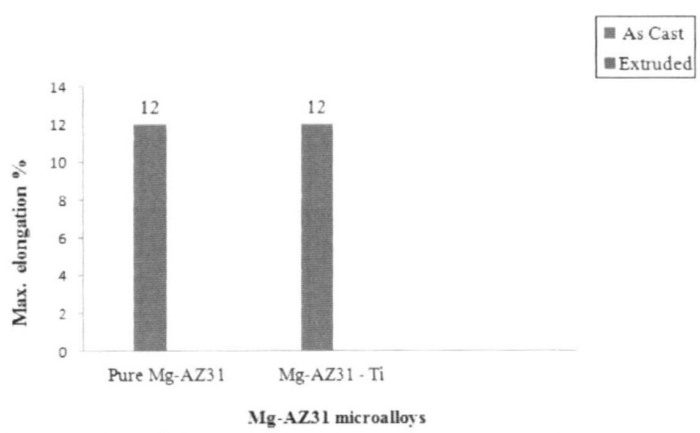

Figure 7. Effect of Ti addition on the maximum elongation % of Mg-AZ31 in the cast condition

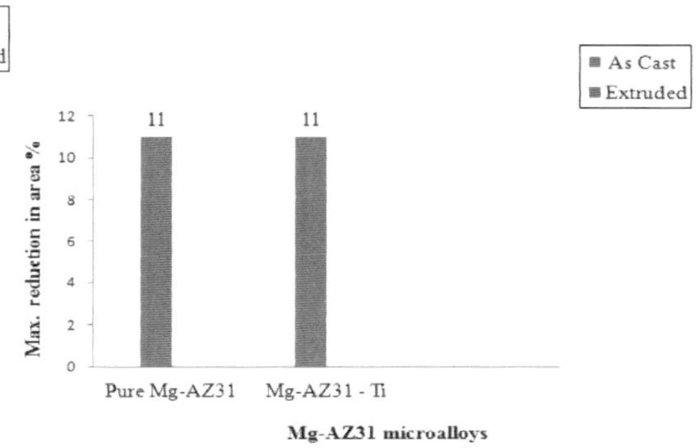

Figure 8. Effect of Ti addition on the maximum elongation % of Mg-AZ31in the cast condition

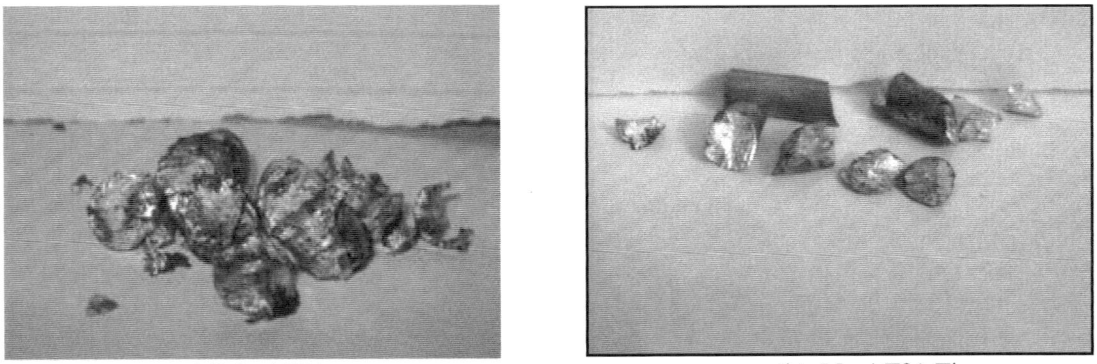

(a)Mg-AZ31 (b) Mg-AZ31-Ti

Figure 9. The fractured Mg-AZ31and Mg-AZ31-Ti (a), (b) during the extrusion process

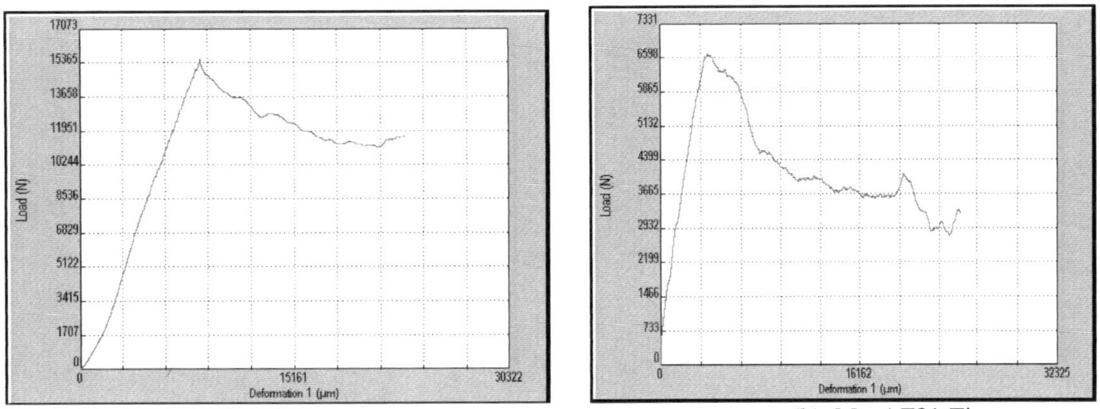

(a) Mg-AZ31 (b) Mg-AZ31-Ti

Figure 10. The autographic records of (a) MgAZ31and (b) MgAZ31-Ti

4. Conclusions

The following points are concluded:
In the cast condition:
Addition of Ti to Mg-AZ31 alloy at 0.15 wt. % resulted in grain refining of its structure and increase of its mechanical strength e.g. its flow stress, strength coefficient and hardness. Furthermore, it resulted in improvement of its ductility and work hardening index ,n, hence improving its formability. Although the addition of Ti to the Mg-AZ31 resulted in increase of its flow stress, it caused decrease in both the extrusion force and energy compared to the Mg-AZ31 alloy.
After extrusion the extrusion process has resulted in further refinement of both Mg-AZ31 alloy and Mg-AZ31-Ti alloy. It also resulted in further enhancement of their flow stress ,work hardening index ,formability, ductility and hardness. Finally it is concluded that the Mg-AZ31 cannot be extruded at room temperature. Furthermore despite the previously gained advantages by the addition of Ti to the Mg-AZ31 it was not possible to be extruded at room temperature.

5. Acknowledgment

The first author, (Adnan I.O. Zaid), is grateful to the Applied Science Private University, Amman, Jordan for partial financial support granted to this research (Grant No.DRGS-2015).

6. References

[1] Kalpakjian S, Scmidt S 2007 *Manufacturing processes of Engineering Materials*, 5thEdition. Prentice Hall.
[2] American Society for Testing Materials 1998 *ASTM*, B-86
[3] Mordike B L, Ebert T 2001 Mater. Sci.Eng.A302,37-45
[4] Cibula A 1949-1950 J. Institute of Metals, 76 321-360
[5] Campbell S A, Sutker GT, 1095The effect of titanium and boron concentrations on aluminum grain refiners
[6] Abdel Hamid A A 1989 Z.Metallkd,80 643-646
[7] ZaidA I O 2003*Grain Refinement of Aluminum and its Alloys* (Proceeding of ISAM-2oo3, Islamabad, Pakistan
[8] Zaid A I O 2006*Grain Refinement of Zinc Alloys* (proceeding of FAEM-06, Lahore, Pakistan
[9] Guang H, Ding H and Xiang L *2009 Grain Refinement of AZ31 Magnesium Alloy by New Al-Ti-CMaster Alloys,*19th International Conference of nonferrous metals society of China, 10th March, 2009
[10] Bryla K, Dutkiewicz J, Malczewski J 2009*International Scientific Journal of materials and manufacturing engineering*, 40 17-22."Grain refinement in Mg AZ31 alloy processed by equal channel angular pressing"
[11] Zhang K, Yin. L and Wu D2006Formability of Mg-AZ31 at warm working conditions *International Journal of machine tools and manufacture*, 1276
[12] Takahashi H,Xu S W, S.Kamado, Hono K, 2013 Effect of extrusion conditions on microstructure and mechanical properties of micro alloyed Mg–Sn–Al–Zn alloys, *Materials Science & Engineering*, A58 318–328
[13] Xuan L, Zhiqiang Z, Wenyi H, Qichi L, Lei BJianzhong C, 2015Effect of Extrusion Speed on the Microstructure and MechanicalProperties of Mg-9Gd—3Y—1.5Zn—0.8Zr alloy,*Journal of Materials Science & Technology*, 221255–268
[14] Lee T J , Kim W J, 2014The significant effect of adding trace amounts of Ti on the high temperaturedeformation behavior of fine-grained Mg–6Al–1Zn magnesium alloys, *Journal of Alloys and Compounds*, 617, 352–358

IOP Conf. Series: Materials Science and Engineering **146** (2016) 012027 doi:10.1088/1757-899X/146/1/012027

Enhanced electrical properties in Nd doped cobalt ferrite nano-particles

S. Abbas, A. Munir, F. Zahra and M. A. Rehman
Applied Thermal Physics Laboratory, Department of Physics,
COMSATS Institute of Information Technology, Islamabad 44000, Pakistan

E-mail: marehman@comsats.edu.pk

Abstract. Spinel ferrites are important class of compounds which has variety of electrical, magnetic and catalytic applications. A small amount of rare earth element causes modification in structural, electrical and magnetic properties of ferrite materials for practical applications. Neodymium doped cobalt ferrites with composition $CoNd_xFe_{2-x}O_4$ where x is 0.1 has been synthesized by sol-gel method. Sol-gel method was preferred because it has good control over stoichiometry, crystallite size and particle size distribution. Characterization was done by using X-Ray Diffraction (XRD) technique for structural analysis and crystal structure was found to be spinel. Particles like morphology was observed in micrographs obtained by Scanning Electron Microscopy (SEM). Thermal analysis of sample has been done which includes Thermogravimetric analysis (TGA) and Differential Scanning calorimetry (DSC). Fourier transform infra-red spectroscopy (FT-IR) of samples was also performed. DC resistivity as a function of temperature has been studied and its shows direct dependence on temperature and inverse dependence on the concentration of Nd dopant. The studied material is a potential candidate for resistive random access memory application.

1. Introduction

Ferrites are material of interest due to unique combination of their electrical and magnetic properties. Nano particles of spinel ferrites are widely used in different application that include microwave devices, biomedical applications and magnetic storage information [1]. Spinel ferrites can be represented by a general formula AB_2O_4 A and B are metal cations that are distributed upon tetrahedral (A) and octahedral (B) interstitials sites that is formed by face centered cubic arrangement of oxygen anions. In the cubic unit cell, to maintain charge neutrality, only 8 out of 64 available tetrahedral A sites and 16 out of 32 available octahedral B sites are occupied. In terms of the cation distribution among these tetrahedral A sites and octahedral B sites, the spinel structure can be normal, inverse, or partially inverse Cobalt ferrites has a special place among all other spinel ferrites due to having unique properties. $CoFe_2O_4$ has high cubic magneto crystalline anisotropy, high coercivity, and moderate saturation magnetization and high chemical stability [2]. In many applications DC resistivity of cobalt ferrites plays important role and desired resistivity can be controlled by addition of a dopant element. Effect of neodymium (rare earth) dopant on electrical resistivity was thoroughly studied in present research work. The results were compared with results of already reported pure cobalt ferrites.

Content from this work may be used under the terms of the Creative Commons Attribution 3.0 licence. Any further distribution of this work must maintain attribution to the author(s) and the title of the work, journal citation and DOI.
Published under licence by IOP Publishing Ltd

2. Experiment

2.1. Synthesis

$CoNd_xFe_{2-x}O_4$ with x=0.1 was synthesized by simplified sol-gel method. The stoichiometric amount of Cobalt Nitrate Hexa Hydrate (Co $(NO_3)_2.6H_2O$), Iron Nitrate Nona Hydrate (Fe $(NO_3)_3.9H_2O$), and Nd_2O_3 were taken. Reagents were of analytical grade and were used without further purification. All precursors were individually mixed into a beaker containing ethylene glycol. The ratio of precursors to ethylene glycol was 1:14 in order for homogeneous mixing of precursors. This solution was heated at 80°C until a wet gel was formed. The gel was then dried at near 250°C which results into burning of the gel. This formed product was then further grinded and a fine powder was obtained. This powder was weighed on a weighing balance. The formed powder sample was then pelletized into a disk shaped pellet using a hydraulic pressure and maintaining a pressure of 1000 psi. The pellet has 13mm diameter with thickness of 2mm and then it was sintered at 550±5 °C for 2 hours. This pellet was then used for DC electrical analysis.

2.2. Characterizations

Nd doped cobalt ferrites sample was characterized by X-ray diffraction (XRD) technique using Cu Kα (λ= 1.5406 Å) radiations in 2θ range 20-80° for structural analysis. Crystallite size $D_{(311)}$ was measured by using Scherrer's formula [3]. Lattice constant a was measured by equation (1)

$$a = d_{hkl}\sqrt{h^2 + k^2 + l^2} \tag{1}$$

Where hkl are miller indices and d_{hkl} is interplanar distance. The Fourier transform infrared (FTIR) absorption spectra was recorded in wave number range of 400-800 cm^{-1}.Thermogravimetric analysis (TGA) and Differential scanning calorimetry (DSC) were used for thermal analysis. DC electrical resistivity as a function of temperature was found by using two-probe method. The sample was heated inside a sample. Temperature range was kept from 40-450°C.

3. Results and discussions

3.1. X-ray diffraction (XRD) analysis

X-ray diffraction of the sintered sample shown in Figure 1 confirmed the formation of face centered cubic (FCC) spinel crystalline structure. X-0.1 showed an increased in lattice constant as compared to pure cobalt ferrites [4]. This was due to larger ionic radii of Nd^{3+} ions (0.983 Å) as compared to Fe^{3+} ions (0.645 Å).Crystallite size was found to be 29 nm. The lattice constant (a), crystallite size ($D_{(311)}$), measured density ρ_m, X- ray density ρ_x and porosity P of x-0.00 and x-0.10 are tabulated in Table 1.

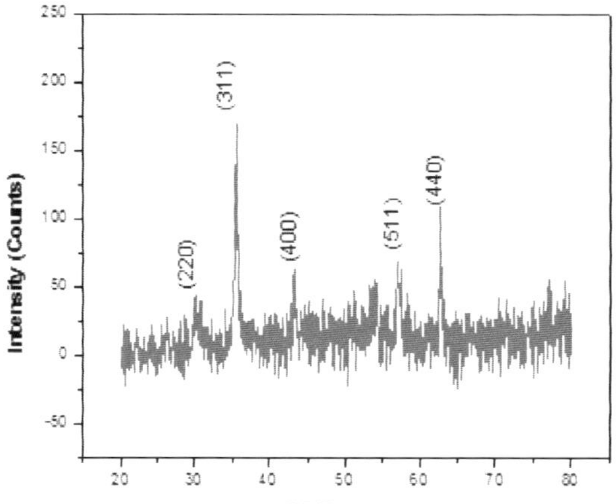

Figure 1. XRD pattern of the sample $CoNd_xFe_{2-x}O_4$ (x-0.10)

Table 1. A comparison of crystallite size $D_{(311)}$, Lattice constant *(a)* , measured density (ρ_m) x-ray density (ρ_x), porosity (P) and DC resistivity (ρ)for the sample $CoNd_xFe_{2-x}O_4$ (x-0.00 and x-0.10)

Parameters	Nd concentration X=0.00	Nd concentration X=0.10
$D_{(311)}$ (nm)	12	29
a (Å)	8.36(1)	8.39(2)
ρ_m g/cm^3	3.278	2.091
ρ_x g/cm^3	5.351	5.490
P (%)	38.7	61.7
ρ (Ω.cm) at 300 °C	2.88×10^4	4.30×10^5

3.2. FTIR spectra analysis

FTIR spectra of x-0.10 is shown in Figure 2. Two prominent bands were observed in the FTIR spectra. The infrared bands in solids are usually due to vibrations of ions in the crystal lattice. The absorption band (υ_1) in the range of 400-450 cm^{-1}. This corresponds to octahedral-metal stretching. The second absorption band (υ_2) was observed in the 550-600 cm^{-1} range. This corresponds to tetrahedral-metal stretching [5]. The formation of spinel structure was confirmed by the presence of characteristics absorption bands in the FTIR spectrum.

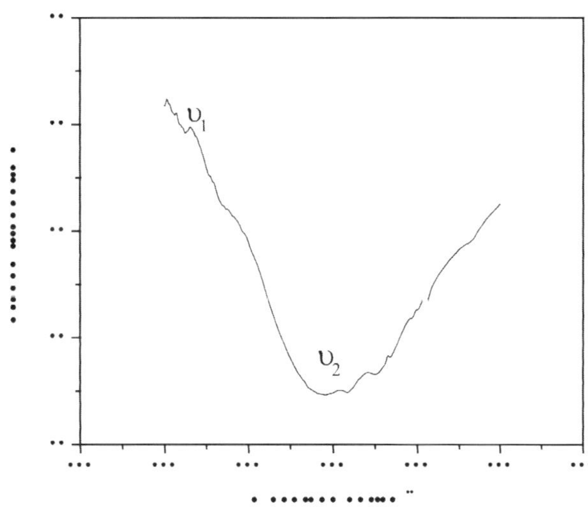

Figure 2. FTIR spectra of the sample $CoNd_xFe_{2-x}O_4$ (x-0.10)

3.3. Thermogravimetric analysis (TGA)/ Differential scanning calorimetry (DSC)

TGA/DSC curve for x-0.10 is shown in Figure 3. The first weight loss was observed at 100°C which corresponds to the removal of moisture. The second weight loss was observed at ~550°C which is due to the removal of by-products that were nitrates in present case and the final weight loss was due to removal of trapped gases within the sample at ~800°C. The crystallization peak was observed at 500°C, after that no further transition was observed for this sample.

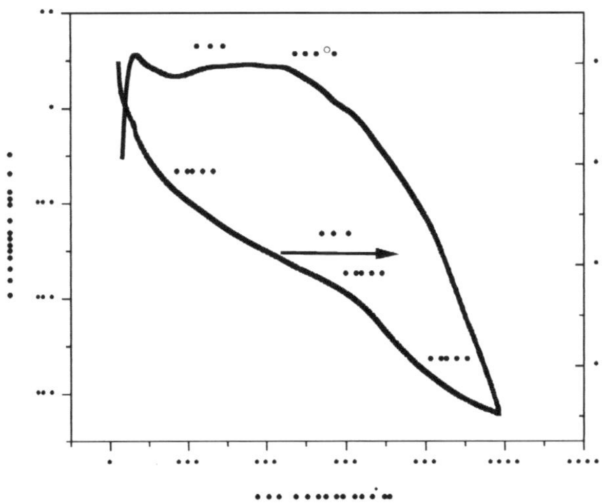

Figure 3. TGA/DSC curves for the sample $CoNd_xFe_{2-x}O_4$ (x-0.10).

3.4. DC electrical properties

The measured DC electrical resistivity of Nd doped cobalt ferrites shown in Figure 4 with x=0.10 was found to be $5.45 \times 10^9\ \Omega$–cm at 40°C. It was observed that electrical resistivity got increased markedly with addition of Nd as compared to x-0.00[6]. Since the conduction in ferrites is considered as the electron hopping between Fe^{2+} and Fe^{3+} at B (octahedral) sites. So the lower Fe^{2+} ions content decreases the conduction and an increase in resistivity. There are many other factor which also influence DC electrical resistivity. There are the microstructural factors that includes grain size, porosity and grain boundaries. Activation energy was also found from linear plots of DC resistivity. Table 1 give a comparison of DC electrical resistivity and activation energy of pure and doped sample. It was also observed that Nd doped cobalt ferrites has increased activation energy as compared to pure cobalt ferrites. The sample showed activation energy greater than 0.4 eV which attribute to polaron hopping mechanism in cobalt ferrites [4].

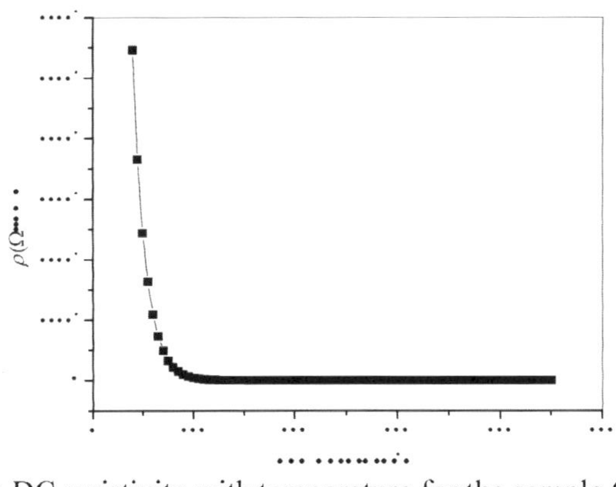

Figure 4. Variation in DC resistivity with temperature for the sample $CoNd_xFe_{2-x}O_4$ (x-0.10)

4. Conclusions

Neodymium doped cobalt ferrites having composition $CoFe_{2-x}Nd_xO_4$where x= 0.10 was successfully synthesized by simplified sol-gel method. XRD confirmed the phase purity with lattice constant of 8.39(2) Å with spinel cubic structure (Fd-3m space group). Neodymium doping has resulted in an increase in lattice constant. Crystallite size was measured by using Scherrer formula. It has value of 29 nm confirming the formation of nanocrystallites. Fourier transform infra-red (FTIR) spectroscopy also confirmed the spinel structure of ferrites. SEM revealed particle like morphology. Thermogravimetric analyses confirmed the thermal stability of the sample in the studied temperature range. DC electrical resistivity (ρ) as a function of temperature showed semiconductor like behavior. Replacement of Fe ions by Nd has caused the higher resistivity than that of undoped sample. Thus electrical properties of cobalt ferrites could be tailored by choice of neodymium ion concentration and hence make it suitable for many technological applications.

5. References

[1] Nongial R, Khan S, Asokan K, Ahmed H and Khan I 2009 *J. Alloys Compd.* 486 640
[2] Zhenfa Z, Yuping S, Xuebin Z, Zhaorong Y and Jianming D 2009 *J. Magn. Magn. Mater.*321 1251
[3] Victoria L, Ddelc C and Rinaldi C 2007*J. Magn. Magn. Mater.* 314 60
[4] Akram M and Anis-ur-Rehman M, 2014 *J. Electron. Mater.* 43 485
[5] Xavier S, Thankachan S, Jacob B P and Mohammed E M *2013 J. Nanomater. Mol. Nanotechnol* 2 1
[6] Gul I H, Abbasi A Z, Amin F, Anis-ur-Rehman and M Maqsood A 2007 *J. Magn. Magn. Mater.* 311 494

Effect of shot peening and grain refinement on the fatigue life and strength of commercially pure Al and two of its alloys: Al-2024-T3 and Al-7075-T6

A. Qandil[1] **and Adnan I. O. Zaid**[2]
[1]Civil Engineering Department, Faculty of Engineering,
Applied Science Private University, Amman 11931 Jordan
[2]Mechanical and Industrial Engineering Department, Faculty of Engineering,
Applied Science Private University, Amman 11931 Jordan

E-mail: adnan_kilani@yahoo.com

Abstract. Aluminum and its alloys are widely used materials in automobile, aircraft and space craft industries due to their high strength- to- weight ratio and corrosion resistance beside their other useful properties. They are the second materials in use after steel alloys. Most of the failures in parts of aircrafts and space vehicles are mainly caused by fatigue and stress corrosion cracking. In this paper, the effect of shot peening on the fatigue life of commercially pure aluminumand two of its alloys namely:Al-2024 and Al-7075-T6 is presented and discussed. Furthermore, the effect of addition of vanadium to Al and Al grain refined by Ti and Ti+Bon Its fatigue life and strengthis also presented and discussed using scanning electron microscope, SEM. It was that shot peening and the addition of V toAl and Al onAl grain refined by Ti and Ti+B have resulted in enhancement of the fatigue life and strength. Ffinally, the effect of shot peening on the surface quality of the peened parts is also presented and discussed.

1. Introduction

Aluminum and its alloys are widely used materials in automobile, aircraft and space industries due to their high strength- to- weight ratio and corrosion resistance beside their other useful properties. They are the second material in use after steel alloys. Most of the failures in parts of aircrafts, space crafts and racing car axes are mainly caused by fatigue failure and stress corrosion cracking. Failure by fatigue is a progressive and localized structural damage which occurs when the part is subjected to cyclic loading. The stress values at which fatigue failures occur are much less than the ultimate tensile strength or even may be well below the yield strength of the material. The failure starts by initiation of cracks at the surface of the part which propagates and extends to cause complete failure of the part.

Content from this work may be used under the terms of the Creative Commons Attribution 3.0 licence. Any further distribution of this work must maintain attribution to the author(s) and the title of the work, journal citation and DOI.
Published under licence by IOP Publishing Ltd

IOP Conf. Series: Materials Science and Engineering **146** (2016) 012028 doi:10.1088/1757-899X/146/1/012028

The corresponding number of load cycles or the time during which the number is subjected to these loads before fracture occurs is referred to as the fatigue life of the member and the maximum stress which can be applied infinite number of cycles to the member without causing its failure,(in case of ferrous materials) is referred to as the endurance limit. However, in case of non-ferrous materials the no of cycles which defines the fatigue life, from which the maximum allowable stress to avoid fatigue failure is defined by the users and design engineers as indicated in Figure 1.

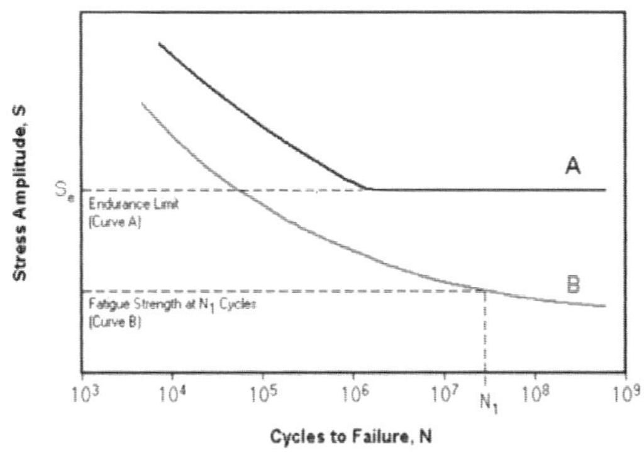

Figure 1. Fatigue S – N curve of A: mild steel and B: aluminum

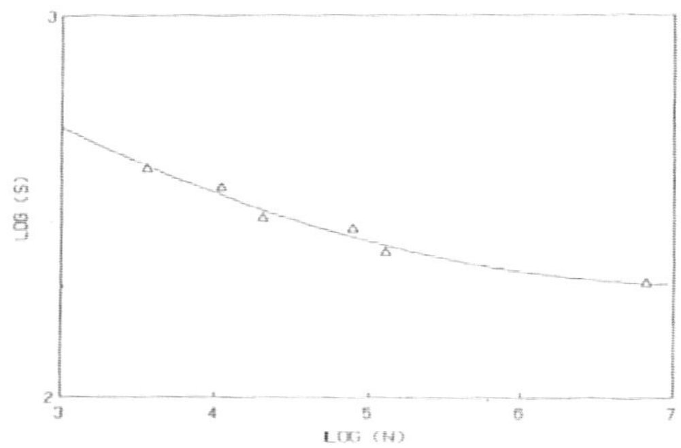

Figure 2. Log stress,(S) versus Log number of cycles to failure,(N)

Also fatigue data may be presented as Log S versus Log N; whereS is the alternating stress and N is the number of cycles.
The total period of fatigue life may be considered to consist of three phases:

i) Initial fatigue damage that produces crack initiation.
ii) Propagation of crack that results in partial separation of cross section of the member.
iii) The last phase when the remaining un-cracked cross section is unable to support the appliedload and then the complete fracture takes place.

1.1. Mechanismand mode of crack initiation and propagation
The position and mode of fatigue crack initiation depends on the microstructure of the material, the type of the applied stress and micro- and macro-geometry of the specimen. The crack initiation period may cover a large percentage of the fatigue life under high-cycle fatigue, i.e. under stress amplitudes just above the fatigue limit. But for larger stress amplitudes the crack growth period can be a substantial portion of the fatigue life. Detailed discussion of the different phases of the fatigue life and the relevant factors associated with them are beyond the scope of this paper.

1.2. Methods of improving fatigue life and strength
In this section two methods (which are used for improving fatigue life and strength) will be dealt with. These are the shot peening and the grain refinement processes.

2. Shot peening
Shot peening is a surface cold working process in which the surface of the component is subjected to a multiple impact by a high velocity stream of hard particles in a defined and controlled manner. This multiple impact of the shots produces a dynamic compressive stress layer in the surface of the component which ranges from 0.25 to 1.2 mm, thereby effectively eliminates cracks and other imperfections; hence its mechanical behavior, fatigue life and strength are improved. Controlled shot peening is an operation which is largely used in the manufacturing of mechanical parts to increase their fatigue life and strength. The kinetic energy of the shot is transformed into plastic deformation of the work piece surface and the steel ball itself. The shot is reflected from the component surface with the remaining kinetic energy. The elastically deformed sub-surface layer tries to resist this surface

expansion, inducing a compressive stress at the surface, balanced by a tensile stress of lower magnitude. If a thin material is peened on one side, a curve will develop in the part and the shot peened side be convex [1,2]. Examination of the available literature on shot peening indicates that the trend has not changed over the subsequent years, [3-16].

2.1. *Parameters affecting shot peening process*

Like any other surface treatment process, the parameters involved in shot peening play a vital role in determining the efficiency, quality and reliability of the process to ensure that the part is correctly treated. From the available literature the main parameters which affect the shot peening process are: shots pressure and velocity, peening time and to a lesser degree the table rotational speed. Their increase will cause increase in peening intensity whereas, increase in the standoff distance results in decrease of the intensity. It is worth mentioning that it is very difficult to modify one of the parameters without having some effect on the others. The culminating effect in the shot peening process is the shot peening intensity and complete coverage of the surface by subjecting or exposing the part to the effect of the shots to a certain time. Hundred percent coverage is reached when the original surface of the material is entirely coveredby the overlapping peening dimples.

2.2. *Peening intensity and saturation*

Peening intensity is the term commonly used to describe the overall effect of shot peening on the treated work piece. Factors that influence the shot peening process are investigated through their effect on peening intensity. These factors include shot size, shape and hardness, shot pressure, velocity, standoff distance, projection angle and exposure time.The Almen strip test method is, so far, the only available non-destructive method for determining peening intensity.The deflection of the Almen strip when subjected to the shot peening process displays the resultant effect of all the process parameters. This deflection is taken as the height of the arc at the center of the strip in thou of an inch and is termed peening intensity provided saturation is achieved. Saturation is obtained when doubling the exposure time will not cause more than 10% increase in the intensity value. Figure 3 shows this variation for 7075-T6 aluminum alloy whereas Figure 4 shows the effect of shot peening on the fatigue life of the 2024-T3aluminum alloy. It can be observed from these two figures that their fatigue lives were enhanced.This is attributed to the grain refinement effect of these two elements on Al which increases the number of grain boundaries which will hinder the fatigue crack propagation; hence resulting in the improvement of their fatigue live and strengths.

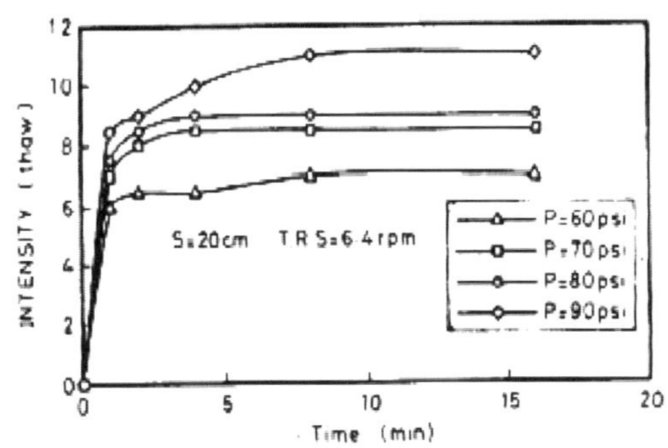

Figure 3. Variation of shot peening ntensity with time at different pressuresfor 7075-T6

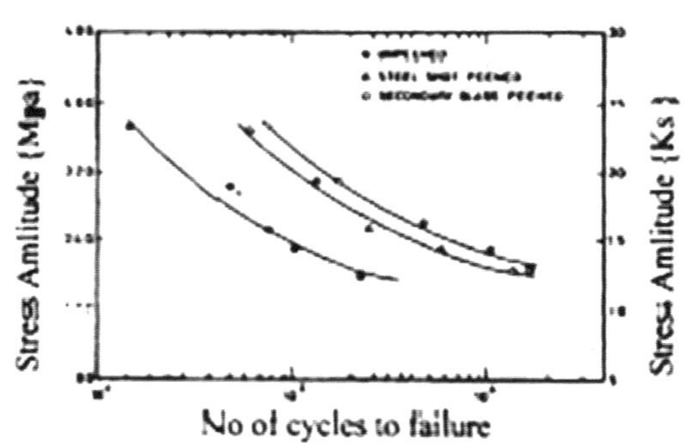

Figure 4. Effect of shot peening on the fatigue life of 2024-T3aluminum alloy

Shot peening intensity optimization to increase the fatigue life of a quenched and tempered structural steel was also investigated and reported in[9].

2.3. *Effect of shot peening on the surface roughness of the peened surface*

Despite the advantages gained by shot peening; it has the disadvantage of increasing the surface roughness of the peened surface as illustrated in Figure 5.

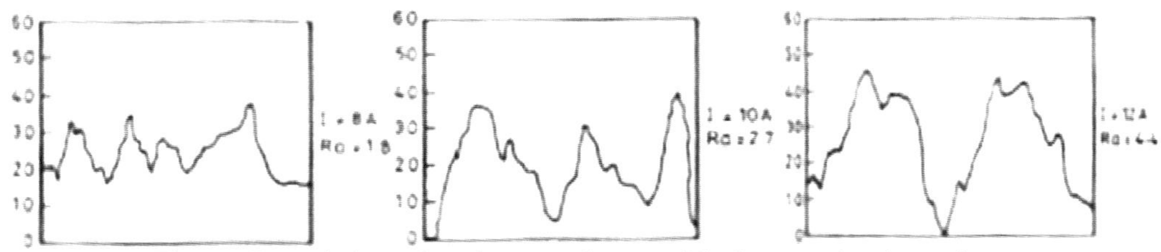

Figure 5. Variation of surface roughness with shot peening intensity

Also surface roughness and X-ray diffraction residual stress measurements were arried out on treated specimens made of X70 micro-alloyed steel (after severe shot peening) using microscopic examinations ann the results of their work is reported in reference [10].

3. Grain refinement

Aluminum and its alloys are the second in use in idustrial applications after ferrous alloys especially in the air craft and automobile industries due to their high strngth to weight ratio, their electrical and thermal conductivities in addition to their good corrosion resistance. However, against these advantages they have the disadvantage of solidifying in a columnar structure with large grains which tends to reduce their mechanical strength, surface quality and their fatigue life. Therefore, it became anecessity to grain refine their structure to overcome these disadvantages. Titanium and titanium + boron are the most widely used as effective refiners. The literature on the subject is voluminus from the metallurgical and mechanical aspects[17-24]. Review of this literature is given by the first author in reference[17].In this section, the effect of the grain refinement on the grain size, hardness, fatigue failure and fracture surface of the commercially pure aluminum is presented and discussed.

3.1. *Effect of the different grain refiners addition to aluminum on its fatigue life*

It can be seen from Figures 6 and 7 that addition of either Ti or Ti+B to commercially pure Al resulted in pronounced enhancement of their fatigue lives. Figure 8 shows comparison between the addition of vanadium to aluminum grain refined by Ti and Ti+B. It can be seen from this figure that the fatigue lives of both alloys had been improved by the vandium addition and is more pronounced in the case of its addition to Ti+B . This is attributed to the grain refinement effect of these two elements on Al which increases the number of grain boundaries which will hinder the fatigue crack propagation hence resulting in improvement of their fatigue lives and strengths.

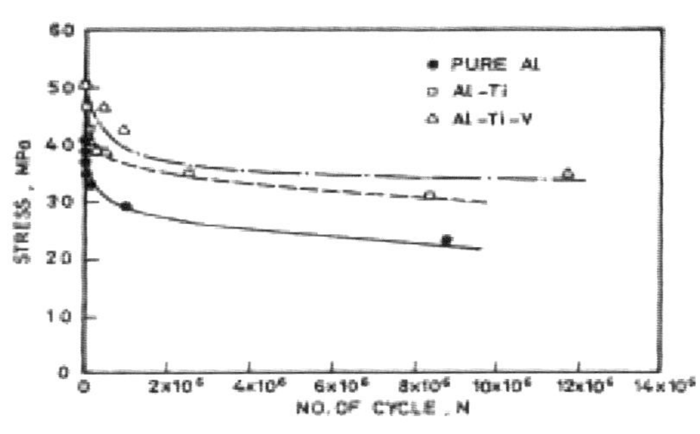

Figure 6. Effect of V on fatigue S-N curves grain refined by Ti

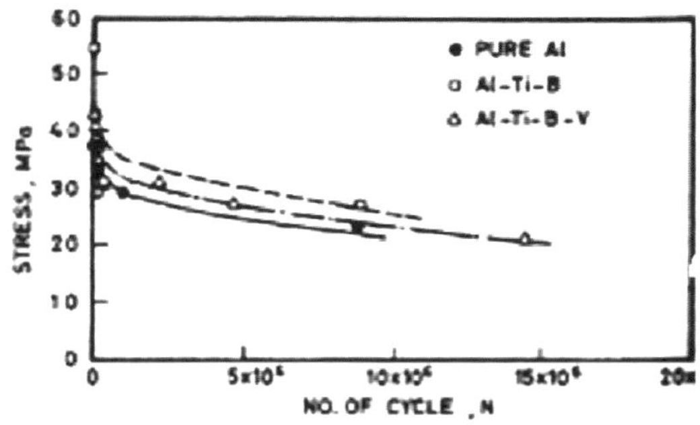

Figure 7. Effect of V on fatigue S-N curves grain refined by Ti+B

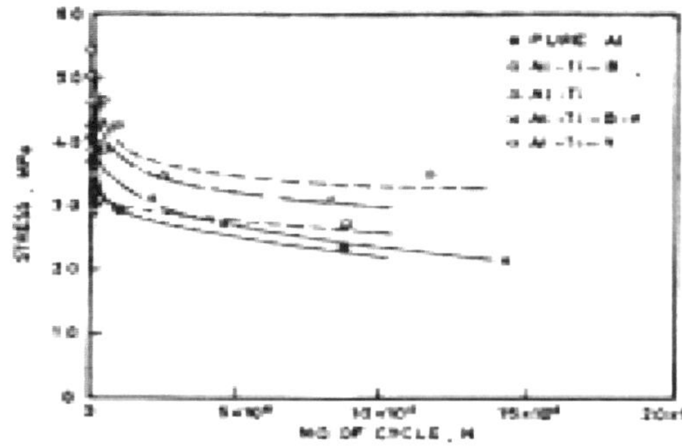

Figure 8. Effect of V on fatigue S-N curves grain refined by Ti and Ti+B

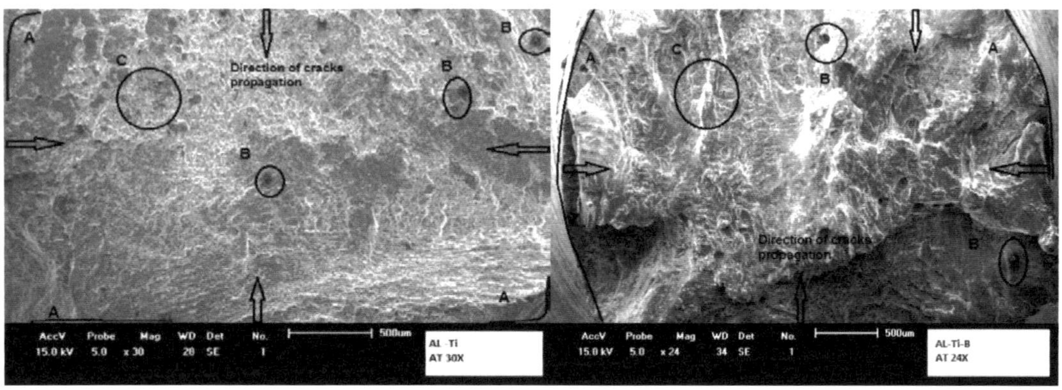

Figure 9. Photo-scans showing fractured surfaces of a).AL-Ti micro-alloy, b). AL-Ti-B micro-alloy

In all the previous photo-scans, it can be seen that there is two stages for the fatigue cracks, first stage is the crack initiation which starts at the surface of the specimen (region A), where it is initiated at a defect like pore or cluster pore located at or below the surface which is described as (region B) in the photo-scans. After the first stage when the micro-cracks have been nucleated, cracks growth starts but very slow and the erratic process occurs, due to the effect of the grain boundaries. Until this moment, fatigue is still material surface phenomenon. However, after some micro-cracks growth has occurred away from the nucleation site, a more regular growth is started, then the initiation period is expected to be completed and a real crack growth begins and the crack propagation takes place through the material until fracture occurs. Consequently, it was observed that the cracks have propagated within the grains (trans-granular), these cracks generate due to the cyclic slip deformation. Or at the grain boundaries (inter-granular) which have a certain angle against the grain boundaries. This may indicate the slip bands blocked by the grain boundary, leading to inter-granular cracks, and both of trans-granular and inter-granular modes occurred consequently.

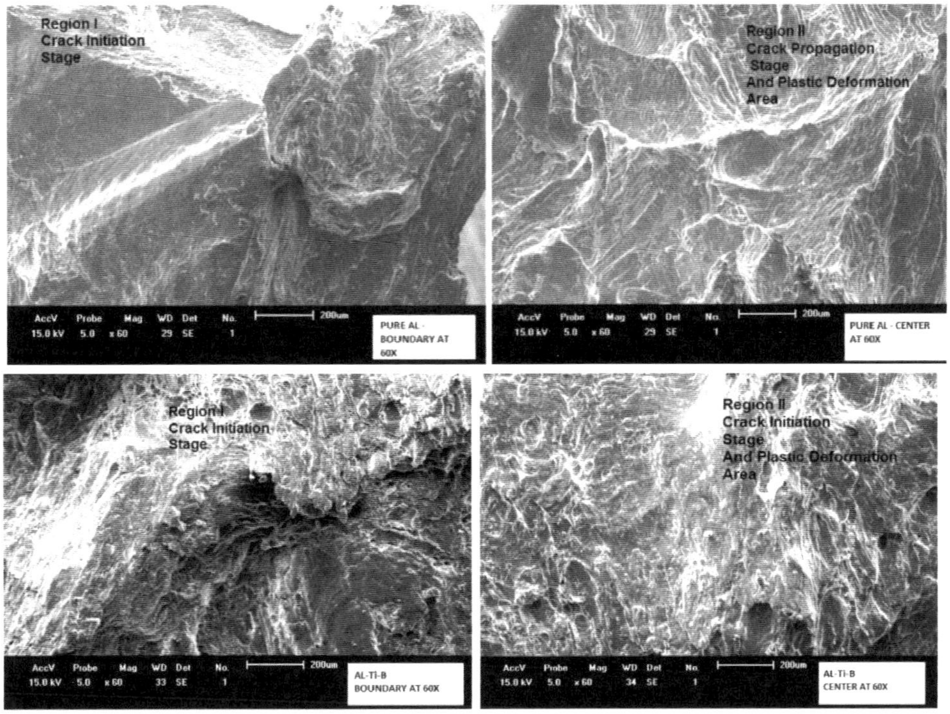

Figure 10. Photo-scans showing the crack initiation regions in Al and Al-Ti-B at 60X

The mode of fatigue failure in the different micro-alloy (to a great extent), alloys is almost identical.It indicates that the specimens suffer a certain amount of plastic deformation prior to fracture, and the amount of the plastic deformation varies with the added grain refiner. The SEM photo scans of Figures 9 and 10, show the fractured surface (At 60X) of pure Al, Al-Ti, Al-Ti-B, Al-Ti-V and Al-Ti-B-V respectively. On the whole, the fatigue failure in aluminum is not sudden as with other materials, e.g. steel, Since the total failure is proceeded with a state of plastic deformation. Examination of the fractured surface reveals that it consists of two regions regarding the crack propagation, which is denoted as (Region I and Region II). Region I contains the fatigue crack initiation as well as initial crack propagation,. It shows that the crack propagates almost is a trans-granular type. As the crack travels through the grains rather than at the grain boundaries. In addition, this region consists of a ductile region at the outer part. The area of the ductile fracture is reduced as the percentage of the hard particles increases in aluminum matrix this was observed in case of addition of vanadium to aluminum grain refined by titanium where a combination of $TiAl3$ and $VAl3$ particles exist in aluminum matrix. Both of them are hard particles. Region II is characteristic of higher rates of crack propagation observed as larger crack size, being much more irregular than Region I. Furthermore some trans-granular cracking is observed in this Region, II. The crack propagation in this region is primarily inter-granular, and the number of the trans-granular cracked grains of the fractured surface decrease as the crack length increases, resulting in more surface irregularity away from the initiation site. It is worth noting, that the fracture surface of the Al-Ti-B-V specimen reflected larger plastic deformation region due to existence of the borides TiB2 and VB2 which are soft particles. Also there is evidence of fatigue striations on the fractured surface of Al-Ti-V described as region D. This may be attributed to the existence of the hard particles TiAl3 and VAl3.

4. Conclusions

The following points are concluded:

4.1. Effect of shot peening:

i. Shot peening intensity increases with the increase of shot pressure and time, whereas the standoff distance has slight effect on intensity and the table speed has particularly no effect on the intensity.
ii. Shot peening resulted in pronounced improvement in the fatigue life and strength of Al and its two alloys7075-T6 and 2024-T3 provided that saturation has been achieved and no over peening has occurred. If over peening occurs it will cause adverse effect on the fatigue life and strength.
iii. Shot peening resulted in deterioration of the surface quality of the peened surface as it caused slight increase of its surface roughness. This can be reduced by decreasing the shots size and controlling the process parameters.

4.2 Effect of grain refinement:

i. Addition of any of the grain refiners Ti, Ti + B or V to Al resulted in reduction of its grain size from columnar structure into equi -axed one. Furthermore, their addition resulted in enhancement of its fatigue life and strength due to the strengthening caused by the increase of the number of grain boundaries.
ii. Addition of any of the used grain refiners used to aluminum resulted in modifying the mode of failure e.g. crack initiation and propagation, being a mixture of trans-granular and inter-granular of different percentages, due to the formation of the hard particles of the intermetallic compounds of titanium and vanadium alum ides within the main matrix.

5. Acknowledgment

The authors are grateful to the Applied Science Private University, Amman, Jordan for partial financial support granted to this research (Grant No.DRGS-2015).

6. References

[1] Zaid A I O 1997 Investigation of the effect of shot peening parameters on peening *Inte. Proc. 5th International Symposiums on advanced materials*, 744-750, Islamabad, Pakistan
[2] Zaid A I O 1999 shot peening: Theory, Application, and Recent Development *6th International Symposiums on advanced materials*, 439-451, Islamabad, Pakistan
[3] Zaid A I O, Ababneh 2000 Effect of shot peening on the fatigue strength of 2024-T3 aluminum welded alloy in the welded conditions current advances in mechanical design, *cario university international MDP conference* 339-347
[4] Zaid A I O 2000 Effect of shot peening on the fatigue life of aluminum alloy 7075-T6. *Proce. International conference on production research.* Bangkok-Thailand
[5] Zaid A I O, Medraj M A 2001 further experiments on the effect of shot peening on the fatigue 4PH *Proceeding of 7th international conference on production Engineering*. **3**, 1537 Alexandaria-Egypt
[6] Zaid A I O 2001 Experimental results on Recent Applications of shot peening *the Symposium on advanced materials* ISAM-7 Islamabad Pakistan
[7] Zaid A I O 2001 Experimental results on recent applications of shot peening *International Symposium on Advanced Materials ISAM-7* Islamabad- Pakistan
[8] Masaki K, Ochi Y,.Matsumura T, 2004 Fatigue Fract.Eng.Mater.Struct.271137–1145
[9] Fitzpatrick M. E. et al., 2005 Determination of residual stresses by X-ray diffraction, National Physics Laboratory, U.K
[10] Mikova A, Bagherifard S, Otakar B, Guagliano M, Trško, L, 2013 Fatigue behavior of X70 microalloyed steel after severe shot peening, International Journal of Fatigue, **55,** 33-42

[11] Vielma, A T, Llanezaand V, Belzunce F J 2014 Shot peening intensity optimization to increase the fatigue life of a quenched and tempered structural steel, XVII International Colloquium on Mechanical Fatigue of Metals (ICMFM17) ,Procedia Engineering 74 273 – 278
[12] Aymen A. Ahmed, Mansour Mhaede, M Basha, Manfred Wollmann, LotharWagner, 2015The effect of shot peening parameters and hydroxyapatite coating on surface properties and corrosion behavior of medical grade AISI 316L stainless steel, Surface & Coatings Technology 280 347–358 Institute of M
[13] Mahmoudi A H, Ghasemi A, Farrahi G H, Sherafatnia K 2016A comprehensive experimental and numerical study on redistribution of residual stresses by shot peening, Materials and Design 90478–487
[14] Kikuchi S, Nakamura Y, Nambu K, Ando M, 2016Effect of shot peening usingultra-fine particles on fatigue properties of 5056 aluminum alloy under rotating bending, Materials Science and EngineeringA652279–286
[15] Barbara Nasiłowska A, Zdzisław Bogdanowicz A, Michał Wojucki B2015Shot peening effect on 904 L welds corrosion resistance, Journal of Constructional Steel Research 115276–282
[16] Yong-Seog Nam A, Yoo-In Jeong B, Byung-Cheol Shin C, Jai-Hyun ByuEnhancing surface layer properties of an aircraft aluminum alloy by shot peening using response surface methodologyEngineering 74273 – 278
[17] Zaid A I O 2001 Effect of some grain refiners on the mechanical properties of aluminum *international symposium on advanced materials ISAM-7* Islamabad- Pakistan
[18] Zaid A I O 2OO5 Metallurgical Aspects of Fatigue Fractured Surfaces of Aluminum Grain Refined by Some Rare Earth Materials", Proceedings of the 9th International Symposium on Advanced Materials ISAM, 19-22 September 2005 Islamabad-Pakistan
[19] Zaid A I O 2OO6 and Al-Dous Effect of Zirconium Addition on the Ductility and Toughness of Cast Zinc-Aluminum Alloy -5 Grain Refined by Ti+ B, Proceedings of the Second International Conference on the Frontier of Advanced Engineering Materials, FAEM 2006, Lahore, Pakistan
[20] Zaid A I O and Al-Dous2OO7 Effect of Zirconium Addition on the Ductility and Toughness of Cast Zinc- Aluminum Alloy- 5 Grain Refined by Ti", Proceedings of the 10th International Symposium on Advanced Materials, ISAM- 2007, pp. 37 – 43, Islamabad, Pakistan
[21] Zaid A I O and Alqawabah S M M 2009 Effect ofZirconium Addition on the Mechanical and Wear Resistance of Aluminum Grain Refined by Titanium", Proc. Of the 12th International Symposium on Advanced Materials , ISAM- 2009, Islamabad, Pakistan
[22] A.I O.Zaid and S.M.A. Al-Qawabah 2012 Refining Aluminum by Titanium or Titanium plus Boron on its Weldability using Gas Tungsten Arc Welding The 2012 International Conference on Industrial Engineering and Operations Management Istanbul, IEOM, Turkey, July 3-6 902-909

Effect of UV radiations to control particle size of Mn–Zn spinel ferrite nano-particles

Ramiza, F. Ameen, S. K. Ajmal[2], M. B. Khan, A. Nasim[1], Y. Jamil, K. Kashif and S. Amira
Department of Physics, Faculty of Sciences, University of Agriculture, Faisalabad, Pakistan
[1]Nuclear Institute of Agriculture and Biology, NIAB, Faisalabad, Pakistan
[2]Department of Physics, University of Sargodha, Women Campus, Faisalabad, Pakistan

E-mail: ramiza_uaf@yahoo.com

Abstract. $Mn_xZn_{1-x}Fe_2O_4$ ($0.0 \leq x \leq 1.0$) ferrite nano particles were synthesized for concentration varying from 0.27 to 0.87 to obtain chemically homogenous powder for obtaining fine particle size by co precipitation technique. Keeping in view the interest of scientists for particle size, the present work focus on the impact of UV radiation to control the particle size of prepared fine magnetic particles. The particles were digested for ninety minutes at a temperature of 90°C. The samples were divided into four equal quantities and were subjected to different doses of UV radiation. The chemically produced samples of Mn–Zn ferrite nano particles were analyzed by XRD which confirmed cubic spinel structure of the material. The average crystallite size (t), lattice parameter (α) and other structural parameters of UV-irradiated $Mn_xZn_{1-x}Fe_2O_4$ spinel ferrite were calculated from XRD data. The spinel peak of the irradiated sample when compared with the control sample, shifted from 35.38 to 35.15. In few samples, additional peaks supporting the ferrite structure were also observed. The variation in the particle sizes observed for various doses of UV irradiation were in the range of 17.6 to 6.2 nm, whereas the particle size of the control was 8.82nm. The experiment was repeated for different concentrations, at the same digestion temperature and time revealed the similar results indicating that UV radiations can have a remarkable effect to control the phase and size of nano size fine magnetic ferrite particles. The present work successfully document the impact of UV to control the particle size.

1. Introduction

The "nano" prefix to technical terms is not simple replacement of the unit of Angstroms; this is a change in Science & Engineering. The world is attain in gun believable achievements because we are capable to figure things that work better on the minimum possibly attained length scales, atom by atom. Nano scaled items in its advanced form has major impact on almost industrial scale and all disciplines of society. It will offer better built, longer lasting, cleaner, safer, and smarter products for home, communications, medicine, transportation, agriculture, and for industry in general [1].
The production of ferrites nanoparticles has driven the research field on contemporary track and these nano-things will revolutionize industries and our lives. Soft ferrite cores are oxides made from Iron (Fe), Manganese (Mn), and Zinc (Zn) which are commonly referred to as manganese zinc ferrites.

Content from this work may be used under the terms of the Creative Commons Attribution 3.0 licence. Any further distribution of this work must maintain attribution to the author(s) and the title of the work, journal citation and DOI.
Published under licence by IOP Publishing Ltd

They have a low coercivity and are also known as soft magnetic ferrites [2]. Because of their comparatively low losses at high frequencies, they are extensively used in switched-mode power supply (SMPS) and radio frequency (RF) transformers and inductors, shielding and linear inductive components [3-5]. Ferrite cores for the high frequency power supply and high quality communication markets are produced in a variety of shapes and sizes for inductors, pulse transformers, high frequency transformers, and noise filters [6]. Notable characteristics of Magnetics ferrite materials are commercially important due to high magnetic permeability and good temperature properties [7].

The magnetic behavior, electrical behavior and structural properties changes with different synthesizing techniques with changing concentrations of cations. The prepared ferrites having same composition of ferrite particles but different preparation techniques show different electromagnetic behavior. Many methods are followed for production of less time consuming and high quality nanoparticles. The processes affect the microstructure, particle size, surface to volume ratio and permeability [1].

For the preparation of ferrites with high values of electromagnetism, many processing techniques are in practice including co-precipitation, sol-gel techniques, spray dying, hydrothermal methods and alcohol dehydration processes [8-13]. The co-precipitation technique has got great attraction for scientists for obtaining good microstructure, better electromagnetic properties and better results at lower temperatures, and not much time consuming property during sample preparation[6].

In the present research work, Manganese Zinc Ferrite nanoparticles ($Mn_xZn_{1-x}Fe_2O_3$) were synthesized by co-precipitation technique to produce chemically homogeneous powder for obtaining fine particles. The concentration (x) were varied from 0.27 to 0.87 in four samples. The samples were prepared at same digestion temperature and then each sample was further divided into four equal quantities. The samples were subjected to electromagnetic (UV) radiations. X-ray diffraction (XRD) was used to determine particle size due to changing concentration of Mn & Zn. The particle size was measured using Scherer's formula. It was observed that X-ray densities were decreased linearly by increasing concentration of Mn. The particle size of the particles was found to be affected by Manganese concentration, but not too extent. The minimum particle size was obtained at x=0.27 when irradiated for 6 hours. In order to get more uniform crystalline structure the calcination of a sample was done after the preparation of samples by co-precipitation method. For lowering the total anisotropy and getting the homogenize material calcination process was used. After calcination the effect of UV irradiation was also observed for getting small particle size [7].

1. Experimental details

Manganese Zinc ferrites ($Mn_xZn_{1-x}Fe_2O_3$) magnetic particles were prepared by using co-precipitation method for getting homogeneous powder of fine particle size. The concentrations of x are 0.27, 0.47, 0.67 and 0.87. The least expensive and readily available salts used for the preparation of Mn-Zn ferrite are $MnCl_2.4H_2O$(Manganese Chloride tetra hydrated), $ZnCl_2$ (Zinc Chloride) and $FeCl_3.6H_2O$ (Ferric Chloride hexahydrated). The [M^+OH^-] was 0.5 for each sample; the concentrations of $FeCl_3$ and NaOH were constant for all the samples. For each sample the ratio of Mn^{+2} and Zn^{+2} ions was taken 1:2. The molecular weights of different chemicals used were 197.9, 136.3, 270.5 and 40 respectively.

For two molar solution of Fe salt, 1 molar solution of combined Mn and Zn salts was used. The next step was the addition of base solution to salt until the pH 12 was attained. 1 mole of $MnCl_2.4H_2O$ contains 1 mole of Mn, 1 mole $ZnCl_2$ contains 1 mole of Zn, 1 mole of NaOH contains 1 mole of Na and 1 mole of $FeCl_3.6H_2O$ contains 1 mole of Fe. All solutions were made in distilled and de-ionized water. All the salts $MnCl_2.4H_2O$, $ZnCl_2$ and $FeCl_3.6H_2O$ with different quantities for different samples were dissolved in distilled water [16]. The volume of this solution was 150 ml. In another beaker 100 ml solution of NaOH was prepared. The morality was taken 1 mol. The beaker containing 150 ml salts of $MnCl_2.4H_2O$, $ZnCl_2$ and $FeCl_3.6H_2O$ was placed on magnetic stirrer of vigorous stirring speed (6o rpm) at room temperature. The aqueous solution of beaker containing base was added drop by drop into beaker containing salt solution. The addition of base solution was continued until the pH 12 was obtained. This process of adding consumed about 60 minutes for each samples. After this addition the final solution was about 250 ml. During the addition dark brown precipitates were obtained. The beaker containing the dark brown precipitates was positioned in a pre-heated water bath containing water. The temperature of water bath was kept constant at 90°C for each sample. Digestion was performed for 90 minutes for each sample. The prepared particles were settled down at the bottom of beaker. The beaker was cooled to the moderate temperature after taking out from water bath. The chemically prepared particles were attained via filtration. The filtered particles were dried at room temperature for 18 hours. Finally, the fine materials were produced and accumulated on the filter paper. Then purified samples were provided heat treatment in an oven at same temperature i.e. 100°C for 24 hours [17-18]. Each sample was then divided into four equal parts. One sample was taken as standard while the others are irradiated for 2 hrs, 4 hrs & 6 hrs respectively by UV source. The Similar process was adopted to prepare all the four samples. The spontaneous magnetization of particles was checked for all the samples with the help of permanent magnet. The prepared samples of manganese zinc ferrites were analyzes with the help of x- ray diffraction technique. The quantities of $FeCl_3.6H_2O$, $ZnCl_2$, $MnCl_2.4H_2O$ and NaOH along with the metal to hydroxide ion concentration ratio are listed in table below:

Table 1. The quantities of salts along with the metal hydroxide ion concentration

Sample No.	$FeCl_3.6H_2O$ (per 50ml)	MnCl2.4H2O (per 50ml)	$ZnCl_2$ (per 50ml)	NaOH (per 100ml)	Digestion Time (Min)
S1	2.70g	0.723g	0.183g	8g	90
S2	2.70g	0.525g	0.320g	8g	90
S3	2.70g	0.327g	0.457g	8g	90
S4	2.70g	0.117g	0.529g	8g	90

Ultra-fine particles of $Mn_xZn_{1-x}Fe_2O_3$ prepared by co-precipitation were analyzed by using X-ray diffraction [16].Powder method was employed for this purpose. In this method, the specimen for diffraction is obtained by sticking some of the powdered materials onto the glass slide with trace of petroleum jelly. A narrow beam of parallel monochromatic X-ray impinges on specimen. The source of X-rays is Coolidge-type X-ray tube with a copper target, equipped with filter to remove all spectral components except desires Kα line. The diffracted radiation is detected by a photo-multiplier tube, which measures the diffracted beam intensities. The diffracted intensity is obtained when Bragg's law is satisfied i.e. $2dSin\theta=n\lambda$ where'd' is inter planer spacing. 'θ' is angle between the incident X-ray beam and diffracting beam plane. 'L' is wavelength of X-rays used.

By using d- values from the peak analysis of samples, the lattice constant were calculated using the relation $a = d\ (h^2+k^2+l^2)^{1/2}$for cubic structure [19]. X-ray density was measured by using formula $D_x = 8\ M/Na^3$. Since each primitive unit cell of the spinel structure contains 8 molecules, the value of the X-ray density, D_x was determined in terms molecular weight(M), Avogadro no.(N)= 6.0225×10^{23} atom/mole and lattice constant(a) [20]. Particle sizes of spinel peaks of all samples were calculated using Scherer's equation [19, 21]. The main characteristic peak (311) for ferrites was used to estimate the particle size. The mean crystalline sizes were calculated from XRD line width of (311) peak using the Scherer equation ($t = 0.9\lambda/B\ \cos\theta_B$), where t is mean crystalline size, λ is X-ray wavelength B is he angular line width of half maxima intensity and θ is Bragg's angle in degrees [21].

2. Results

Spinel structure ferrite belongs to technologically important materials. Fine Manganese Zinc Ferrite powders were prepared and then exposed to UV irradiation. The XRD analysis was made in order to determine the structure characterization, particle size, lattice constant and density.

Table 2. Showing the effect of UV irradiation on Particle Size, Lattice Constant, Density(X)

UV Treatment Time (hrs)	Sample Code	Mn (X) Contents	Particle Size (nm)	Lattice Constant (A^0)	Density (gm/cm^3)
Un-Treated	S1-A	0.27	8.82	8.43	5.28
	S2-A	0.47	17.64	8.47	5.17
	S3-A	0.67	6.57	8.43	5.19
	S4-A	0.87	8.82	8.40	5.19
2 hr treatment	S1-B	0.27	17.64	8.41	5.32
	S2-B	0.47	8.86	8.44	5.22
	S3-B	0.67	10.59	8.43	5.20
	S4-B	0.87	13.32	8.48	5.05
4 hr treatment	S1-C	0.27	7.23	8.40	5.34
	S2-C	0.47	13.22	8.45	5.21
	S3-C	0.67	8.83	8.42	5.26
	S4-C	0.87	7.24	8.42	5.16
6 hr treatment	S1-D	0.27	6.19	8.43	5.28
	S2-D	0.47	10.59	8.42	5.26
	S3-D	0.67	8.82	8.40	5.19
	S4-D	0.87	10.58	8.47	5.07

3. Result discussions

4.1 X-Ray diffraction studies

From values given in Table 2 it could be observed that by increasing the time of ultra violet radiation, 2θ angle for characteristic spinel peak shifts showing precise spinel structure. It is concluded that ultra violet radiation have a remarkable effect on manganese zinc ferrites when the concentration of content is 0.27 & 0.67, while ultra violet radiation have no remarkable effect on manganese zinc ferrites when the concentration of content is 0.47& 0.87,but it was experienced that increase exposure to ultra violet radiation effect remarkably effect the structural properties of ferrites. The X-ray diffraction patterns for sample S3-A, S3-B, S3-C & S3-D are given as Figure 1 (a-d),

4.2 Lattice parameters

The results shown from Figure 2 indicate that the lattice parameter increases with increasing Mn ion contents. The increase of the lattice parameter can be attributed to the ionic radii of the ions. The value of the lattice constant for $ZnFe_2O_4$was found to be 8.42 A^o which is in good agreement with that obtained by JCPDS card where a = 8.44A^o, while the value of lattice constant for $MnFe_2O_4$was found to be 8.48A^o.The measured lattice parameter for all samples changed from 8.40 to 8.48A^o.

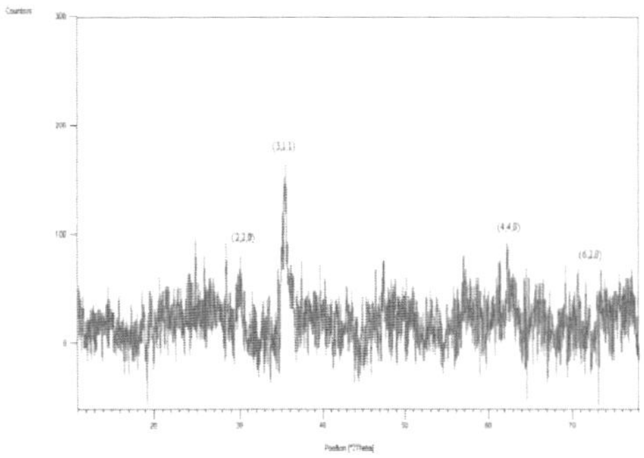

Figure 1 (a). XRD pattern of Mn-Zn ferrite without UV treatment

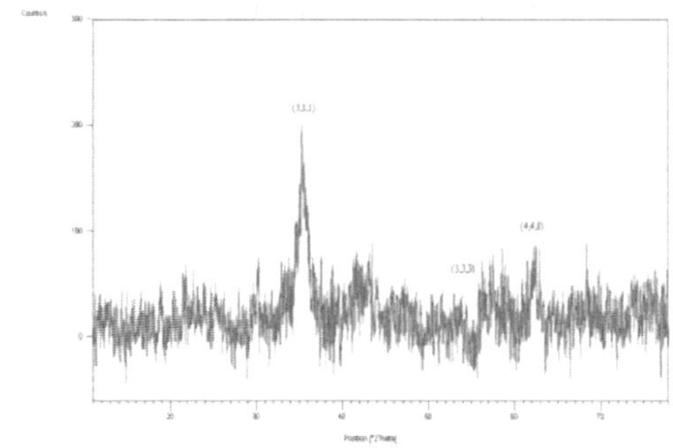

Figure 1 (b). XRD pattern of Mn-Zn ferrite after 2 hrs UV treatment

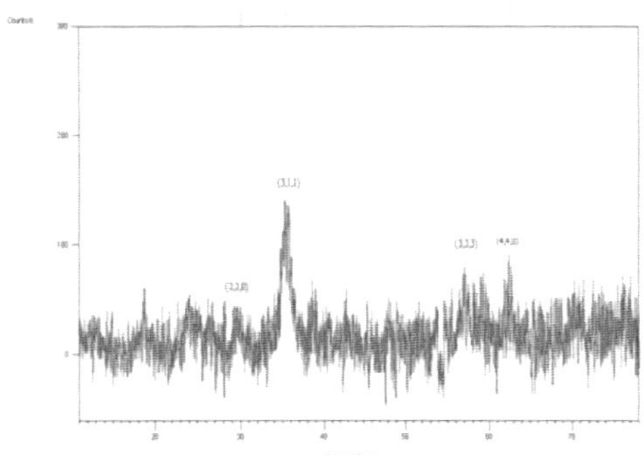

Figure 1 (c). XRD pattern of Mn-Zn ferrite after 2 hrs UV treatment

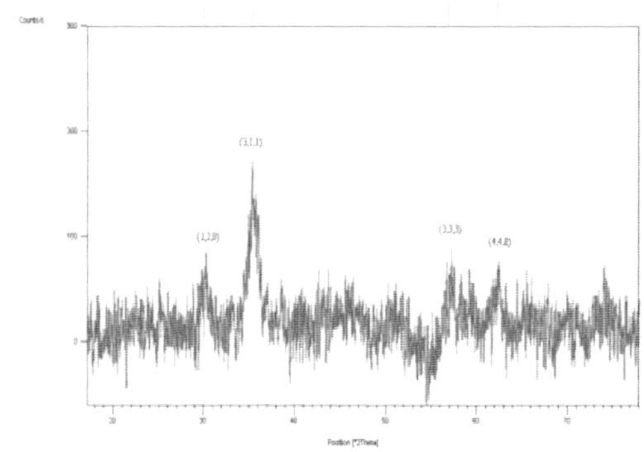

Figure 1 (d). XRD pattern of Mn-Zn ferrite after 2 hrs UV treatment

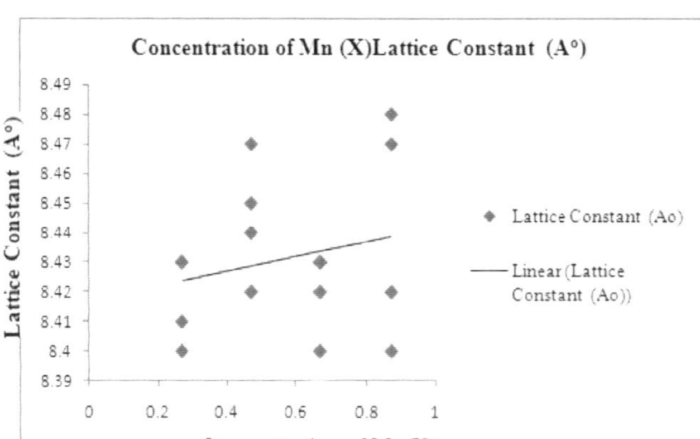

Figure 2. Graph between lattice constant and content (x)

4.3 X-Ray density

From the above data, it is clear that D_x decreases with increasing x. The value of D_x for $Mn_xZn_{(1-x)}Fe_2O_4$ was found to be 5.34 to 5.07gm/cm^3. The decrease of D_x with Mn ion and Zn ion content could be attributed to the atomic weight and the radii of constituent ions, Figure 3.

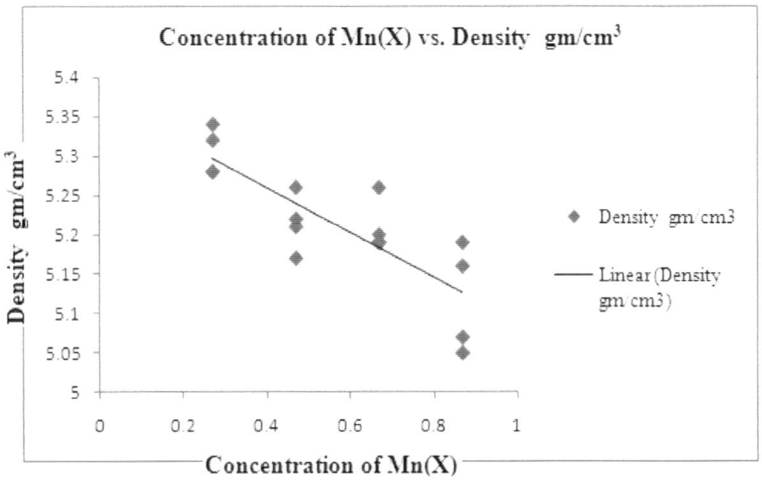

Figure 3. Graph between X-ray density and content (x)

4.3 Particle size

The X-ray densities decrease linearly by Mn concentration. The particle sizes of the samples were also affected by Manganese concentration. The minimum particle size (6.198) was obtained at X = 0.27 when it irradiated for 6 hours. The results showed that the lattice parameter increases with increasing Mn ion content Figure 4.

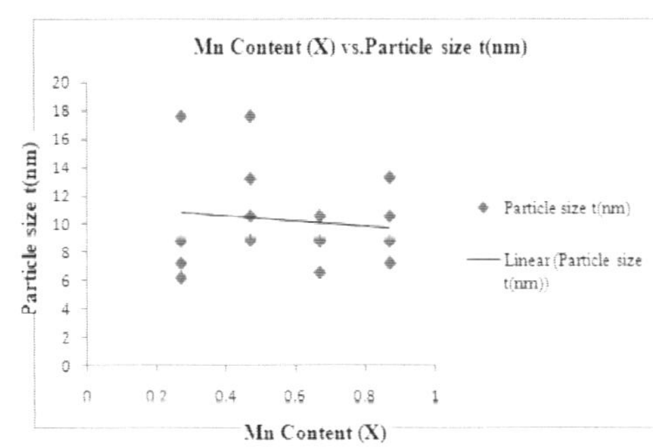

Figure 4. Graph between particle size t (nm) and content (x)

The increase of the lattice parameter can be attributed to the ionic radii of the ions. The results showed that co-precipitation technique was successfully used for the preparation of Manganese Zinc ferrites ($Mn_XZn_{1-X}Fe_2O_4$) magnetic particles where the concentrations (X) are varying from 0.27 to 0.87. The samples were irradiated by ultra violet source and particle size was measured by Scherer's formula. The XRD patterns of samples were found to be in agreement with the original patterns. In most samples Ultra violet showed the remarkable effect in improving the properties of Mn-Zn ferrites.

Effect of calcination and UV irradiation

The samples S3-A was calcined at 900°C and then divided into two equal parts. One part treated as standard and other was irradiated for 2 hours. The XRD pattern of pre irradiated and post irradiated samples after calcination are shown in Figure 5 (a) and 5(b) respectively indicating the reduction in particle size from approximately 90 nm to 50 nm.

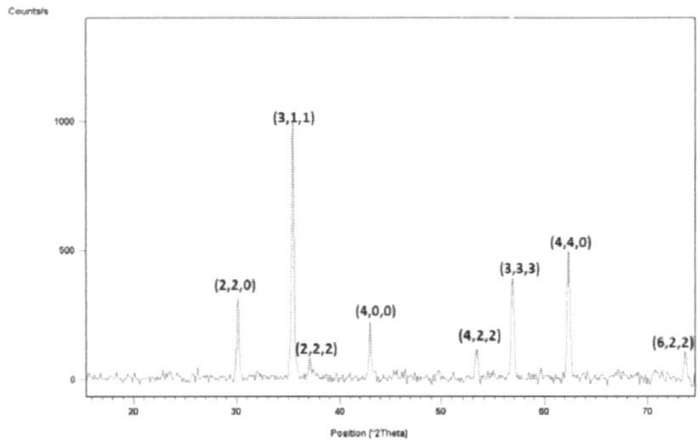

Figure 5(a). XRD pattern of calcined and untreated (UV) samples

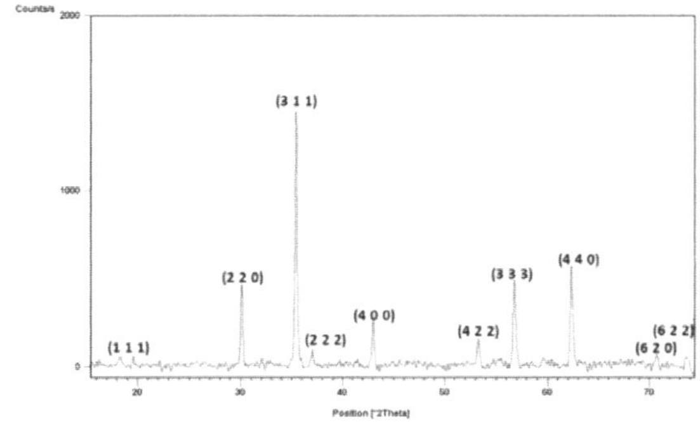

Figure 5(b). XRD pattern of calcined and post Irradiated sample for 2 hrs

4. Conclusions

The consumption of ferrites is rapidly increasing with new advancement in electronics industry specifically power conversion and communication. The purpose of current study was to investigate the effect of UV radiations on the fabrication of Manganese Zinc Ferrite prepared using co-precipitation technique and achievement of lowest particle size along with better magnetic properties. Following are the findings:

1. The improvement in the existing techniques for fabrication of ferrites could be made by using UV treatment of samples and will also be helpful for exploration of new areas of research.
2. The single phase crystallite structure could be achieved with reduced particle size by increasing the time for UV treatment. By calcination and UV treatment, the counts of main peaks have been increased which shows improved crystalinity.
3. The unwanted peaks in XRD pattern analysis corresponds to impurities and secondary phases. With the help of UV treatment these could be reduced which alternatively means increase in phase formation.
4. The increase in Mn(x) concentration makes it possible to achieve high density Manganese Zinc ferrites.
5. The height of spinel peak of concentration achieved after calcination and then UV treatment compared to non-calcined and untreated sample shows the better magnetic qualities and better crystal structure.

6. References

[1] Anwar H, Maqsood A 2010 Temperature dependent structural and electrical analysis of Mn-Zn Nano Ferrites, *J. Pak. Mater. Soc.*4(2) 81-94

[2] Ghazanfar U, Siddiqi SA, Abbas G 2005 Structural analysis of the Mn-Zn ferrites using XRD technique, *Structural Materials Science and Engineering*, 118:84-86

[3] Ott G, Wrba J, Lucke R 2003 Recent developments of Mn–Zn ferrites for high permeability applications. *Journal of Magnetism and Magnetic Materials*, 254/255: 535–537

[4] Snelling EC. (Edn) 1988*Soft Ferrites – Properties and Applications*, (Butterworths: London)

[5] Goldman A, Modern Ferrite Technology 2006 2nd Edn. (Springer: Berlin)

[6] Ahmed M A, Okasha N and El-Sayed MM. 2007 Enhancement of Physical Properties of rare-earth-substituted Mn-Zn ferrites prepared by flash method, *Ceramic International*, 33 (1): 49-58

[7] Dahotre S G and Singh L N 2014 Supermagnetism and FMR study of Nano Mn-Zn Ferrite, *Advances in Applied Science Research*, 5(1):146-149

[8] Hu P, Yang H, Pan D, Wang H, Tian J, Zhang S, Wang X and Volinsky A 2010 Heat treatment effects on microstructure and magnetic properties of Mn–Zn ferrite powders, *Journal of Magnetism and Magnetic Materials*,322: 173–177
[9] Ozgur U2009Microwave ferrites: fundamental properties, *Journal of Materials Science-materials in Electronics*, 20: 789–834
[10] Collin R E 2000*Foundations for Microwave Engineering*, 2nd Edn, (IEEE Press,New York)
[11] Verma A, Alam MI, Chatterjee R, Goel TC and Mendiratta R G 2006 Development of a new soft ferrite core for power applications, *Journal of Magnetism and Magnetic Materials*, 300: 500–505
[12] Mathew D S, Juang R S 2007An overview of the structure and magnetism of spinel ferrite nanoparticles and their synthesis in microemulsions, Chemical Engineering Journal, 129:51–65
[13] Kosak A, Makovec D 2004 Drofenik M, *Phys. Stat. Sol.*, C112: 3521–3524
[14] Paevatheeswara Rao, O Caltun, WS Cho, Chong-Oh Kim and CheolGi Kim, 2007, Synthesis and characterization of mixed ferrite nanoparticles, *Journal of magnetism and magnetic materials,*Vol. 310:812-814
[15] Xuan Y, Qiang Li, Gang Yang 2007 Synthesis and magnetic properties of Mn-Zn ferrite nanoparticles, *Journal of Magnetism and Magnetic Materials*, 312: 464-46
[16] Arulmurugana R, Jeyadevanb B, Vaidyanathana G, Sendhilnathanc S, 2005, Effect of Zinc substitution on Co-Zn and Mn-Zn ferrite nanoparticles prepared by co-precipitation, *Journal of Magnetism and Magnetic Materials,*Volume 288, 470-477
[17] Wahab M A 2015 Solid State Physics Structure and Properties of Materials, 3rd Edn. (Narosa Publishing House: India)
[18] Manzoor A, K. Hasnain S Mumtaz A, F Bertino M and Franzel L 2012 Effects of size and oxygen annealing on the multiferroic behavior of bismuth ferrite nanoparticles, *Journal of Nanoparticle Research*, 14:1310
[19] Cullity B D, 1978, *Elements of X-Ray diffraction*, 2nd Edn (Wesley Publishing Company: Canada)
[20] IslamMU, F Aen, SB Niazi, MA Khan, M Ishaque, T Abbas and MU Rana 2008Electric transport properties of CoZn ferrite-SiO_2 composites prepared by co-precipitation technique, *Materials Chemistry and Physics*, 109:482-487
[21] JamilY, Ahmad M R, Hafeez A, Zia-ul-Haq and AminN 2008 Microwave assisted synthesis of fine magnetic Mn ferrite particles using co-precipitation technique, *Pak. J. Agri. Sci.*, 45 (3):59-64
[22] Abbas T, Khan Y, Ahmad M and Anwar S 1992 X-ray diffraction study of the cation distribution in the Mn-Zn-ferrites, *Solid State Communication,*82:701-703

IOP Conf. Series: Materials Science and Engineering **146** (2016) 012030 doi:10.1088/1757-899X/146/1/012030

Bio-molecular sensors based on guided mode resonance filters

M. R. Saleem[1], R. Ali[1], S. Honkanen[1] and J. Turunen[1]
Institute of Photonics, University of Eastern Finland,
P.O. Box 111, FI-80101, Joensuu, Finland
[1]Center for Advanced Studies in Energy (CAS-EN), National University of Sciences and Technology, postal code 44000, Sector H-12, Islamabad, Pakistan

E-mail:rizwanphy@gmail.com

Abstract. In this work a low surface roughness and homogenous, high refractive index, and amorphous TiO_2 layer on corrugated structures of diffractive optical element is coated by Atomic Layer Deposition (ALD) for biosensors. The design of Guided Mode Resonance Filters (GMRFs) is based on refractive indices and thicknesses of the waveguide biomolecular layers. The designed spectral shifts are calculated by Fourier Modal Method (FMM) and depend on the magnitude of the variations in refractive index of the biomolecular layer on waveguide structures. Furthermore, the sensitivity of the biomolecular sensors depends on the thickness of biomolecular layer and periodicity of the structures. The waveguide structures designed for larger periods show an enhancement in the sensitivity (nm/RIU) of the biomolecular sensor at longer wavelengths. The periodicities of nanophotonic structures are varied from 300 to 500 nm in design calculations with predominance of increase in effective index of the structure to support leaky waveguide modes.

1. Introduction

Metamaterials encountered great interest in several applications of precise optical sensors and other elements. The development of perfectly tuned optical components is a result of highly efficient and sophisticated modeling tools and modern fabrication of nanostructures by accurate experimental lithographic, coating, and etching techniques. Metallic sub wavelength nanostructures show useful optical properties, Surface Plasmon Resonance (SPR) to measure minute changes in the refractive index of materials in regions of nanostructures referred as the refractive index of the cavity n_c. The spectral shift of central resonance peak is a measure of the sensitivity of the sensor in nanometers per Refractive Index Unit (RIU) [1-4]. Generally, a high efficiency and spectral resolution optical sensor with larger figure of merit detects minor changes in refractive index of cavity [5].

The physics of metallic nanostructures is based on scattering phenomena while sub wave length dielectric nanostructures support Wood's anomalies of Guided Mode Resonance Filters (GMRFs) [6]. Incident light couple to the excited leaky waveguide modes guided by the waveguide nanostructure under resonance conditions [7]. This results in narrowband of reflected or transmitted wavelengths through the sensor at resonance [8]. GMRFs are quite often used to target several promising applications such as pressure, humidity, or protein detections. The precise sensing with highest resolution of such sensor elements is achieved and facilitated by rapid and cost-effective productions through accurate control of structural and optical parameters at resonance wavelengths [9-11].

The practical realization of such sensors and corresponding nanostructures require precisely uniform depositing techniques compatible with the interacting wavelengths. Traditionally, optical

Content from this work may be used under the terms of the Creative Commons Attribution 3.0 licence. Any further distribution of this work must maintain attribution to the author(s) and the title of the work, journal citation and DOI.
Published under licence by IOP Publishing Ltd

coatings of metals and dielectrics are obtained by Physical Vapor Deposition (PVD) using solid source materials through resistive heating, electron-beam evaporation, sputtering, laser-assisted evaporation, or ion-assisted deposition at high enough deposition rates. To confine and smooth propagation of waveguide modes through dielectric films, the control of film homogeneity, composition, thickness uniformity, adhesion with underlying substrate, and better microstructure control can be achieved by Atomic Layer Deposition (ALD) [12]. ALD method exhibits angstrom level resolution, layer-by-layer growth of ultra-thin compound films on planar and high-aspect ratio nanostructures. ALD operates in cycles consisting of four steps: (1) exposure of first precursor material, (2) evacuation or purging of chamber, (3) exposure of second precursor material, and (4) evacuation or purging of reaction byproducts from the chamber [13-15].

In this paper we report on the design and optimization of optical properties of GMR bio-molecular sensors. The sensing properties of the sensors have been optimized theoretically with increase in the thickness of the analyte layer. The localized refractive index is tailored by the incremental increase in the thickness of analyte layer within the cavity. Figure 1 shows the structure of a TiO_2 based GMRF bio-molecular sensor. A polarized light (s and p) illuminates the replicated grating [16-17] at an angle of incidence φ_i with respect to the normal to the grating and outgoing light makes an angle φ_o with the normal to the grating.

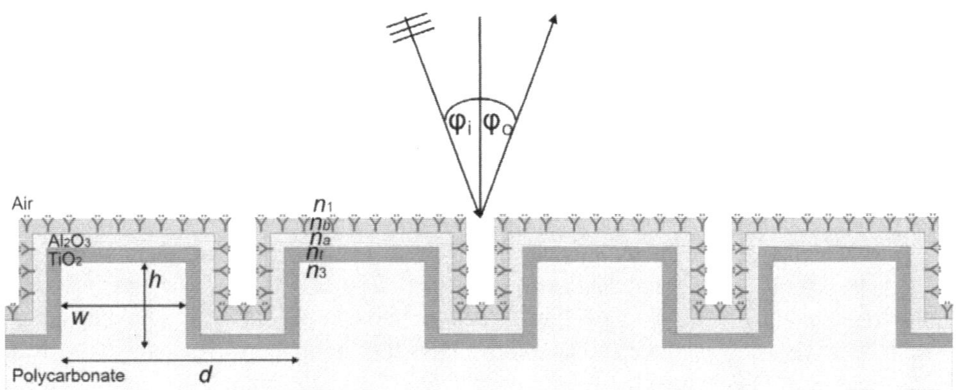

Figure 1. Ideal grating profile of bio-molecular sensor functionalized with an analyte in the presence of coated TiO_2 and Al_2O_3 thin layers

2. Results and discussion

2.1. Design and simulation

Consider 1D grating structure, the grating can couple light into and out of the waveguide by diffraction phenomena. The grating structure can slightly modify the envelope and propagation constant of the waveguide propagating mode. The height of the grating structure is typically much smaller than the incident wavelength λ_o. The grating period $d \sim \lambda_o/n_b$, where n_b is refractive index of analyte, so the first diffracted orders into the structure are propagating, whereas rest of the higher diffracted orders are evanescent. The first order diffraction occurs at: $\frac{2\pi}{d} = \pm n_1 k_o \mathrm{Sin}\varphi_i + \beta_r$, where $k_o = \frac{2\pi}{\lambda_o}$ and β_r is real part of propagation mode constant $\beta = \beta_r + i\beta_i$ [18].

The scope of this work is to develop highly sensitive and efficient GMR structures for bio-molecular sensors. For this purpose, we systematically calculated the spectral shift in resonance wavelength by gradually increasing the analyte thickness. A schematic presentation of such sensors is shown in Figure 1. Design was carried out using rigorous coupled wave analysis through Fourier Modal Method (FMM). Design was performed for both TE- and TM-polarized lights. Figure 1 shows a three-layered structure of TiO_2/Al_2O_3/analyte layers on a replicated grating in polycarbonate substrate. Film thicknesses of TiO_2 and Al_2O_3 materials are kept fixed at 40 and 20 nm, respectively while the thickness of analyte layer varies from 5 nm onward. Structural parameters of the grating are: periodicity d = 325 nm, line width w = 205 nm, ridge height h = 100 nm, angle of incidence φ_i = 18° using both TE- and TM-polarized light. Refractive indices of: air n_1 = 1, polycarbonate n_3 = 1.53, bio-molecule n_b = 2.2, $Al_2O_3 n_a$ =1.67, and $TiO_2 n_t$ = 2.4.

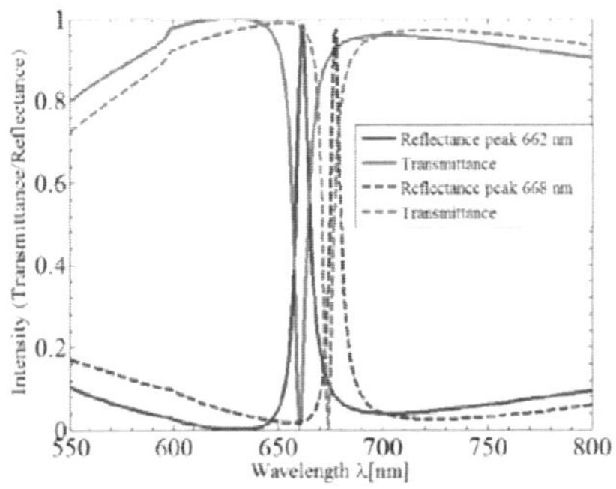

Figure 2. Transmittance/reflectance spectra of GMR sensor with peak positions at 662 nm and 668 nm at 5 and 15 nm thickness of analyte layers, respectively

2.2. Wavelength shift of bio-molecular sensor with incremental increase in analyte thickness

Figure 2 shows transmittance/reflectance spectra of GMR sensor for TE-mode with resonance wavelength peak positions at 662 nm and 668 nm for analyte layer thicknesses of 5 nm and 15 nm, respectively. Resonance peaks show a shift around 6 nm on increasing analyte thickness by 10 nm. Figure 3 shows calculated reflectance spectra for TE-mode of bio-molecular sensor with 5 nm increase in analyte thickness in each spectral calculation. The wavelength shifts towards longer wavelengths by 3 nm on increasing analyte thickness by 5 nm as shown in each spectral image. As a result, peak shifts from 662 nm to 684 nm by increasing analyte thickness from 5 to 40 nm. The shift in peak wavelength results in due to increase in effective index of analyte layer owing to the increase in the optical thickness of analyte layer. One can estimate a linear change in the wavelength shift as a function of analyte thickness with low sidebands and almost a symmetric spectral profile. It shows an average shift of 0.6 nm per nm increase in analyte thickness.

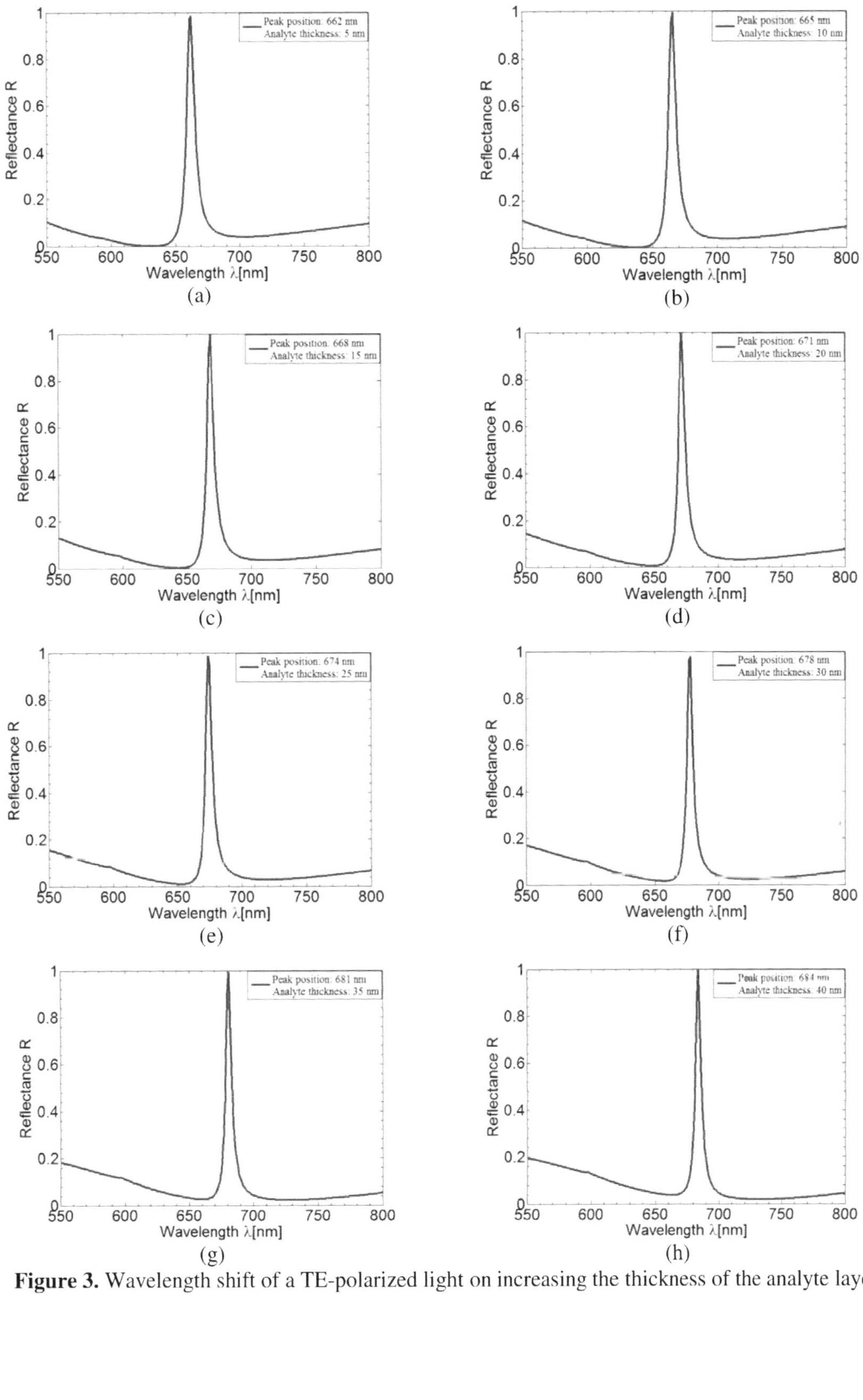

Figure 3. Wavelength shift of a TE-polarized light on increasing the thickness of the analyte layer

IOP Conf. Series: Materials Science and Engineering **146** (2016) 012030 doi:10.1088/1757-899X/146/1/012030

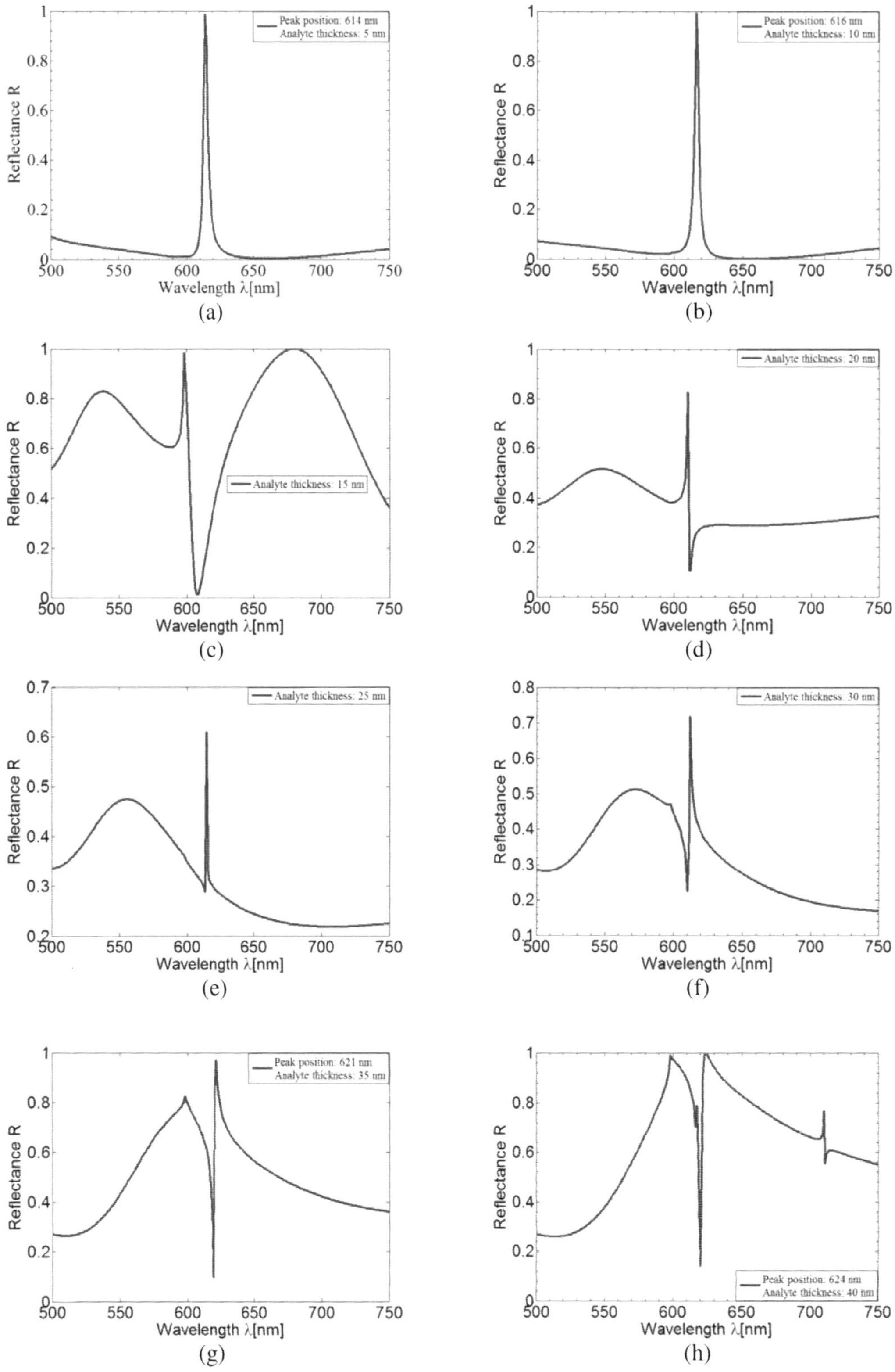

Figure 4. Wavelength shift of a TM-polarized light on increasing the thickness of the analyte layer

IOP Conf. Series: Materials Science and Engineering **146** (2016) 012030 doi:10.1088/1757-899X/146/1/012030

Figure 4 shows calculated reflectance spectra for TM-mode by increasing the analyte layer thickness from 5 to 40 nm. It is evident from the spectra that narrow line width of spectral waveform with lower sidebands and symmetric profile appears at relative low analyte thicknesses (5 to 15 nm). However, more waveguide modes appear on increasing the thickness beyond 15 nm. In the context of ultra-thin layers, we introduce the spectral shifts in response to the thickness of the analyte layer. For simplicity in the calculations, the thickness of the analyte layer is considered; however, a detailed analysis can be made using optical thickness. For TM-modes the peak shifts are not linear with the thickness of the analyte layer which shows that the sensor is more efficient with layer thicknesses around 5 to 15 nm rather than with thicker layers.

The designed biomolecular sensor is also optimized in a flow cell configuration to investigate the kinetics of a chemical reaction for which such sensors can be used in microfluidic applications. Figure 5 shows TE- and TM-polarized reflectance spectra calculated in a flow cell arrangement filled with either air (refractive index: n_1 = 1 or water: refractive index n_3 = 1.33). In the case of structure with air filling scheme, the calculated spectra for TE- and TM-polarized lights shows resonance peak positions at 662 and 614 nm, respectively. However, all spectra show shifts towards longer wavelengths once the refractive index of the cavity is changed with water. The TE- and TM-shifted resonance peaks show spectral positions at 694 and 651 nm, respectively. The calculated long-range sensitivities obtained for TE- and TM-modes are 97 and 82 nm/RIU, respectively. It is worthwhile to describe that the peaks in water environment are much narrower than those calculated for air which facilitate to design sensors at higher figure of merit value due to low refractive index contrast between the cavity and optical materials (polycarbonate, TiO_2, Al_2O_3 etc.,).

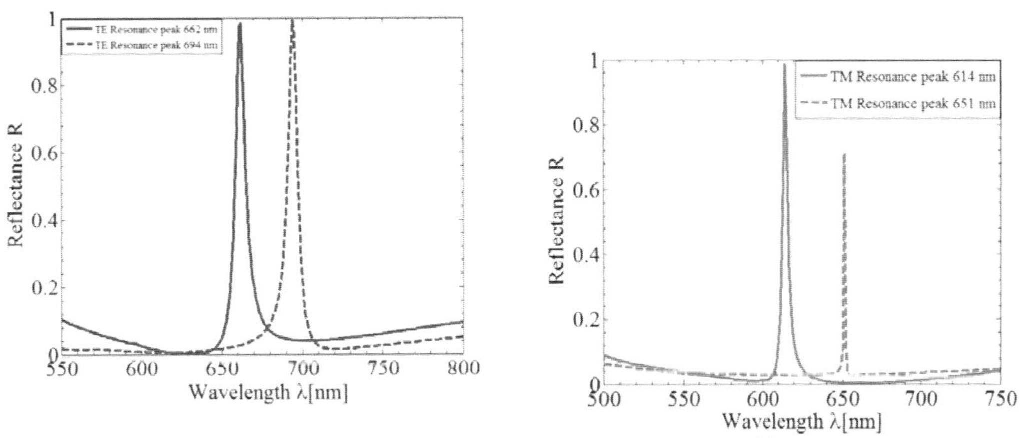

(a) (b)

Figure 5. Reflectance spectra in a flow cell configuration of (a) TE-Mode and (b) TM-Mode

3. Conclusion

Guided mode resonance biomolecular sensors have been designed with respect to the thickness layer of the analyte. The 3-layer structure design is made for a replicated grating in polycarbonate, with thin films of metallic oxides (TiO_2 and Al_2O_3) and analyte layer as bio-molecule. The calculated TE-polarized spectra showed a linear shift in resonance peak wavelength with the increase in analyte layer thickness from 5 to 40 nm while TM spectra followed linearity only for very thin layers (5 to 15 nm) of analyte and showed generation of new modes with more narrowed peaks. The sensitivities of TE- and TM-modes are calculated in a flow cell configuration filled with either air or water. In the case of cavity filled with water, significant peak shifts have been observed for both TE- and TM-modes which resulted in long range high sensitivities of 97 and 82 nm/RIU, respectively. The calculated high sensitivities are attributed to low refractive index contrast between the cavity and grating's optical materials.

4. Acknowledgement
Financial support from Strategic funding of the University of Eastern Finland, Institute of Photonics and USAID program for Center for Advanced Studies in Energy (CAS-EN), NUST, Islamabad is highly appreciated.

5. References
[1] R. Magnusson, D. Wawro, S. Zimmerman, Y. Ding, "Resonant photonic biosensors with polarized-based multiparametric discrimination in each channel," Sensors 11, 1476–1488 (2011)
[2] X. Fan, I.M. White, S.I. Shopova, H. Zhu, J.D. Suter, Y. Sun, "Sensitive optical biosensors for unlabeled targets: a review, " Anal. Chim. Acta 620, 8–26 (2008)
[3] X. Wei, S.M. Weiss, "Guided mode biosensor based on grating coupled porous silicon waveguide," Opt. Express 19, 11330–11339 (2011)
[4] H.N. Daghestani, B.W. Day, "Review:Theory and Applications of Surface Plasmon Resonance, Resonant Mirror, Resonant Waveguide Grating, and Dual Polarization Interferometry Biosensors," Sensors 10, 9630–9646 (2010)
[5] A. Szeghalmi, E.B. Kley, M. Knez, "Theoretical and experimental analysis of the sensitivity of guided mode resonance sensors," J. Phys. Chem. C 114, 21150–21157 (2010).
[6] M.R. Saleem, "Resonant waveguide gratings by replication and atomic layer deposition," PhD thesis, University of Eastern Finland, ISBN: 978–952–61–0884–1 (2012)
[7] B.T. Cunningham, P. Li, B. Lin, J. Pepper, "Colorimetric resonant reflection as a direct biochemical assay technique," Sens. Actuators B81, 316–328 (2002)
[8] B.T. Cunningham, P. Li,S. Schulz, B. Lin,C. Baird, J. Gerstenmaier, C. Genick, F. Wang, E. Fine, L. Laing, "Label-Free Assays on the BIND System,"J. Biomol. Screen. 9, 481–490 (2004)
[9] M.R. Saleem, D. Zheng, B. Bai, P. Stenberg, M. Kuittinen, S. Honkanen, J. Turunen, "Replicable one-dimensional non-polarizing guided mode resonance gratings under normal incidence," Opt. Express 20, 16974–16980 (2012)
[10] M.R. Saleem, S. Honkanen, J. Turunen, "Effect of substrate overetching and heat treatment of titanium oxide waveguide gratings and thin films on their optical properties," Appl. Opt. 52, 422–432 (2013)
[11] J-N. Yih, Y-M. Chu, Y-C. Mao, W-H. Wang, F-C. Chien, C-Y. Lin, K-L. Lee, P-K. Wei, S-J. Chen, Optical waveguide biosensors constructed with subwavelength gratings. Appl. Opt. 45, 1938–1942 (2006)
[12] M.R. Saleem, R. Ali, M.B. Khan, S. Honkanen, J. Turunen, "Impact of atomic layer deposition to nanophotonic structures and devices," J. of Frontiers in Materials 1, 1–15 (2014)
[13] M.R. Saleem, P. Silfsten, S. Honkanen, J. Turunen, " Thermal properties of TiO_2 films grown by atomic layer deposition," Thin Solid Films 520, 5442–5446 (2012)
[14] M.R. Saleem, R. Ali, S. Honkanen, J. Turunen, " Thermal properties of thin Al_2O_3 films and their barrier layer effect on thermo-optic properties of TiO_2 films grown by atomic layer deposition," Thin Solid Films 542, 257–262 (2013)
[15] R. Ali, M.R. Saleem, P. Pääkkönen, S. Honkanen, "Thermo-Optical Properties of Thin-Film TiO_2–Al_2O_3 Bilayers Fabricated by Atomic Layer Deposition," *Nanomaterials* 5, 792–803 (2015)
[16] D. Rosenblatt, A. Sharon, A.A. Friesem, "Resonant Grating Waveguide Structures," IEEE J. Quantum. Electron. 33, 2038–2059 (1997)
[17] M.R. Saleem, P. Stenberg, M.B. Khan, S. Honkanen, J. Turunen, "Hydrogen silsesquioxane resist stamp for replication of nanophotonic components in polymers," J. Micro/Nano lith. MEMS MOEMS 11, 013007 (2012)
[18] S. Glasberg, A. Sharon, D. Rosenblatt, A.A. Friesem, "Spectral shifts and line-shapes asymmetries in the resonant response of grating waveguide structures," Opt. Comm. 145, 291–299 (1998)

The effect of anisotropy on the intermediate and final form in deep drawing of SS304L, with high draw ratios: Experimentation and numerical simulation

F. Qayyum, M. Shah, A. Muqeet and J. Afzal
Fracture Mechanics and Fatigue Laboratory, Mechanical Engineering Department, University of Engineering and Technology, Taxila, Pakistan

E-mail: faisal.qayyum@uettaxila.edu.pk

Abstract: High deep draw ratio of metal sheets is often required in industry to produce complex structural components and is considered to be problematic beyond a draw ratio of 1.7. In addition, anisotropy of material plays a great role in the final form of deep drawn products. In this research AISI SS304 L high draw ratio deep drawing (HDR) cups have been developed by using multiple intermediate annealing steps. The form of flange produced at each draw step is recorded. Numerical simulation of the process is carried out using ABAQUS Standard™. Accurate modelling of the process requires anisotropic parameters. Tensile tests at 0°, 15°, 30°, 45°, 60°, 75°, 90° to the rolling direction are carried out to determine the Lankford coefficients and hardening coefficients in each direction. The detailed material data obtained is then applied to simulate the HDR process. The forms recorded experimentally are compared with simulation results and conclusion is drawn as to the accuracy of the simulation.

Notations

wo	Width of specimen before test
wf	Width of specimen after test
lo	Length of specimen before test
lf	Length of specimen after test
rx	Lankford's r value in rolling direction
ry	Lankford's r-valuein trasverse direction
R22	Anisotropy coefficient in x-direction to simulate anisotropy in ABAQUS
R33	Anisotropy coefficient in y-direction to simulate anisotropy in ABAQUS
R12	Anisotropy coefficient in x-y plane to simulate anisotropy in ABAQUS

Content from this work may be used under the terms of the Creative Commons Attribution 3.0 licence. Any further distribution of this work must maintain attribution to the author(s) and the title of the work, journal citation and DOI.
Published under licence by IOP Publishing Ltd

1. Introduction

A. S. Korhonen[1] devised a technique for estimating the maximum drawing force in the deep drawing of cylindrical cups. He approximated that at the limiting drawing ratio maximum force will act at the punch nose rounding and that there will be no contact between the sheet and the punch at the point of necking, so that frictional force cannot assist in carrying any drawing load. He proposed simple charts for determining the maximum drawing force as a function of anisotropy and tooling geometry. M. Ahmetoglu, et al. [2] found the wrinkling and fracture boundaries of aluminum alloy 2008-T4 to remove defects, increase part quality and rise the draw depth. They performed experiments on oval, oblong and rectangle specimens by varying the Blank Holding Force (BHF). They concluded that BHF as a function of time increases the formability and quality of final part. Xi Wang and Jian Cao [3] based on simplified flat or curved sheet models with approximate boundary conditions devised a modified energy approach utilizing energy equality and the effective dimensions of the region undergoing circumferential compression. Xi Wang and Jian Cao [4] devised a method is based on the wrinkling principle suggested by Cao and Boyce [5] for forecasting the buckling performance of sheet metal under normal limitation. They used a mixture of energy conservation and plastic bending theory. Their analysis provided the serious buckling stress and wavelength as affected by normal pressure. The results obtained were in excellent arrangement with those obtained from Cao and Boyce's numerical method [5], and also matched well with the experimental outcomes of a square cup development. They also discussed that how wrinkling behavior is affected by material properties.Anupam Agrawal, N. Venkata Reddy and P.M. Dixit [6] predicted the minimum blank holding pressure compulsory to evade wrinkling in the flange region during axisymmetric deep drawing procedure. They equated the energy accountable for wrinkling to that which overpowers the wrinkles. The model was verified by comparing with already published data.S. Han, M. Bruhis, and M. K. Jain [7] developed a new FE model for deep drawing and redrawing, which also accounted for the stiffness of working machine. a mathematical model to determine the limiting drawing ratio (LDR) of deep drawing and redrawing procedures was resulting based on the extension of an existing analytical model proposed by Leu [8] and Hill's anisotropic criterion [9]. The consequences of the scientific model were validated by consistent experimental and FE simulation work in relationships of punch load and clamping force against punch displacement and thickness allocations along the product profile.R. Padmanabhan, M.C. Oliveira, J.L. Alves and L.F. Menezes [10] worked on the effect of process parameters on the deep drawing of stainless steel. They used Taguchi method to recognize the relative effect of each process restriction. They found out that die radius has the highest effect on the deep drawing of stainless steel blank sheet followed by the effect of blank holder force and the friction coefficient. They also found out that blank holder force and local lubrication arrangement improves the superiority of the formed part.M. Kadkhodayan and F. Moayyedian [11] based on the Tresca yield criterion, bifurcation functional and Tresca yield criterion along with the assumption of perfectly plastic material developed a closed-form answer for the critical drawing stress. In their study they used nonlinear plastic stress field and the deformation theory of plasticity. For larger width of flanges they successfully predicted effect of a blankholder upon wrinkling and on the number of waves produced. It was also demonstrated that using large deflection theory has the same result as the using small deflection theory. Wang Wu-rong, et.al. [12] worked on the limiting drawing ratio (LDR) and formability forecast of progressive high strength dual-phase steels (DP-AHSS). They conducted experiments to recognize the maximum blank diameter with onset crack, then simulated the same problem using three yielding models (Hill [9], Batlat[13] and Banabic [14]). The investigation showed that a Swift and Hockett–Sherby joint formula is in good contract with the flow curve of the tensile test and Batlat-89 yield model positively forecasts the beginning of shear crack of DP AHSS.

In the current research cups with high deep draw ratios (HDR) have been developed using multiple deep draw punches with intermediate annealing steps. Experimentations was done using customized die assembled on MTS 810 system to determine various parameters affecting the deep drawing process i.e. force applied, punch displacement and gripping pressure on sheet blank. Evolution of flange shape from circular to complex polygonal shape was also studied by recording the flange shape after each step in pictorial form. Detailed material testing was done to determine plastic properties, strain hardening coefficients and Lankford's r-values at 0°, 15°, 30°, 45°, 60°, 75°, 90° of sheet where 0° is rolling direction and 90° is transverse direction. This material model was used while defining numerical simulation model. Results obtained from numerical simulation were compared with actual experimentation, it was observed that due to good definition of numerical model it predicted the experimental trends very closely with some errors.

2. Experimentation

2.1. Materials

2.1.1. AISI D2

Die for hydroforming was manufactured using AISI D2. The chemical composition [15] of this material is shown in Table 1, and mechanical properties of AISI D2 [16] are shown in Table 2. The material was heat treated by using ASTM B661-12 [17].

Table 1. Chemical properties of AISI D2

Component	C	Si	Mn	Cr	Mo	V	Ni	Co	Cu
Wt. %	1.55	0.3	0.4	11.8	0.8	0.8	0.3	1	0.25

Table 2. Mechanical properties of AISI D2

Density Kg/m3	Young's Modulus GPa	Poisson's Ratio	Compressive Yield Strength MPa	Thermal Conductivity W/m.K	Specific Heat KJ/Kg.K
7700	210	0.3	2200	20	0.46

2.2. AISI SS304

Chemical composition of AISI SS304 [15] is shown in Table 3 and mechanical properties of AISI SS304 [16] are shown in Table 4.

Table 3. Chemical properties of AISI SS304 [15]

Component	C	Cr	Fe	Mn	Ni	P	S	Si
Wt. %	0.08	18-20	66.3-74	Max. 2	8-10.5	0.045	0.03	Max. 1

Table 4. Mechanical properties of AISI SS304

Density Kg/m3	Poisson's Ratio	Young's Modulus GPa	Yield Strength MPa	UTS MPa	Rockwell Hardness
8000	0.3	200	215	505	70

To determine detailed plasticity data tensile tests 0.5mm thick SS304L sheet were done using ASTM E8 Standard [18] at different deformation speeds. Results of these tests are shown in Figure 1. It can be observed that the material is highly stain rate dependent. At deformation speed of 0.48 mm/min, it shows the yield strength of 223 MPa, at deformation speeds of 4.8 mm/min and 48 mm/min, the material shows yield strength of 273 MPa and at speed of 480 mm/min it has very high value of yield strength which is close to 400 MPa. Hence it can be concluded that AISI SS304L shows greater strain hardening at higher strain rates, hence it should be deep drawn at relatively lower strain rates.

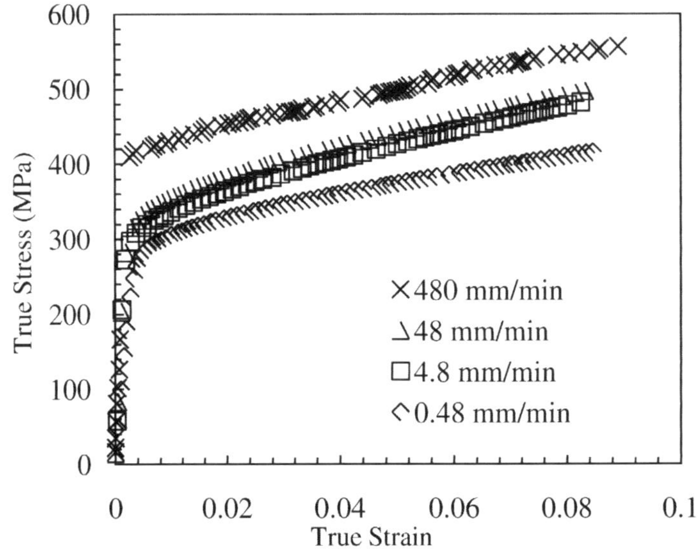

Figure 1. Stress Strain diagram for 0.5mm thick sheet at different deformation speeds

Commercially available rolled sheets show anisotropic behaviour while forming. To determine the Lankford r-values for anisotropy, specimens were prepared and tested according to the ASTM E517 Standard [19]. As for sheet only planar anisotropy is considered, specimens were cut in 3 directions which have been represented in Figure 2.

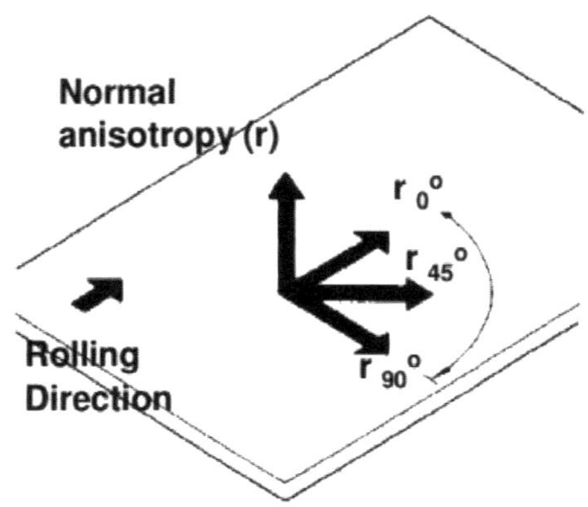

Figure 2. Extraction of 3 specimens for the test and determination of r-values according to ASTM E517 Standard

According to standard they were elongated 20% to their original length and the data shown in Table 5 was recorded for further calculations. Specimens before testing and after testing have been presented in Figure 3.

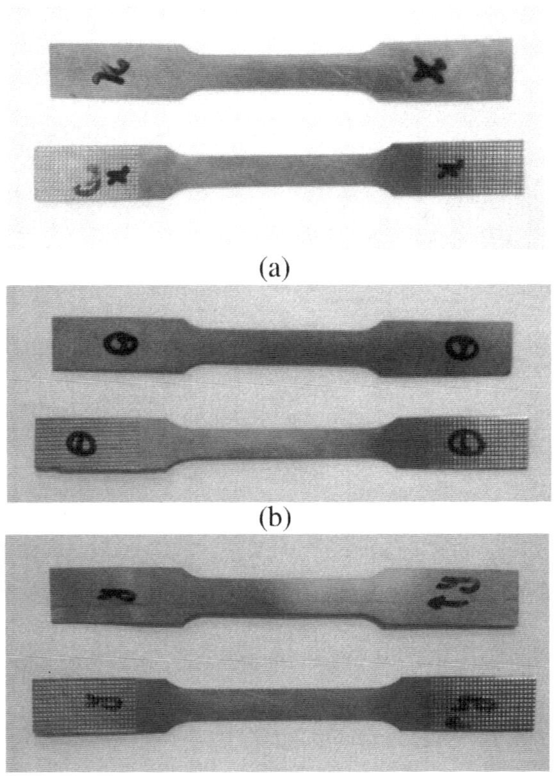

(a)

(b)

(c)

Figure 3. Specimens (undeformed & deformed) prepared for determination of Lankford r-values (a) rolling direction (b) medial direction (c) perpendicular to rolling direction

The r value is considered a measure of sheet metal draw ability, Δr is measure of the tendency to form ears in the flange of deep drawn cylindrical parts in the direction of higher r values [19]. The results of these tests are discussed later.

3. Experimental setup

Die was designed for blank of 70mm diameter to get a deep draw ratio of 2.5 during deep drawing, and internal bulge was introduced to hydroform the final cup into a relatively complex shape. The schematics of the deep drawing process has been shown in Figure 4. Clamping rods, spring and flanges were used to incorporate load cell into the assembly, so that the gripping force on sheet blank could be estimated. Different modes of final assembly have been shown in Figure 5.

Before drawing, sheet was lubricated using graphite grease, and gripping pressure of 20kN was applied on the gripper to eliminate wrinkling, these gripping pressures were kept constant throughout the experiment. It was observed that the flange becomes hard due to excessive deformation radially and tangentially. The sheet was fully drawn into the die using multi step deep drawing with intermediate annealing steps. The process was conducted using force control technique to avoid the sheet from reaching the limiting force, which was 34kN in our case.

Die, grabber and punch were manufactured using AISI D2. After machining these parts were heat treated by following ASTM B661-12 [17], to make these parts hard enough so that they might not deform at very high pressures which were to be applied during the hydroforming process. External surface of punch and internal surface of grabber were ground to attain tight fit of hydraulic grade.

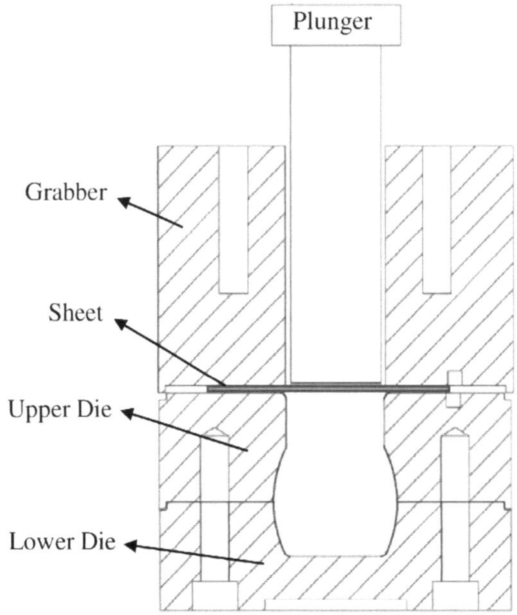

Figure 4. Schematics showing the basic deep drawing process

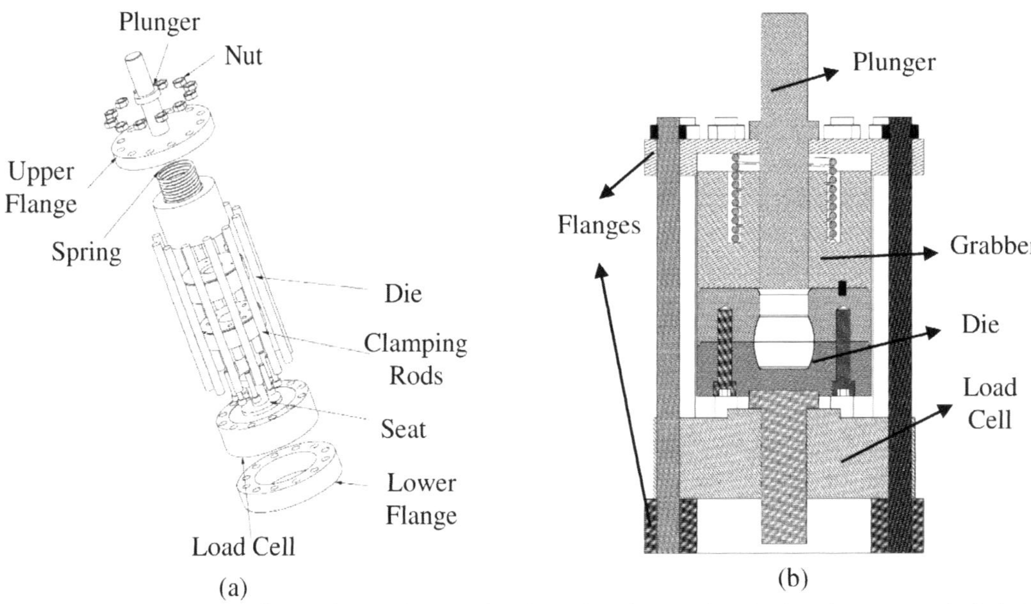

Figure 5. Final assembly of experimental setup for deep drawing and hydroforming (a) exploded view showing all components (b) sectioned view

Fully assembled die was mounted on MTS 810 system to operate the whole process in a controlled manner, and to get a detailed set of data which can be further processed to develop accurate relationships between above mentioned boundary conditions affecting the outcome of the process. We were able to set exact gripping force due to incorporated load cell into the assembly, and get force and displacement data from the LVDT and load cell of MTS 810 system. Analysis of data has been presented in results and discussion section.

4. Image processing

Commercially available metal sheets behave anisotropically when drawn into cups. This behaviour is due to the longer metal grain development in rolling direction and smaller metal grains in transverse direction. This anisotropic behaviour affects the final shape of part. In this research, we closely observed the experimental behaviour of material flow into the die cavity. Shape of flange after each step was recorded, using NIKON D600 with standard 55mm lens of 5 aperture, at ISO-100, exposure time of 0.5 seconds and resolution of 300 dpi in horizontal/vertical directions, to produce a set of images as shown in Figure 6.

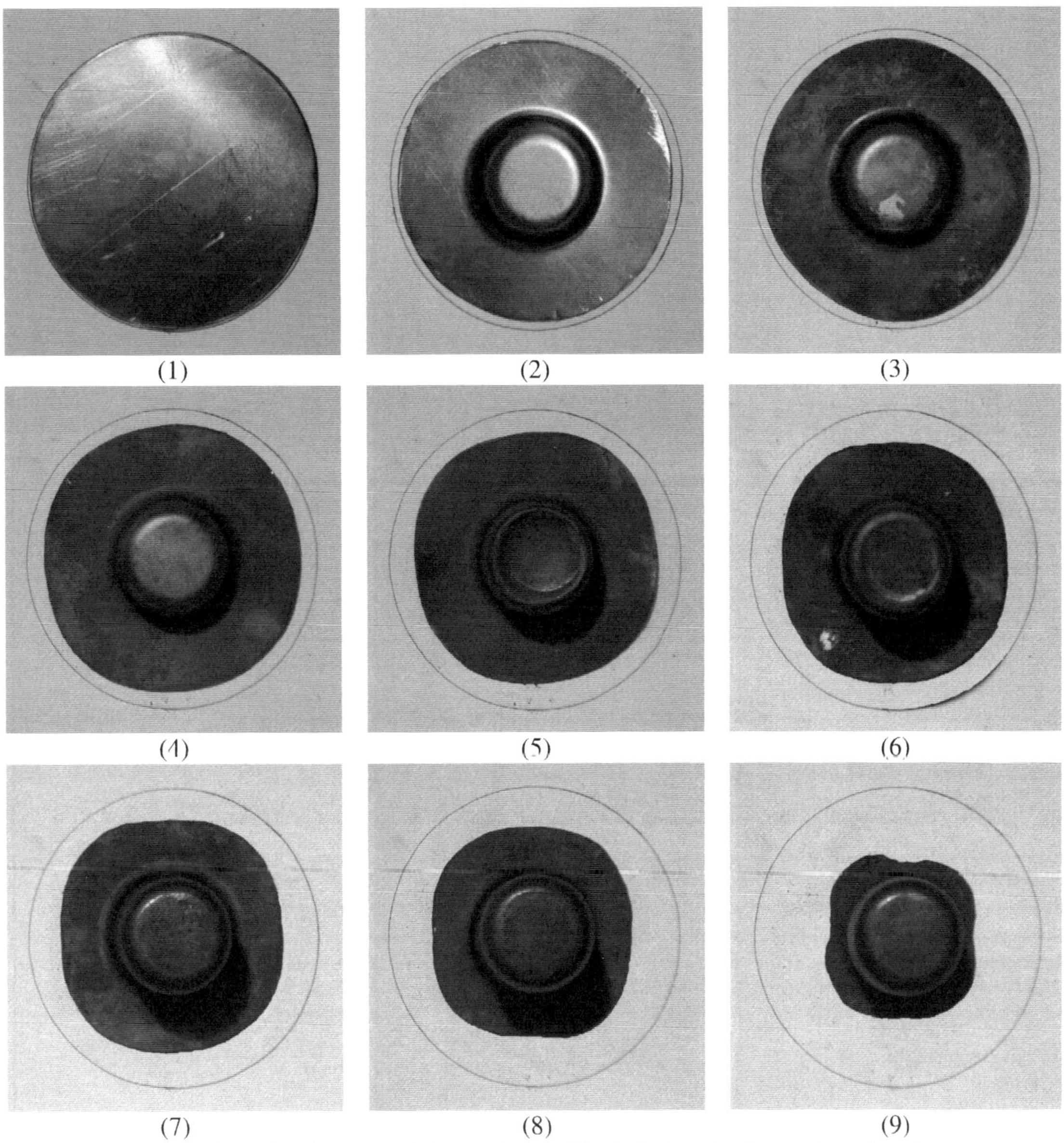

Figure 6. Evolution of earing at periphery of sheet blank during the 8 step deep drawing process

These images were then further processed using commercially available java 1.6.0_20 based code for image processing "Image J 1.47v" developed by National Institute of Health USA. The processed set of images is shown in Figure 7.

Figure 7. Processed images showing the evolution of flange shape after each step

These images were then processed in MATLAB to trace boundaries. Centroids of each polygon could be selected manually or automatically. Manual selection was preferred as it produced better results. Radial distance of boundary from centroid after specific interval of angel was recorded and processed to get desired radial plots, which were then compared with the simulation results.

5. Finite element analysis

The detailed FE analysis of the procedure was carried out on ABAQUS Standard TM 6.12. The problem was simplified by only considering deformable sheet blank, rest all the parts including plunger, die and grabber were considered to be rigid. Detailed finite element analysis was performed on the final assembly of all the components as shown in Figure 8.

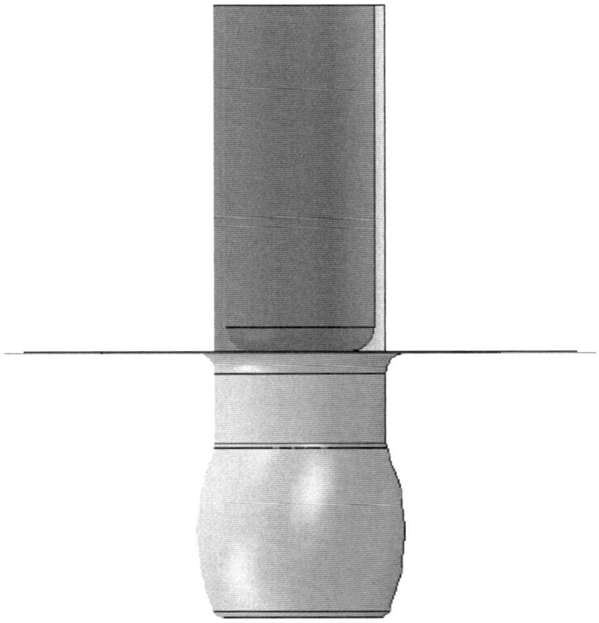

Figure 8. Full assembly of all parts used in simulation

Die, grabber and punch were modeled to be as 3D analytically rigid revolved shells as per actual drawing. While sheet blank was modeled to be as 3D deformable solid revolved part of 70mm diameter and of 0.5mm thickness. The blank was then sectioned to assign appropriate material properties and mesh to specific areas. Sectioned sheet blank has been shown in Figure 9.

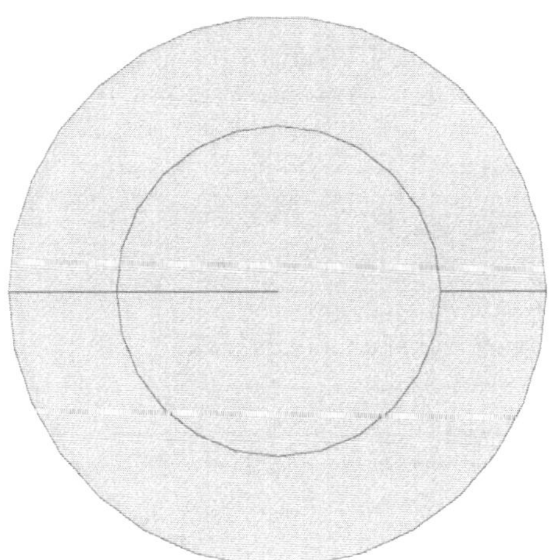

Figure 9. Sectioned sheet blank used in simulation

Material properties of elastic modulus, poisson's ratio and density were assigned to the sheet blank as shown in Table 4, while the plastic properties of SS304L were determined from tensile tests and were defined as shown in Figure 10.

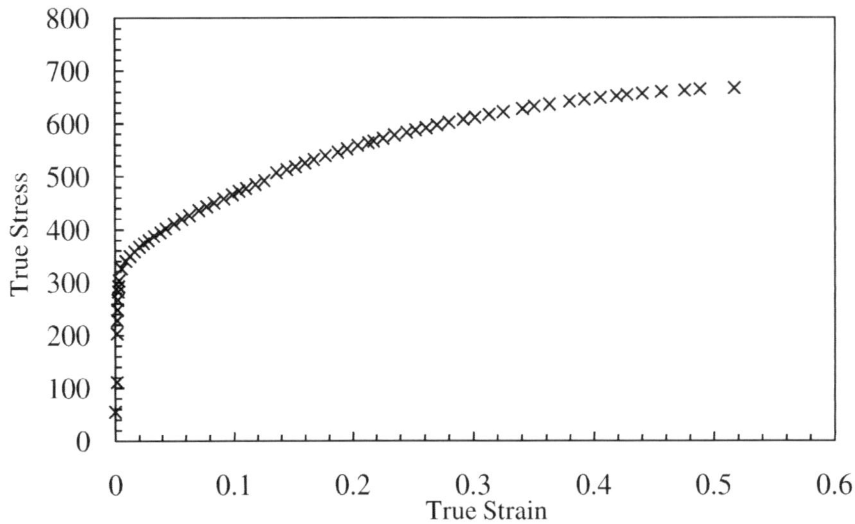

Figure 10. Plastic properties of 0.5mm thick sheet of SS304L, 25°C and 48mm/min strain rate

Anisotropic material properties play a great role in the final shape of the formed part. To keep this effect under consideration during simulation standard anisotropic tests using ASTM E517 [19] were conducted for 0°(rolling direction), 45°(medial direction), 90°(transverse direction). Three tests in each direction were carried out and the results were averaged to minimize the error. Observation of these tests are shown in Table 5.

Table 5. Average change in respective length and width of specimens and their corresponding r-values in 3 directions

	Thickness (t) (mm)		**Length (l) (mm)**		**Lankford's r-values**
	Initial (t_0)	Final (t_f)	Initial (l_0)	Final (l_f)	
Rolling (0°)	19.81	18.31	20	24	0.760165
Medial (45°)	19.76	17.78	20	24	1.375956
Transverse (90°)	19.76	18.19	20	24	0.8317542

(1) was used to calculate the r-values using the data obtained from tests. Calculated r-values at 0°, 45° and 90° are shown in Table 5.

$$r = \left(\frac{\ln \frac{w_0}{w_f}}{\ln \frac{l_f w_f}{l_0 w_0}} \right) \quad (1)$$

These r-values were converted to anisotropic coefficients of ABAQUS using (2), (3) and (4) which have been discussed in software documentation respectively.

$$R_{22} = \sqrt{\frac{r_y(r_x+1)}{r_x(r_y+1)}} \quad (2)$$

$$R_{33} = \sqrt{\frac{r_y(r_x+1)}{r_x+r_y}} \quad (3)$$

$$R_{12} = \sqrt{\frac{3r_y(r_x+1)}{(2r_{45}+1)(r_x+r_y)}} \quad (4)$$

Values of R11, R13 and R23 have been kept 1 because of the planar anisotropy. Above mentioned values of R22, R33 and R12 are applicable when the sheet lies in plane 2-3. The values will be defined in the respective boxes if the plane of sheet changes. All the corresponding values of constants have been presented in Table 6.

Table 6. Input coefficients for simulating anisotropy in ABAQUS

Coefficient	$\mathbf{R_{11}}$	$\mathbf{R_{22}}$	$\mathbf{R_{33}}$	$\mathbf{R_{12}}$	$\mathbf{R_{13}}$	$\mathbf{R_{23}}$
Value	1	1.023	0.959	1	1	0.858

In simulation, all parts were assembled together using coaxial position constraints and using displacement command. To exactly simulate the actual process 8 dynamic explicit steps considering nonlinear geometry were defined with consecutive annealing steps, to soften the material after each punch. Isothermal frictionless surface to surface contacts were assumed between die-sheet pair and grabber-sheet pair, while rough surface contact was defined for punch-sheet pair. Die, grabber and punch were constrained to be rigid along their respective reference points. While defining the boundary conditions, die and grabber were fixed initially to grip the sheet, and punch was displaced in each step to comply with the actual experimental data.

The sheet was meshed using approximate global size of 0.0014 with total 10832 nodes and 7989 elements. 7875 3D explicit linear hexahedral C3D8R (8-node linear brick, reduced integration, hourglass control) elements and 114 3D explicit linear wedge C3D6 (6-node linear triangular prism) elements were defined to homogenously mesh the blank. Global material orientation was defined using global coordinates of the part. Final meshed sheet blank is shown in Figure 11.

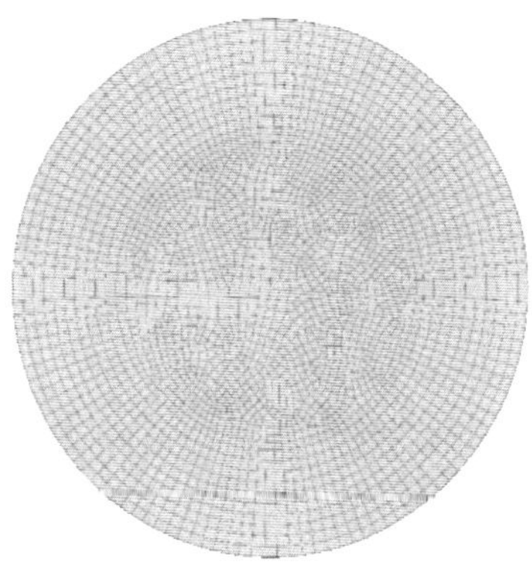

Figure 11. Final mesh of sheet blank

Figure 12 shows the shape of flange after each deep drawing step as predicted by numerical simulation model. It can clearly be observed that numerical simulation adapted the anisotropic deformation in flange and predicted the evolution of flange shape from circular to complex polygonal shape till the end of experiment. Detailed results obtained from numerical simulation and its comparison with experimental data have been discussed later.

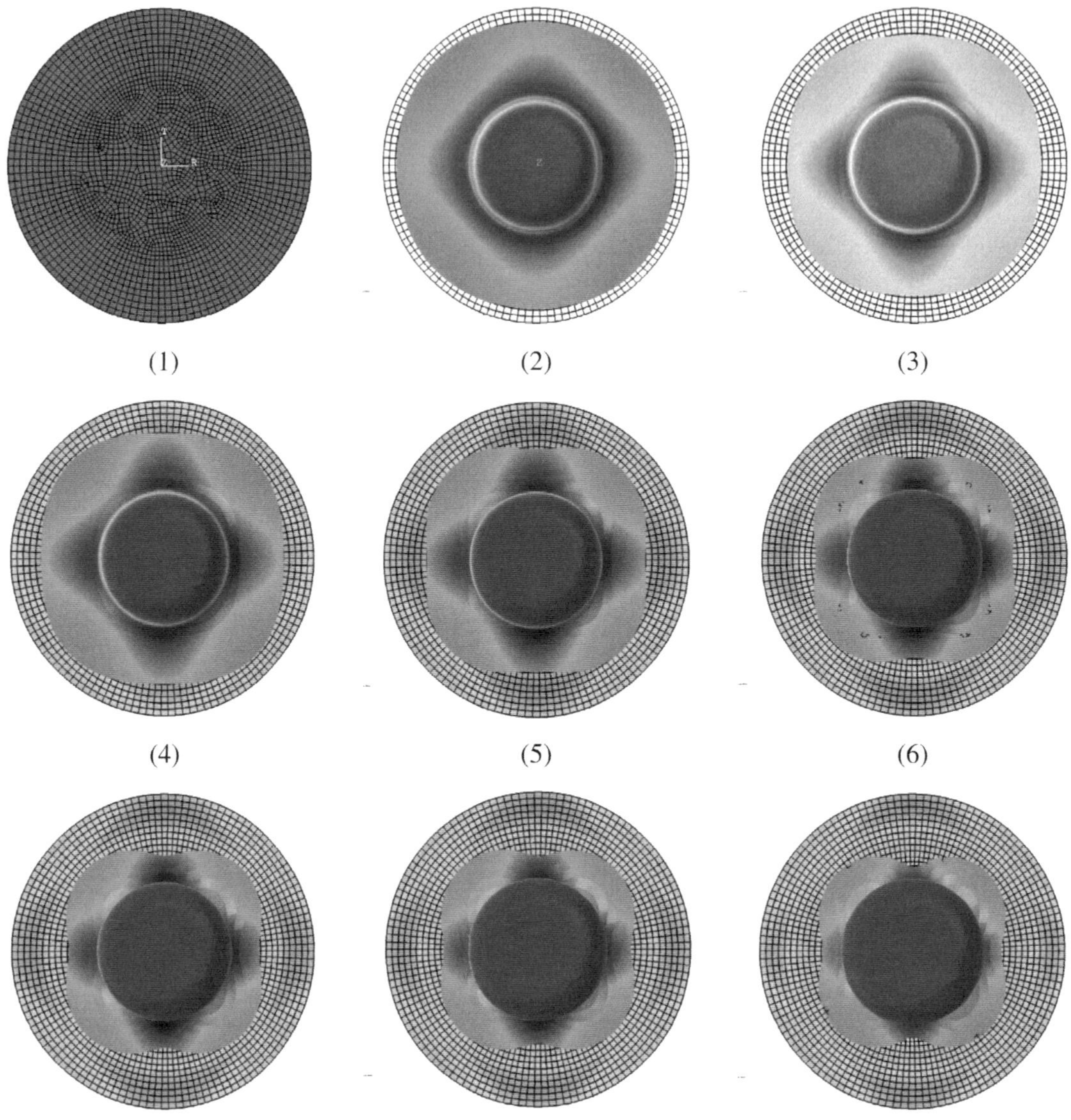

(1) (2) (3) (4) (5) (6) (7) (8) (9)

Figure 12. Shape of flange after each step presented on top of initial circular blank for comparison purpose

6. Results and discussion

High draw ratio deep drawing (HDR) can be a useful tool to manufacture relatively complex parts, which are difficult to manufacture otherwise. Simulation model with detailed material properties, directional strain hardening coefficients and Lankford's r-values predicts the experimental parameters and geometric outcomes. It was observed during experimentation that it is impossible to achieve draw ratios of 2.46 in single punch, hence multiple deep drawing steps with intermediate annealing steps to remove residual stresses in the flange and soften the material are necessary.

It was observed that anisotropic hardening properties of sheet play a great role in the development of final shape of finished product. To predict the actual shape of flange with the help of simulation, detailed material model should be defined, tensile tests in which each specimen was elongated 20% were done. Figure 13 shows the graph of strain hardening coefficients at different directions. 0° is the rolling direction of sheet while 90° is the transverse direction of sheet. Strain hardening coefficient (n) for our sheet varies from 0.258 to 0.267. It is interesting to note that material is soft at 0° (rolling direction) and shows constant increase in hardening till 60° (medial direction) after which the hardening coefficient decreases until we reach 90° (transverse direction). As discussed in previous section, tests were conducted for first and third quadrant only at 0°, 15°, 30°, 45°, 60°, 75°, 90°, 180°, 195°, 210°, 225°, 240°, 255°, 270° (filled markers) and the results were extrapolated for 360° circle (hollow markers) assuming symmetric strain hardening coefficients for second and fourth quadrant.

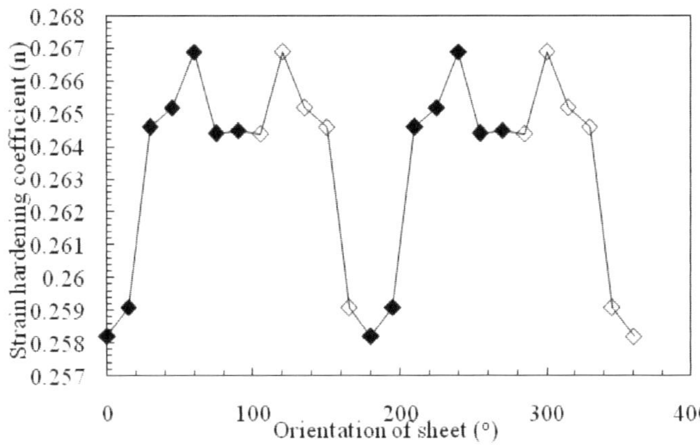

Figure 13. Strain hardening coefficients at respective orientation of commercially available SS304L sheet

It was observed during experimentation that it is impossible to achieve draw ratios of 2.46 in single punch. During deep drawing the flange deforms and hence becomes hard and stops flowing into the die cavity. Necking at wall of deep drawn cup occurs and eventually the part fails. Figure 14 shows the force displacement curve of one step deep drawing process during which the part fails. Similar curve of failure for cup wall was obtained multiple times during experimentation. Depth of drawing is directly proportional to the force applied. It can be observed that specifically for our case, maximum or limiting drawing force (LDF) which can be applied during deep drawing is 34kN. In one step deep drawing the part fails at 16mm depth.

Similar problem was simulated in ABAQUS Standard TM as discussed previously, and the simulation results are shown in Figure 14. Simulation model predicted the limiting drawing force to be 25kN with 13mm depth of drawing. While comparing both curves it can be seen that the overall trend of both curves is very similar, yet numerical simulation model under predicts the force and displacement when compared with experimental values.

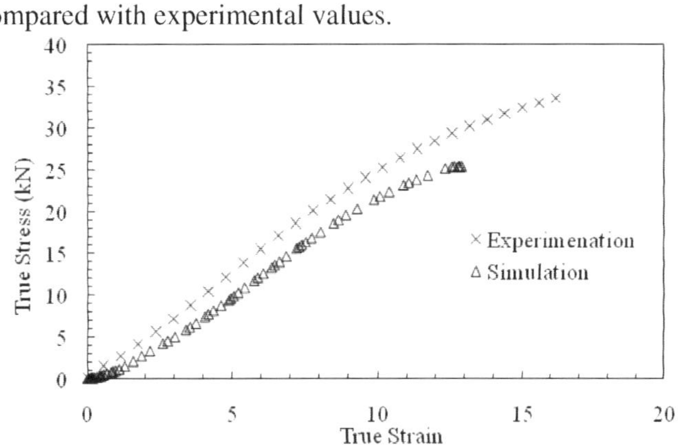

Figure 14. Comparison of experimentation and simulation results for tearing of sheet wall during one step deep drawing

As single step HDR is not possible, it was carried out using multiple deep drawing steps with intermediate localized annealing of flange to remove residual stresses and soften the flange before the wall reached the limiting force which it can carry. Figure 15 shows the force displacement curves for multi-step deep drawing process. Complete process was carried out in 8 steps. Process was conducted using force control technique and limiting force was set to be 28kN. As the force reached the limiting value after each step, part was removed from the die, flange was quick annealed using oxy-acetylene torch. 2nd and 6th punch of experimentation show less deep drawing for exactly the same amount of force applied in all steps, it is because of relatively higher gripping pressures in these 2 steps. There is a drop of force in 8th step which shows that the part has been fully drawn into the die. In Figure 15 experimental results (hollow makers) have been compared with the simulation results (solid markers) as well. It can be seen that simulation model predicts the trends of force displacements curves very closely, with the fact that initially the predicted force is highest and it gradually decreases with decrease in flange area and in final step the prediction of force is lower than the actual value. Ideally it should be the case when the contacts are frictionless.

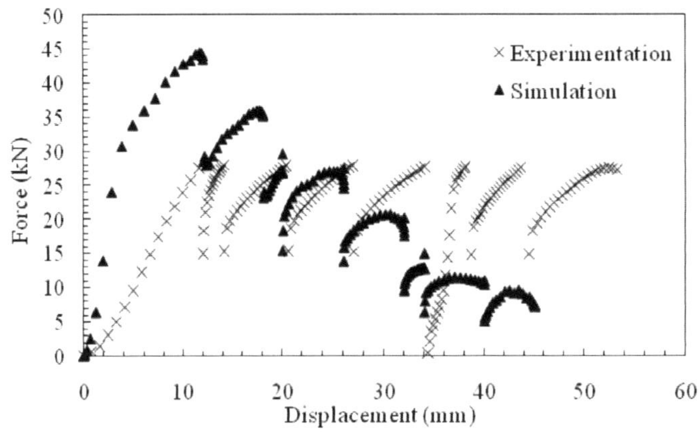

Figure 15. Comparison of experimentation and simulation results for multi-step deep drawing with intermediate annealing

Flange shape after each punch was recorded as shown in Figure 6 and was compared with the flange shape predicted by the simulation model which has been discussed in the previous section. Comparison of results for 1st, 4th and last steps are shown in Figure 16-18 correspondingly. Simulation model predicts greater flow of material into the die cavity, it is because of the frictionless model which has been assumed during simulation.

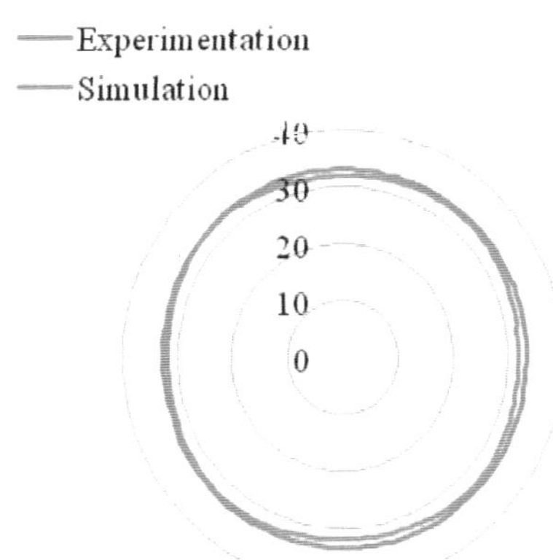

Figure 16. Comparison of flange shapes from experimentation and simulation after 1st step

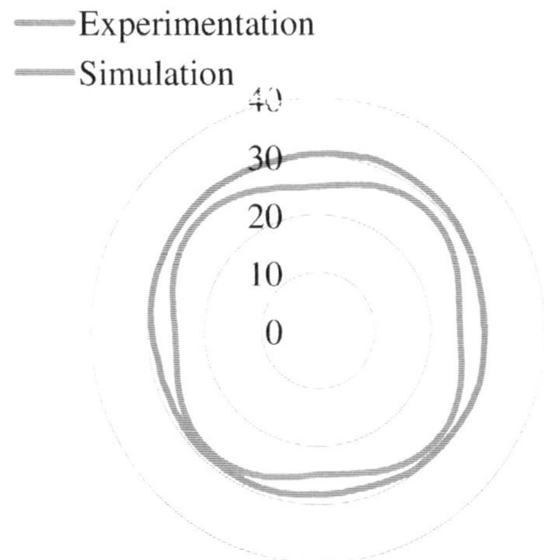

Figure 17. Comparison of flange shapes from experimentation and simulation after 4th step

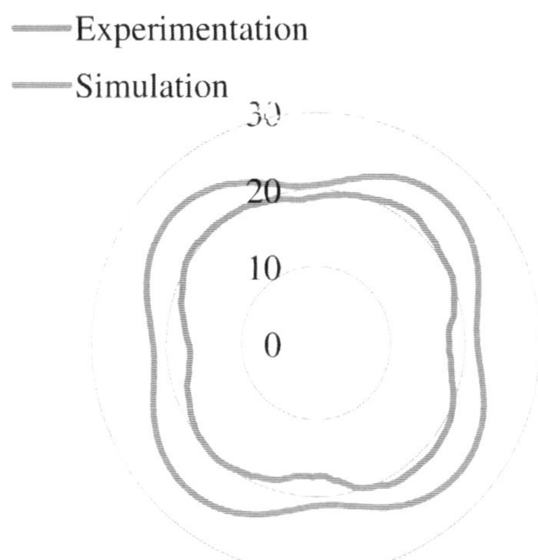

Figure 18. Comparison of flange shapes from experimentation and simulation after 8th step

Error between flange shapes obtained from experimentation and those predicted by simulation was calculated at 0°, 45° and 90°. Error plot after each step has been shown in Figure 19. Initially the blank is circular with radius of 35mm in experimentation and simulation, still there is up to 4% error, it might come due to the processing of flange shape during edge detection in MATLAB. It can be observed from the figure that there is an increasing trend in error up till 6th punch, which then decrease in the next punch, it is because simulation predicts greater material flow into the die cavity during each step and the error tends to increase.

It can be observed that error in all directions follows same trends, but at 45° there is least error (max 12%) in the prediction of flange shape, at 90° the error in prediction of flange shape is intermediate (max 17%) and maximum error is seen at 0° orientation of flange (max 22%).

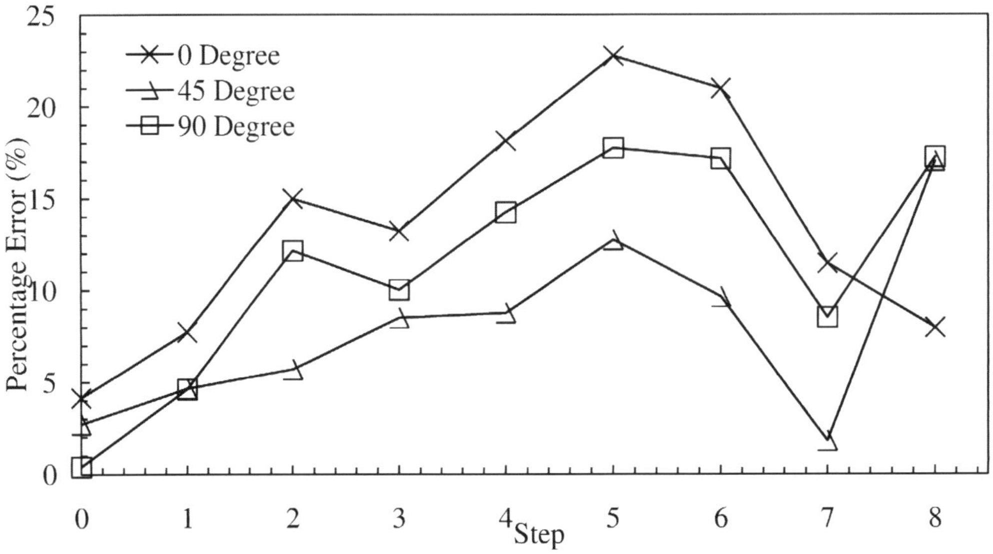

Figure 19. Comparison of average percentage error in prediction of flange shape

7. Conclusion

Effect of anisotropy on the final shape of deep drawn parts has been investigated. Material was tested to determine detailed material data and Lankford's anisotropic coefficients, which was further used to develop detailed numerical simulation model to exactly predict the final shape of the flange. Detailed material model with Lankford's anisotropic coefficients was used to define detailed numerical simulation model to precisely predict the force-deformation curve and final shape of flange after each deep draw step. Experimental results were compared with numerical simulation.

By observing Lankford's r-values in Table 5, it can be concluded that rolled material is most ductile in rolling direction (0°), 8.4% less ductile in transverse direction (90°) and 44.7% less ductile in medial direction (45°). Which is the exact trend of material while flowing into the die cavity. Deformation is maximum in rolling direction and least in medial direction as shown in Figure 7.

It was observed that it is impossible to attain HDR of 2.46 in single step. Excessive deformation of flange hardens the flange which hinders its flow into the die cavity and the cup wall fails as depicted by force-deformation curve of single punch in Figure 14. It is recommended to anneal the flange of cup as soon as the limiting value of force is achieved which in our case was 28 kN.

Impossibility of attaining HDR of 2.46 for 0.5mm thick sheet of AISI SS304 was also predicted by numerical simulation model with an average error of 14% as shown in Figure 14. Numerical simulation predicts on average 7% to 12% more material flow into the die cavity than actual experimental results because of the frictionless model defined, the detailed error plot in 3 directions for each step has been presented in Figure 19.

8. Acknowledgement

This study was funded and supported by Department of Mechanical Engineering, University of Engineering and Technology Taxila, Pakistan, which is gratefully acknowledged.

9. References

[1] Korhonen, A.S., *Drawing force in deep drawing of cylindrical cup with flat-nosed punch.* Journal of Engineering for Industry, 1982. 104: p. 29

[2] Ahmetoglu, M., et al., *Control of Blank Holder Force to Eliminate Wrinkling and Fracture in Deep-Drawing Rectangular Parts.* CIRP Annals - Manufacturing Technology, 1995. 44(1): p. 247-250

[3] Wang, X. and J. Cao, *On the prediction of side-wall wrinkling in sheet metal forming processes.* International Journal of Mechanical Sciences, 2000. 42(12): p. 2369-2394
[4] Wang, X. and J. Cao, *An Analytical Prediction of Flange Wrinkling in Sheet Metal Forming.* Journal of Manufacturing Processes, 2000. 2(2): p. 100-107
[5] Cao, J. and M. Boyce, *Wrinkling behavior of rectangular plates under lateral constraint.* International journal of solids and structures, 1997. 34(2): p. 153-176
[6] Agrawal, A., N.V. Reddy, and P.M. Dixit, *Determination of optimum process parameters for wrinkle free products in deep drawing process.* Journal of Materials Processing Technology, 2007. 191(1–3): p. 51-54
[7] Han, S., M. Bruhis, and M. Jain. *Some Considerations In Modeling Axisymmetric Deep Drawing And Redrawing Process And LDR Prediction.* in *AIP Conference Proceedings*. 2007
[8] Leu, D.-K., *The limiting drawing ratio for plastic instability of the cup-drawing process.* Journal of Materials Processing Technology, 1998. 86(1): p. 168-176
[9] Hill, R., *The mathematical theory of plasticity*. Vol. 11. 1998: Oxford university press
[10] Padmanabhan, R., et al., *Influence of process parameters on the deep drawing of stainless steel.* Finite Elements in Analysis and Design, 2007. 43(14): p. 1062-1067
[11] Saxena, R.K. and P.M. Dixit, *Prediction of flange wrinkling in deep drawing process using bifurcation criterion.* Journal of Manufacturing Processes, 2010. 12(1): p. 19-29.
[12] Wu-rong, W., et al., *The limit drawing ratio and formability prediction of advanced high strength dual-phase steels.* Materials & Design, 2011. 32(6): p. 3320-3327.
[13] Barlat, F. and K. Lian, *Plastic behavior and stretchability of sheet metals. Part I: A yield function for orthotropic sheets under plane stress conditions.* International Journal of Plasticity, 1989. 5(1): p. 51-66
[14] Banabic, D., et al., *An improved analytical description of orthotropy in metallic sheets.* International Journal of Plasticity, 2005. 21(3): p. 493-512
[15] Volume, A.H., *1: Properties and Selection: Irons, Steels, and High-Performance Alloys.* ASM International, 1990
[16] Handbook, M., *Desk edition.* ASM, Metals Park, 1985
[17] ASTM, *B661-12 Standard Practice for Heat Treatment of Magnesium Alloys.* ASTM Standards, 2012
[18] E8, A., *Standard test methods for tensile testing of metallic materials.* Annual book of ASTM standards, 1997. 3.
[19] Standard, A., *E517-00.* ASTM Standard Annual Book Vol. 0.3, 2000. 1

Thermal decomposition and kinetic evaluation of decanted 2,4,6-trinitrotoluene (TNT) for reutilization as composite material

M. F. Ahmed, A. Hussain and A. Q. Malik
School of Chemical and Materials Engineering,
National University of Sciences and Technology Islamabad-44000, Pakistan

E-mail: farooqahmed@scme.nust.edu.pk

Abstract. Use of energetic materials has long been considered for only military purposes. However, it is very recent that their practical applications in wide range of commercial fields such as mining, road building, under water blasting and rocket propulsion system have been considered. About 5mg of 2,4,6-trinitrotoluene (TNT) in serviceable (Svc) as well as unserviceable (Unsvc) form were used for their thermal decomposition and kinetic parameters investigation. Thermogravimetric/ differential thermal analysis (TG/DTA), X-ray diffraction (XRD) and Scanning electron microscope (SEM) were used to characterize two types of TNT. Arrhenius kinetic parameters like activation energy *(E)* and enthalpy (ΔH) of both TNT samples were determined using TG curves with the help of Horowitz and Metzger method. Simultaneously, thermal decomposition range was evaluated from DTA curves. Distinct diffraction peaks showing crystalline nature were obtained from XRD analysis. SEM results indicated that Unsvc TNT contained a variety of defects like cracks and porosity. Similarly, it is observed that thermal as well as kinetic behavior of both TNT samples vary to a great extent. Likewise, a prominent change in the activation energies *(E)* of both samples is observed. This in-depth study provides a way forward in finding solutions for the safe reutilization of decanted TNT.

1. Introduction

2,4,6-trinitrotoluene (TNT), being one of the most widely used military high explosives, has extensively been investigated for its thermal behaviour and decomposition kinetics. Since 1870, it is available in pure form as an explosive substance for use in various ways [1-4]. TNT forms part of aromatic nitro compounds with three nitro groups attached to carbon atoms. Out of all constitutional isomers of TNT, only α-TNT also known as Tolite or Trotyl, is used for military grades with 99% of purity and a small fraction of impurity of ~1% mainly due to 2,4-DNT.A huge quantity of unwanted and unserviceable (Unsvc) munitions containing TNT has been discarded through conventional disposal techniques i.e. Open Burn Open Detonation (OB/OD), incineration, demolition, etc. after WW1 [5-6]. Since all of these methods produced adverse climatic and health hazards, any significant move towards the safest disposal of these Unsvc energetic materials is well appreciated all around the globe. In this paper, structural analysis as well as kinetic study of TNT thermo chemical decomposition has been carried out to examine its feasibility for further use. About 5mg of decanted TNT Unsvc has been used for comparative analysis with serviceable (Svc) TNT. Both the samples were investigated for their thermal and kinetic behaviour with the help of available techniques such as

Content from this work may be used under the terms of the Creative Commons Attribution 3.0 licence. Any further distribution of this work must maintain attribution to the author(s) and the title of the work, journal citation and DOI.
Published under licence by IOP Publishing Ltd

Thermo gravimetric/differential thermal analysis (TG/DTA), X-ray diffraction (XRD) and Scanning electron microscope (SEM). Samples for this study were arranged through defence organizations. TG curves were used for the determination of Arrhenius kinetic parameters, like activation energy *(E)*and enthalpy *(ΔH)*of both Svc and UnsvcTNT samples using Horowitz and Metzger method. Simultaneously, weight loss/gain and thermal decomposition range were evaluated from TG/DTA curves. For structural investigation of the samples, XRD was used which gave distinct diffraction peaks showing crystalline nature. Similarly, molecular structures of both the samples were examined through SEM. SEM analysis revealed a variety of defects present in Decanted TNT Unsvc sample such as voids, porosity, cracks, etc. Interestingly, it was observed that thermal as well as kinetic behavior of both TNT samples vary to a great extent, keeping in view various aspects such as shelf life, environment conditions, manufacturing, filling and formulation processes. Additionally, the decomposition temperature of Decanted TNT Unsvc sample increased substantially. A prominent change in activation energies *(E)* of both the samples under investigation was also observed.

2. Experimental procedure

2.1 Material and methods

TNT Svc sample was arranged from National Defence Organizations. However, Decanted TNT Unsvc sample was recovered from Unsvc munitions containing TNT through detailed decanting plant available in Pakistan army institute. In this process, hot water spray was injected through shell nozzle resulting into melting of TNT. Decanted TNT Unsvc sample was then collected and dried in open air for subsequent use in chemical laboratory.

2.2. Analytical techniques

Samples were analysed through various instrumental techniques. Multiple tests were conducted for the identification of various parameters such as thermochemical decomposition, structural analysis and kinetic evaluation. Instruments used in the present work include SEM-6490A JEOL made in Japan, TG/DTA-Perkin Elmer Model Pyris Diamond and XRD- made by Theta-Theta Toe, Germany.

2.2.1. SEM characterization

SEM is used for the extraction of structural and chemical information of both the samples under investigation through topographic imaging, e.g. fracture surface, cracks etc.SEM version 6490A JEOL (made in Japan) was used for the microscopic study of samples from various dimensions. Different electron beam energies were applied to obtain maximum output from the SEM images with magnifications ranging from 10000X to 1500X.

2.2.2 TG/DTA characterization

Wide utility of TG/DTA makes it one of the most versatile tools for the verification and thermal cum kinetic studies of material sample using a controlled temperature environment. In order to obtain thermal data about the samples, Perkin Elmer Model Pyris Diamond TG/DTA instrument was used. Nitrogen (N_2) gas was used as purge gas for maintaining inert environment for the samples. A heating rate of 10^oC/min was kept constant throughout the analysis along with temperatures ranging between room temperature to 340^oC.

2.2.3 XRD analysis

XRD analysis, being a widely used method for the crystallographic identification of material sample, was selected to identify crystal structure of both the samples. XRD analysis gave useful information about the samples which includes their crystalline phases, atomic arrangement, orientation of a single crystal, lattice parameters, density etc.

IOP Conf. Series: Materials Science and Engineering **146** (2016) 012032 doi:10.1088/1757-899X/146/1/012032

2. 3 Kinetic Method

Literature finds various methods for the measurement of activation energy *(E)*and enthalpy *(ΔH)* of explosive material. Few of the most commonly known methods are; Doyle method, Coats and Redfern method, Horowitz and Metzger method, Free man and Carroll method, Newkirt method. Although all of these methods are quite accurate and reliable in use, however, present study for the calculation of activation energy *(E)* and enthalpy *(ΔH)* of TNT explosive samples was carried out with help of Horowitz and Metzger Method using a Curve Fitting Program.

3. Results and discussions

3.1 SEM analysis

3. 1.1 Decanted TNT unsvc

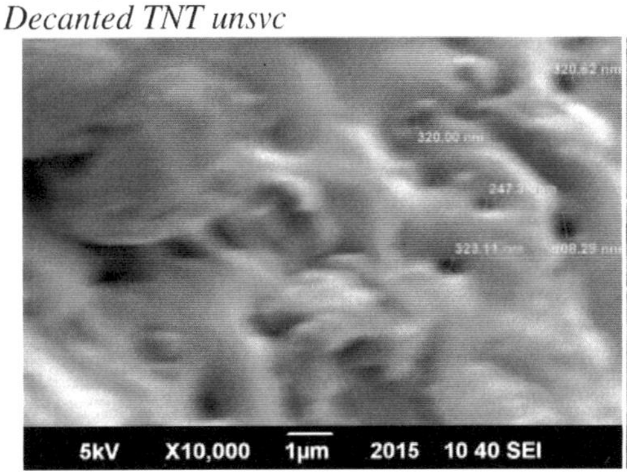

(a) Image at 5kV and 10000x magnifications

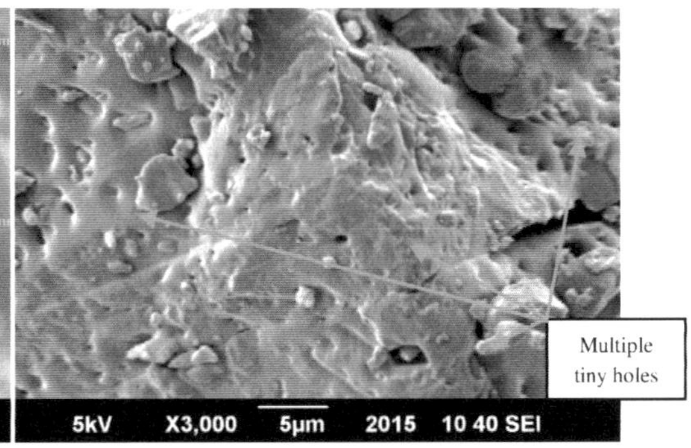

(b) Image at 5kV and 3000x magnifications

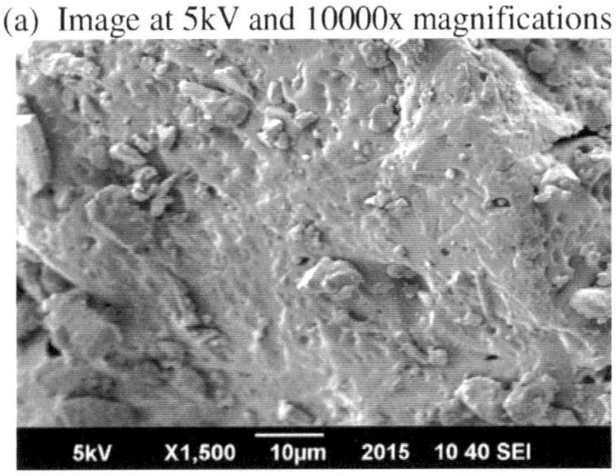

(c) Image at 5kV and 1500x magnifications

Figure 1. SEM images of Decanted TNT Unsvc sample

For an in-depth analysis of Decanted TNT Unsvc, about 5mg of sample was examined at5kV electron beam energy and with various magnifications through SEM. All three images at different magnifications show an irregular dough shaped morphology with no definite orientation of TNT sample Figure 1. A predictable number of pores in the size range of 247-608 nm are visible on the surface, giving an indication of less possibility for potential use as military grade high explosive. Similarly, a wide range of cracks on the sample surface renders it unsuitable for use in pure form. One of the primary reasons for these pores and cracks may be linked to treatment with hot water spray during decanting process along with storage conditions and handling processes.

3.1.2 TNT Svc

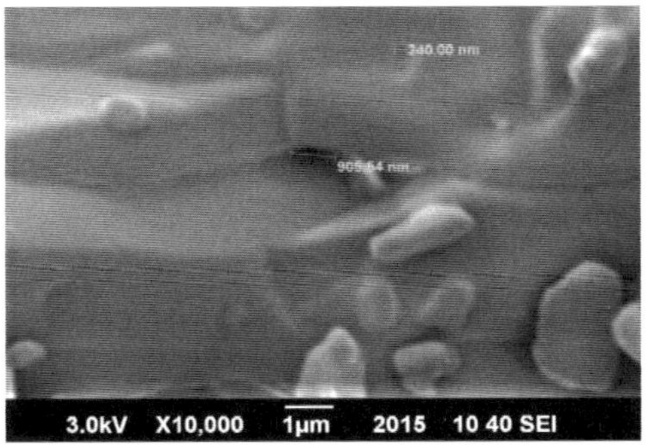

(a) Image at 3kV and 10000x magnifications

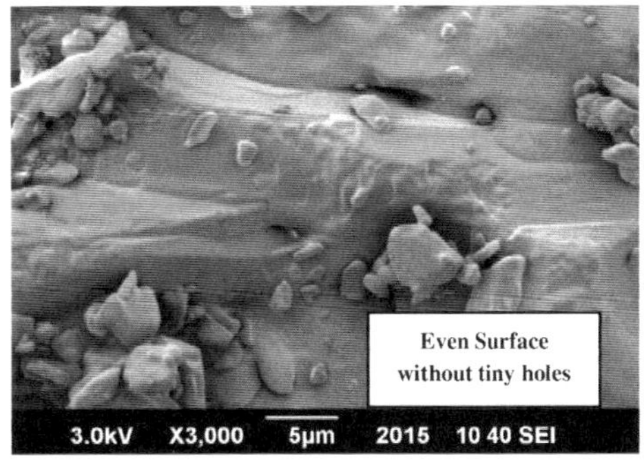

(b) Image at 3kV and 3000x magnifications

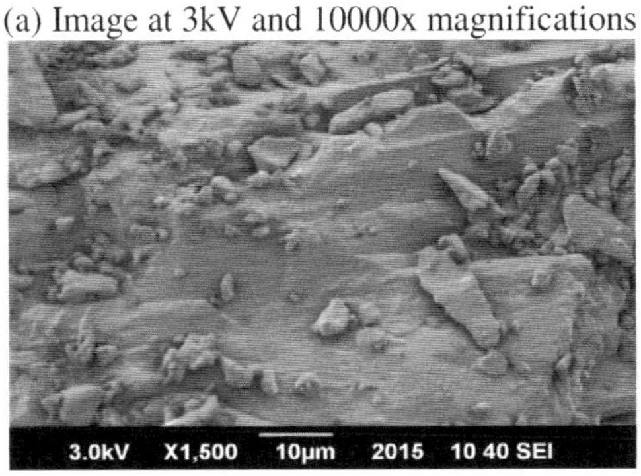

(c) Image at 3kV and 1500x magnifications

Figure 2. SEM images of TNT Svc sample

About 5mg of TNT Svc in powder form was utilized for SEM analysis. This particular sample was selected for comparison of structural changes occurring in Decanted TNT Unsvc *vis-a-vis*TNT Svc. Once exposed for investigation, it was revealed that TNT Svc sample has a smooth surface, with irregular pebble like shape and no distinct orientation Figure 2. The number and size of pores (between 240.00 nm to 905.54 nm) visible on sample surface are fewer than those on Decanted TNT Unsvc sample surface. Similarly, less number of cracks can be found on the microstructure. It can be deduced from both the samples under investigation that TNT Svc sample covers the basic requirement of pure explosive use for military grade because of its clear, compact, smooth and homogeneous microstructure; in contrast with the Decanted TNT Unsvc sample. Variation in storage conditions, handling procedure and environmental conditions greatly affect the explosive's physical as well as chemical properties and its shelf life.

3.2 TG/DTA analysis

3.2.1 TG analysis

For thermal analysis, Decanted TNT Unsvc and TNT Svc were examined through Perkin Elmer TG/DTA instrument. Figure 3 gives TG curves for both the samples. Study of TG curves shows that weight loss occurred in both samples in one step with both the samples diminishing nearly 98% around 285°C which is in agreement with literature data [7-8]. Interestingly, onset of weight loss of both the samples seems to occur around 152-155°C. A rapid change in mass loss is generally observed between 175-266°C, which can be related to thermal decomposition of both TNT samples. Consequently, total weight loss for Decanted TNT Unsvc sample under inquiry over prescribed temperature range of 155- 266°C is observed to be nearly 98%, whereas for Svc TNT sample, the decline in weight loss seems to progress further and comes closer to 95% exhaustion around 285°C. Thisimplies that Decanted TNT Unsvc sample decomposes earlier than TNT Svc sample, which could be due to impurities, prolonged exposure to high temperature environment, a long shelf life or even treatment with hot water spray during decanting process.

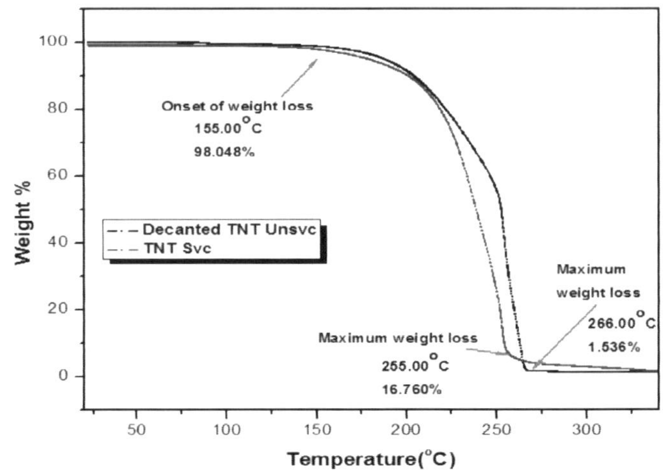

Figure 3. TG Curves of Decanted TNT Unsvc and TNT Svc samples

3.2. DTA

DTA curves for Decanted TNT Unsvc and TNT Svc are observed in Figure 4. DTA thermogram indicates that both the samples nearly exhibit similar curves until 1st sharp endothermic peak for melting point of TNT between 81-83°C is visible [9-12]. On closer observation, it is revealed that melting of TNT Svc samples takes place earlier than Decanted TNT Unsvc sample. Similarly, second endothermic event with slightly broad peaks between 245-255°C corresponds to endothermic decomposition of both TNT samples [13-15]. A slightly earlier decomposition of Decanted TNT Unsvc in contrast to TNT Svcis also observed, which may be linked primarily to the existence of the impurities in Decanted TNT Unsvc sample. However, presence of moisture content in this particular sample may also contribute as impurity. However, results shows that Decanted TNT Unsvc can be effectively used with number of other energetic materials as composite materials for energy blast applications as well as for use as commercial blasting agents (commercial mining, metal brazing, cutting etc) [16].

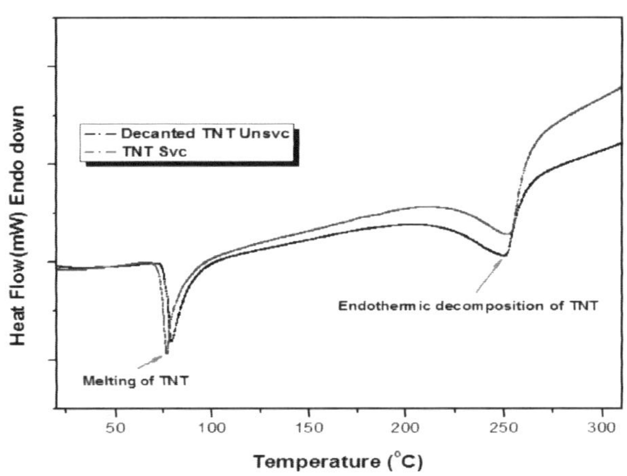

Figure 4. DTA Curves of Decanted TNT Unsvc and TNT Svc samples

3.3 Horowitz and metzger method

Kinetic parameters like activation energy *(E)* and enthalpy(ΔH) for Decanted TNT Unsvc *vis-à-vis* TNT Svc explosive samples were determined with the help of Horowitz and Metzger method using TG curves. Graphs used for the calculation of kinetic parameters are shown in Figure 5(a) and 5(b) for Decanted TNT Unsvc and TNT Svc samples respectively.

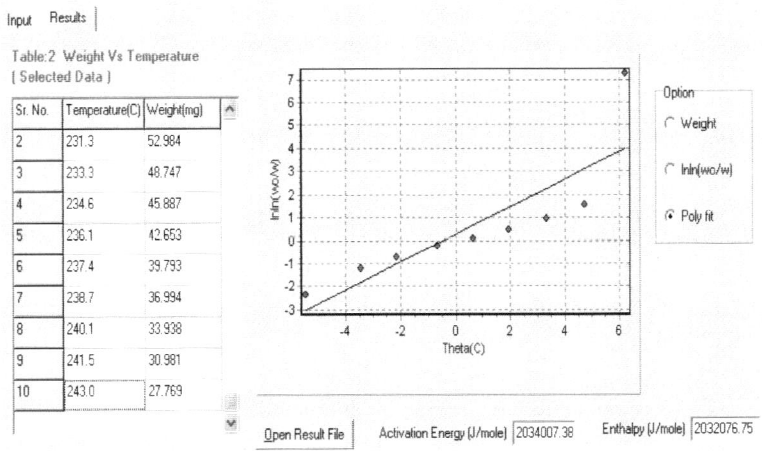

(a) Graph for calulation of kinetic parameters of Decanted TNT Unsvc

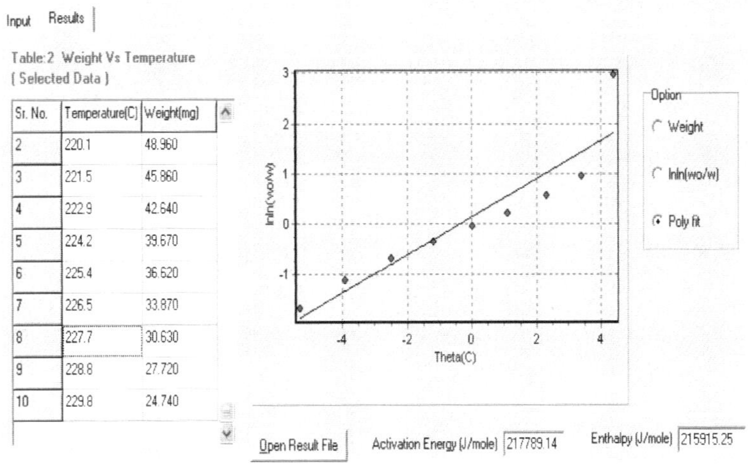

(b) Graph for calulation of kinetic parameters of TNT Svc

Figure 5. Kinetic parameters measured for Decanted TNT Unsvc and TNT Svc sample

It is revealed in Figure 5 that activation energy *(E)*of Decanted TNT Unsvc sample under controlled temperature program is lower i.e. 203.40kJmol^{-1}(48.613 kcalmol^{-1}) than literature values for the pure TNT i.e. 222 kJmol^{-1} (53.06 kcalmol^{-1}). However, activation energy of TNT Svc measured with the help of Horowitz and Metzger method is 217.78 kJmol^{-1} (52.05 kcalmol^{-1}), which is in close agreement to the literature value for pure TNT [15,17]. On the other hand, enthalpy (ΔH) of Decanted TNT Unsvc sample is determined to be 203.20kJmol^{-1}(48.56 kcalmol^{-1}), whereas, the same is 215.91 kJmol^{-1} (52.08 kcalmol^{-1}) for TNT Svc sample. It is deduced from comparative analysis that TNT Svc sample will result into better performance in terms of its thermal stability and sensitivity. Its rate of thermal decomposition and activation energy *(E)* are well defined. On the contrary, Decanted TNT Unsvc shows slight variation in its chemical properties which may be due to multiple reasons such as ingress of moisture contents, impurity, voids etc. All or some of these properties may have affected serviceability of TNT and resulted into change of its chemical or physical properties for instance, decrease in activation energy *(E)* and enthalpy (ΔH). This decrease in activation energy *(E)* of the Decanted TNT Unsvc has negative impact on its thermal stability and shock sensitivity because thermally stable compounds are more difficult to detonate. The summary of TNT samples resultis presented in Table 1.

Table 1. Summary of TNT samples result

S.No	Type of Sample	Melting point(°C)	Activation energy Ea (kJmol^{-1})	[a]Literature Value of Ea for pure TNT(kJmol^{-1})	EnthalpyΔH (kJmol^{-})	Decomposition Range(°C)
1.	Decanted TNT Unsvc	81.30	203.40	222	203.20	155– 266
2.	Svc TNT	80.81	217.78	222	215.91	155– 285

[a]Ahmed S R and Cartwright M 2014 Laser Ignition of Energetic Materials *John Wiley & Sons*1161-62

3.4 XRD analysis

XRD results of Decanted TNT Unsvc and TNT Svc shown in Figure 6 reveals several different distinctive diffraction peaks with varying intensity. XRD spectra exhibited these peaks in the range of 20 to 40 (2 theta degree). The diffraction pattern of TNT Svc can be easily distinguished from Decanted TNT Unsvc. XRD Spectra reveals that both the TNT samples exist in monoclinic structure possessing more stability than meta-stable state [18]. Miller indices are identified on XRD spectra with number of h, k, l values of (004), (021),(-315), (1021) for both the TNT samples. Similarly, TNT peak with h, k, l value of (004) seems to be the most crystalline peak of XRD spectra. From XRD spectra, it is revealed that experimental results for Decanted TNT Unsvc sample are in agreement with literature values reported for pure TNT sample and thus prove its existence in its real state even after aging and treatment with hot water spray.

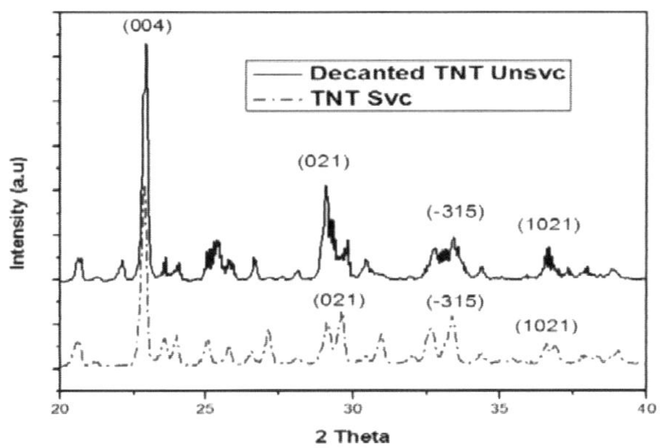

Figure 6. XRD spectra of Decanted TNT Unsvc and TNT Svc samples

4. Conclusion

Present research work was aimed at finding out various characteristics of two types of TNT samples which includes thermal decomposition, microstructure analysis, surface morphology etc. From experimental results, it is revealed that Decanted TNT Unsvc has gained impurity during manufacturing, storage, handling and transportation. Presence of impurities adversely affected its thermal properties, where its activation energy *(E)* became lower than the intended value thus making it more sensitive to accidental initiation by friction or shock. Surface morphology of Decanted TNT Unsvc sample also reveals that the surface is rougher with visible cracks and porosity thus giving rise to the creation of hotspot. Similarly earlier decomposition of Decanted TNT Unsvc sample reveals that the sample contains impurities thus making it decompose prematurely. Although, numerous changes were observed for the Decanted TNT Unsvc sample during analytical study, but interestingly it has maintained its crystalline nature even though after a long shelf life, aging and exposure to variant environmental condition. This in-depth study has opened up a new dimension for reutilization of Decanted TNT Unsvc as composite material with metals for high energy blast applications. It can also be used as composite material for formulation of composite explosive materials. Similarly its conversion into commercial explosive for later use in mining, quarrying or even rock blasting cannot be ruled out.

5. References

[1] Cohen R, Zeiri Y, Wurzberg E, and Kosloff R 2007 Mechanism of Thermal Unimolcular Decomposition of TNT (2,4,6-Trinitrotoluene): A DFT Study*J. Phys. Chem. A*11111074-83
[2] Brill T B and James K J1993 *Chem. Rev.*93 2667-92
[3] Brill T B and James K J 1993 *J. Phys. Chem.*97 8759
[4] Brill T B, James K J, Chawla R, Nicol G, Shukla A and Futrell I H 1999*J. Phys. Org. Chem.*12819
[5] Becanova J,Friedl Z and Simek Z 2010 Identification and determination of trinitrotoluene and their degradation products using liquid chromatography–electrospray ionization mass spectrometry *Int. J. Mass Spectrom.* 291 133–39
[6] Pennington J C andBrannon J M 2002 Environmental fate of explosives *Thermochim.Acta* 384 63–72
[7] Ingale S V, Wagh P B, Tewari R and Gupta S C 2010 Nano crystalline trinitrotoluene (TNT) using sol–gel process *J. Non-Cryst. Solids* 356 2162–67
[8] Makashir P S and Kurian E M 1991Spectroscopic and thermal studies on 2,4,6 Trinitrotoluene (TNT) *J. Therm. Anal. Calorim.* 55 173-85
[9] Gibbs T R and Popolato A 1980 LASL Explosive Property Data*University of California Press*Barkeley and Los Angeles, California281-82
[10] Ingale S V, Wagh P B , Tewari R and Gupta SC 2010 Nano crystalline trinitrotoluene (TNT) using sol–gel process*J. Non-Cryst. Solids* 3562162–67
[11] Srihakulung O and Soongsumal Y 2011 Improving TNT curing process by using Infrared camera World Academy of Science Engineering and Technology54
[12] Lixia S, Rongzu Hu and Jiamin Li 1995 Phase diagram and compatibility of the binary systems TNT-polyester and TNT-picric acid*Thermochim. Acta* 253111-16
[13] Yinon J and Zitrin S 1984 The Analysis of explosives *Pergamon series in analytical chemistry* 3136
[14] Long G T, Brems B A and Wight C A 2002 Autocatalytic thermal decomposition kinetics of TNT *Thermochim. Acta* 388 175-81
[15] Brill T B and James K J 1993 *Chem. Rev.*93 2667-92.
[16] Horst Krause 1994 Proceedings of the NATO Advanced Research Workshop on Conversion Concepts for Commercial Applications and Disposal Technologies of Energetic Systems Moscow, Russia 14 xi-xii.
[17] Ahmed S R and Cartwright M 2014 Laser Ignition of Energetic Materials *John Wiley & Sons* 1 161-62
[18] Miller G R and Garroway A N 2001 A Review of the Crystal Structures of Common Explosives Part I: RDX, HMX, TNT, PETN, and Tetryl *Naval Research Laboratory*4555 Overlook Avenue, SW Washington, DC 20375-5320

Effect of micro-structural modifier on the morphology of silicon rich secondary phase and strain hardening behavior of eutectic Al-Si alloy

M. Mansoor, I. Salam and A. Tauqir
Institute of Industrial Control Systems, P. O. Box 1398, Rawalpindi, Pakistan

E-mail: malik01677@yahoo.com

Abstract. Eutectic Al-Si alloys find their applications in moderate to severe tribological conditions, for example: pistons, casings of high speed pumps and slide sleeves. The higher hardness, so the better tribological properties, are originated by the formation of a silicon rich secondary phase, however, the morphology of the secondary phase drastically influence the toughness of the alloy. Microstructural modifiers are used to control the toughness which modifies the Si rich secondary phase into dispersed spherical structure instead of needle-like network. In the present study, a mixture of chemical fluxes was used to modify the Si phase. The alloy was cast into a sand mold and characterized by scanning electron microscopy, energy dispersive spectroscopy, hardness testing and tensile testing. It was found that the morphology of the Si phase was altered to acicular structure due to the modification process. In comparison, the un-modified alloy contained Si phase in needle-like structure. The effect of modifier was also pronounced on the mechanical properties, where increase of 50% in yield strength, 56% in tensile strength and 200% in elongation occurred. A discernable raise in strain hardening component indicated the improved strain harden ability and formability of the modified alloy.

1. Introduction

While research have been performed recurrently on Al–Si alloys since mid of nineteen century, it is only after 1921 that the alloys have attracted commercial attention after the introduction of "Alpax" a commercial alloy containing ~15% silicon by Pacz [1]. The Al-Si alloy is a simple binary eutectic system with about 12% silicon being the eutectic composition. Al-Si alloys are called eutectic alloys when silicon content is 11% - 13% of the alloys. These alloys are frequently used in aeronautical and automobile industries for cylinders, pistons, guiderails, etc. The composition of most Al-Si foundry alloys is in the vicinity of the eutectic point in order to take full advantage of its excellent castability, and these alloys normally contain about 50 - 90 vol.% (Al-Si) eutectic phase. Hence, the (Al-Si) eutectic phase is important in determining the mechanical properties of these alloys.

In unmodified Al-Si alloys, the eutectic silicon phase has brittle, coarse and needle-like morphology. The sharp ends of this needle-like silicon phase serve as stress concentrators and promote crack initiation and propagation, and ultimately results in poor mechanical properties of the alloy. Through the process of modification, this brittle, coarse and needle-like eutectic silicon structure can be transformed to fine fibrous acicular eutectic silicon structure with much enhanced mechanical properties [2-5]. Modification is the process of adding trace amounts of certain elements to an alloy to change the morphology of the phase(s) present in the alloy.

Content from this work may be used under the terms of the Creative Commons Attribution 3.0 licence. Any further distribution of this work must maintain attribution to the author(s) and the title of the work, journal citation and DOI.
Published under licence by IOP Publishing Ltd

Research work have been done on eutectic modification of Al-Si alloys with various modifiers [6-8] but work incorporating eutectic modification of Al-Si alloys along with its effect on strain hardening have hardly been reported so far. In present study modification was carried out for a melt of 1Kg alloy using mixture of fluxes (NaF, KCl and NaCl). The alloy was cast into sand molds and evaluated for its microstructural and mechanical characteristics.

2. Experimental

Generally, eutectic Al-Si alloys have a range of silicon contents i.e. from 7.5 to 12 percent. In present work the proposed chemical composition the alloy is mentioned in Table 1.

Table1. Chemical composition of the eutectic alloy

	Si	Mg	Mn	Fe	Cu	Zn	Al
LM9	8.0 ~ 10.5	0.17 ~ 0.3	0.2 ~ 0.5	≤ 0.60	≤ 0.10	≤ 0.20	bal.
Achieved	8.5	0.15	0.38	≤0.32	≤0.21	≤0.16	bal.

To incorporate various alloying additions in the melt of aluminum various addle materials in the form of Al-Si, Al-Mn, and Al-Mg were used. The raw materials used for the melting were cleaned manually. It was assured that any kind of dirt, dross, grease or organic material was not present on the raw materials. After cleaning, raw materials were placed in box oven for pre-heating to eliminate any sort of moisture present in the charge. Any moisture present on the surface of the materials could be a potential source of hydrogen dissolution during melting, which may result in subsequent hydrogen embrittlement. For Al-Mg, the material was heated at 150~200 °C for 20 minutes, while rest of the materials were heated at 25~300 °C for 90 minutes. Pre-heated raw materials were then charged to the electric resistance-heating furnace. Following steps were involved during heating and casting operation:

A graphite crucible was preheated to 600 ~ 700 °C in an electric resistance furnace. At the preheating temperature, pure aluminum was poured in the crucible followed by the addition of melting flux i.e. 0.5 wt. percent of the melt. After complete melting of aluminum, pre-weighed Al-Mn alloy was added. Subsequently, Al-Mg was added. The melt temperature was maintained at 735 ~ 745 °C. To degas the melt, two stage degasing was performed: using Degaser-190 tablets (provided by FOSECO, UK) and vacuum degasing at 10^{-2} ~ 10^{-3} mbar pressure. After degasing chemical modification of the eutectic alloy was carried out using a physical mixture of KCl, NaCl and NaF. The modification salt contained 63%NaCl, 12%KCl and 25%NaF and was added in 3% by weight, while maintaining melt temperature between 730 ~ 740°C. After the addition of salt modifier the melt was kept for 10~15 minutes. Meanwhile, the melt was stirred slowly with a steel plunger without breaking the surface of modifier flux. At 710°C the flux was rabbled into the surface of the melt until the exothermic reaction has removed the metal from the flux. The powdery dross was skimmed and alloy was poured into the sand molds.

Energy dispersed spectrometry, optical spark emission and inductively coupled plasma techniques were carried out to analyze the composition of the alloy (table-1).Further, the alloys (with and without modification) were characterized for their microstructural features using scanning electron microscope (SEM) in secondary electron mode. While for mechanical properties Brinell hardness tester was used to determine hardness of the alloys. A steel ball of 2.5mm diameter with a testing load of 30Kg was used for indentation. The value of the hardness was calculated by taking the average of five indentations. An INSTRON universal testing machine was used for tensile properties with a strain rate of $0.02S^{-1}$. Two sets of mechanical test specimens (comprising three specimens each) were prepared and tested in accordance with ASTM E8 standard. To determine strain hardening component ASTM E646 standard was followed, according to which:

$$\delta = k\varepsilon^n \quad (1)$$

where, δ is true stress, ε is true strain, k is strengthening coefficient and n is strain hardening component. To calculate true stress (δ) and true strain (ε) from engineering stress-strain curves, Eq. 2 and 3 were used.

$$\delta = S(1 + e) \quad (2)$$
$$\varepsilon = \ln(1 + e) \quad (3)$$

where, S and e are engineering stress and strain, respectively. The seven data points (N) of engineering stress (S) and strain (e) were taken from the region of engineering stress-strain curves between yield point and ultimate tensile strength. Whereas, values of strain rate component and strengthening coefficient were calculated using Eq. 4 to 6.

$$n = \frac{N\sum_{i=1}^{N}(\log\varepsilon \log\delta) - (\sum_{i=1}^{N}\log\varepsilon)(\sum_{i=1}^{N}\log\delta)}{N(\log\varepsilon)^2 - (\sum_{i=1}^{N}\log\varepsilon i)^2} \quad (4)$$

and

$$k = \exp\{b\} \quad (5)$$

where;

$$b = \frac{\sum_{i=1}^{N}\log\delta - n\sum_{i=1}^{N}\log\varepsilon}{N} \quad (6)$$

The strain hardening component (n) and strengthening coefficient (k) were calculated for three representative engineering tensile curves of the each alloy and averaged out.

3. Results and discussion

3.1 Microstructural Characteristics

The specimens were sectioned from the core area of the castings (both unmodified and modified) and prepared for microstructural investigations. The prepared specimens were intentionally deep etched to reveal the secondary network. Figure 1 shows SEM micrograph of unmodified alloy featuring a typical unmodified secondary silicon phase having needle like morphology. The silicon phase seemed to be disconnected when viewed at low magnification however, at higher magnification, the silicon rich phase appeared as 3-D network (Figure 1b). Additionally, it had flake like morphology (arrows in Figure 1b).

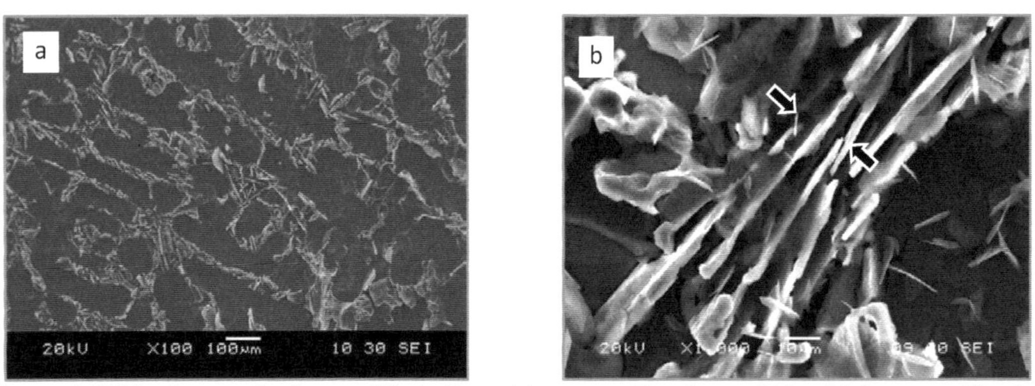

Figure 1.SEM micrograph of unmodified alloy, a) low magnification and b) at higher magnification

The effect of chemical modification was apparent (Figure 2a), where silicon rich secondary phase was transformed into acicular spherical structure. The otherwise needle-like network structure of secondary silicon rich phase had been fully transformed to fine eutectic dispersion. The phase appeared as isolated spherical crystals at high magnification (Figure 2b). The spherical dispersed structure was fairly uniform in size and needle-like structures of silicon rich phase were completely transformed after the chemical modification process. Moreover, the dendritic structure of the matrix material (aluminum) was refined after the addition of modifier. The refinement of the matrix dendrites was effectuated by the uniform and prolific dispersion of the silicon rich phase. Such type of structural refinement could also be achieved by rapid solidification rate, however it is worth mentioning that the mechanism behind the two structural refinement processes (i.e. rapid solidification and chemical modification) are quite different with their effects on mechanical properties.

In modified alloys solidifying at large cooling rates there exist a balance between the rates of heat flow from the liquid to the solid through the interface and released latent heat of fusion, which play a vital role. Aluminum and silicon have large differences in their thermal conductivities and latent heats of fusion i.e. 0.53&0.20 cal. cm^{-2}. $°C^{-1}$ and 94.6 &337 cal. g^{-1}, respectively. Due to the considerable differences in the thermal conductivities and latent heats of fusion of the two constituent of the alloy, aluminum solidifies rapidly than silicon (figure3a). At higher cooling rate the lead of aluminum becomes larger causing encasement of lagging silicon crystals (Figure 3b & c).

In chemically modified alloys, the impurity induced twinning theory proposed by S. Lu et al [10] explains the mechanism by which chemical modifiers transform the morphology of silicon phase. It is the more widely recognized eutectic modification theory, according to which, chemical modifiers are impurities that poison already growing atomic silicon layers by getting adsorbed onto surface steps and kinks thus preventing the attachment of silicon atoms to the crystal. Furthermore, the adsorbed impurity atoms induce twinning of silicon crystals by altering the stacking sequence of atomic layers as the newly added layers seek to grow around the adsorbed impurity atom.

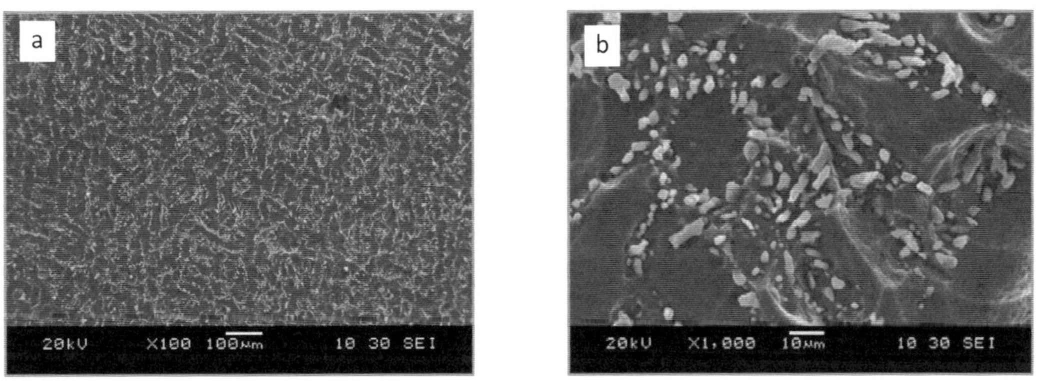

Figure 2. SEM micrograph of unmodified alloy, a) low magnification and b) at higher magnification

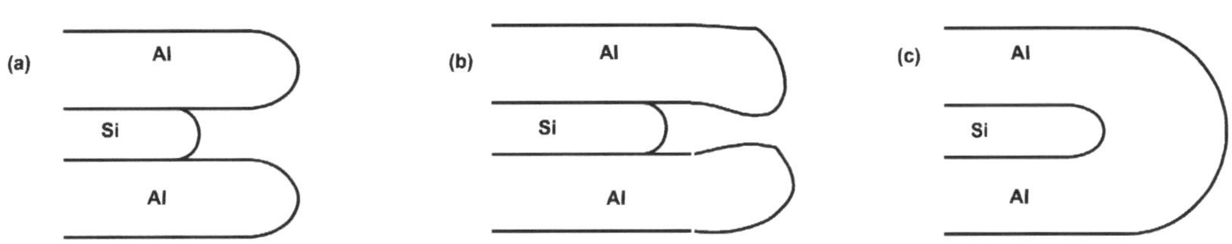

Figure 3. Eutectic solidification in unmodified chill cast Al–Si alloys [9].

3.2 Mechanical characteristics

To understand the effect of chemical modification on the hardness of alloys Brinell hardness tester was used. It was observed that the hardness of modified samples was increased as compared to that of

unmodified alloy. Such improvements in hardness can be attributed to the basic factor of modification of eutectic silicon needle into fine fibrous form. The results of hardness tests are given in Table 2.
The mechanical properties of Al-Si alloys in unmodified and modified conditions are shown in Table 2. The yield strength increased approximately 50% from 60 MPa for the unmodified cast alloy to 90 MPa for chemically modified alloy, and similarly tensile strength of modified alloy increased by 56% as compare to unmodified alloy i.e. from 82MPa to 128MPa. Furthermore, the elongation of modified alloy increased approximately thrice than that of unmodified alloy i.e. 0.9% to 2.76%.

For unmodified alloy, it was believed that under loading conditions the needle-like structure of the silicon rich phase not only provided crack initiation but also caused early failure due to the premature fracture of silicon. Hence, compromising on overall strength and ductility of the alloy. In case of modified alloy, where silicon rich phase was present as acicular form, no premature failure of the silicon rich phase occurred, therefore improved strength and ductility were observed[11, 12].

Table 2. Results of hardness and tensile tests

Specimen	Yield Strength (MPa)	Ultimate Tensile Strength (MPa)	Elongation (%)	Brinall Hardness (HB)
Unmodified	59.7 ±2.4	81.6±2.9	0.9 ± 0.1	103.1±2.2
Modified	89.3 ± 1.8	127.6±3.3	2.7 ± 1.4	118.7±1.5
Change (%)	~50	~56	~200	~15

The σ-ε curves of the tensile test were, subsequently, used for determination of strain hardening component and strengthening coefficient of the two alloys (Table 3). It is known that due the morphological modification of the silicon rich phase higher strengths (or strengthening coefficient, k) are achievable. However, it was observed that besides the strengthening coefficient (k), the modified alloy had higher strain hardening component (n) than the unmodified alloy.

Generally speaking, strain hardening component is a measure of increase in strength of the materials due to plastic deformation. Or in other words, as the strain hardening component is calculated from the data points between yield point and ultimate tensile strength; it is the degree of material's deformability prior to the necking or thinning. It determines how the metal behaves when it is being formed. Materials that have higher n values have better formability than those with low n values [13, 14].

Table 3. "n" and "k" values of unmodified and modified alloys.

	Unmodified Alloy	Modified Alloy
Strain Hardening Component (n)	0.17±0.02	0.22±0.03
Strengthening Co-efficient (k)	511±12	988±17

4. Conclusions

Eutectic modification of Al-Si alloy was successfully carried out using chemical flux. The needle-like structure of silicon rich phase was completely transformed into acicular fibrous structure. Moreover, the strain harden ability was improved besides tensile strengths and hardness of the modified alloy.

5. Reference

[1] Pacz, US Patent No. 1387900, 1921

[2] M. D. Hanna, S. Z. Lu and A. Hellawell, Metallurgical and Materials Transactions A, Vol. 15, No. 3, 1984, pp. 459-469. Doi: 10.1007/BF02644969

[3] C. B. Kim and P. W. Heine, Journal of the Institute of Metals, Vol. 92, 1963, pp. 367-376

[4] F. Yilmaz and R. J. Elliott, Journal of Materials Science, Vol. 24, No. 6, 1989, pp. 2065-2070. Doi: 10.1007/BF02385422

[5] B. S. Pena and J. A. Lozano, Materials Characterization, Vol. 57, No. 4-5, 2006, pp. 218-226. doi:10.1016/j.matchar.2006.01.015

[6] Stuart D. McDonald, Kazuhiro Nogita, Arne K. Dahle, Acta Materialia 52 (2004) 4273–4280.

[7] Barbora Bryks, Stunov, Acta Polytechnica Vol. 52 No. 4/2012

[8] A.K. Dahle, K. Nogita, S.D. McDonald, C. Dinnis , L. Luc, Materials Science and Engineering A 413–414 (2005) 243–248

[9] B.M. Thall, B. Chalmers, J. Inst. Met. 78 (1949) 79

[10] S. Lu, A. Hellawell, Met. Trans. 18A (1987) 1721–1733

[11] Shi W, Gao B, Tu G, Li S, Hao Y, and Yu F, J Rare Earth 28 (2010) 367-370

[12] Kilicaslan MF, Lee W, Lee T, Sohn Y, and Hong S, Mater Lett 71 (2012) 164-167

[13] Hosford, William R. & Caddell, Robert M. Metal Forming: Mechanics and Metallurgy,Prentice-Hall Inc, Englewood Cliffs, ©1983

[14] Richard Gedney, Sheet Metal Formability, Advanced Materials and Processes, 8:2002: 33-36

AFM investigation and optical band gap study of chemically deposited PbS thin films

S. Zaman, M. Mansoor, Abubakar and M. M. Asim
Institute of Industrial Control Systems, P. O. Box 1398, Rawalpindi, Pakistan

E-mail: superbs_1@yahoo.com

Abstract. The interest into deposition of nanocrystalline PbS thin films, the potential of designing and tailoring both the topographical features and the band gap energy (*Eg*) by controlling growth parameters, has significant technological importance. Nanocrystalline thin films of lead sulfide were grown onto glass substrates by chemical bath deposition (CBD) method. The experiments were carried out by varying deposition temperature. We report on the modification of structural and optical properties as a function of deposition temperature. The morphological changes of the films were analyzed by using SEM and AFM. AFM was also used to calculate average roughness of the films. XRD spectra indicated preferred growth of cubic phase of PbS films in (200) direction with increasing deposition time. Optical properties have been studied by UV-Spectrophotometer. From the diffused reflectance spectra we have calculated the optical Eg shift from 0.649-0.636 eV with increasing deposition time.

1. Introduction

Semiconductor nano-materials are important due to their unique optical and electronic properties originating from quantum confinement and they exhibit size dependent band gap energies [1-3]. A nano-crystalline material contains high concentration of grain boundaries as in case of porous films, i.e. a high internal surface compared to the bulk material. These properties are of considerable importance for electrical transport, optical and mechanical properties [4-9]. Most of the work on nano-semiconductor particles reported so far has been restricted to optical absorption. The absorption edge shifted to blue as the particles size reduces. Most studied crystalline semiconductors, in recent years, belong to the II-VI and IV-VI groups, since they are relatively easy to synthesize and are generally prepared as nano particles or in the form of thin [10].

PbS thin film is an important binary IV–VI semiconductor with a narrow band gap (0.41 eV) and relatively large excitation Bohr radius (180 nm) [11], which results in strong quantum confinement of electrons and holes presents into the crystal lattice of PbS. Band gap of the PbS thin films can be controlled simply by modifying particle size and shape [12-14]. Structural, optical and electrical properties of PbS nanoparticles have been studied extensively [15-18]. Nano-crystalline PbS is important due to its potential application for infra-red region, however, reduction of grain size extends its application over wide spectral range. Grain boundaries in compound semiconductor films are different from those in elemental semiconductors. Boundaries for larger crystallite films differ from those for small-grained ones. In a polycrystalline semiconductor thin film, factors like film defects, surface scattering and grain boundaries complicate identification of electrical properties. Among the

Content from this work may be used under the terms of the Creative Commons Attribution 3.0 licence. Any further distribution of this work must maintain attribution to the author(s) and the title of the work, journal citation and DOI.
Published under licence by IOP Publishing Ltd

different PbS thin film fabrication methods, chemical bath deposition CBD technique requires relatively mild conditions, is cost effective, scalable and technically straight-forward [19].By CBD the dimensions of the crystallites can be varied by controlling deposition parameters. This paper focuses on the effect of temperature and dipping time on the growth of grains during deposition in a chemical bath and its influence on the morphology and size of grains and optical band gap of grown films.

2. Experimental

The thin films of PbS were deposited onto chemically cleaned glass substrates using AR grade chemicals by CBD process. Thiourea (Fluke)was used as a source of sulphide ions and lead acetate (Alfa Acer) as Pb^{+2} precursor in alkaline medium, using hydrazine hydrate (Aldrich) as base instead of commonly used basis such as sodium hydroxide. Chemical bath was prepared by mixing aqueous solutions of thiourea, lead acetate and hydrazine hydrate in 0.7M:0.7M:5M respectively. The substrates were mounted on a special holder that was immersed inside the chemical bath. The starting solution had 12 pH. The temperature of bath was raised to the boiling point of water and during this process substrates were removed from bath at different temperatures according to Table 1. After deposition, films were rinsed in de-ionized water and dried in oven at 60-80°C. Figure 1 shows the experimental setup for chemical bath deposition. The formation of PbS is governed by the equation (1),

$$[Pb\,(HH)^{+2}] + NH_2CSNH_2 + H_2O \rightarrow PbS\downarrow + NHCNH + [HH.2H]^{+2} + H_2O \qquad (1)$$

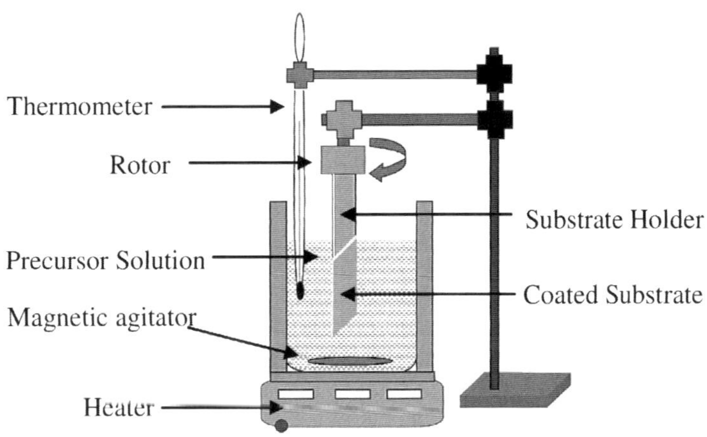

Figure 1. Experimental Set Up for CBD

Table 1. Details of samples collected during CBD process

S. No.	Sample name	Deposition temperature	Deposition time
1	A	30°C	10 min
2	B	50°C	15 min
3	C	70°C	20 min
4	D	90°C	25 min
5	E	100°C	30 min

The obtained samples were characterized to see their morphology and optical properties. The structure and phase studies were done by Jeol X-ray diffractometer. Topography and surface roughness was determined with help of easyscan2 Nanosurfatomic force microscope. Film thickness was determined by gravimetric method. The surface morphology of films was studied by scanning electron microscope. The dark resistance of film was measured with a two probe method and the optical properties were studied on Shimadzu UV/VIS/NIR spectrophotometer.

3. Results and discussion

3.1 XRD studies

X-ray diffraction patterns of PbS thin films were obtained using Copper Kα source. Figure 2 shows XRD graph of samples prepared at different temperatures, while going from room temperature to 100°C.In the start of deposition, up to 50°C, the PbS films are amorphous in nature, and on further increase in temperature some crystalline peaks starts to grow. At 90°C well defined crystalline structure is formed. A further 10 degree rise in temperature leads to increase in intensity due to increased number density of grains and preferential growth in one direction. The 2θ diffraction peaks observed are at 26, 30 and 43 degrees correspond respectively to (111), (200), and (220) planes. The XRD pattern confirms cubic structure of PbS (JCPDS reference no. 78-1901) with slightly preferential (200) orientation of grains.

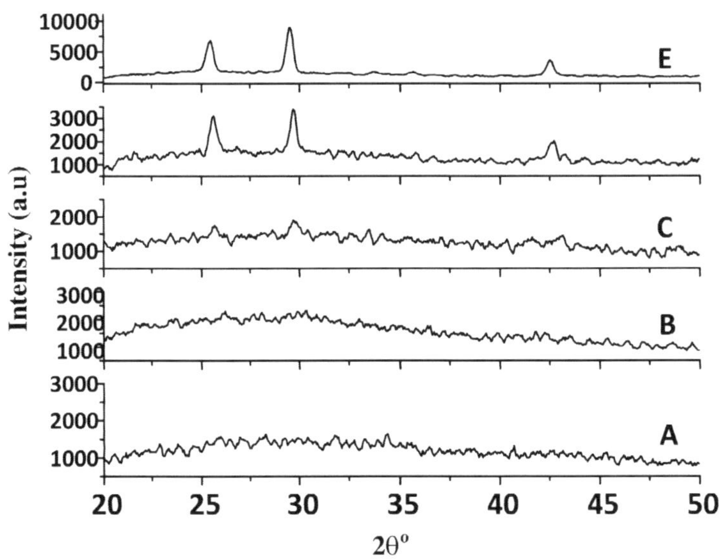

Figure 2. XRD comparison of PbS samples A, B, C, D and E showing evolution of crystalline structure with increasing deposition temperature and Td

It was verified that the relative intensity of XRD peaks changes with respect to the temperature and dipping time.

3.2. AFM Studies

The surface topography and surface roughness of the deposited films was determined by atomic force microscopy. AFM measurements were performed in tapping mode in air. These studies revealed that there is a considerable change in morphology and surface roughness as the temperature and deposition time increases. Figure 3 shows that when deposition temperature is increased from 30 to 100°C, PbS films become more rough due to increasing grain size. The grains are spherical up to 70°C and start merging with neighboring grains on further increase in temperature. Some elongated grains are also seen at 100°C.

IOP Conf. Series: Materials Science and Engineering **146** (2016) 012034 doi:10.1088/1757-899X/146/1/012034

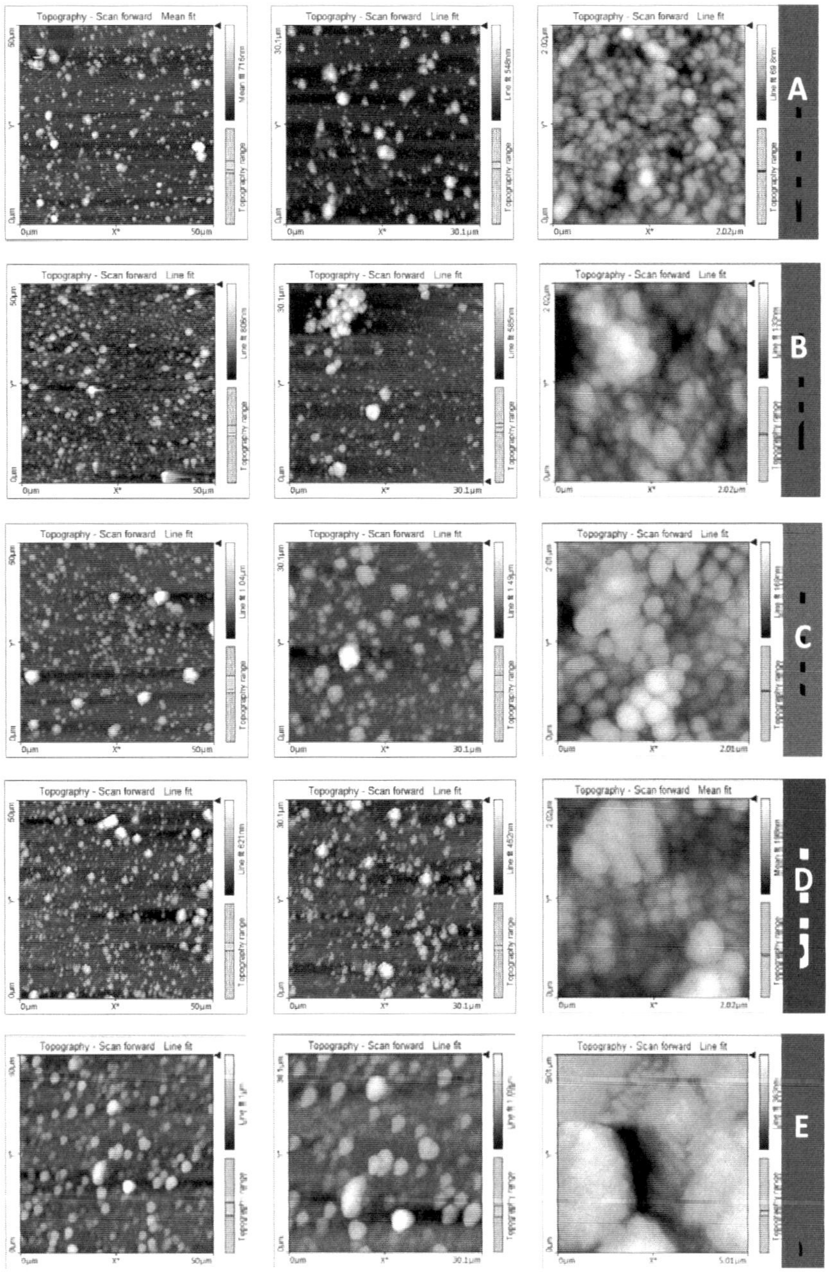

Figure 3.AFM scans of PbS films A, B, C, D and E showing morphology evolution and topography as a function of temperature

Surface roughness was calculated by using AFM measurement software. Increase in deposition temperature and time leads to growth of crystallites and film thickness that increases the surface roughness and average surface roughness (S_a) increase from 59.55nm to 103.54 nm as shown in Table 2.

Table 2. Change in Surface roughness of deposited films with increasing deposition temperature

S. No	Sample	Average roughness Sa (nm)
1	A	59.55
2	B	72.43
3	C	83.78
4	D	94.17
5	E	103.54

3.3. SEM analysis
The surface morphology of deposited thin films was determined by scanning electron microscopy (SEM). Fig. 4 shows the SEM images of films obtained at different temperatures and time.SEM images, from A to E, show film growth sequence during whole deposition process. At 30°C, nucleation and coalesces of grains is visible. However, the grains are isolated and porosity in film is apparent. At 50°C and 70°C, the process of coalesces dominates and film starts becoming more compact. In the range of about 90°C the grains start to diffuse and grain growth dominates. Here the porosity of film diminishes. At 100°C the grain growth forms into pyramidal shape crystals. The porosity in film reaches to its minimum value.

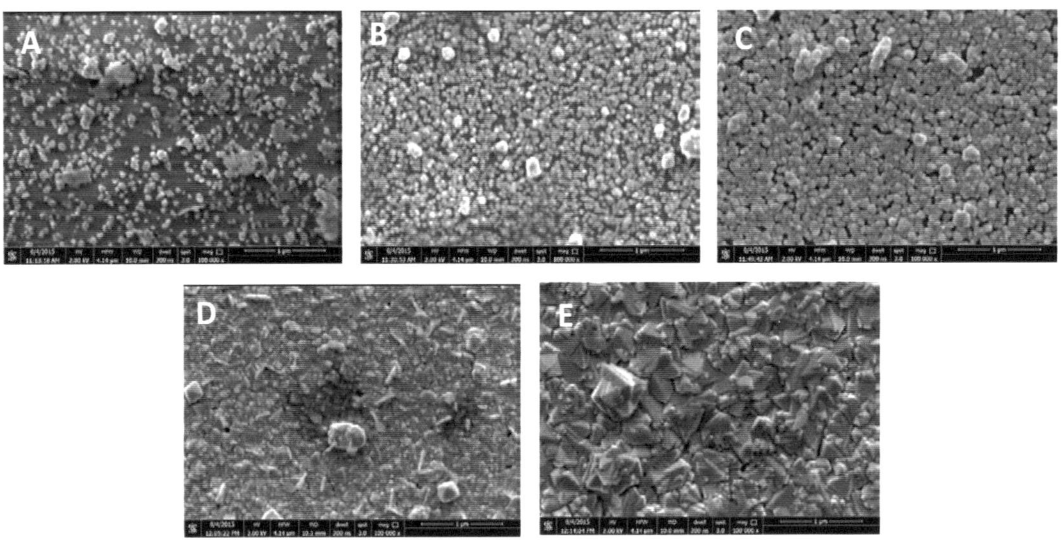

Figure 4.SEM micrographs of PbS films A, B, C, D and E showing morphology evolution and grow that as mentioned deposition temperatures

3.4. Optical studies
Optical properties of the deposited films were determined with the help of their diffused reflectance graphs in the range of 1500nm to 3000nm, taken by using UV-3600 spectrophotometer. The diffused reflectance data was used to calculate absorptivity, and band gap by using Kubalka Munk function i.e.

$$F[R] = (1-R)^2 /2R$$

Where F[R] denotes absorptivity i.e. α^n, for direct band gap materials n=2. Plot of α^2 versus energy gives the band gap of materials as shown in Figure 5.

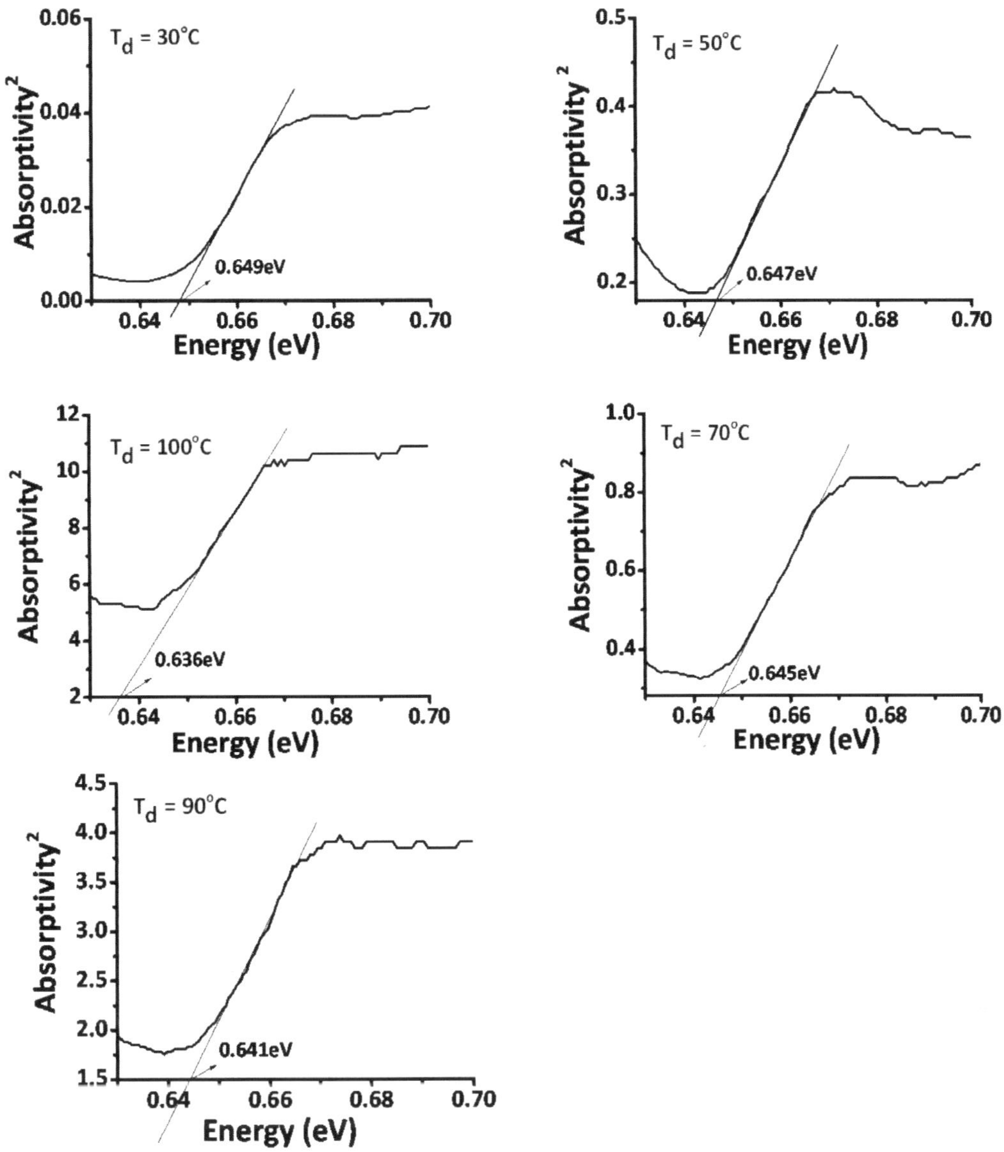

Figure 5. Absorptivity (α^2) versus Energy graph of PbS films at different deposition temperatures and Td. Abscissa intercept gives the band gap of PbS films

These graphs show that optical band gap is sensitive to deposition temperature. The band gap of PbS films decreases from 0.649eV to 0.636eV while going from 30°C to 100°C.The band gap varies with increasing grain size, although the variation is not as high as observed by Wang et al [20]. This can be attributed to larger grain size, as the grains grow bigger; decrease in band gap becomes less effective. The value of $F[R]^2$ i.e. absorptivity^2 also increased with increasing deposition temperature and time as shown in Figure 6.

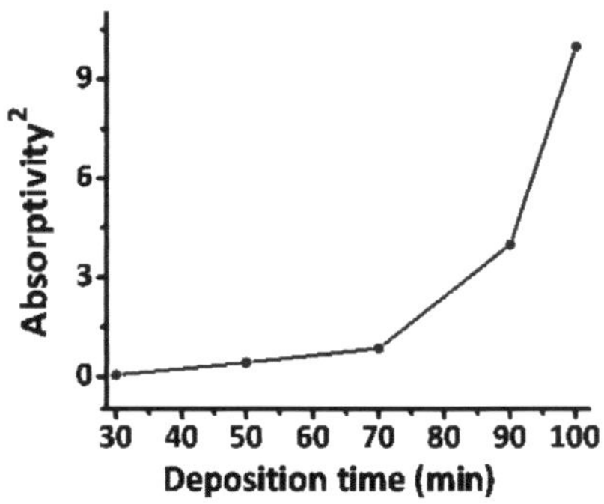

Figure 6. Effect of deposition temperature on absorptivity of deposited films

4. Conclusion

Structural and optical properties were determined and correlated with the different deposition temperatures. Some conclusions are made on the basis of the experimental data.

I. Growth of PbS films is greatly affected by deposition temperature. Films are amorphous at the start of growth but as the temperature increases and films become thick the crystallinity also increases and we get well defined peaks of nanocrystalline PbS films in XRD patterns.

II. AFM and SEM results revealed that temperature plays significant role in film growth and grain modification. The grains are spherical in start with a size of about 100nm. As the time passes and temperature rises, grains diffuse into larger pyramidal shape that are up to 400-500nm size. The surface roughness also increases with temperature, many facets are observed in AFM and SEM scans.

III. Optical studies revealed that on increasing deposition temperature there is a decrease in the optical band gap of PbS thin films as a result of increasing grain size. Absorptivity of films is also enhanced as a result of increased deposition temperature.

5. References

[1] Srinivasan P and Rajesh P 2012 *Elixir Thin Film Tech.* 42 6206
[2] Godea F, BaglayanbO andGunericE 2015 *Chalcogenide Letters*12(10) 519
[3] Moreels I et al 2009 *American Chem. Soci.Nano*3(10) 3023
[4] Tang F and Okuda S 1996 *Mater. Trans.*37 1813
[5] Lang M J, Duarte-Dominguez M, Birringer R, Hampelman R, Natter H and Arnold W 1999*Nanostruct. Mater.*12 1811
[6] Manouchehri S. et al2014*The Experiment*26 (3)1816
[7] EzekoyeB A, EmeakarohaT M, EzekoyeV A, Ighodalo K O and Offor P O 2015*Int. J. Phys. Sci.*10(13) 385
[8] Tanimoto H, Sakai S and Mizubayashi H 1999 *Nanostruct. Mater.*12 951
[9] Andrierskii R A and Glozer G M 2000 *Phys. Met. Metallogr.*89 83
[10] Pentia E, Pintilie L, Matei I, Botila T and Ozbaya E 2001 *J. Optoelectronics and Adv. Mat.* 3,2 525
[11] Saloniemi H 2000 *Academic dissertation universityof Helsinki*
[12] Wang Y, Suna A, Mahler W, Kasowski R J 1987 *J. Chem. Phys.*87 7315
[13] Gao F, Lu Q Y, Liu X Y, Yan Y S and Zhao D Y 2001 *Nano Lett.*1 743
[14] Nanda K K, Kruis F E, Fissan H and Acet M 2002 *J. Appl. Phys.*91 2315
[15] Nanda K K, Kruis F E, Fissan H 2002 *Phys. Rev. Lett.*89 256103
[16] Joshi R K, Sehgal H K 2004 *Physica* E 23 168
[17] Kuang D B, Xu A W, Fang Y P, Liu H Q, Frommen C and Fenske D 2003 *Adv. Mat.*15 1747
[18] Joshi R K, Kanjilal A and Sehgal H K 2004 *Appl. Surf. Sci.*221 43
[19] Rempela A, Kozhevnikovab N, Leenaersa A and van den Berghe S 2005 *J. Cryst. Growth*280300
[20] Kotani K 1990*J. of Non-Crystalline Solids* 126 87

Humidity effect on organic semiconductor NiPc films deposited at different gravity conditions

N. Fatima, M. M. Ahmed, Kh. S. Karimov[1,2] and Kh. Ahmedov[2]
Department of Electrical Engineering, Capital University of Science and Technology, Islamabad Expressway, Kahuta Road, Zone-V, Islamabad,44000, Pakistan
[1]GIK Institute of Engineering Sciences and Technology, Topi, District Swabi, KPK, 23640, Pakistan
[2]Center for Innovative Development of Science and Technologies of Academy of Sciences, Aini 299/2, Dushanbe, 734063, Tajikistan

E-mail: fatima_yusufzai@yahoo.com

Abstract. In this study, thin films of Nickel Phthalocyanine (NiPc) were deposited by centrifugation at high gravity (70*g*), and also at normal gravity (1*g*) conditions to fabricate humidity sensors. Ceramic alumina sheet, coated with silver electrodes, having interelectrode distance of 0.21 mm were used to assess the electrical properties of the sensors. Room temperature capacitance and impedance variations were measured as a function of relative humidity ranging from 25 % ~ 95 % at 1 kHz frequency. It was observed that sensors fabricated at 70*g* were more sensitive compared to sensors fabricated at 1*g*. Sensors fabricated at 70*g* exhibited 1.8 times decrease in their impedance and1.5 times increase in their capacitance at peak ambient humidity. SEM images showed more roughness for the films deposited at 70*g* compared to films deposited at 1*g*. It was assumed that surface irregularities might have increased active surface area of 70*g* sensors hence changed the electrical response. Impedance-humidity and capacitance-humidity relationships were modeled and a good agreement was observed between experimental and modeled data. Experimental data showed that NiPc films could be useful for instrumentation industry to fabricate organic humidity sensors.

1. Introduction

Humidity sensors play an important role in today hi-tech era, especially in the field of metrology and environmental assessment [1-6]. On the basis of their measuring standards, humidity sensors are categorized as resistive, capacitive, gravimetric, or hydrometric [7]. To fabricate a humidity sensor, diverse methodologies and materials have been reported in the literature [8]. A plainer construction of capacitive humidity sensors, using copper phthalocyanine (CuPc) films, was reported in [9]. Humidity effects on electrical conductivity of carbon nanotubes (CNTs) were studied by many researchers, and it has been reported that the water molecules act as a donor and modify the electrical properties of single-walled carbon nanotubes (SWNTs) [10]. Whereas, investigation pertaining to the electrical transportation of double-walled carbon nanotubes (DWNTs) showed that water molecules act as an acceptor in DWNTs [11]. Humidity sensors, based on Ce-doped nanoporous ZnO, Fe_2O_3 and polyaniline were also investigated and reported in [12-14]. Asa result of these investigations, it can be said that different materials with different compositions have a useful potential for humidity sensing applications.

Nickel phthalocyanine (NiPc) is an organic semiconductor that was considered a promising phthalocyanine for optoelectronic devices and gas sensing applications [15-24]. NiPc has a charge carrier mobility of 0.1 $cm^2V^{-1}s^{-1}$, which is 1000 times greater than copper phthalocyanine (CuPc) [24].

Content from this work may be used under the terms of the Creative Commons Attribution 3.0 licence. Any further distribution of this work must maintain attribution to the author(s) and the title of the work, journal citation and DOI.
Published under licence by IOP Publishing Ltd

Its energy band gap is 2.24 eV and 3.2 eV for indirect and direct allowed transitions, respectively [25]. Zubair [26] and Shah [27] fabricated and investigated NiPc based Schottky barrier diodes, and their studies confirmed the potential of NiPc as an organic semiconductor material to be used in electronic devices and sensors technology.

It is an established fact that the performance of organic semiconductor devices is heavily dependent on the chosen technology. Thin films of organic semiconductors, which usually act as a conducting channel, can be deposited by: a) vacuum evaporation; b) drop-casting;orc) spin-coating. At the same time, in the recent past,organic semiconductor thin films deposition by centrifugation at high gravity was also reported [28]. In [28], heterojunctions of silicon and thin layer of poly-N-epoxypropylcarbazole with tetracyanoquinodimethane (TCNQ) impurities were fabricated. The poly-N-epoxypropylcarbazole films on silicon wafers were deposited at normal temperature, butat various gravity conditions: 1*g*, 123*g*, 277*g*, and also at 1107*g*. A smooth layer and strong adhesiveness was obtained by this process of deposition. Current-voltage characteristics of the deposited hetrostructures, as a function of temperature ranging from 20—60 °C, were assessed. It was observed that almost every sample exhibited p-p isotype heterojunctions. Temperature dependent junction's parameters such as reverse saturation current, rectification ratio, junction resistance and threshold voltage were measured.

In 2009, Tahir*et.al.* manufactured many organometallic vanadium complexes such as VO_2(3-fl), VO $(PBO)_2$ and VO (DBM), and deposited the thin films by centrifugation process at 183*g*, 733*g* and 1650*g* [29]. As a result, sandwich type samples were fabricated based on Al/VO_2(3-fl)/Ga, Cu/VO_2(3-fl)/Ga, Al/Vanadyl(acac)/Ga, Cu/Vanadyl(acac)/Ga, Al/VO$(PBO)_2$/Ga, Cu/VO$(PBO)_2$/Ga, Al/VO (DBM)/Ga, Cu/VO(DBM)/Ga, Al/Vanadyl(acac)/TiO_2/Ga, and Cu/VO_2(3-fl)/TiO_2/Ga. Current voltage characteristics of the finished devices showed non-linear rectification behavior. Samples fabricated using VO_2(3-fl) and Vanadyl (acac) showed switching effect. These samples also exhibited two distinct conductance states referred to as low and high conductance states. Such behavior is typically known as memory cell characteristics of the type called as write-once-read-many-times (WORM). It is worth mentioning that memory cell characteristics were observed only in those devices which were fabricated using Cu electrodes.

This article deals with fabrication and characterization of NiPc based humidity sensors and their optimization as a function of fabrication parameters. Finished devices were characterized using a humidity chamber fitted with LCR meter to assess the change in the device impedance and capacitance by varying ambient humidity. The composition of the rest of the article is that Section-2 deals with experimental techniques used to fabricate NiPc thin films sensors at 1*g* and also at 70*g*.Section-3 describes experimental results and their plausible explanation followed by the conclusions drawn from this study.

2. Experimental

Figure 1 illustrates molecular structure of NiPc organic semiconductor. For thin films deposition commercially available NiPc (C_{32} $H_{16}N_8N_i$) of Sigma-Aldrich was used. NiPc films were deposited at 1*g* and 70*g* by drop costing and centrifugation respectively from 5wt. % solution prepared in chloroform. A table-top centrifuge commercially known as HETT ICH EBA-20S was used. Acceleration (*a*) was calculated using Equation (1):

$$a = R\omega^2 \tag{1}$$

where *R* is radius and *ω* is angular frequency.

IOP Conf. Series: Materials Science and Engineering **146** (2016) 012035 doi:10.1088/1757-899X/146/1/012035

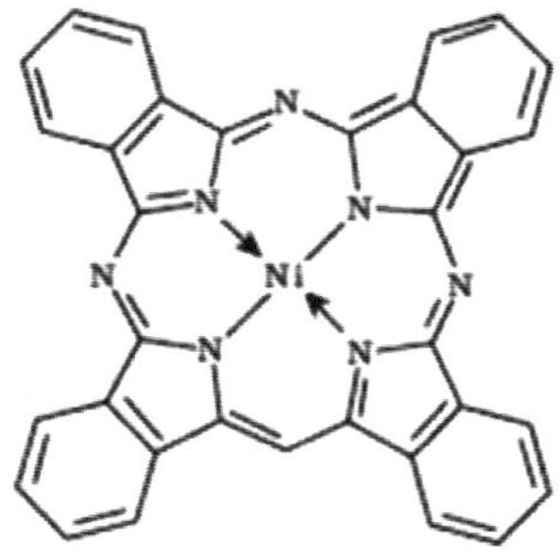

Figure 1. Molecular structure of Nickel phthalocyanine (NiPc)

Figure 2 shows schematic diagram of the substrate used for the deposition of NiPc thin films. It is made of ceramic alumina sheet having 14 mm length and 7 mm width. It contains surface-type interdigitated silver electrodes having 0.21 mm interelectrode distance, fabricated by screen printing and followed by chemical etching [12].

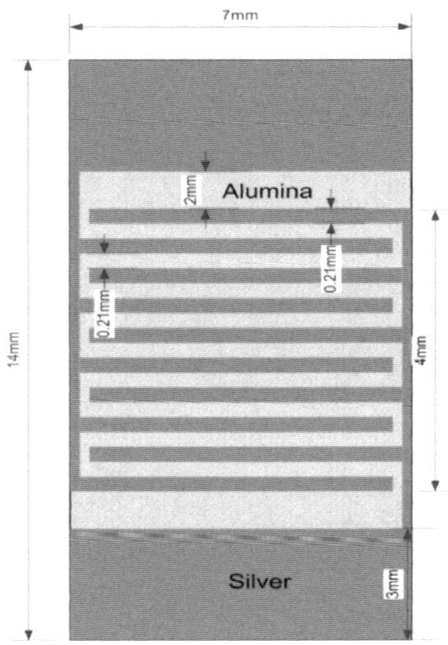

Figure 2. Ceramic alumina substrate with surface type silver electrodes

During centrifugation process, each experiment was performed using two symmetrically installed glass tubes filled with equal volume of solution (0.5 ml). The tube had internal diameter of 12 mm and length of 95 mm. The centrifuge's rotation speed was set at 5000 rpm. The achieved thickness of NiPc grown films were 7—10μm. It was observed that the layers achieved at 1*g* were usually thicker compared to those grown at 70*g*, which means that the acceleration and thickness have an inverse relationship. The same behaviour was observed by us earlier and was reported in [28, 29].

As a norm during centrifugal procedure, solution evaporation was carried out at room temperature and at atmospheric pressure. It was noted that during centrifugation process,10-15 min were required to get films deposited on the substrates. Figure 3 shows SEM images of NiPc films deposited at 1*g*. Moreover, Figure 4 represents SEM micrographs of NiPc films deposited at 70*g* from solution prepared in chloroform.

The obtained SEM images of NiPc films showed that the films deposited at 1*g*appearedto be more uniform and smooth but with larger grain size. On the other hand, SEM micrographs of the films deposited at 70*g* showed relatively rough surface morphology, but with smaller grain size. The impact of roughness of the films and unit grain size on the humidity sensing characteristics is discussed in Section 3. Electrical properties of the sensors were investigated in a humidity chamber. Impedance and capacitance variation were assessed by using LCR meter MT 4090 by setting frequency at 1 kHz and voltage level at 1 V.

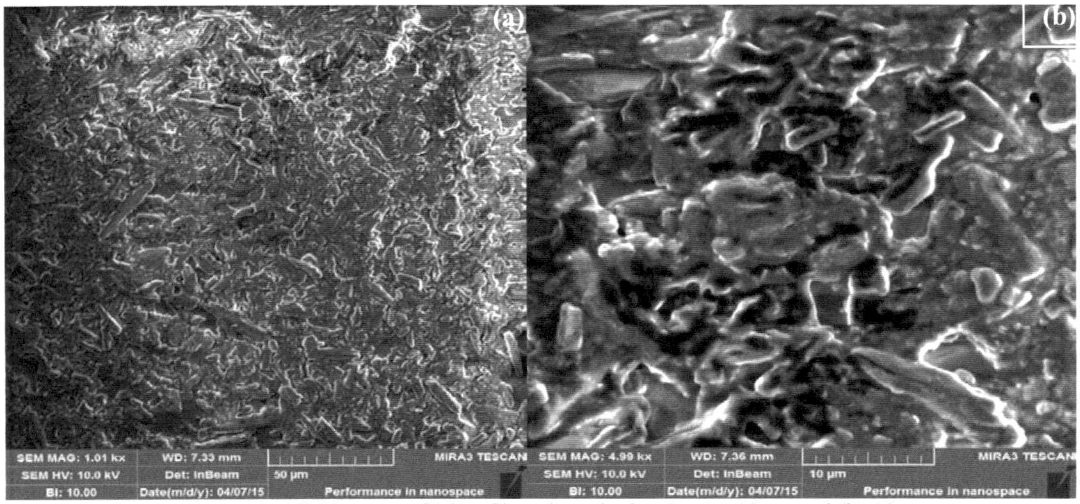

Figure 3. SEM images of NiPc films deposited at 1 *g* (a) 50μm and (b) 10μm

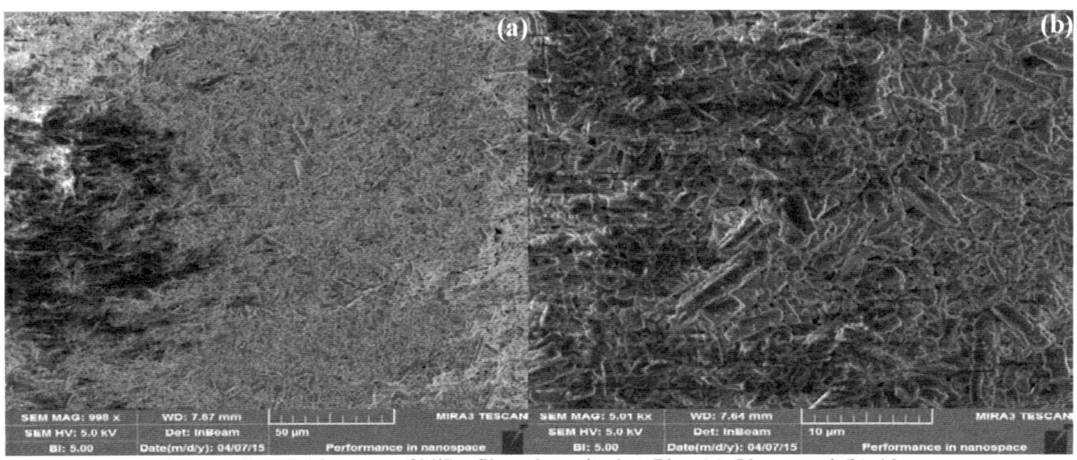

Figure 4. SEM images of NiPc films deposited at 70 *g* (a) 50 μm and (b) 10 μm

3. Results and discussion

Figure 5 (a) and (b) show impedance-humidity and capacitance-humidity relationships for NiPc samples deposited at 1*g* by drop-casting method, respectively. It can be seen that the impedance decreased1.5 times; whereas, the capacitance exhibits an increase of 1.1 times in its magnitude while changing the ambient humidity from 25% to 95%.

In Figure 6 (a) and (b) impedance-humidity and capacitance-humidity profiles are shown for NiPc samples deposited at 70*g* by a centrifugation process, respectively. In the figure, the impedance of the sample shows 1.8 times decrease in its magnitude while changing the humidity from 25 % to 95 %; whereas, for the same change in humidity the capacitance exhibited an increase of 1.5 times with respect to its initial values.

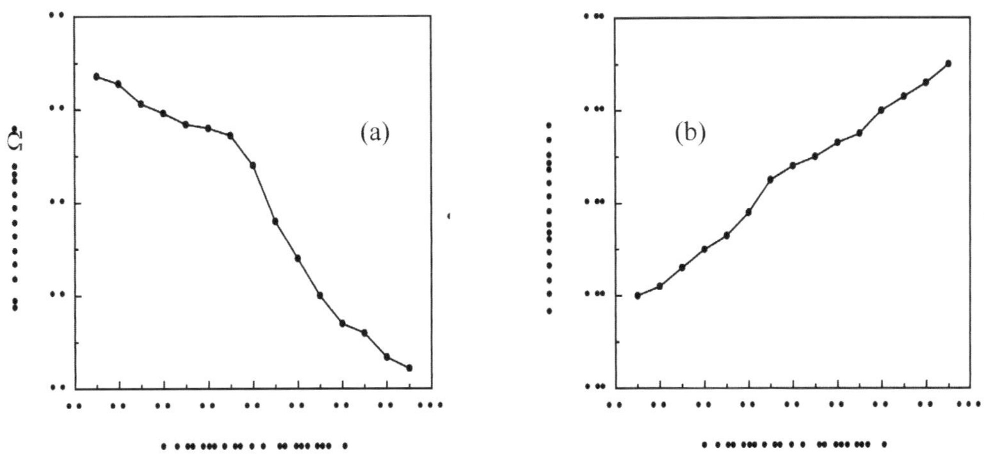

Figure 5. Experimental data showing (a) impedance-humidity and (b) capacitance-humidity relationships for NiPc samples deposited at 1g.

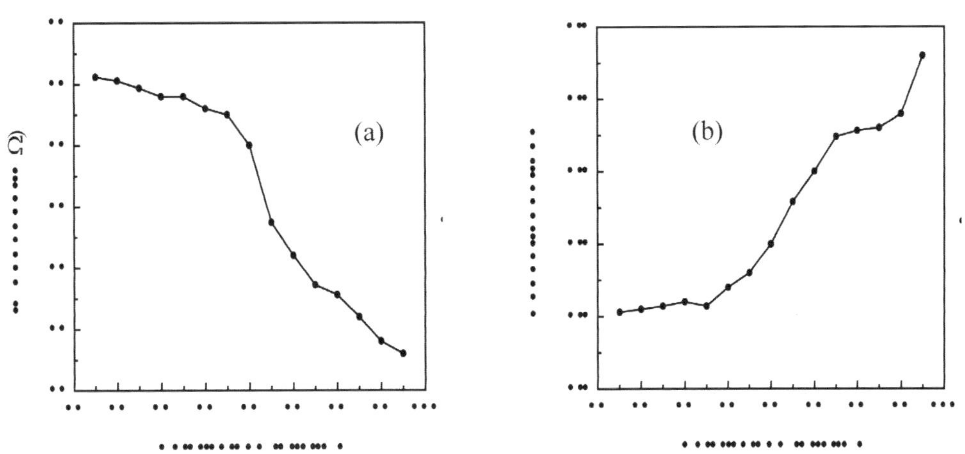

Figure 6. Experimental data representing (a) impedance-humidity and (b) capacitance-humidity relationships for NiPc samples deposited at 70g.

The impedance-humidity and capacitance-humidity relationships shown in Figures 5-6 can be modeled by placing a resistance (*R*) and a capacitance (*C*) in parallel [30, 31]. From circuit point of view, decrease in impedance (*Z*) with incremental increase in humidity can be explained by using an expression involving *R* and *C* as given below [30, 32, 33]:

$$Z = \frac{R}{(1 + j\omega RC)} \qquad (2)$$

Equation (2) describes that the observed change in the magnitude of *Z* because of the changing value of ambient humidity can be associated with the collective response generated by both *R* and *C*, but with dominating *R* contribution.

It is an established fact that capacitance of a material is dependent on its polarizability which has got numerous basic constituents like dipolar (α_{dip}), ionic (α_i) and electronic (α_e) polarizability [32]. In NiPc,α_{dip} could be associated with H_2O dipoles absorbed by the films, and appear to be a dominating factor in determining the characteristics of NiPc. On the other hand, α_e is common and arises due to the relative displacement of the orbital electrons. There is a good possibility that organic molecules may interact with water molecules to define charge-transfer complexes and as a result, the presence of α_i in NiPc can safely be assumed. Furthermore, it can be assumed, that α_{dip} and α_i polarizabilities affect the capacitive properties at lower frequencies (120 Hz-1 kHz), where α_e dominates at high frequencies. In [34], it was discussed that the transfer of charge carriers, i.e., electrons and holes, can originate another type of polarizability referred to as charge transfer polarizabilty, α_{tn}. Hence, we can possibly write the total polarizability, α_n for NiPc film:

$$\alpha_n = \alpha_{dip} + \alpha_i + \alpha_e + \alpha_{tn} \quad (3)$$

According to Clausius-Mosotti, the relative permittivity, ε and electrons/holes concentration, N_n under normal conditions, can be associated with α_n as [34, 35]:

$$\frac{\varepsilon - 1}{\varepsilon + 2} = \frac{N_n \alpha_n}{3\varepsilon_0} \quad (4)$$

where $\varepsilon_0$is permittivity of free space. Considering that capacitance is proportional to permittivity on the basis of equation (4), one can write[36]:

$$\frac{C_H}{C_n} = \frac{[1 + 2N_n \alpha_n (1 + k\Delta RH)/3\varepsilon_0]}{[1 - N_n \alpha_n (1 + k\Delta RH)/3\varepsilon_0]\varepsilon} \quad (5)$$

In Equation (5),C_H and C_n are capacitances at RH>25 % and RH =25 % respectively k is a constant defining humidity based capacitive factor (or fitting parameter) of the sample, and $\Delta RH=RH-RH_o$ is the percentage increase of relative humidity with respect to initial value RH_o = 25 %. Assuming relative permittivity of NiPc equal to 4, one can evaluate the value of $N_n\alpha_n$ by employing Equation (4). The evaluated value $N_n\alpha_n$ can then be used in Equation (5) and assuming C_H/C_n = 1.13 at RH=95 %, the value of k was found to be 0.013 $(\%)^{-1}$. Thus, Equation (5) can be used as first approximation to simulate capacitance-humidity response of NiPc samples. As Z includes both capacitance and resistance (equation-2), the simulation of the capacitance thus allows us to simulate impedance as well.

Change in capacitance due to the change in humidity can possibly be associated with the increase in dielectric permittivity due to the absorption of water molecules by the surface of the NiPc, which in turn increases its capacitance. Further, absorbed water molecules also generate more displacement current, which, in turn, reduces resistance of the sample. Another plausible explanation of increased capacitance and reduced resistance at higher humidity could be associated with the doping of NiPc film by the absorbed water molecules. This, in turn, increases the polarizability and also the concentration of charges which inherently increases capacitance and reduces resistance of the sample [36].

Sensitivity of a humidity sensor can be evaluated using incremental change in Z and C (S_z and S_c) as a function of changed humidity conditions and can be estimated as [30]:

$$Sz = \frac{\Delta Z}{\Delta RH} \quad (6)$$

$$Sc = \frac{\Delta C}{\Delta RH} \quad (7)$$

where ΔZ, ΔC and ΔRH are increments of impedance, capacitance and relative humidity (in %). It can be shown that in average S_z= -0.32 MΩ/%, S_c= 0.025 pF/% for samples deposited at 70g, and S_z= -0.22 MΩ/%, S_c = 0.007 pF/% for samples deposited at 1g at 1 kHz frequency. Thus, NiPc samples can potentially be used to fabricate humidity sensors for environmental monitoring and assessment.

To evaluate the dynamics of the water vapor absorption and desorption processes of NiPc samples, the average response and recovery time were evaluated by impedance-time measurements at the change of humidity from 25 % to 95 % and from 95 % to 25 % which were found to be 1.0 min and 2.5 min, respectively. The response and recovery times were measured accordingly, at the change of impedance from 90 % to 10 % and vice versa.

4. Conclusion

To fabricate humidity sensors, Nickel phthalocyanine's (NiPc) thin films were deposited by centrifugation at high gravity (70*g*), and also by drop-casting at normal gravity (1*g*) on surface-type ceramic substrate having interdigitated silver electrodes. The sensitivity of the sensors was then assessed by evaluating impedance and capacitance variations as a function of ambient humidity. It was found that NiPc sensors fabricated at 70*g* showed larger response to the humidity than 1*g*devices. It was also observed that by varying ambient humidity impedance of the samples decreases whilst the capacitance increases, which could be associated with the diffusion of water molecules into NiPc films which presumably had increased charge concentration, and also the permittivity of the NiPc films. This effect was higher for devices fabricated at high gravity condition, contrary to those madeat 1*g*, due to increased surface area as evident from SEM micrographs. It has been demonstrated that NiPc films can potentially be used in organic semiconductor industry especially, in instrumentation technology.

5. Acknowledgement

The authors are thankful to Capital University of Science and Technology and Ghulam Ishaq Khan Institute of Engineering Sciences and Technology for their support to this research work.

6. References

[1] Patowary B B 2014 *International Journal of Advanced Research in Electrical Electronics and Instrumentation Engineering* 3 9351-9360
[2] Dzugan T, Kroupa M and Reboun J 2014 *Procedia Engineering* 69 962-967
[3] Yang M Z, Dai C L and Lin W L 2011 *Sensors*11 8143-8151
[4] Weng Q and Yang S 2006 *Environ Monit Assess*117 463
[5] Anjaneyulu Y, Jayakumar I, Bindu V, Rao P, Sagareswar G, Ramani K and Rao T 2007 *Environ Monit Assess*124 371
[6] Joyce A, Adamson J, Huntley B, Parr T and Baxter R 2001 *Environ Monit Assess*68 127
[7] Chen Z and Lu C 2005 *sensor letters*3 274-295
[8] Park S, Kang J, Park J and Mun S 2001 *Sensors Actuators B* 76 322
[9] Karimov Kh S, Khan I Q, Khan T A and Draper P H 2008 *Environ Monit. and Assess*141 323-328
[10] Rinkio M, Zavodchikova M Y, Torma P and Johansson A 2008 *Physica Status Solidi B*245 2315–2318
[11] Tang D, Ci L, Zhou W and Xie S 2006 *Carbon*44 2155-2159
[12] Anbia M and Fard S E M 2012 *Journal of Rare Earths*30 38-42
[13] Chani M T S, Karimov Kh S, Khan S B, Asiri A, Saleem M and Bashir M 2013 *Optoelectronic and Advanced Materials-Rapid Communication* 7 861-865
[14] Chani M T S, Karimov Kh S, Khalid F A and Moiz S A 2013 *Solid State Sciences* 18
[15] Shafai T S and Anthopoulos T D 2001 *Thin Solid Films*361 389-399
[16] Anthopoulos T D and Shafai T S 2004 *Journal of Physics and Chemistry of Solids*65 1345-1348
[17] Chaabane R B, Ltaief A, Dridi C, Rahmouni H, Bouazizi A and Ouada H B, 2003 *Thin Solid Films* 427 371-376
[18] Binu P, Joseph C, Shreekrishnakumar K and Menon C 2003 *Materials Chemistry and Physics* 80 591-594
[19] Liu C, Shih J and Ju Y 2004 *Sens. Actuators B*99 344

[20] El-Nahass M, Abd-El-Rahman K, Farag A and Darwish A 2005 *Org. Elect.*6 129
[21] El-Nahass M, El-Deeb A and Abd-El-Salam F 2006 *Org. Elect.*7 261
[22] El-Nahass M, Abd-El-Rahman K and Darwish A 2007 *Journal of Microelect*38 91-95
[23] Anthopoulos T and Shafai T 2003 *THIN SOLID FILM*441 207
[24] El-Nahass M and Abd-El-Rahman K 2007 *Journal of Alloys Comp* 430 194
[25] Petraki F, Papaefthimiou V and Kennou S 2007 *Org. Elect*8 522
[26] Ahmad Z, Sayyad M, Wahab F, Sulaiman K, Shahid M, Chaudry J, Munawar M and Aziz F 2013 *Physica B: Condensed Matter*413 21-23
[27] Shah M, Sayyad M H, Karimov Kh S and Tahir M M 2010 *PhysicaB*405 1188
[28] Ahmed M M, Karimov Kh S and Moiz S A 2004 *IEEE Transactions on Electron Devices* 51 121-126
[29] Tahir M M and Karimov Kh S 2009 *Journal of Optoelectronics and Advanced Materials* 11 83-88
[30] Dally J W, Riley W F and McConnell K G 1993 *Instrumentation for Engineering Measurements, Second Edition* (New York: John Wiley & Sons, Inc)
[31] Boylestad R L 2010 *Introductory Circuit Analysis, 12th Edition* (Prentice Hall)
[32] Khan S B, Chani M T S, Karimov Kh S, Asiri A M, Bashir M and Tariq R 2014 *Talanta120,* 443–449
[33] Chani M T S, Karimov Kh S, Khalid F A, Abbas S Z and Bhatty M 2013 *Chin. Phys. B* 22
[34] Boguslavsky S and Vannikov V V 1968 *Organic Semiconductors* (Moscow)
[35] M. Omar 2002 *Elementary Solid State Physics: Principles and Applications (*Singapore: Pearson Education Pte. Ltd)
[36] Karimov Kh S, Akhmedov K, Qazi I and Khan T 2007 *Journal of Optoelectronics and Advanced Materials* 9 2867-2872

Polymerization of acrylic acid using atmospheric pressure DBD plasma jet

M. Bashir and S. Bashir[1]
Department of Physics, COMSATS Institute of Information Technology, Islamabad, Pakistan
[1]Department of Physics and Applied Mathematics, Pakistan Institute of Engineering and Applied Sciences, P. O. Nilore, Islamabad, Pakistan

E-mail: dr.bashir77@gmail.com

Abstract. In this paper polymerization of acrylic acid was performed using non thermal atmospheric pressure plasma jet technology. The goal of this study is to deposit organic functional coatings for biomedical applications using a low cost and rapid growth rate plasma jet technique. The monomer solution of acrylic acid was vaporized and then fed into the argon plasma for coating. The discharge was powered using a laboratory made power supply operating with sinusoidal voltage signals at a frequency of 10 kHz. The optical emission spectra were collected in order to get insight into the plasma chemistry during deposition process. The coatings were characterized using Fourier transform infrared spectroscopy, atomic force microscopy and growth rates analysis. A high retention of carboxylic functional groups of the monomer was observed at the surface deposited using this low power technique.

1. Introduction

Recently the use of atmospheric pressure non thermal plasmas for coating of organic materials has gained much attention due to their low cost and low temperature operation [1-4]. Atmospheric pressure cold plasmas can be generated using different type of sources: pulsed DC, inductively coupled plasma, radio frequency AC and dielectric barrier discharge (DBD) [5-7]. Among these the dielectric barrier discharges are the most suitable sources for the generation of non thermal plasmas at atmospheric pressure due to their cold, robust and low frequency operation. Plasma jet devices based on DBD can produce non thermal atmospheric pressure plasma which is low cost, safe to use and extremely stable for longer operation.

Acrylic acid (AA) is a commonly used precursor for plasma polymerization as it is easily available and low cost material. Thin films of the plasma polymerized acrylic acid (PPAA) have been a dynamic research for applications in biomedical, biosensors and micro fluidics [8-9]. Plasma deposition of acrylic acid using low pressure vacuum systems has been extensively studied [10-11]. However, low pressure coating systems are associated with practical limitations due to the expensive vacuum technology and their low deposition rates. The development of new technology based on atmospheric pressure deposition has gained much interest due to its flexibility in terms of low cost and ease of operation. Few studies have been made that address the use of atmospheric pressure plasma for polymerization of acrylic acid [12-13].

Content from this work may be used under the terms of the Creative Commons Attribution 3.0 licence. Any further distribution of this work must maintain attribution to the author(s) and the title of the work, journal citation and DOI.
Published under licence by IOP Publishing Ltd

Table 1. Deposition conditions for plasma polymerization of acrylic acid

Processing parameters	Quantity
Ar gas flow rate (sccm)	500
Monomer flow rate (sccm)	10 - 20
Monomer temperature (°C)	80
Deposition time (min)	5- 15
Plasma power (W)	25

This paper presents the deposition and characterization of polyacrylic acid films using an atmospheric pressure DBD plasma jet. The aim of the current investigation is to study the potential of this specific DBD plasma technique for deposition of organic functional films. The properties of the deposited surfaces were studied using atomic force microscopy, Fourier transform infrared spectroscopy and growth rate analysis. The paper is structured as follows. The experimental methods are described in the section 2. The results with discussion are presented in the section 3 and the conclusions of this study are drawn in section 4.

2. Experimental methods

2.1. Polymerization setup

The experimental setup used for atmospheric pressure plasma deposition is shown in Figure 1. A 99.9% pure argon (Ar) gas was used for plasma generation and to carry the monomer vapours in reactor. The setup consists of two mass flow controllers to control the flow rate of Ar gas. A custom made glass bubbler was heated at 80 °C for vaporizing the monomer solution. A low flow rate Ar gas was passed through the bubbler and plasma deposition was achieved using vapours. A 50 Ω resistor was connected from the ground side of the reactor for measuring plasma power.

2.2. Electrodes configuration for plasma jet

The DBD plasma jet was generated in a flask made from Pyrex glass (Figure 1). The high voltage (HV) electrode is a steel tube with inner diameter of 5 mm and outer diameter of 7 mm. The steel tub was fixed at the centre of the flask by inserting into a rubber bang at the top of the flask. The rubber bang seals the flask in order to create the vacuum. The steel tube serves as a HV electrode for plasma generation and also as an inlet for gas flow. The ground electrode, made of aluminium sheet, was placed underneath the flask. The bottom glass wall of the flask acts as a dielectric barrier for plasma generation. The distance from the lower end of the steel tube to the bottom of the flask was fixed to 2.5 cm in order to accommodate the substrate. A concentrated high voltage electric field that breaks down the gas in the flask is emitted from the lower end of the high voltage electrode. Plasma was produced in the form of a jet. The glass substrate was placed under the HV electrode in the flask before starting the polymerization process. In the next step the flask was evacuated to a pressure of 100 mTorr using a rotary pump. The discharge flask was then filled with Ar gas to atmospheric pressure. When the pressure in the flask is reached to atmospheric the high voltage electric field was applied for generating plasma in the flask. Finally a mixture of gas containing Ar and monomer vapours was introduced to Ar plasma for deposition.

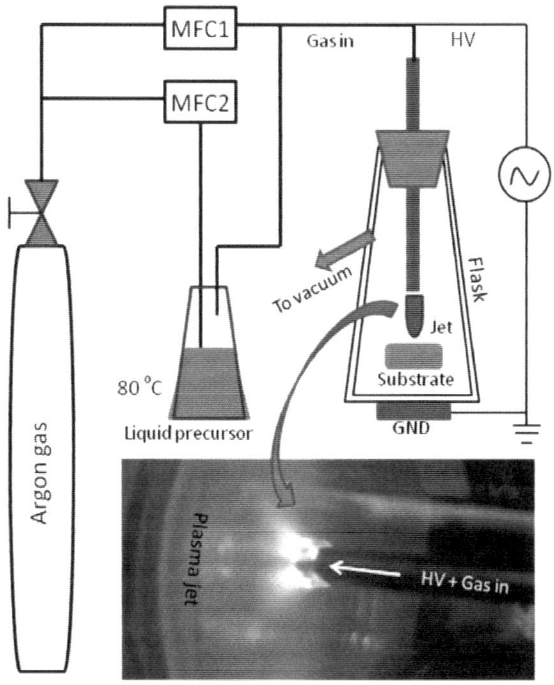

Figure 1. Schematic diagram of the experimental setup used for plasma polymerization at atmospheric pressure. Photograph of DBD plasma jet generated in a Pyrex flask (bottom)

3. Results and discussion

Plasma deposition was performed using a laboratory made DBD jet plasma configuration as described in section 2.2. A circular area with an average diameter of 1.5 cm was deposited on the substrate. The commercially available glass microslides were used as a substrate for deposition. They were cleaned using methanol and deionized (D.I.) water. The substrates were pre-treated with Ar discharge for 2 minutes. The deposition conditions for plasma polymerization are summarized in Table 1. The monomer flow rate of 10 sccm, Ar gas flow rate of 500 sccm, deposition time of 10 minutes and plasma power of 25 W were optimized and kept constant throughout the observation until otherwise stated.

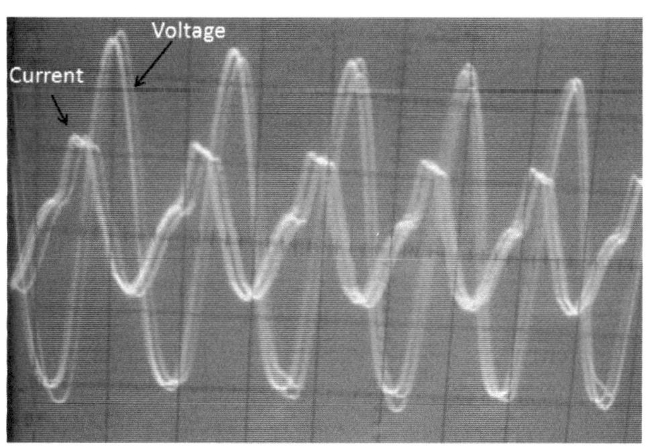

Figure 2. Typical voltage and current signals of plasma jet. Voltage Vp = 8.4 kV, current Ip = 6 mA

3.1 Electrical characterization of plasma

The electrical characterization of plasma was performed by measuring the applied high voltage and the resultant current. An oscilloscope, a high voltage probe and a current probe was used to perform the electrical characterization of the discharge. The characteristics current-voltage signals were recorded simultaneously on the oscilloscope and are shown in Figure 2. Typically the atmospheric pressure DBD plasmas operating in filamentary mode consist of a large number of current spikes appearing on each half cycle of voltage [14]. These current spikes are diminished in the case of glow discharge plasma and get the shape close to the sinusoidal waveform which corresponds to the capacitive current [15]. Also, the current signal is leading the voltage with a phase shift of < 90° (Figure 2) which characterizes the capacitive load. Hence, the DBD jet plasma used here for polymerization of acrylic acid operates in glow mode.

3.2. Plasma emission spectroscopy

The optical emission spectroscopy (OES) was used in order to get insight into the plasma chemistry during deposition process. The emission spectra were collected using a USB2000 spectrometer (Ocean Optics). The signals were recorded by placing the lens of fibre optic cable close to the plasma jet. Figs. 3a and 3b show the emission spectra of argon discharge before deposition and argon mixed with acrylic acid during deposition process respectively. The discharge was generated at a power of 25 W and a monomer flow rate of 10 sccm. The emission peaks were identified and assigned with reference to previous studies [16-17] and to the online data available from the National Institute of Standards and Technology (NIST) [18]. The spectrum of the Ar discharge shows the presence of OH (309 nm), N_2 band (310 nm - 440 nm) and Ar band (695 nm - 850 nm). The spectral lines intensities of pure Ar were decreased by introducing the monomer precursor into the plasma jet as shown in Figure 3b. The decrease in the emission intensities of Ar plasma by introducing monomer vapours is most probably due to the utilization of plasma energetic species (Ar metastables etc.) for dissociation of organic molecules. This reduction in Ar metastables is caused by the plasma quenching due to the additional reactions related to the increase of monomer content into the discharge volume. The peaks of NH at 337 nm, CN at 359 nm and CO at 457 nm appeared in the emission spectrum when the acrylic acid was mixed with Ar plasma in order to achieve deposition. Similar peaks of NH and CN have been reported by Tran et al. [19] during plasma deposition at atmospheric pressure. The presence of strong peaks corresponding to NH, CN and CO indicated that the monomers were fragmented during the plasma polymerization process. Moreover, the emission of the lines from hydroxyl (OH) radicals and N_2 indicates the presence of excessive air in open to atmosphere plasma jet.

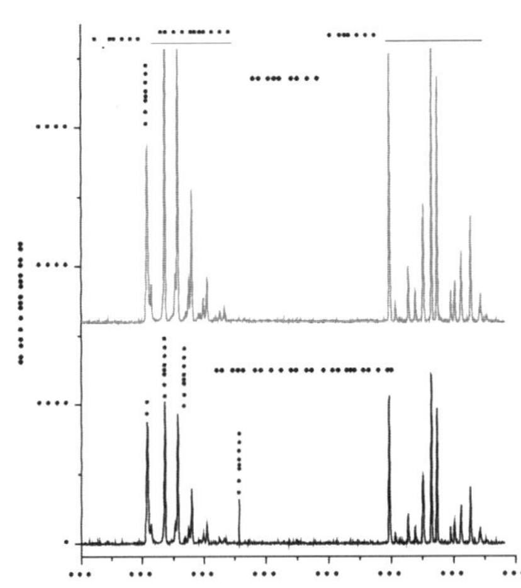

Figure 3. Optical emission spectroscopy. (a) Ar plasma and (b) Ar plasma mixed with acrylic acid

3.3. FTIR spectroscopy

The chemical properties of the plasma polymerized acrylic acid films have been studied using Fourier transform infra red (FTIR) spectroscopy. The spectra were obtained by using FTIR spectrometer (Perkin Elmer USA) at a resolution of 4 cm^{-1}. Multiple scans (32) were performed and then averaged for each sample. Since the absorption depth of infrared beam is typically 1 μm the relatively thick samples were selected for FTIR analysis. Therefore, the samples for FTIR spectrum were prepared at a monomer flow rate of 20 sccm and deposition time of 10 minutes. Figure 4 shows the FTIR spectra of AA monomer and plasma polymerized acrylic acid. The comparison of the monomer spectrum with plasma polymers provide the evidence that the monomers have actually been polymerized in the plasma and are deposited on the glass substrate. The bands in the monomer spectra are well resolved compared to the broad bands

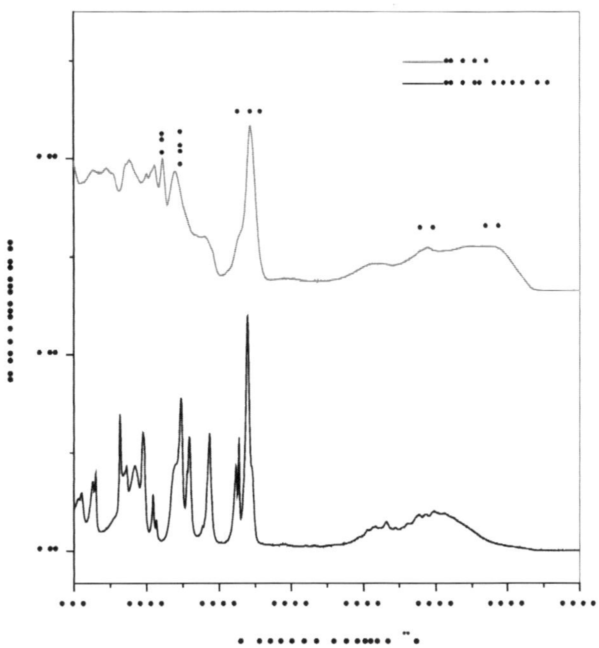

Figure 4. Fourier transform infrared spectra of acrylic acid monomer (bottom) and plasma polymerized acrylic acid (top)

obtained from the plasma polymers [20]. The spectral peaks were identified using data previously reported in the literature [12, 21, 22]. The spectrum of PPAA exhibits a dominant absorption band at 1720 cm^{-1} which can be assigned to stretching vibration of the C=O group. The spectrum contains a very broad band ranging from 3600 cm^{-1} to 2400 cm^{-1} which can be assigned to the stretching vibrations of the hydroxyl group OH. A smaller peak at 2945 cm^{-1} which can be attributed to stretching vibrations of the CH bond is also superimposed on a broad band of OH. At low wave numbers, two other significant bands appeared at 1200 cm^{-1} and 1112 cm^{-1} which can be assigned to the stretching vibrations of C-OH and C-O groups respectively [22]. Hence, a high retention of carboxylic functional groups of the monomer was observed at the surface deposited using this low power technique.

3.4. Coating morphology

The surface morphology of plasma polymerized acrylic acid films was observed using atomic force microscopy (AFM). The images of surface were recorded by adopting the taping mode of AFM. The scans were performed by selecting an area of 5 μm2 at the surface of the PPAA films. Figures 5a, 5b show 3D and 2D AFM images for PPAA surfaces respectively. AFM analyses showed that the surfaces deposited using jet plasma were continuous and free from pin holes. However, the average value of the surface roughness (20.5 nm) was found to be higher than the previous reports [20] according to which the films prepared using plasma enhanced chemical vapour deposition (PECVD) method are smooth with lower surface roughness. This roughness is an indication of the surface etching due to bombardment of energetic species at the substrate in the plasma jet. The continuity of the plasma coatings was further confirmed under an optical microscope using CCD camera (Figure 5c).

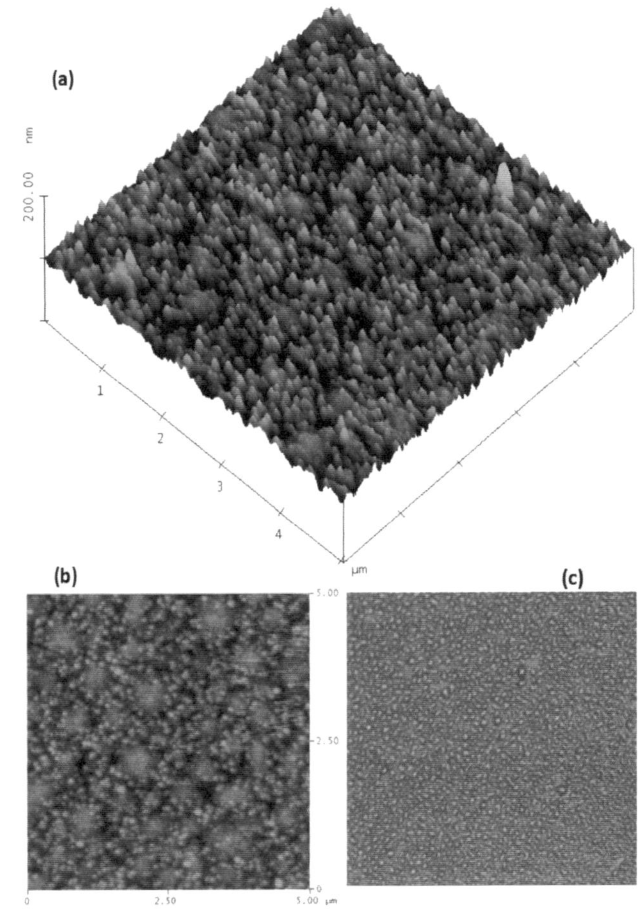

Figure 5. Atomic force microscopy of the films, (a) 3D and (b) 2D images of the PPAA films. (c) CCD image under optical microscope with 10X resolution

3.5. Thickness and growth rate measurements

The film thickness was determined using step height method in atomic force microscopy (AFM). The microscope glass slides were masked during jet plasma deposition in order to measure the film thickness. Figure 6a shows the film thickness as a function of deposition time at two different flow rates of 10 sccm and 20 sccm of the monomers. A linear relationship was observed for a polymerization time of 15 minutes. The plasma power was kept constant at 25 W during this analysis. The average growth rates were calculated from the slop of the linear curves in Figure 6a. The average growth rates of 69.3 nm min^{-1} and 150.8 nm min^{-1} were found for monomer flow rates of 10 sccm and 20 sccm respectively. The increase in monomer concentration increases the number of monomer molecules in the plasma and hence, enhances the rate of reactive species reaching to the substrate surface. Thus, the film growth rate in plasma polymerization process depends mainly on the plasma processing conditions. Figure 6b shows the step height for measuring thickness of the PPAA film.

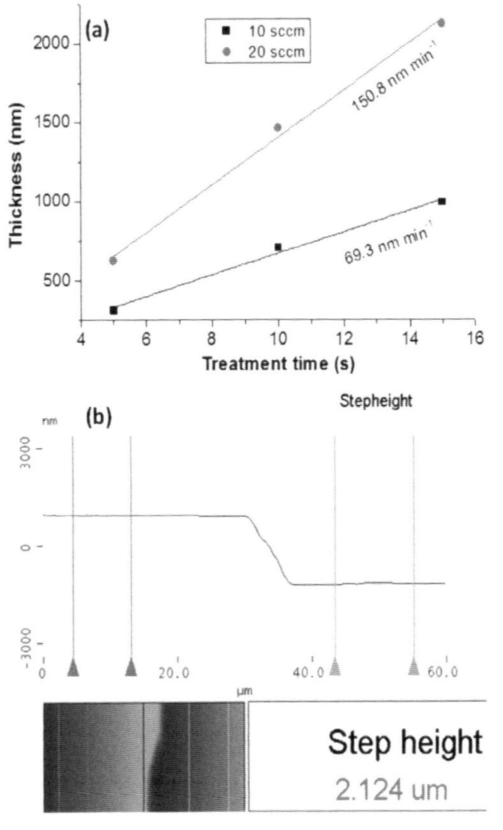

Figure 6. (a) Thickness of the films as a function of deposition time for two different concentrations of monomers.(b) The image shows the step height for measuring film thickness through AFM

4. Conclusions

The plasma polymerized acrylic acid films deposited using an atmospheric pressure plasma jet are presented in this paper. The FTIR spectrum indicated a high retention of carboxylic functional groups of the monomers at the surface when the films were deposited at a discharge power of 25 W, monomer flow rate of 20 sccm and a treatment time of 10 minutes. AFM results showed that the films were continuous and free from pin holes. An average growth rate of 150.8 nm min^{-1} was obtained for a monomer flow rate of 20 sccm. Hence, the films with promising characteristics can be obtained by the technique of atmospheric pressure DBD plasma jet.

5. Acknowledgment

M.B. would like to acknowledge the Higher Education Commission of Pakistan for financial support through its TTS program.

6. References

[1] Asandulesa M, Topala I, Pohoata V and Dumitrascu N 2000 J. Appl. Phys. 108 093310
[2] Bradley J W, Oh J-S, Olabanji O T, Hale C, Mariani R and Kontis K 2011 IEEE T Plasma Sci. 39 2312

[3] Bashir M and Bashir S 2015 Plasma Chem. Plasma Process. 35 739
[4] Bashir M, Bashir S, Rees J M and Zimmerman W B 2013 Surf. Coat. Technol. 234 82
[5] Li B, Chen Q and Liu Z W 2010 Appl. Phys. Lett. 96 041502
[6] Xu Gui-M, Ma Y and Zhang G-J 2008 IEEE Trans. Plasma Sci. 36 1352
[7] Lee S-Y, Gho J-S, Kang B-H, and Cho J-S 2008 34th annual conference of IEEE. doi: 10.1109/IECON.2008.4758014 556
[8] Siow K S, Britcher L, Kumar S and Griesser H J 2006 Plasma Process. Polym. 3 392
[9] Bashir S, Casadevall i Solvas X, Bashir M, Rees J M, and Zimmerman W B 2014 Bio Chip J.8(2) 122
[10] Detomaso L, Gristina R, Senesi G S, Agostino R d and Favia P 2005 Biomaterials 26 3831
[11] Jafari R, Tatoulian M, Morscheidt W and Arefi -Khonsari F 2006 React. Funct. Polym. 661757
[12] Morent R, Geyter N D, Vlierberghe S V, Vanderleyden E, Dubruel P, Leys C and Schacht E 2009 Plasma Chem. Plasma Process. 29 103
[13] Carton O, Salem D B, Bhatt S, Pulpytel J and Arefi-Khonsari F 2012 Plasma Process. Polym. 9 984
[14] Tang J, Duan Y and Zhao W 2010 Appl. Phys. Lett. 96 191503
[15] Park J, Henins I, Herrmann H W, Selwyn G S, and Hicks R F 2001 J. Appl. Phys. 89 20
[16] Gordillo-Vazquez F J, Camero M and Gomez-Aleixandre C 2006 Plasma Sources Sci.Technol. 15 42
[17] Bashir M, Rees J M, Bashir S and Zimmerman W B 2014 Phys. Lett. A 378 2395
[18] See http://physics.nist.gov/PhysRefData/ASD/lines form.html
[19] Tran D T, Mori S, Tsuboi D and Suzuki M 2009 Plasma Process. Polym. 6 110
[20] Bashir M, Rees J M and Zimmerman W B 2013 Surf. Coat. Technol. 234 82
[21] Wan J, Wang S, Song M, Jia X and Yang J 2009 Plasma Process. Polym. 6 825
[22] Wan J, Wang S, Song M, Jia X and Yang J 2009 Plasma Chem. Plasma Process. 29 103

IOP Conf. Series: Materials Science and Engineering **146** (2016) 012037 doi:10.1088/1757-899X/146/1/012037

Effects of nano-sized boron nitride (BN) reinforcement in expandable graphite based in-tumescent fire retardant coating

E. S. Zulkurnain, F. Ahmad and Q. F. Gillani
Universiti Teknologi PETRONAS, Bandar Seri Iskandar, Malaysia

E-mail: qandeelgillani@gmail.com

Abstract. The purpose of in-tumescent fire retardant coating (IFRC) is to protect substrate from fire attack by limiting heat transfer. A range of coating formulations have been prepared using Bisphenol A epoxy resin BE-188 and polyamide solidifier H-2310 as two-part binder, ammonium polyphosphate (APP) as acid source, melamine (MEL) as the blowing agent, expandable graphite (EG) as carbon source and nano-boron nitride (BN) as inorganic nano filler. The filler was used to improve the performances of the APP-EG-MEL coating. The effects of nano-BN on the char morphology and thermal degradation were investigated by fire test, thermo gravimetric analysis (TGA), Fourier transform infrared spectroscopy (FTIR), X-ray photoelectron spectroscopy (XPS) and field emission scanning electron microscopy (FESEM). The results showed that by substituting or reinforcing of 4% weight percentage of nano-BN, residual weight of the char increases by 23.82% compared to APP-EG-MEL coating without filler. Higher carbon content was obtained in the char and a more compact char was produced. The results indicated that nano-BN could be used as a filler to improve thermal stability of the APP-EG-MEL coating.

1. Introduction

Fire can results in loss of lives and property. It can be controlled using fire protection system classified into two: active and passive. Active systems such as fire extinguishers, fire alarms and water sprinklers are commonly installed. However, passive systems such as fire-rated doors and fire retardant coatings are often overlooked and disregard. Their purpose are to contain fire, delay building collapse and provide safe evacuation of occupants, and to limit the destructive effects of fire [1].

Steel is a strong material and the need to protect it may not be apparent. The exposure of steel structural members to high temperature results in buckling leading to structural collapse when it reaches 500ºC, its critical temperature [2]. To ensure the safety of steel structure during fire, its temperature must be maintained below the critical temperature. This can be achieved by using intumescent fire retardant coatings [3].

Intumescent is defined as the swelling up of certain substances when they are heated and it comes from latin word 'intumescere' which means 'to swell up' [4]. The coating comprises of three main ingredients: an acid source, a carbon source and a gas source [5]. Past researches commonly used ammonium polyphosphate, pentaerythritol and melamine [6]. When the temperature rises in a fire, intumescent fire retardant coating swells many times of its original thickness to form carbonaceous foam charred layer that limit heat transfer to the substrate [7].

The expansion of the charred layer often depends on the coating ingredients as well as the ratio of carbon, nitrogen and phosphorus atoms in the compound [8]. The char layer is the key parameter to

Content from this work may be used under the terms of the Creative Commons Attribution 3.0 licence. Any further distribution of this work must maintain attribution to the author(s) and the title of the work, journal citation and DOI.
Published under licence by IOP Publishing Ltd

ensure the coating provides full protection of substrate. The relationship between flammability and charring process of intumescent coating explained by Joseph and Ebdon [9] is as follows: the higher the amount of residual char after combustion, the lower the amount of combustible material to propagate the flame thus increasing the flame retardancy of the material. Existing intumescent fire retardant coatings reported that the char layer would damage easily by the turbulence during flame spread. The residual weight of the coating is also low as it is easily oxidized at high temperature [10].

Several studies [11-13] have indicated the usage of fillers to improve the performances of the coatings. Incorporating nano-sized fillers specifically into coatings can greatly enhance its fire resistance and anti-oxidation properties [14]. Boron compounds are one of the widely used fire retardant fillers for example: borax, zinc borate, barium metaborate and ammonium fluoroborate. These type of compound are non-toxic to human and act as fire suppressant during condensed and vapour phase [15].

A study by Jimenez, Duquesne and Bourbigot stated that borates and boric acid are well established as flame retardants; and addition of boric acid to the epoxy-based intumescent fire retardant has improved its performance by increasing the time of failure. Furthermore, the formation of boronphosphate during the degradation of boric acid and ammonium polyphosphate together has been suggested to provide superior mechanical resistance of the char and also promotes the adhesion between char and substrates [16].

Boron nitride (BN) is an inorganic boron compound that demonstrates excellent chemical and thermal stability, hence indicates it can be used as a filler to improve the performance of intumescent fire retardant coatings [17]. Nowadays, it has been utilized in many applications specifically in polymer composites [18], refractory materials [19] and coatings. Boron nitride will act as a thermal insulator for the underlying unburned coating when the temperature reaches above 600°C [19]. This will result in mass loss suppression and stabilizes the resulting char [20] thus increasing the fire performance of intumescent coating.

The objective of this study were to investigate the performance of intumescent fire retardant coating reinforced with nano-BN by determining its optimum thermal performance, residual weight, functional group of the char and char morphological structure of intumescent fire retardant coating.

2. Materials and methods

Both Bisphenol A epoxy resin BE-188 and polyamide solidifier H-2310 were obtained from Mc-Growth Chemical Sdn. Bhd., Malaysia. Ammonium polyphosphate was supplied by Clariant (Malaysia) Sdn. Bhd., Malaysia. Melamine, expandable graphite, and boric acid were purchased from Sigma-Aldrich (M) Sdn. Bhd., Malaysia. Boron nitride was supplied by Xuzhou Jiechuang New Material Technology Co., Ltd., China. Structural steel A36M was obtained from TSA Industries (Ipoh) Sdn. Bhd. Malaysia.

2.1. Coating preparation

All intumescent ingredients were weighed according to their respective weight percentages as in Table 1 and pulverized using ROCKLABS Bench Top Ring Mill for 60 sec. Binder and intumescent ingredients were then mixed using a shear mixer at 40 rpm for 30 min. The coatings were applied on sandblasted structural steel plate of 50x50x1.5mm and cured for 24 hours at room temperature before characterization tests.

Table 1. The weight % of materials in intumescent fire retardant coating.

Formulation	EG	APP	MEL	BA	Epoxy	Hardener	BN
BN-0	5.85	11.76	5.76	11.5	43.42	21.71	0
BN-1	5.85	11.76	5.76	11.5	42.75	21.38	1
BN-2	5.85	11.76	5.76	11.5	42.09	21.04	2
BN-3	5.85	11.76	5.76	11.5	41.42	20.71	3
BN-4	5.85	11.76	5.76	11.5	40.75	20.38	4
BN-5	5.85	11.76	5.76	11.5	40.09	20.04	5

2.2. Characterization techniques

Fire test was carried out using Bunsen burner for 60 min and the distance between Bunsen burner and coating was set at 7cm. The coating's backside temperature was recorded using three thermocouples type K with miniature connector and Anritsu Data Logger AM-8000K. Thermal gravimetrical analysis (TGA) of intumescent fire retardant coating samples was performed on Pyris 1 TGA (Perkin Elmer) at 10ºC/min under nitrogen over temperature range of 25ºC–800ºC. Fourier transform of Infra-red Radiation (FTIR) for functional groups of char residual were determined by Spectrum One (Perkin Elmer) in range of 4000–400cm^{-1}. X-ray photoelectron spectroscopy (XPS) of char residual was performed using K-alpha XPS System (Thermo Scientific) with 14.0 kV of voltage and 300 W of power. The char layer was characterized by FESEM using SUPRA 55VP (Carl Zeiss AG) to physically observe its morphological structure.

3. Results and discussion

3.1. Analysis of char expansion

Furnace test was performed to burn the coating samples using CWF 13/5 Standard Chamber Furnace (Carbolite) at 6ºC/min and held at 500ºC for 1 h. The initial thickness of the coating was measured using Positector 6000 FS1 (DeFelsko). The char thickness after furnace test was obtained using digital Verniercaliper.

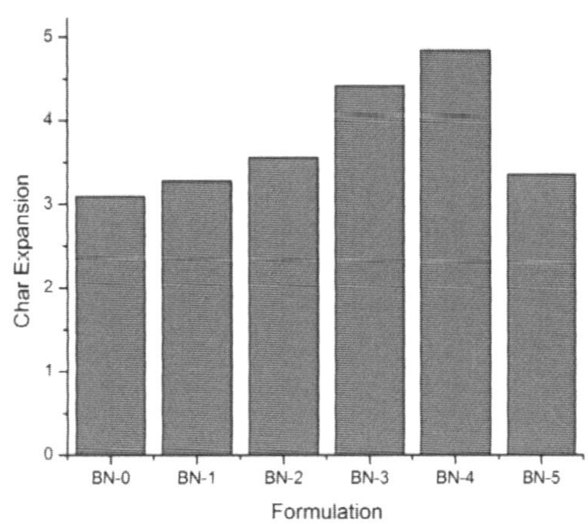

Figure 1. Expansion of intumescent coating after furnace test

The char expansion results are as follows: 3.1 (BN-0), 3.28 (BN-1), 3.56 (BN-2), 4.42 (BN-3), 4.84 (BN-4), and 3.36 (BN-5) times, from their original coating thickness as shown in Figure 1. Meanwhile Figure 2 shows the resulting char of BN-1, BN-2, BN-3, BN-4 and BN-5 after the furnace test. By comparing formulation BN-4 to BN-0, BN-1, BN-2, BN-3 and BN-5, BN-4 gives better expansion.

This is because of higher weight percentage of BN while APP, melamine, EG and boric acid used are kept constant in all formulations.

The role of APP, EG, melamine and boric acid are as acid source, carbon source, blowing agent and additive, respectively. Both boric acid and boron nitride will decompose into boric oxide that contributes to high residue and resulting in high amount of char [21]. Additionally, both boron compounds act as adhesion agent to provide strong bonding between the substrate and the char [22].

Initially the formulation expanded uniformly, but with the progressing fire over time, the remaining boron nitride continue to balance the expansion [17]. In previous studies, Duquesne et al. stated that a certain degree of expansion is required to get high thermal performance, but the highest expansion does not necessarily providing the highest efficiency [5]. Intumescent coating reinforced with boron nitride has moderate char expansion but it can limit heat transfer to the substrate effectively. With the presence of boron nitride, the resulting char does not contain any appalling cracks or holes that can cause heat to penetrate towards the substrate. Even though boron nitride used in BN-5 is the highest, it cannot contribute to higher expansion. This indicates that in any formulation with reinforced fillers, there is certain limitation in weight percentage where it provides optimum results [16].

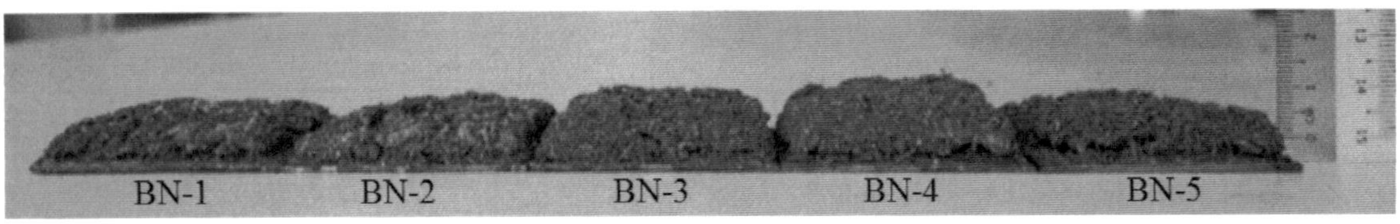

Figure 2. Physical char expansion comparison of BN-1, BN-2, BN-3, BN-4 and BN-5 from left to right respectively, after furnace test

3.2. Analysis of thermal insulation

Fire test was carried out using Bunsen burner for 60 min and the distance between Bunsen burner and coating was set at 7cm. The coating's backside temperature was recorded using three thermocouples type K with miniature connector and Compact Thermologger AM-8000K (Anritsu Meter). Time versus substrate temperature curve for heat shielding/insulation is illustrated in Figure 3.

Six samples of coating with different compositions of in tumescent ingredients were burnt using portable Bunsen burner for 60 minutes. The substrate temperature of BN-0 after 30 minutes is 232°C, and after 60 minutes the temperature reached 261°C, respectively as reported in Figure 3. BN-0 contains no boron nitride and the char structure could not sustain the fire for 60 minutes as there were no further reaction products which can trim down the substrate temperature.

Figure 4 (a, b, c, and f) shows the char photograph of BN-0, BN-1, BN-2 and BN-5 with cracked char surface. The substrate temperature of BN-1, BN-2 and BN-5 was 234°C, 178°C and 193°C after 60 minutes respectively. The temperature at 60 minutes for BN-1, BN-2 and BN-5 is 27°C, 83°C and 68°C less than BN-0 respectively. Figure. 4(b, c, and f) shows that the surface of BN-1, BN-2 and BN-5 char is also cracked and oxidized.

Figure 4 (d, e) shows a better char surface of BN-3 and BN-4 char compared to BN-0, BN-1, BN-2 and BN-5. The substrate temperature of BN-3 and BN-4 formulation reached 148°C and 145°C after 30 minutes and 162°C and 138°C after 60 minutes. This substrate temperature is significantly lesser than BN-0, BN-1, BN-2 and BN-5 after 60 minutes. Figure 2(d) of char photograph of burnt BN-3 formulation shows the existence of cracks on the surface. The cracks gradually increased the substrate temperature to 162°C after 60 minutes. In general, the addition of boron nitride increases adhesion of char to substrate [17]. From the literature [7], it was reported that when APP and boric compound were combined into the resin; the temperature of steel substrate reached 400°C in 29.5 minutes in contrast to 4 minutes when epoxy resin was used alone. Increasing the wt% of boron nitride results in heat shielding effects increases as shown in Figure 4 and surpasses the performance of coating without the usage of boron nitride as filler.

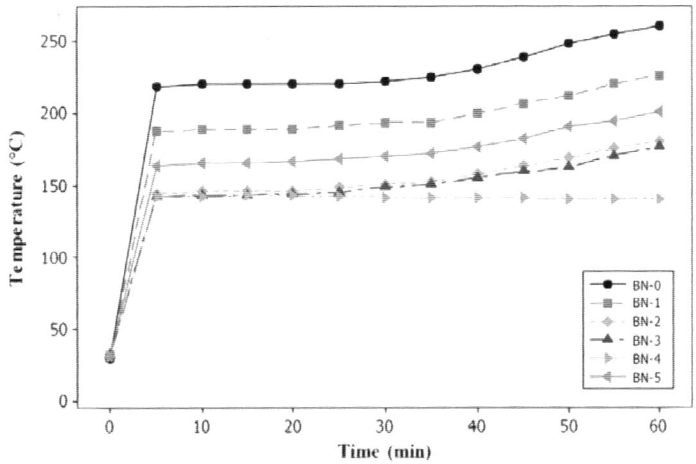

Figure 3. Time and substrate temperature curve of intumescent fire retardant coating

3.3. Analysis of char morphology

The char layer was characterized by FESEM using SUPRA 55VP (Carl Zeiss AG) to physically observe its morphological structure. The outer surface micrographs of char layer from furnace test of BN-0, BN-1, BN-2, BN-3, BN-4 and BN-5 are shown in Figure. 5 (a, b, c, d ,e f) respectively. It was noted that the morphological structure of the outer surface of the char layers were compact with tiny holes and cracks. Large crack surface and voids can be seen from BN-0, BN-1, BN-2 and BN-3. Char layer with large cracks could not provide good fire protection as heat can passes through char thus penetrating the substrate [7]. The char layer of BN-4 and BN-5 shows more refined compact surfaces with no crack and small voids that sized less than 50 μm. The compact outer surface char structure can provide resistance and limit heat transfer to the substrate hence protecting the substrate against fire. Thus, both well-expanded char and defined char structure are very crucial to general fire resistance properties of coatings [11].

Although nano filler was not incorporated in BN-0, it also formed bubbles but the distribution is quite uneven and there were some crack in the structure. Therefore, char morphological structure of BN-4 and BN-5 with 4% and 5% nano-BN respectively are better than BN-0 without nano filler. The char morphology of BN-4 formulation in Figure 5 (e) shows a thick and porous inner surface with the bubbles formation which is formed due to emission of gaseous compounds during the fire test. These gaseous compounds are produced from the thermal degradation of the coating during fire test [23].

3.4. Functional groups in char

FTIR for functional groups of char residual were determined by Spectrum One (Perkin Elmer) in range of 4000–400cm^{-1}. The FTIR spectra of BN-0, BN-1, BN-2, BN-3, BN-4 and BN-5 char are presented in Figure. 6. The FTIR spectra of BN-0 char shows two regions of absorption. In the 1st region, the three bands at 545, 620, 925 and 1085cm^{-1} are attributed to ν (B–O–H) stretching mode, ν (C–H) stretching mode, δ (B–O–H) deformation bond, and ν (B–O) stretching bond, respectively [24]. In the 2nd region above 2100 cm^{-1}, two weak bending peaks occurred at 2257 and 3215 cm^{-1} which represent the weak stretching vibrations of (-CH_3- CH_2-) and O-H groups due to cured epoxy resin.

The FTIR spectrum of BN-1 char shows three regions of absorption. In the low frequency side, at 494 cm^{-1} of the spectrum is attributed to fourth regions of ν is PO_4 [25]. The two broad bands centered on 1151 and 970 cm^{-1} are characteristic of phosphate anion [24, 26, 27]. In the region from 1380 – 1700cm^{-1}, the peaks at 1633 and 1402 cm^{-1} are due to amino groups (NH_2) and ring(C,N), respectively. In the 3rd region above 2100 cm^{-1}, one stretching peak at 2627 cm^{-1} vibrations of

(-CH_3-CH_2-) and two weak bending peaks at 3154 and 3450 cm^{-1} are identified the O-H groups. The spectra of BN-2, BN-3, BN-4 and BN-5 look almost similar as BN-1 because the ingredients and testing environment during fire test is the same. The difference in the formulations is in the amount of boron nitride.

(a) BN-0 showing hole and cracked char

(b) BN-1 showing cracked surface of char

(c) BN-2 showing small crack and char oxidation

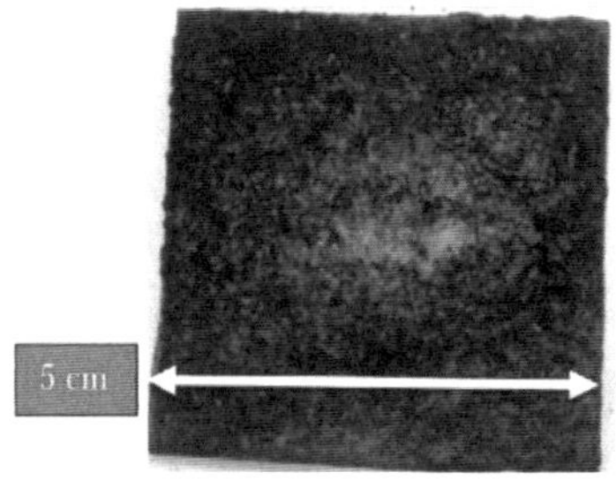

(d) BN-3 showing thick char with no crack

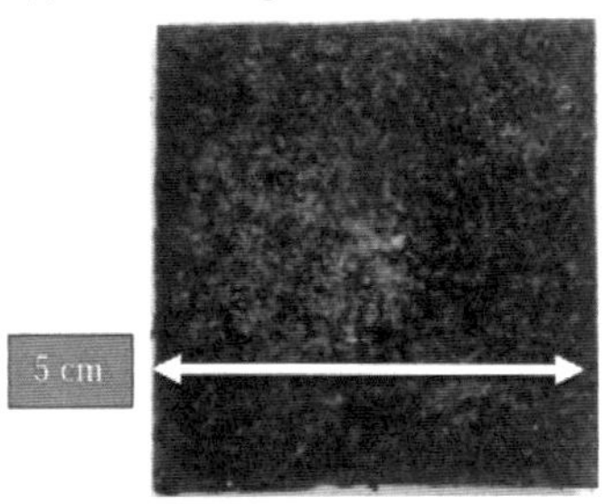

(e) BN-4 showing well expanded char with little amount of oxidation

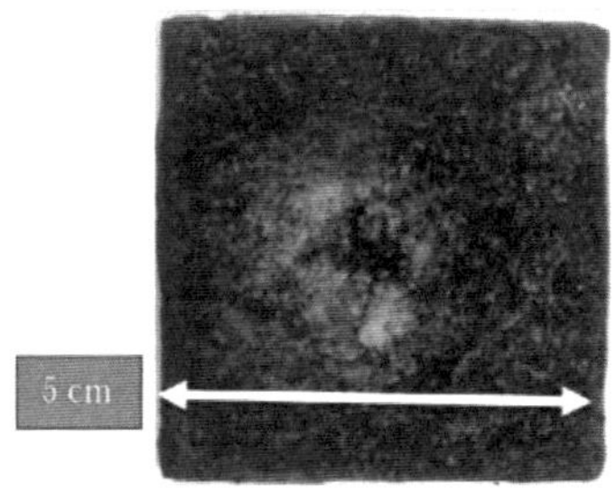

(f) BN-5 showing hole at the centre due to oxidation of char

Figure 4. Char photographs (a, b, c, d, e, f) after fire test

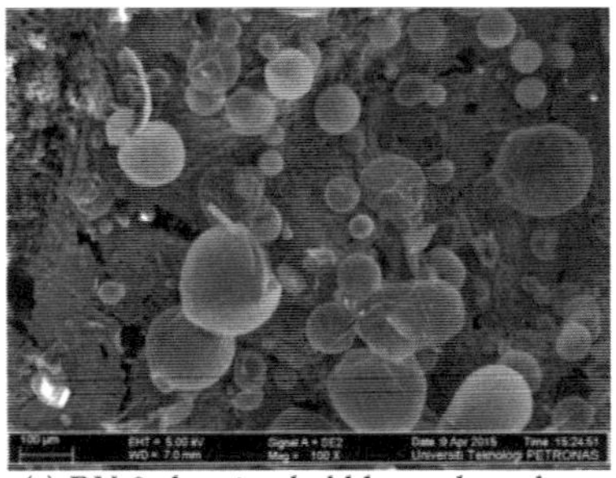

(a) BN-0 showing bubbles and cracks,

(b) BN-1 showing rough and cracked surface

(c) BN-2 with holes and cracks,

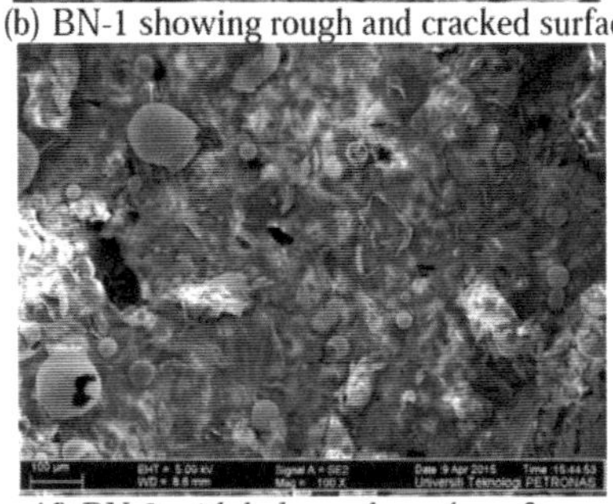

(d) BN-3 with holes and rough surface

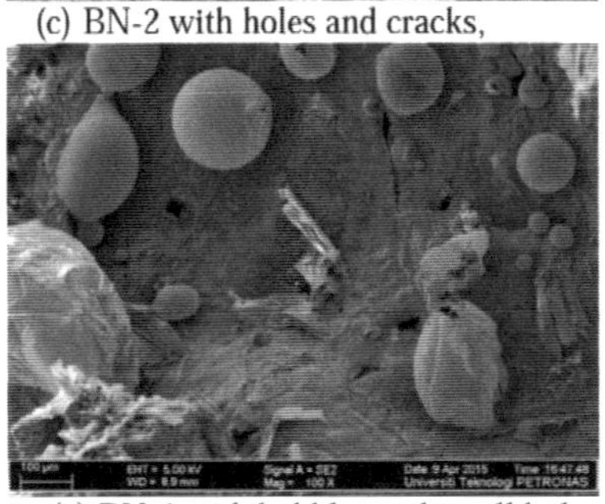

(e) BN-4 with bubbles and small hole

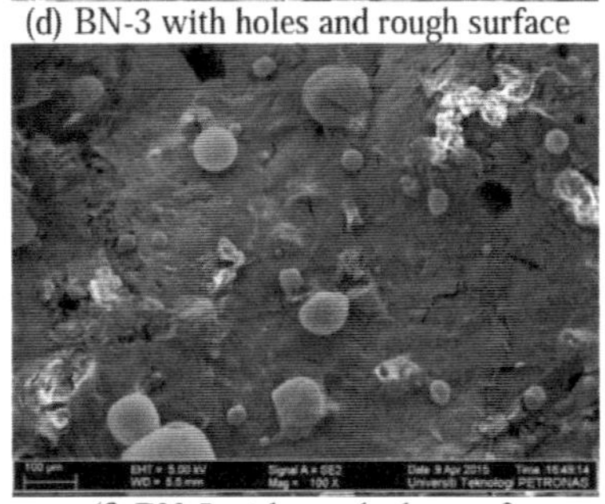

(f) BN-5 with rough char surface

Figure 5.FESEM images (a, b, c, d, e, f) showed the char morphology of intumescent fire retardant coating

In the Figure 6, the B-O-P bending motions are observed at 613, 618, 644 cm^{-1} [25, 26] and O-P-O are observed at 548, 537 and 539 cm^{-1}, respectively. The peak from 548-537 cm^{-1} in B_2O_3 is assigned to δ(O–B–O) mode. The two bands in the range of 975-824 cm^{-1} and 1089-1185 cm^{-1} are the stretching mode of ν(B–O–P) [24]. The bending vibrations observed at 1412, 1436 and 1446 cm^{-1} of CH_2 or CH_3 are due to epoxy binder, while the C≡C stretching vibration at 1630 cm^{-1} is due to epoxy binder and APP derivative in the char. In the second region around 2257-2348 cm^{-1} a strong stretching vibration characteristic of C=N is due to polyamide hardener. In the region above 2400 cm^{-1}, a single broad peak at 2495–2895 cm-1 is a weak bending peak due to the bending vibration of CH_3-CH_2 and a double peak at 3213, 3405, 3467 cm^{-1} is identified as the weak bending peaks of O-H group. Both types of peak originate from the epoxy binder. The functional group of residual char compounds is identified by FTIR spectroscopy which revealed that the char contains functional groups of borophosphate, phosphate oxide, boron oxide, -CH_3-CH_2 and aromatic compounds obtained from the reaction between EG, APP, boric acid, melamine, boron nitride and cured epoxy resin during the fire test.

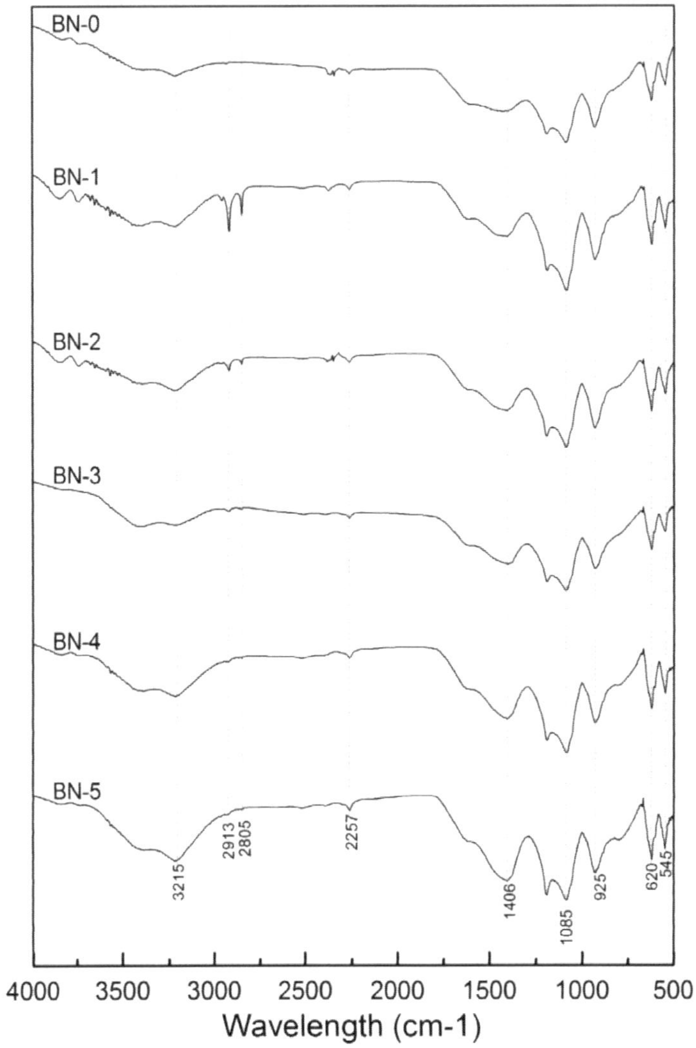

Figure 6. FTIR spectrum (a, b, c, d, e, f) showed the functional groups of residual char

3.5. Elemental analysis of char

XPS of char residual was performed using K-alpha XPS System (Thermo Scientific) with 14.0 kV of voltage and 300 W of power. XPS analysis provides detailed information for the elemental composition of charred intumescent fire retardant coating. Table 2 lists elemental contents of the char residue of BN-0, BN-4and BN-5after burned in the furnace. Their C1s spectra is shown in Figure 7 (a, b, c) respectively. The amount weight percentage of element C in char residue of BN-5surpasses BN-0 and BN-4 by 18.04% and 10.18% respectively. Higher carbon content generally indicates higher level of char accumulation and higher level of cross-link [20]. This means there are more atoms and cross-link per unit area in char residue of the coating when boron nitride is incorporated. This will results in stronger and more compact char.

The peak from C1s spectra at around 285 eV is assigned to C–H and C–C in aliphatic and aromatic species in the sample while the peak at around 288.2 eV is ascribes to C–O bonds [19]. It can be seen that C1s peak intensity of BN-5is higher compared to BN-0 and BN-4. The area of 285 eV in BN-5increases from 18.46% and 33.73% to 47.81% compared to BN-0 and BN-4respectively. The result further shows that nano-BN increases the carbon accumulation in the coating which is favorable to its fire performance.

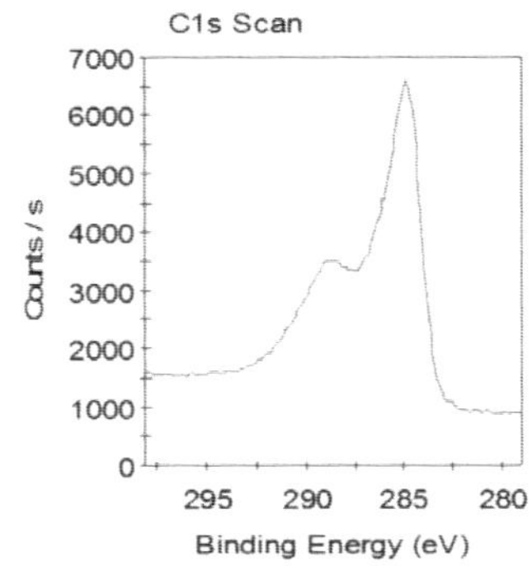

(a) C1s spectrum of BN-0

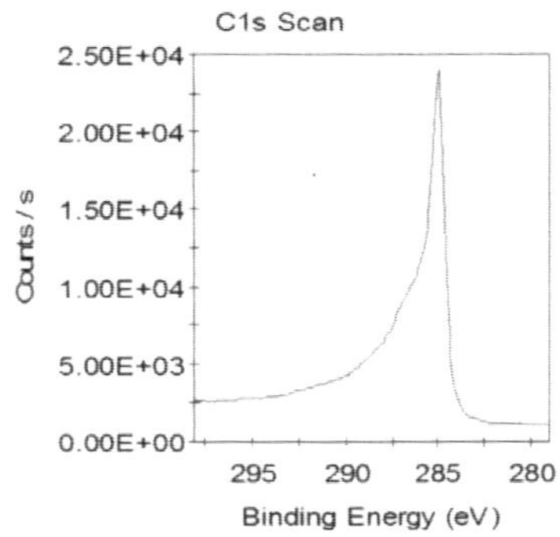

(b) C1s spectrum of BN-4

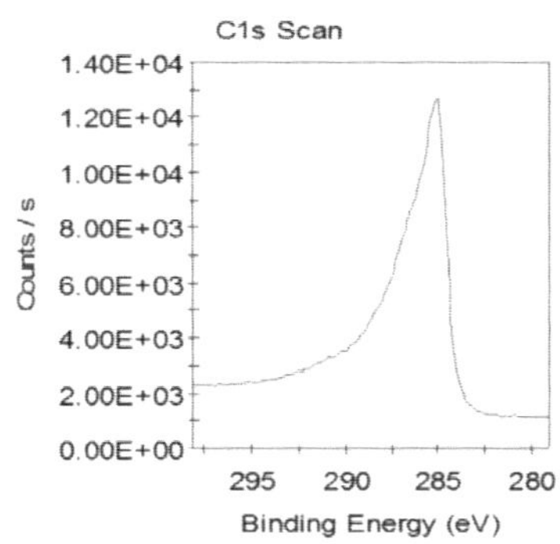

(c) C1s spectrum of BN-5

Figure 7. C1s XPS spectra (a, b, c) from char residue of intumescent fire retardant coatings

3.6. Analysis of char residual weight

TGA of intumescent fire retardant coating samples was performed on Pyris 1 TGA (Perkin Elmer) at 10ºC/min under nitrogen over temperature range of 25ºC–800ºC. The TG curves of the intumescent fire retardant coatings are shown in Figure 8. It shows the final residual weight of the coatings before and after nano-BN was incorporated. The residual weight of each formulation at 800ºC is: 34.64 wt% (BN-0), 37.85 wt% (BN-1), 39.62 wt% (BN-2), 41.38 wt% (BN-3), 44.50 wt% (BN-4) and 40.02 wt% (BN-5). The higher weight percentage of boron nitride contributed to an increase in the residual weight of BN-1 to BN-5 compared to BN-0. The TG curve shows that increasing the weight percentage of boron nitride increases the total residual weight. The residual weight of BN-4 formulation is 44.50%, which are 22.15 percent higher than that of BN-0. Based on these results, it is possible that boron nitride alongside boric acid are main components in the final protective layer developed in the char when the intumescent coating is exposed to fire.

The TG curves have four decomposition stages: stability stage (29 – 200 ºC), formation of char layer (200 – 350 ºC), char loss stage (350 – 550 ºC) and inorganic layer formation stage (550 –800 ºC). The curves suggested there was little difference between the intumescent fire retardant coating without filler and with nano-BN in the first and second stage. In third stage however, BN-4 and BN-5 with 4% and 5% nano-BN respectively shown little weight loss compared to other coatings. This suggested that nano-BN leads to the formation of a more thermally stable coating when it is incorporated at right weight percentage. In the first stage, weight loss was very small due to decomposition of volatile solvent and softening of the binder; and between 100–200 °C, boric acid degraded into metaboric acid and boron oxide [6]. During second stage, char began to form from the reaction of APP, EG and MEL. At 250 °C, APP began to decompose releasing ammonia and water and EG began to decompose releasing carbon dioxide [7]. MEL decomposed at 200–375 °C eliminating ammonia to further blows the charring layer [8] formed between APP, EG and boric acid. In the third stage, significant weight loss occurred due to further decomposition of APP into polyphosphoric and metaphosphoric acid [9]. In the final stage, only non-combusted char layer and inorganic substances were remained.

Nano fillers have superior surface-to-volume ratio but they are extremely difficult to disperse uniformly [10]. Thus, the weight percentage of nano fillers plays an important role in enhancing the intumescent fire retardant coating.

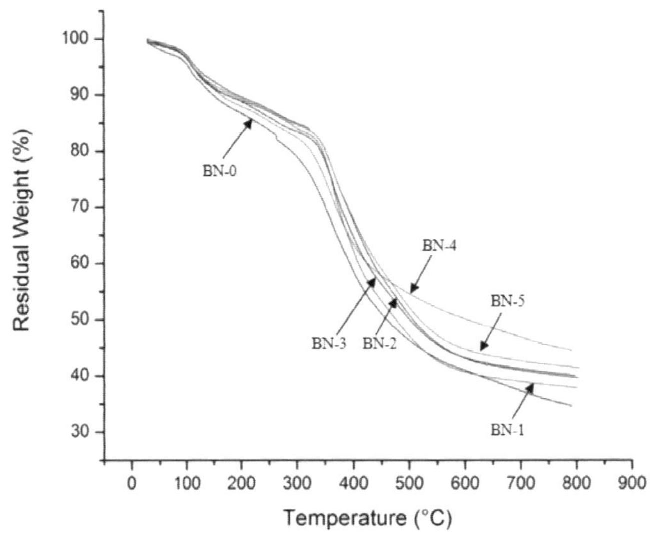

Figure 8. TGA of intumescent fire retardant coating with various percentages of boron nitride.

4. Conclusion

The effect of nano-boron nitride on thermal degradation of intumescent fire retardant coating were studied. Results show that by incorporating small quantities of nano-BN; it improves fire retardant properties of the intumescent fire retardant coating. This is evident from the measurement of substrate temperature which shows that the substrate temperature has been lowered by various degrees. The formulations BN-4 and BN-3 have shown better results on the substrate temperature, 138°C and respectively after 60 minute fire test. SEM results show improvements in the structure of residue char flakes when the wt% of boron nitride was increased. The TGA results demonstrate higher residual weight of intumescent fire retardant coating when nano-BN was introduced in the coating. As the weight percentage of nano-BN increased from 1%, 2%, 3%, 4% and 5%; the residual weight was 37.85%, 39.62%, and 41.38%, 44.50% and 40.02% respectively. This concluded that the increase of nano-BN will increase the percentage residual weight of the coating. XPS results show that nano-BN increases the carbon content and increases the C1s spectrum intensity. This will be beneficial in enhancing the strength of the char layer and improving the structure of the char layer itself. Thus, nano-BN as a filler could enhance the performance of intumescent fire retardant coating.

5. References

[1] A. R. Horrocks, "Developments in flame retardants for heat and fire resistant textiles—the role of char formation and intumescence," *Polymer Degradation and Stability,* vol. 54, no. 2-3, pp. 143-154, 1996

[2] G. Wang and J. Yang, "Influences of binder on fire protection and anticorrosion properties of intumescent fire resistive coating for steel structure," *Surface and Coatings Technology,* vol. 204, no. 8, p. 1186–1192, 2010

[3] Z. Wang, E. Han and W. Ke, "Fire-resistant effect of nanoclay on intumescent nanocomposite coatings," *Journal of Applied Polymer Science,* vol. 103, no. 3, p. 1681–1689, 2007

[4] D. Enescu, A. Frache, M. Lavaselli, O. Monticelli and F. Marino, "Novel phosphorous–nitrogen intumescent flame retardant system. Its effects on flame retardancy and thermal properties of polypropylene," *Polymer Degradation and Stability,* vol. 98, no. 1, p. 297–305, 2013

[5] S. Duquesne, S. Magnet, C. Jama and R. Delobel, "Intumescent paints: fire protective coatings for metallic substrates," *Surface and Coatings Technology,* Vols. 180-181, p. 302–307, 2004

[6] J. Gu, G. Zhang, S. Dong, Q. Zhang and J. Kong, "Study on preparation and fire-retardant mechanism analysis of intumescent flame-retardant coatings," *Surface and Coatings Technology,* vol. 201, no. 18, p. 7835–7841, 2007

[7] M. Jimenez, S. Duquesne and S. Bourbigot, "Characterization of the performance of an intumescent fire protective coating," *Surface and Coatings Technology,* vol. 201, no. 3-4, p. 979–987, 2006

[8] S. Bourbigot, M. Le Bras, R. Delobel, R. Decressain and J. Amoureux, "Synergistic effect of zeolite in an intumescence process: study of the carbonaceous structures using solid-state NMR," *Journal of the Chemical Society, Faraday Transactions,* vol. 92, pp. 149-158, 1996

[9] J. Ebdon, B. J. Hunt and P. Joseph, "Thermal degradation and flammability characteristics of some polystyrenes and polymethyl methacrylates chemically modified with silicon-containing groups," *Polymer Degradation and Stability,* vol. 83, no. 1, p. 181–185, 2004

[10] G. J. Griffin, "The Modeling of Heat Transfer across Intumescent Polymer Coatings," *Journal of Fire Sciences,* vol. 28, no. 3, pp. 249-277, 2010

[11] T. Olcese and C. Pagella, "Vitreous fillers in intumescent coatings," *Progress in Organic Coatings,* vol. 36, no. 4, p. 231–241, 1999

[12] S. Almeras, M. L. Bras, F. Poutch, S. Bourbigot, G. Marosi and P. Anna, "Effect of fillers on fire retardancy of intumescent polypropylene blends," *Macromolecular Symposia,* vol. 198, no. 1, p. 435–448, 2003

[13] G. Bertelli, G. Camino, E. Marchetti, L. Costa, E. Casorati and R. Locatelli, "Parameters affecting fire retardant effectiveness in intumescent systems," *Polymer Degradation and Stability,* vol. 25, no. 2-4, p. 277–292, 1989

[14] Z. Wang, E. Han and W. Ke, "Influence of nano-LDHs on char formation and fire-resistant properties of flame-retardant coating," *Progress in Organic Coatings,* vol. 53, no. 1, p. 29–37, 2005

[15] J. Green, "Mechanisms for Flame Retardancy and Smoke suppression -A Review," *Journal of Fire Sciences,* vol. 14, no. 6, pp. 426-442, 1996

[16] S. Ullah, F. Ahmad and P. S. M. Megat Yusoff, "Effect of boric acid and melamine on the intumescent fire-retardant coating composition for the fire protection of structural steel substrates," *Journal of Applied Polymer Science,* vol. 128, no. 5, p. 2983–2993, 2013

[17] Y. Dong and G. Wang, "Influence of nano-boron nitride on fire protection of waterborne fire-resistive coatings," *Journal of Coatings Technology and Research,* vol. 11, no. 2, pp. 265-272, 2014

[18] C. Zhi, Y. Bando, T. Terao, C. Tang, H. Kuwahara and D. Golberg, "Towards Thermoconductive, Electrically Insulating Polymeric Composites with Boron Nitride Nanotubes as Fillers," *Advanced Functional Materials,* vol. 19, no. 12, p. 1857–1862, 2009

[19] J. Eichler and C. Lesniak, "Boron nitride (BN) and BN composites for high-temperature applications," *Journal of the European Ceramic Society,* vol. 28, no. 5, p. 1105–1109, 2008

[20] Y. Dong, G. Wang and Q. Su, "Influence of nano-boron nitride on anti-aging property of waterborne fire-resistive coatings," *Journal of Coatings Technology and Research,* vol. 11, no. 5, pp. 805-815, 2014

[21] M. Jimenez, S. Duquesne and S. Bourbigot, "Intumescent fire protective coating: Toward a better understanding of their mechanism of action," *Thermochimica Acta,* vol. 449, no. 1-2, p. 16–26, 2006

[22] M. Jimenez, S. Duquesne and S. Bourbigot, "Kinetic analysis of the thermal degradation of an epoxy-based intumescent coating," *Polymer Degradation and Stability,* vol. 94, no. 3, p. 404–409, 2009

Development and characterization of cobalt based nano-structured super hydrophobic coating

H. Mohsin, U. Sultan, Y. F. Joya, S. Ahmed, M. S. Awan[1] and S. N. Arshad[2]
Faculty of Materials and Chemical Engineering, Ghulam Ishaq Khan Institute of Engineering Sciences & Technology, Topi-23640, Khyber Pakhtunkhaw, Pakistan
[1]Ibn-e-Sina Institute of Technology H-11/4 Islamabad, Pakistan
[2]Department of Chemistry, Syed Babar Ali School of Science & Engineering, Lahore University of Management Sciences, Lahore, Punjab, Pakistan

E-mail: yasir.joya@giki.edu.pk

Abstract. A super hydrophobic coating on the surface of glass substrate has been developed using chemical bath deposition (CBD) process. A water contact angle (WCA) greater than 150° has been achieved. Cobalt Chloride ($CoCl_2$) has been used as the main precursor to investigate optimum composition and high superhydrophobicity. The water droplet has been observed to slide with a sliding angle less than ~3°. This effect is particularly due to the surface morphology (roughness) and low surface energy that causes water droplet to form a large contact angle thus allowing the surface to show water-repellent properties. Deposition time is the primary parameter affecting the coating properties and a different WCA value has been observed by increasing time. Scanning Electron Microscopy (FE-SEM) images show the presence of a nano flower-like morphology that helps in imparting superhydrophobic behavior. Energy Dispersive X-ray Spectroscopy (EDX) indicate the coating to be composed of cobalt as the main constituent. Contact Angle Measurement confirms the contact angle value to be greater than 170°.

1. Introduction

Superhydrophobicity is defined as the state in which water shows a contact angle greater than 150° with the surface. Non-wetting surfaces having high contact and low sliding angles have received significant attention in recent years. Properties such as self-cleaning [1], corrosion resistance [2-3], resistance to frost formation and accumulation [4] and oleophobicity [5] have been studied. Naturally, plants such as lotus have been found to exhibit superhydrophobic behavior [6].

High surface roughness and low surface energy [7-9] are the requirements for a surface to be superhydrophobic. The surface roughness is provided by the morphology of the coating deposited on the surface while low surface energy is due to the attachment of certain molecules onto the tips of the surface. Cassie/Baxter model can be used to determine the water contact angle (WCA) with the surface [9]. This is given by (1) as:

$$\cos\theta_c = \varphi_s(\cos\theta_e) + (1 - \varphi_s)\cos\theta_x \quad (1)$$

Content from this work may be used under the terms of the Creative Commons Attribution 3.0 licence. Any further distribution of this work must maintain attribution to the author(s) and the title of the work, journal citation and DOI.
Published under licence by IOP Publishing Ltd

Where, θ_c is the WCA, φ_s (less than 1) is the fraction of area covered by solid at the top of the surface, θ_e represents the WCA with the solid, $(1 - \varphi_s)$ represents the fraction of area covered by air gaps at the surface and θ_x shows the WCA with the entrapped air (within the surface grooves).θ_x is normally taken to be 180° due to which the contact angle increases. In this case, the water drop remains in contact with the tops of the surface leading to the flow of air beneath.

To develop superhydrophobic coatings, both bottom up (sol-gel [10], layer by layer deposition [11], chemical vapor deposition [12]) and top down (plasma treatment of surfaces [13], templating [14], photolithography [15]) approaches can be used.

Cobalt based (cobalt hydroxide and cobalt oxide) coatings have been found to exhibit water-repellent properties. Saravani et al. [10] used cobalt chloride as the precursor for producing a superhydrophobic surface based on cobalt oxide (Co_3O_4) nanoparticles and reported a WCA of 153.45° along with a sliding angle of 2.5°.In a comparative study, Ho et al. [16] deposited layered cobalt carbonate hydroxide on a substrate using different precursors. The cobalt oxide films synthesized from these precursor salts were in shape of straight acicular nanorods, bending acicular nanorods, nanosheets, and net-shaped nanosheets. Chemical bath deposition (CBD) is an economical route for depositing oxide based coatings [17].A WCA of 178° has been reported by Zhou et al. [18] using CBD by developing cobalt hydroxide based coating. Joo et al. [19] synthesized cobalt oxide based coating with a thickness between 437-843nm.

In this study, cobalt based coatings have been developed by CBD using cobalt chloride as the main precursor. Moreover, cobalt nitrate and cobalt sulfate salts have also been used as precursors to see the difference in surface morphology of the coating. Stearic acid treatment has been used to impart superhydrophobic behavior. Deposition time has been varied to see the effect on WCA values.

2. Experimental

2.1 Materials

Cobalt chloride ($CoCl_2$), cobalt nitrate hexahydrate ($Co(NO_3)_2.6H_2O$) and cobalt sulfate heptahydrate ($CoSO_4.7H_2O$) were procured from Sigma-Aldrich. Ethanol was procured from Scharlau and stearic acid was procured from BDH.

2.2 Development of the superhydrophobic coating

Cobalt based coatings were developed on a glass slide using CBD. The glass slide was first cleaned by rinsing with ethanol and drying in air. The solution for CBD was prepared by mixing known amount of $CoCl_2$salt (0.1 mol/dm^3) and CH_4N_2O(20 g) in 100 ml water. $CoSO_4.7H_2O$ and $Co(NO_3)_2.6H_2O$were also used as precursors. The solution was transferred to a sealed glass bottle (glass slide was immersed in the solution)and placed on a hot plate at 60°C for varying amount of time. The substrate was then removed, rinsed with ethanol and dried in air at room temperature. The dried coated glass slide was then immersed in an ethanolic solution of stearic acid ($C_{18}H_{36}O_2$) for 10 min and dried in air.

2.3 SEM analysis

The surface morphology of the coating was observed using a field emission scanning electron microscope (FEG-SEM).

2.4 EDX analysis

The composition of the coating was determined using energy dispersive X-ray spectroscopy (EDX) with a FEG-SEM equipment.

2.5 Contact angle measurements

The water contact angle with the surface was measured by sessile drop mode of Drop Shape Analyzer (KRUSS DSA-30).

3. Results and discussion

CBD process is highly favorable due to its low temperature and ease of operation. It is a two-step process involving nucleation and growth of the nuclei. The coating developed on glass substrate by chemical bath system (precursor plus activating agent i.e. urea) does not show superhydrophobic behavior unless further treated. The nuclei grow in the form of a flower composed of very fine nanopinswhen precursor is $CoCl_2$ as shown in Figure 1 in comparison to other precursor salts.

Figure 1. SEM image of Co based coating using (a), (b) $CoCl_2$, (c) $CoSO_4.7H_2O$ and (d) $Co(NO_3)_2.6H_2O$ precursor.

The WCA at this stage depends on the deposition time. A deposition time as high as 22 h causes the water droplet to spread leading to a contact angle value ~0°. At lower deposition times, the drop shows hydrophilic to hydrophobic behavior but tends to spread continuously making contact angle measurements extremely difficult as compared to an uncoated glass surface that exhibits hydrophilicity (WCA equal to ~55°) as shown in Figure 2.

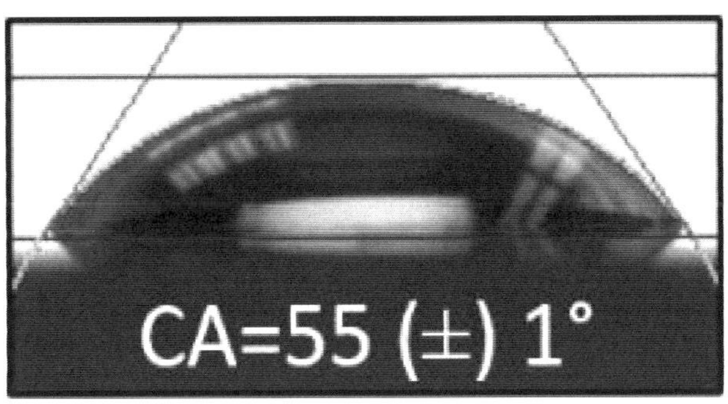

Figure 2. Water contact angle on an uncoated glass slide is ~55°

At this stage, one requirement of superhydrophobicity i.e. surface roughness is fulfilled. For satisfying the second condition, necessary treatment is required to ensure the lowering of surface energy. Treating the surface with ethanolic solution of stearic acid makes the coating superhydrophobic. Stearic acid molecule like all the fatty acid molecules is made up of two parts: a polar head which is hydrophilic in nature and a non-polar tail that has a hydrophobic nature. These molecules arrange themselves in such a way that the hydrophilic head attaches itself onto the surface features (nanopins in this case) while the hydrophobic tail is aligned away from the surface. These non-polar tails effectively decrease the surface energy so that when a water droplet comes into contact with the surface it forms a sphere and a high WCA is obtained.

Figure 3. SEM image of Co based superhydrophobic coating using 0.1 M $CoCl_2$ at 22 h

Figure 3 shows the morphology of a superhydrophobic coating on glass (the pins become finer – rod like at the bottom and very fine at the tip). The fineness of the coating features is due the stearic acid treatment that deposits itself as a monolayer over the pins thus lowering the surface energy and allowing the surface to show water-repellent properties. A WCA value of 171.7° has been achieved at a deposition time of 22 h using 0.1 mol/dm^3 $CoCl_2$ as precursor. The water drop tends to roll off the surface even at very slight tilting of the glass slide. This indicates the sliding angle to be less than ~3°. The EDX (area scan) results indicate the coating to be composed of cobalt and oxygen as the main constituents with cobalt having the highest percentage (80 wt% and 52.93 at%) as shown in Figure 4. Oxygen also has a relatively higher proportion because of the stearic acid molecule that is rich in oxygen. Chlorine is also observed in the EDX scan that is indicative of the precursor ($CoCl_2$).

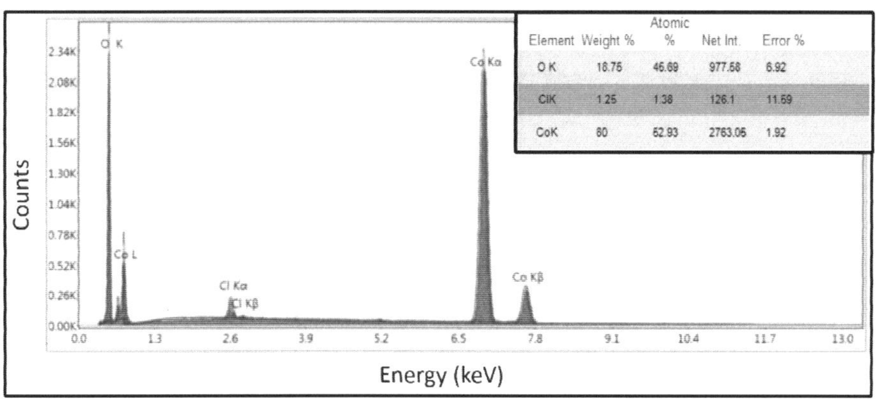

Figure 4. EDX scan indicating the relative proportions of the constituents in the coating

Deposition time is an important parameter that affects the coating properties. The WCA values tend to increase as the deposition time is increased and depends on the surface morphology (see Figure 5). This is particularly due to the increase in surface roughness i.e. formation of greater number of pins with time. A coating developed at 6 h has very few pins and the dominant morphology is bead type (interconnected flakes) but the coating shows superhydrophobic behavior with a WCA of 164.2°. On increasing the deposition time, pins begin to form and grow in the form of a flower-like arrangement. At 8 h, the coating is composed of maximum number of pins rather than beads but these pins appear to be in the initial stages of growth. Further increase in deposition time leads to further growth of the pins thus leading to an increase in contact angle values. The WCA does not increase very sharply with time as only a little increase in values is observed. Superhydrophobicity has been observed at a time as less as 6 h and the sliding angle in all cases was observed to be the same i.e. < ~3°. Increase in deposition time also tends to improve adhesion of the coating as observed by finger nail test. Figure 5 shows the effect of increase in deposition time on coating (stearic acid treated) morphology.

Figure 5. Evolution of superhydrophobic coating morphology with deposition time; (a) 6 h, (b) 8 h, (c) 10 h and (d) 12 h (insets show the images of water droplet on respective glass slide)

4. Conclusions

Superhydrophobic cobalt based coatings on glass exhibited a very fine nanopin flower-like morphology. A water contact angle of 171.7° was successfully achieved by the application of stearic acid. Stearic acid makes the pins very fine at the tip with rod like features at the bottom. The water droplet tends to roll off the surface at a sliding angle < ~3°. Increasing the deposition time tends to increase the contact angle due to increase in surface roughness by the formation of nanopins and their growth. EDX analysis confirms the coating to be composed of cobalt as the main constituent.

5. Acknowledgements

The authors are extremely grateful to Dr. Salman Noshear Arshad (Lahore University of Management Sciences, Pakistan) for his assistance in conducting FE-SEM analysis and contact angle measurements. We are also thankful to Dr. M. Saifullah Awan (ISIT, Pakistan) for providing the FE-SEM images and EDX analysis of research samples.

6. References

[1] Nishimoto S and Bhushan B 2013 Bioinspired self-cleaning surfaces with superhydrophobicity, superoleophobicity, and superhydrophilicity *RSC Adv.* 3 671–90
[2] Liu C, Su F, Liang J and Huang P 2014 Facile fabrication of superhydrophobic cerium coating

with micro-nano flower-like structure and excellent corrosion resistance *Surface & Coatings Technology* 258580–6
[3] Li W and Kang Z 2014 Fabrication of corrosion resistant superhydrophobic surface with self-cleaning property on magnesium alloy and its mechanical stability *Surface & Coatings Technology* 253 205–13.
[4] Xu Q, Li J, Tian J, Zhu J and Gao X 2014 Energy-effective frost-free coatings based on superhydrophobic aligned nanocones *ACS Applied Materials and Interfaces* 6 8976–80
[5] Lakshmi R V, Bharathidasan T, Bera P and Basu B J 2012 Fabrication of superhydrophobic and oleophobic sol–gel nanocomposite coating *Surface & Coatings Technology* 206 3888–94
[6] Barthlott, W and Neinhuis C 1997 Purity of the sacred lotus, or escape from contamination in biological surfaces *Planta.* 202 1–8
[7] Nosonovsky M and Bhushan B 2009 Superhydrophobic surfaces and emerging applications: nonadhesion, energy, green engineering *Current Opinion in Colloid & Interface Science* 14 270–80
[8] Zhao X D, Xu G Q and Liu X Y 2012 *Superhydrophobic Surfaces: Beyond Lotus Effect* (*Bioinspiration: From Nano to Micro Scales)* ed X Y Liu (New York: Springer) pp 331–78
[9] Roach P, Shirtcliffe N J and Newton M I 2008 Progess in superhydrophobic surface development *Soft Matter* 4 224–40
[10] Moulapanah-Konaroi M, Ahmad M A and Saravani H 2013 Fabrication of superhydrophobic surface by Co_3O_4 nanoparticles *Indian Journal of Physics* 87 211–5
[11] Zhai L, Cebeci F C, Cohen R E and Rubner M F 2004 Stable superhydrophobic coatings from polyelectrolyte multilayers *Nano Lett.* 4 1349–53
[12] Ma M, Mao Y, Gupta M, Gleason K K and Rutledge G C 2005 Superhydrophobic fabrics produced by electrospinning and chemical vapor deposition *Macromolecules* 38 9742–8
[13] Fresnais J, Benyahia L and Poncin-Epaillard F 2006 Dynamic (de)wetting properties of superhydrophobic plasma-treated polyethylene surfaces *Surf. Interface Anal.* 38 144–9
[14] Yoshimitsu Z, Nakajima A, Watanabe T and Hashimoto K 2002 Effects of surface structure on the hydrophobicity and sliding behavior of water droplets *Langmuir* 18 5818–22
[15] Fürstner R, Barthlott W, Neinhuis C and Walzel P 2005 Wetting and self-cleaning properties of artificial superhydrophobic surfaces *Langmuir* 21 956–61
[16] Kung C, Lin C, Li T, Vittal R and Ho K 2011 Synthesis of Co_3O_4 thin films by chemical bath deposition in the presence of different anions and application to H_2O_2 sensing *Procedia Engineering* 25 847–50
[17] Pawara S M, Pawara B S, Kima J H, Joob O and Lokhandeb C D 2011 Recent status of chemical bath deposited metal chalcogenide and metal oxide thin films *Current Applied Physics*11 117–61
[18] Hosono E, Fujihara S, Honma I and Zhou H 2005 Superhydrophobic perpendicular nanopin film by the bottom-up process *Journal of the American Chemical Society* 12 13458–9
[19] Kandalkar S G, Gunjakar J L, Lokhande C D and Joo O 2009 Synthesis of cobalt oxide interconnected flacks and nano-worms structures using low temperature chemical bath deposition *Journal of Alloys and Compounds* 478 594–8

Experimental thermal transport evolution of silane activated nano-clay reinforced styrene butadiene elastomeric nanocomposites

S. S. Iqbal, N. Iqbal[1], T. Jamil, A. Bashir[2] and M. Shahid[2]
Department of Polymer Engineering and Technology, University of the Punjab, Lahore, Pakistan
[1]Center of Undergraduate Studies, University of the Punjab, Lahore, Pakistan
[2]School of Chemical and Materials Engineering, National University of Sciences and Technology, Islamabad, Pakistan

E-mail: sadiasagariqbal.pu@gmail.com

Abstract. In this study, silane activated nanoclay was reinforced in styrene butadiene rubber (SBR) to enhance the thermal resistance/stability and mechanical properties of SBR. silane activated nanoclay with variant concentrations was impregnated in the rubber matrix to fabricate polymer nanocomposites under control processing conditions. Experimental thermal transport, thermal oxidation, phase transition study, and mechanical properties of the nanocomposite specimens were carried out. Thermal insulation, thermal stability, and heat flow response were remarkably enhanced with the addition of nanokaolinite in the polymer matrix. Phase transition temperatures, their corresponding enthalpies, tensile strength, elastic modulus, elongation at break and hardness of the rubber composites were positively influenced with the filler incorporation into the host matrix. The Even dispersion of nanoreinforcements, morphological and compositional analyses of the thermal transport tested specimens were performed using scanning electron microscopy and energy dispersive spectroscopy, respectively.

1. Introduction

The clays study has become a prominent topic of the present era due to their easy and abundant availability, low cost, and environment friendly nature. These clays have multidisciplinary applications, i.e., construction, polymer, medicine, ceramic, and metal processing industries [1-3]. Montmorillonite, kaolinite, bentonite, etc were used in polymer matrices to enhance their thermal, physical, mechanical and fire retardant characteristics. Particle size of clay is an important parameter that influences the thermomechanical character of the base polymer. Nanoclays have several advantages over micro size clays due to their large surface area, nanoscale interaction with the polymeric chains and low density [4-6]. Kaolinite is an aluminosilicate mineral clay having layered structure used in paints, medicines, rubbers, ceramics, etc. Nanokaolinite remarkably influences the physical and chemical characteristics of the base polymer matrix. Filler-matrix compatibility and its uniform dispersion within the host polymer are the major reinforcing parameters that drastically change the characteristics of the fabricated composites. Functionalization of the incorporated nanofillers by using appropriate chemical method is one of the procedure viz. efficiently used in academia and industry to enhance filler to matrix compatibility. Silane functionalization route in

Content from this work may be used under the terms of the Creative Commons Attribution 3.0 licence. Any further distribution of this work must maintain attribution to the author(s) and the title of the work, journal citation and DOI.
Published under licence by IOP Publishing Ltd

which adequate silane coupling agent is used to attach silane moiety is effective, easy, economical and environment friendly method to functionalize the incorporated reinforcements compared to other established chemical processes. Nanokaolinite (NK) with low density, high aspect ratio, large surface area and soft nature has affirmed the promising reinforcing effects if uniformly dispersed in the rubber matrices, i.e., ethylene propylene diene monomer rubber, styrene butadiene rubber (SBR), acrylonitrile butadiene rubber, neoprene rubber, etc [7-10]. SBR is a synthetic rubber, synthesized through emulsion or solution polymerization. It has good abrasion, aging and impact strength. The main application area of this rubber is tire, shoes and gasket industries. Usually carbon black is reinforced in the SBR rubber to attain the required level of mechanical strength but thermal resistance suffers with the incorporation of this filler [11-14]. During the tire operation, enormous heat is generated due to road friction which may degrade the mechanical characteristics of the rubber [15-18]. So, the filler that improves thermal endurance, thermal resistance and mechanical strength, synchronously is required for SBR to perform better during its usage [19].

Thermal conductivity, thermal impedance, thermal degradation, differential thermal analysis, differential scanning calorimetric study and mechanical properties are evaluated for NK/SBR composites in this novel research. In addition, the effect of nanokaolinite loadings on the glass transition, crystallization, melting temperatures and their associated specific enthalpies are also executed herein.

2. Experimental

2.1. Materials

Emulsified Styrene butadiene rubber (INDOPOL 1502) was bought from Evergreen Global Pte. Singapore. Silane coupling agent (SCA), 3-aminopropyl-trimethoxysilane (APTMS) and cetrimonium bromide (CTAB) were purchased from Sigma Aldrich, Germany. Kaolinite, nanosilica, sulphur, zinc oxide and stearic acid were purchased from Merck, Germany. Aromatic oil and wax were received from International petrochemicals (Pvt) Ltd, Pakistan. Mercaptobenzthiazole Disulphide (MBTS) and Cyclohexyl Benzthiazyl Sulphenamide (HBS) were supplied by Dalian Richon Chemical Co. Ltd, China.

2.2. Synthesis of nanokaolinite with silane coupling agent

A silane coupling agent (SCA), 3-aminopropyl-trimethoxysilane (APTMS) along with cetrimonium bromide (CTAB) was used for the surface treatment to modify the layered silicate mineral clay (kaolinite). A specific amount of the unmodified kaolinite was added in ultra pure water in a ratio 10:100 and the mixture was sonicated in ultra-sonication bath at 70°C and 40 KHz frequency for 30min. After that, 1mass% CTAB was introduced in the preset mixture followed by further sonication under similar conditions for next 30min. Subsequently, 5 volume % SCA was added in the mixture and sonicated it for next three hours. Consequently, magnetic stirring followed at 1000rpm for an hour with 90°C. The sonicated solution was diluted five times to maintain its pH at 7. Centrifuge (Centurion Scientific C2 series) at 4000rpm was used to recover the nano alumino-silicate clay. The silane functionalized nanokaolinite (NK) sample was dried for 12h at 80°C to eradicate moisture from the specimen.

2.3 Fabrication of NK/SBR nanocomposites

A preset amount of nano-reinforcement (nanosilica), lubricant (aromatic oil + wax), surface modifier for nanoclay (stearic acid) and four progressive loadings (0–15wt %) of NK were inserted in the dispersion kneader chamber along with the SBR matrix at 110°C for 30 minutes to disperse nanofillers properly in the host matrix. Cross linker (sulphur), primary accelerator (MBTS), secondary accelerator (HBS), and activator (zinc oxide) were incorporated into the rubber matrix in the post mixing stage on a two roller mixing mill at 60°C and 30rpm roller's speed for twenty minutes to disperse the reinforcements and processing aids homogeneously within the polymer matrix [20]. Table 1 illustrates the basic composition of SBR composite (SK1). The four silane functionalized nanoclay concentrations, i.e., 0, 5, 10, and 15 wt% were impregnated in the rubber matrix and the four rubber nano compositions are nominated as SK1, SK2, SK3, and SK4, correspondingly. The base formulation of the polymer nanocomposite is SK1 and is illustrated in Table 1.

Table 1. Base formulation of SBR nanocomposite

SBR (wt%)	Nano Silica (wt%)	ZnO (wt%)	Stearic Acid (wt%)	Aromatic Oil (wt%)	Sulphur (wt%)	[a]HBS (wt%)	[b]MBTS (wt%)	Wax (wt%)
100	40	5	2.5	10	2.5	2	2.5	2.5

[a] HBS:Cyclohexyl Benzthiazyl Sulphenamide
[b]MBTS: Mercaptobenzthiazole Disulphide

2.1. Characterization
Scanning electron microscopy (SEM, JSM 6940A, Jeol, Japan) along with the energy dispersive spectroscopy was used to characterize the even dispersion of nanoclay, surface morphology of the post thermal conductivity/impedance tested nanocomposite specimens and their composition analysis.
Thermal conductivity (λ_{SN}) of the polymer nanocomposite specimens were carried out using ASTM E1225-99. Experimental setup design to evaluate λ_{SN} of the composite specimens in the temperature range 323K to 523K is depicted in Figure 1.

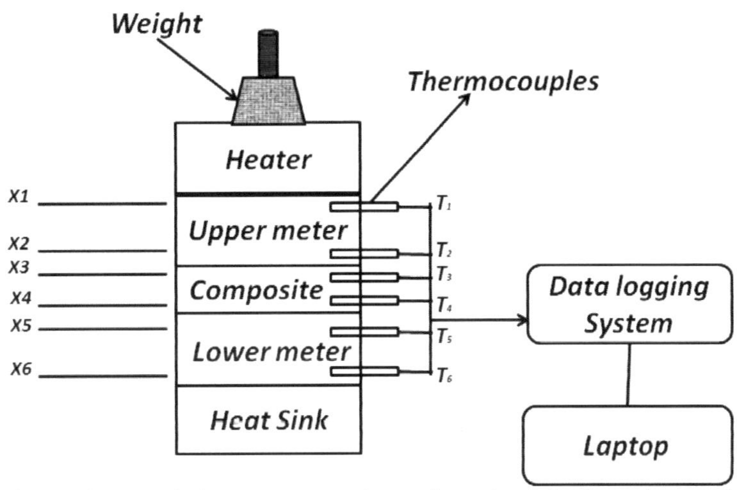

Figure 1. Experimental setup design to execute thermal conductivity/impedance of the fabricated rubber nanocomposites

Schematic illustration of the comparative guarded longitudinal heat flow system shows the feasible locations of K-type thermocouples, heating source, water heat sink, temperature data logger, copper upper and lower meter bars and laptop [21]. Time–temperature profiles of all thermocouples located at specific positions were displayed on the laptop display using Squirrel OQ610 series temperature data logger. Thermal conductivity of the composite specimens was measured at different input temperatures using equation 1.

Thermal conductivity of specimen= $\lambda_{SN} = ((Q'_T + Q'_B)(d_4 - d_3)/2(v_4 - v_3))$ (1)

Heat flow at the top bar = $Q'_T = \lambda_c * (v_2 - v_1)/(d_2 - d_1)$

Heat flow at the bottom bar $= Q'_B = \lambda_c * (v_6 - v_5)/(d_6 - d_5)$

v_1, v_2, v_3, v_4, v_5, and v_6 are the temperatures of the six thermocouples and d_1, d_2, d_3, d_4, d_5, and d_6 are their positions. λ_c is thermal conductivity of the copper bar.

Thermal impedance of a material is its ability to resist thermal/temperature fluctuations in a variable heating environment. For thermal impedance four thermocouples have been used instead of six in afore mentioned thermal conductivity design. Thermal impedance of the nanocomposite specimens were measured according to the ASTM D5470-03 and by following equation 2.

$$\text{Thermal impedance } (m^2\text{-K/W}) = \zeta = \lambda_c (V_A - V_D)*(A/\Omega) \quad (2)$$

Ω = V (Voltage) x I (Current)

Temperature of upper bar surface in contact with the sample (K) = $V_a = v_2 - ((v_1 - v_2)*(d_b/d_a))$
Temperature of lower bar surface in contact with the sample (K) = $V_d = v_3 - ((v_3 - v_4)*(d_d/d_c))$

v_1 (K) = Upper temperature of upper bar
v_2 (K) = lower temperature of upper bar
v_3 (K) = Upper temperature of lower bar
v_4 (K) = Lower temperature of lower bar
d_a (m) = Distance between T_1 to T_2
d_b (m) = Distance between T_2 to S Sample
d_c (m) = Distance between T_3 to T_4
d_d (m) = Distance between Sample to T_3

Thermal degradation and endothermic/exothermic capability of the rubber nanocomposites were carried out using Perkin Elmer Diamond TG/DTA, Japan. Heating rate and temperature range during the thermogravimetric and differential thermal analyses was 10°C/min and 25-850°C, respectively. Perkin Elmer Diamond DSC, Japan was used to analyze the heat flow response of the polymer composites from -75°C-450°C to determine physical transformations such as phase transitions (glass, crystallization, and melting), and their specific enthalpies with a heating rate (10°C/minute) of polymer nanocomposites. Both of these studies were performed in the air atmosphere. Tensile strength, elongation at break, and modulus of elasticity of the polymer nanocomposite specimens were executed using universal tensile testing machine (AG-20KNXD Plus, Shimadzu) according to the ASTM standard D412-98A. Shore A hardness of polymer nanocomposites were measured from Torsee, Tokyo testing machine. Scanning electron microscopy (SEM, JSM 6940A, Jeol, Japan) along with the energy dispersive spectroscopy was used to analyze the surface morphology of the post thermal conductivity tested specimens and compositional analysis of the rubber nanocomposite. Fourier Transformation In-fared (FT-IR) spectroscopy in which Infrared spectrum measures the quantity of radiation absorbed as a function of its frequency. FT-IR (Perkin-Elmer Diamond 1000) was used to analyze the changes taken place on the nanokaolinite surface due to the silane treatment in the frequency range 400–4000 cm^{-1}. 5200 Jeol Atomic force microscopy, Japan was used to examine the surface topography of the functionalized nanokaolinite.

3. Results and discussion

3.1. Confirmation of Silane functionalization

Figure 2 shows the FTIR spectra of unmodified/modified kaolinite. The peaks at 2922.91, 2849.14, and 1463.25cm^{-1} indentifies the presence of trimethoxy silane moiety at the surface of nanokaolinite Experimental results reveal that the functionalization of nanokaolinite due to the attachment of silane moiety in the presence of APTMS. The average particle size observed in Figure 3a is <1µm for the unmodified kaolinite. The synthesis and modification process has successfully reduced the particle size of the clay up to 1000 times. The average particle size noticed in Figure 3(b, c) for modified kaolinite is 15nm. Another interesting thing in the morphological analysis of the nanokaolinite is that all particles have approximately similar spherical shape and size. Another advantage of this route is that both the nano-particle synthesis and modification have taken place synchronously. The EDS analysis in Figure 3d, illustrates the presence of Al (21.05), Si (32.86), and O (46.09) present in nanokaolinite ($Al_2Si_2O_5(OH)_4$) that means synthesized nanoparticles maintain their original composition (unmodified kaolinite). AFM images clear the particle size fineness (1-5nm) of modified

IOP Conf. Series: Materials Science and Engineering **146** (2016) 012039 doi:10.1088/1757-899X/146/1/012039

nanokaolinite in Figure 4(a, b). The approximate average particle size observed in AFM and SEM images is less than 10nm. The measured average particle size of the fabricated modified nanokaolinite is around 5nm using SEM software. The morphology of the synthesized modified nanokaolinite is spherical.

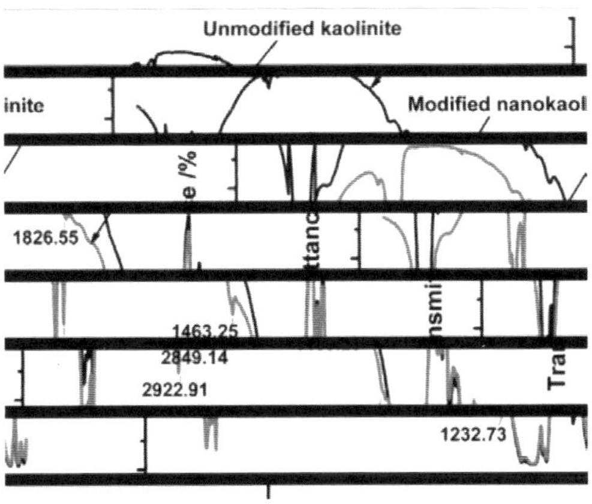

Figure 2. FTIR Spectra of unmodified/ modified kaolinite

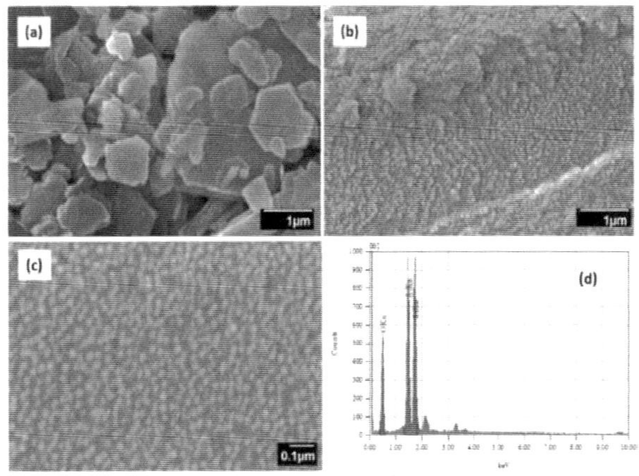

Figure 3. Surface morphology of unmodified (a) and modified (b, c) kaolinite along with its compositional analysis (d)

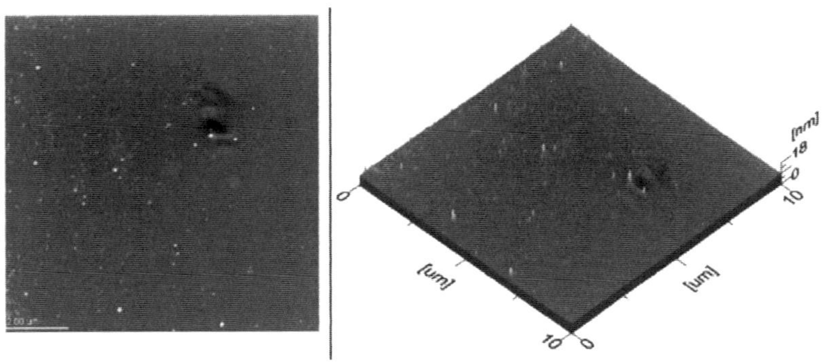

Figure 4. Surface topography of modified Kaolinite

3.2. Thermal conductivity and thermal impedance analyses

Thermal conductivity contours of the rubber nanocomposites in the temperature range 323K to 523K are illustrated in Figure 5. Initially, thermal conduction through the SK1 composite is minimum compared to the clay nanocomposites. A drastic decline is observed in the λ_{SN} values of the composite in the temperature range 333-373K, after that each rubber nanocomposite has established a separate level of λ_{SN} in the proceeding TR. The SK4 nanocomposite having 15 wt% NK, has the least level of λ_{SN} due to enormous heat quenching by the nanoclay during its phase changes and water evaporation. It elucidates that λ_{SN} is diminished with the progressive addition of NK in the rubber matrix. The utmost NK loaded nanocomposite formulation has 41%, 22%, and 41% less λ_s compared to 0 wt% concentrated SK1 composite specimen at 373, 423, and 473K, respectively due to the high surface area, excellent heat absorbing capability, and nanoscale interaction of NK with the polymer matrix[22-25].

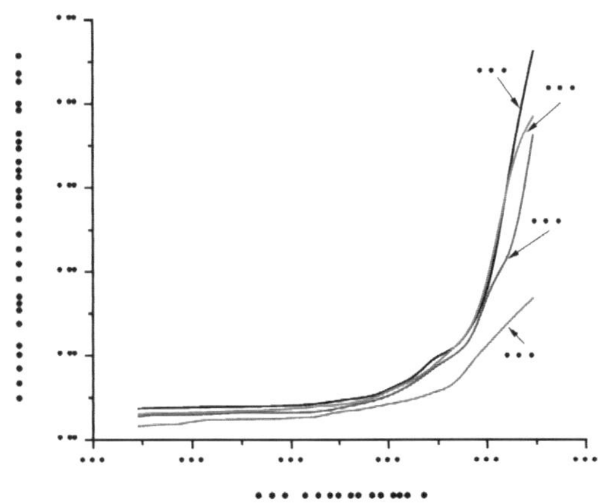

Figure 5. The variation of thermal conductivity with the progressive addition of nanoclay in the SBR matrix at three diverse temperatures

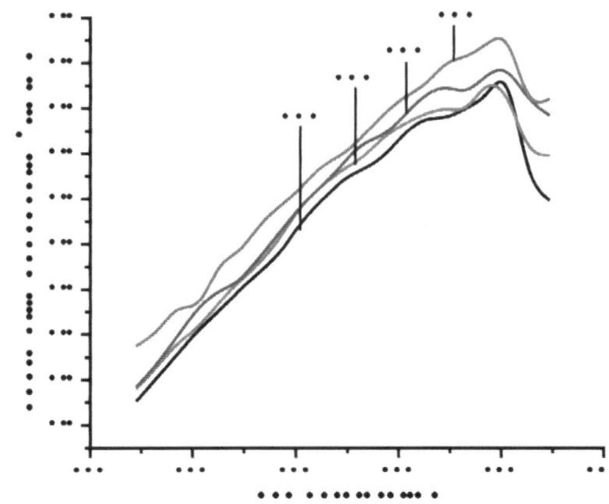

Figure 6. The effect of nanokaolinite on the thermal impedance of the rubber nanocomposites in the temperature range 323K-523K

Thermal impedance (ζ) is an important characteristic of a polymer composite which ensures the capability of a material to resist thermal variation encounter during a specific application. Figure 6 simulates the effect of NK on the thermal impedance of the nanocomposite specimens. An elevation in ζ values up to 493K is observed in the graph and then a reduction is scrutinized in the further temperature line due to transpirational cooling effect associated with evaporation of aromatic oil within the nanocomposite specimens.

The nanocomposite SK4 has the maximum ability to resist incoming heat flux within the complete tested temperature range due to the maximum nanoclay incorporation into the polymer matrix [26, 27]. Figure 7 (a, b, c, and d) simulates the surface morphological analysis of the post thermal conductivity tested nanocomposite specimens in the temperature range 323-523K. A large number of micro voids are observed in SKI while the size and quantity of theses pores are remarkably reduced with increasing NK concentration in the rubber matrix. It means that NK impregnated rubber nanocomposites has less surface rupture in the heating environment compared to the base polymer due to the excellent thermal stability and heat absorbing capability on the nanokaolinite. Figure 7e elaborates the compositional analysis of the 15 wt% loaded rubber composite. The elements C (59.99%), Al (5.69%), Si (12.55%), O (14.52%), S (1.10%), Ca (1.96%), Mg (2.58%) and Na (1.62%) are found in the energy dispersive spectroscopic analysis of the polymer nanocomposite.

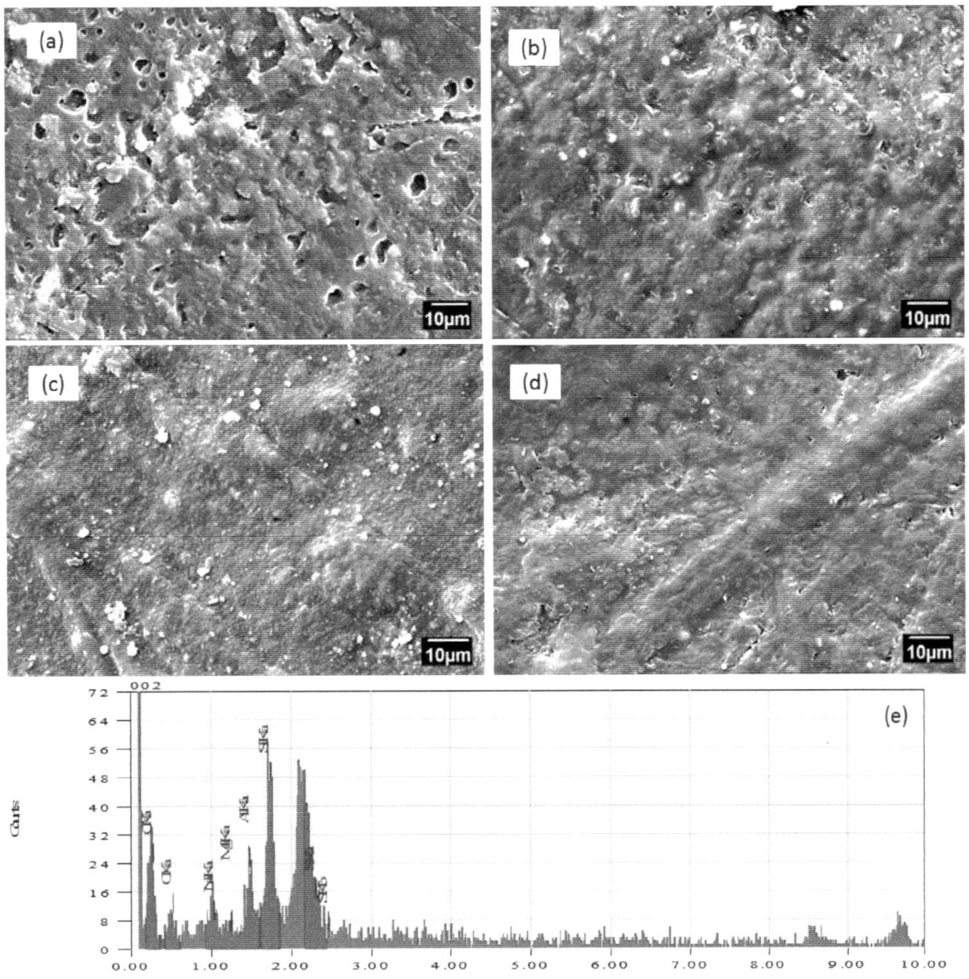

Figure 7. SEM micrographs of the post thermal conductivity tested SC1 a), SC2 b), SC3 c), SC4 d) rubber nanocomposites and compositional analysis of 15 wt% loaded composite specimen

3.3. Thermal decomposition and heat response of the polymer nanocomposites

Thermal endurance of the rubber nanocomposites were analyzed in air atmosphere from room temperature to 850°C and portrayed in Figure 8 and Table 2. An insignificant weight loss is observed up to 425°C and then the maximum thermal degradation is scrutinized in the proceeding 100°C due to polymer pyrolysis and polymeric molecular chain cession. From 550°C to 850°C, there is no significant change in weight loss is observed for the nanocomposite specimens due to the remaining carbon char and the nano inclusion.

A momentous enhancement in thermal endurance is noticed at all temperatures with the NK concentration elevation in the SBR matrix [28]. Figure 9 demonstrates the heat quenching potential of the NK/SBR nanocomposite. The composite specimens absorb enormous heat up to 300°C and the upward plateaus are observed up to 600°C due to the exhaust of volatile products included in the rubber nanocomposites.

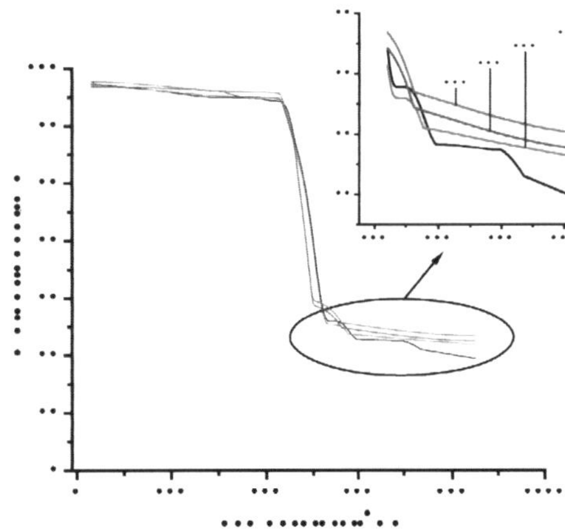

Figure 8. Thermal degradation analysis of the nanokaolinite impregnated SBR nanocomposites in the temperature range 25°C – 850°C in air environment

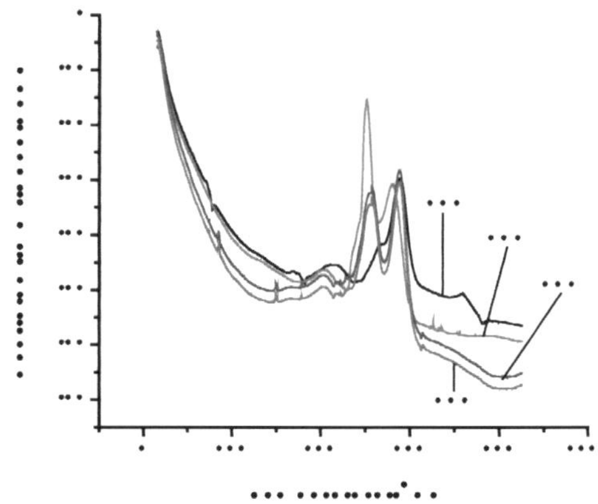

Figure 9. Thermal absorbance capability of the rubber nanocomposites in the temperature range 25–850°C

Table 2. Weight loss data of the nanocomposite specimens at various temperatures

Sample ID	Weight loss (%) at 200°C	Weight loss (%) at 400°C	Weight loss (%) at 600°C	Weight loss (%) at 800°C
SC1	1.274±0.01	3.271±0.01	65.843±0.01	69.887±0.01
SC2	1.088±0.01	3.064±0.01	64.682±0.01	66.728±0.01
SC3	0.259±0.01	2.591±0.01	63.432±0.01	66.064±0.01
SC4	0.011±0.01	1.291±0.01	62.132±0.01	64.764±0.01

The heat absorbing capability of the nanocomposites is augmented with the progressive incorporation of NK in the rubber matrix due to the excellent thermal stability, high surface area, and nanoscale interaction of the silane functionalized nanokaolinite (NK) with polymeric molecular chains [29].

2.1. Differential scanning calorimetric study

Figure 10 illustrates the DSC analysis of the rubber nanocomposites in the temperature range -74°C to 450°C to analyze the effect of nanoclay concentration on the phase transition temperatures and their specific enthalpies. Glass transition temperature (T_g), crystallization temperature (T_c), onset melting temperature (T_{m1}), and offset melting temperature (T_{m2}) of the NK/SBR nanocomposites are compiled in Table 3.The T_g of the polymer nanocomposites is reduced with increasing filler concentration in the host matrix and a 10°C T_g reduction is noticed with the 15 wt% incorporation of NK in the rubber matrix. It means that the nano inclusion remarkably enhances the rubber phase of SBR nanocomposite by interacting with polymeric molecular chains at the nanoscale, reducing the crosslinking density of the base polymer, and due to the soft nature of the filler.

Crystallization temperature of the SN4 nanocomposite is more than the SK1 composite due to the excess NK loading in the former mentioned formulation. The progressive addition of the nanoclay has enhanced the T_c of the composite specimens due to the crystalline nature of the NK.

Table 3. The measured data of the phase change temperatures of the NK impregnated rubber composites

Sample ID	Glass transition temperature (T_g, °C)	Crystallization temperature (T_c, °C)	Melting temperature (T_{m1}, °C)	Melting temperature (T_{m2}, °C)
SC1	-32.73±0.1	76.23±0.1	210.01±0.1	372.05±0.1
SC2	-38.94±0.1	77.89±0.1	209.58±0.1	383.17±0.1
SC3	-39.35±0.1	78.88±0.1	210.16±0.1	388.65±0.1
SC4	-42.70±0.1	79.14±0.1	211.60±0.1	391.83±0.1

The effect of the filler concentration on the melting characteristics of the rubber nanocomposite is opposite to T_g and T_c. The melting temperatures T_{m1} and T_{m2} proceed in the following direction and the utmost wt% concentration of NK elevates T_{m2} up to 19°C due to high thermal stability, heat absorbing capability, surface area and even dispersion of the nanoclay in the host polymer matrix [30-32]. The specific enthalpies of glass transition temperature and crystallization temperature phases are diminished while a significant evolution is noticed in the onset and offset melting phase transitions due to the additional heat absorbing capacity of the NK as tabulated in Table 4.

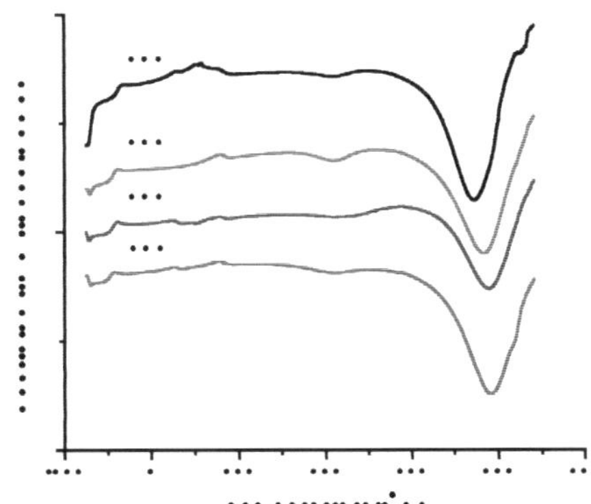

Figure 10. Heat flow response of the nanocomposite specimens in the temperature range -74 to 450°C

Table 4. Specific enthalpies of the phase change temperatures of the polymer nano composites

Sample ID	SC1	SC2	SC3	SC4
Specific Enthalpy of Glass transition (J/g)	3.227+0.01	2.975±0.01	2.966±0.01	2.941±0.01
Specific Enthalpy of Crystallization (J/g)	3.412±0.01	3.991±0.01	9.812±0.01	11.777±0.01
Specific Enthalpy of First Melting Phase (J/g)	4.369±0.01	11.073±0.01	21.605±0.01	37.191+0.01
Specific Enthalpy of Second Melting Phase (J/g)	182.610±0.01	204.756±0.01	220.499±0.01	257.184±0.01

The SK4 nanocomposite has 74 J/g more specific enthalpy compared to the SK1 formulation due to higher concentration of the reinforcing filler in the aforementioned composition compared to afterward, counterpart.

3.5. Mechanical properties

The stress–strain contours of the polymer nanocomposites are displayed in Figure 11, which simulates a remarkable evolution in toughness of the composite specimens with increasing filler incorporation into the rubber matrix. The effect of NK loading on the tensile strength, elongation at break, and modulus at 100% elasticity of the fabricated composite formulations is compiled in Figure 12a. The ultimate tensile strength is progressed from 2.78-5.10MPa, elongation at break is elevated from 137-161%, and modulus of elasticity at 100% elongation is promoted from 2.04-3.40Mpa with the 15 wt% addition of the silane functionalized nanokaolinite (NK) in the polymer matrix due to the high surface area, layered structure, even dispersion, and nanoscale interaction of the nanofiller with the polymeric molecular chains of the host matrix. Shore A hardness comparison of the rubber nanocomposites is displayed in Figure 12b [33-34].

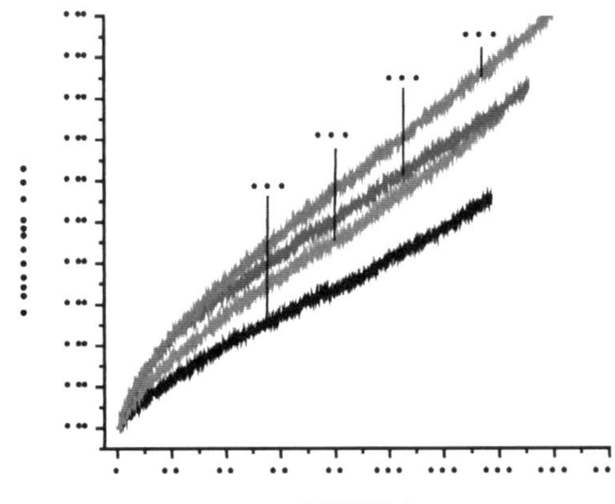

Figure 11. Stress–strain contours of the nanoclay/SBR nanocomposites

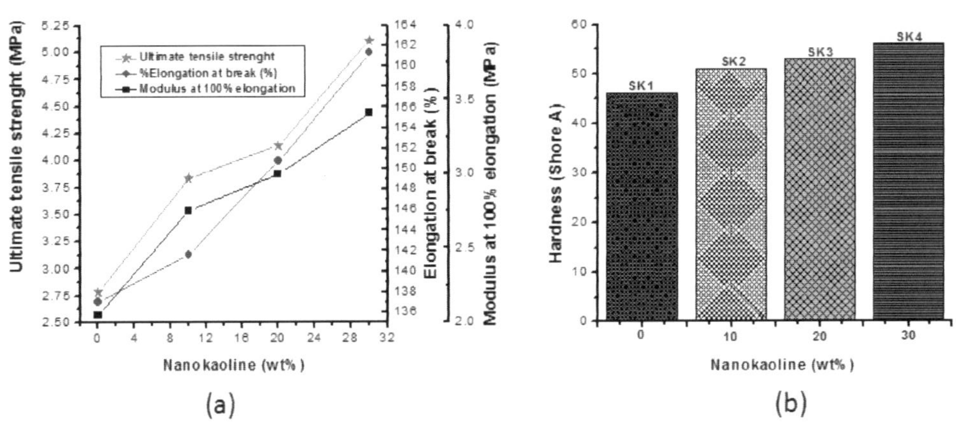

Figure 12. The effect of nanokaolinite concentration on the ultimate tensile strength, elongation at break, and modulus 100% elasticity (a) and Shore A hardness (b) of the polymer nanocomposites with diverse nanoclay loadings

Shore A hardness is augmented from 46 to 51, 53, and 56 with the 5, 10, and 15 wt%, respectively progressive impregnation of nanokaolinite in the SBR base formulation (SK1) due to the enhancement in polymeric chain flow resistance offered by the silane functionalized nanokaolinite-SBR linkage at nanoscale.

4. Conclusion

Silane functionalized nanokaolinite, a layered silicate mineral is evenly dispersed in the styrene butadiene rubber matrix using dispersion kneader and two roller mixing mill. Thermal conductivity of the rubber nanocomposites is reduced up to 41%; thermal impedance is elevated up to 41%; thermal stability is promoted up to 5%; glass transition is diminished up to 10°C; crystallization temperature, first and second melting phase temperatures are increased up to 3°C, 1.5°C and 19.7°C, correspondingly; ultimate tensile strength, rubber hardness, elongation at break, and modulus of elasticity at 100% elongation are augmented up to 2.37MPa, 10 Shore A, 24%, and 1.3MPa, respectively with 15 wt% nanoclay impregnation in the base composite formulation. Heat absorbing capability and toughness of the polymer nanocomposites are remarkably enhanced with the progressive incorporation of the nanofiller in the host rubber matrix. SEM analysis of the post thermal transport tested specimens reveals the micro void's size and quantity reduction with increasing filler to matrix ratio.

5. Acknowledgements

The authors would like to greatly acknowledge Start Up Research Grant program (SRGP) from Higher Education Commission (HEC) of Pakistan (No: 21-433 SRGP/R&D/HEC/2014) for providing sufficient funds to execute this research; School of Chemical and Materials Engineering (SCME), National University of Sciences and Technology (NUST) Longman Mills, Lahore to facilitate us with the best of their rubber nanocomposite fabrication and testing expertise and facilities.

6. References

[1] Conzatti L, Stagnaro P, Colucci G, Bongiovanni R, Priola A, Lostritto A, and Galimberti M 2012 *Appl. Clay Sci.* 61 14-21
[2] Bahramian A R, Kokabi M, Famili M H N, and Beheshty M H dsxsx2008 *J. Hazard.Mater.* 150 136-145
[3] Diez J, Barral L, Bellas R, Lopez J, Ramirez C, and Rodriguez A 2012 *J. Appl. Polym. Sci.* 125 E705-E713
[4] Vijayan P P, Puglia D, Jyotishkumar P, Kenny J M, and Thomas S 2012 *J.Materi. Sci.,* 47 5241-5253
[5] Salehi M and Razzaghi-Kashani M 2012 *J. Appl. Polym. Sci.* 126 253-259
[6] Kaneko M L Q A, Romero R B, Gonçalves M C, and Yoshida I V P 2010 *European Polym. J.* 46 881-890
[7] Panda A K, Mishra B, Mishra D, and Singh R 2010 *Colloid Surf. A: Physicochemical Eng. Aspects,* 363 98-104
[8] El-Sabbagh S H, Ahmed N M, and Ward A A 2012 *Mater. Desig.* 40 343-355
[9] Al-Safy R, Al-Mahaidi R, Simon G P, and Habsuda J 2012 *Const. Build. Mater.* 28 769-778.
[10] Chen S, Yu H, Ren W, and Zhang Y 2009 *Thermochimica Acta,* 491 103-108
[11] Noriman N Z, Ismail H, and Rashid A A 2012 *Adv. Polym. Tech.* 31 100-108
[12] Wang Q G, Gao W X, and Zhang L Q 2011 *J. Appl. Polym.Sci.,* 120 3196-3203
[13] Salgueiro W, Somoza A, Silva L, Consolati G, Quasso F, Mansilla M A, and Marzocca A J 2011 *Phy. Rev. E,* 83
[14] Tian M, Cheng L J, Liang W L, and Zhang L Q 2006 *J. Appl.Polym. Sci.* 101 2725-2731.
[15] Zhang H L, Liu N A, Ran X H, Han C Y, Han L J, Zhuang Y G, and Dong L S 2012 *J. Appl.Polym. Sci.* 125 E550-E561
[16] Lei Y D, Tang Z H, Zhu L X, Guo B C, and Jia D M 2012 *J. Appl.Polym. Sci.* 123 1252-1260.
[17] Zanchet A, Carli L N, Giovanela M, Brandalise R N, and Crespo J S 2012 *Mater. Desig.,* 39 437-443
[18] De Falco A, Marzocca A J, Corcuera M A, Eceiza A, Mondragon I, Rubiolo G H, and Goyanes S 2009 *J. Appl.Polym. Sci.* 113 2851-2857
[19] Zhou X W, Zhu Y F, and Liang J 2007 *J. Appl.Polym. Sci.* 106 1836-1841
[20] Iqbal N, Khan M B, Sagar S, and Maqsood A 2013 *J. Appl.Polym. Sci.* 128 2439-2446
[21] Yau Y H, Wong H F, and Ahmad N 2012 *Int. J.Heat Mass Trans.* 55 2879-2888
[22] Pham D C May 2008 *Int. J.Heat Mass Trans.* 51 3355-3361
[23] Mu L W, Shi Y J, Feng X, Zhu J H, and Lu X H 2012 *Tribology Int.* 53 45-52

[24] Li Z G, Chen H, Zhu Z H, and Zhang Y 2011 *Polym. bull.,* 67 1091-1104
[25] Ali Raza M, Westwood A, Stirling C, Brydson R, and Hondow N 2012 *J. Appl.Polym. Sci.* 126 641-652
[26] Liu K, Takagi H, Osugi R, and Yang Z M 2012 *Compos. Sci.Tech.* 72 633-639
[27] Yu H, Li L L, Kido T, Xi G N, Xu G C, and Guo F 2011 *J. Appl.Polym. Sci.* 124 669-677
[28] Zulkepli N N and Ismail H 2012 *Polym. Plast. Tech. Eng.* 51 350-357
[29] Tasdemir M and Ulug E 2012 *Polym. Plast. Tech. Eng.* 51 164-169
[30] Cibulkova Z, Cerna A, Simon P, Uhlar J, Kosar K, and Lehocky P 2011 *J. Therm. Anal. Calorim.* 105 607-613
[31] Jayasree T K and Predeep P 2012 *J. Therm. Anal. Calorim.* 108 1151-1160
[32] Cibulkova Z, Cerna A, Simon P, Uhlar J, Kosar K, and Lehocky P 2012 *J. Therm. Anal. Calorim.*108 415-419
[33] Qu L L, Yu G Z, Wang L L, Li C Q, Zhao Q S, and Li J 2012 *J. Appl.Polym. Sci.* 126 116-126.
[34] Vasilakos S and Tarantili P 2011 *J. Appl.Polym. Sci.,* 125 E548-E560

Effect of sintering on structural and electrical properties of co-precipitated Mn-Zn ferrites

S. Mehmood, F. Zahra and M. A. Rehman
Applied Thermal Physics Laboratory, Department of Physics,
COMSATS Institute of Information Technology, Islamabad 44000, Pakistan

E-mail: marehman@comsats.edu.pk

Abstract. Mn-Zn ferrite is one of the important class of soft ferrites. These are famous for possessing high initial permeability. In the present work, we have studied the effect of sintering on Mn-Zn nano particles. The particles were synthesized using co-precipitation method. The structural characterization of the prepared sample after each sintering step were done by using XRD. The XRD analysis showed the spinel structure. The electrical properties were studied as a function of temperature. It was observed that dielectric constant, loss tangent and AC conductivity varies with respect to temperature. The prepared composition is useful in microwave devices.

1. Introduction

Ferrites nanoparticles are ceramic ferromagnetic materials. These materials are famous due to their wide field of technological applications in microwave setup, TV, and radio sets [1].Mn-Zn ferrites belong to the soft ferrites. These are renowned for their utilization in modern electronic devices such as EMI suppression, switch mode power supplies and data transmission circuitry [2,3]. There are number of methods by means of which these nano ferrite particles can be prepared such as co-precipitation, hydrothermal and sol-gel mechanical method [4,5]. In order to control the structural and electrical properties of ferrites there are certain important factors, one of them is the method of preparation [6] and the other one is sintering temperature [7]. By controlling these parameters the structural and the electrical properties of a material can be changed.

In the present work $Mn_{0.5}Zn_{0.5}Fe_2O_4$ nanoparticles were synthesized by using co-precipitation method. The effect of sintering on the structural and the electrical properties of $Mn_{0.5}Zn_{0.5}Fe_2O_4$ nanoparticles were studied as a function of temperature. The prepared sample was sintered from 350 ^oC to 500 ^oC with the step of 50 ^oC for 2 hours. For a fixed frequency of 1MHz and the temperature ranged from room to 500 ^oC. AC electrical properties including dielectric constant (ε'), loss tangent ($\tan\delta$) and AC conductivity (σ_{AC})were measured.

2. Experiment

Mn-Zn ferrite of composition $Mn_{0.5}Zn_{0.5}Fe_2O_4$was prepared by the co-precipitation method. The precursors of iron nitrate nonahydrate$Fe(NO_3)_3.9H_2O$, zinc nitrate hexahydrate $Zn(NO_3)_2.6H_2O$ and manganese nitrate tetrahydrate $Mn(NO_3)_2.4H_2O$ were weighed in stoichiometry. These precursors were dissolved in de-ionized water through magnetic stirring. The molarity of the solution was kept 0.4M. The solution of NaOH was used to start the process of precipitation. These solutions were heated for 30 minutes using hot plate magnetic stirrer. The temperature was kept constant at 75±5 ^oC

Content from this work may be used under the terms of the Creative Commons Attribution 3.0 licence. Any further distribution of this work must maintain attribution to the author(s) and the title of the work, journal citation and DOI.
Published under licence by IOP Publishing Ltd

and the pH was set to 12.5-13. Reaction vessel was covered with a plastic cover to diminish the evaporation of the solution. Finally the de-ionized water were used for washing the precipitates. The particles were dried in an oven at 105 °C and then grinded into fine powder. Pellets were prepared and then sintered at different temperatures.

3. Results and discussion

3.1. Structural analysis

The X-ray diffraction (XRD) of the prepared pellets was done by using an X' pertPro X-ray diffractometer with CuKα radiation (λ=1.5406Å). The analysis of as prepared and sintered pellets were studied. By using the two probe method the dielectric constant (ε'), loss tangent (tanδ) and AC conductivity (σ_{AC}) were studied. The XRD pattern of the prepared sample sintered at different sintering temperature were studied as shown in Figure 1.

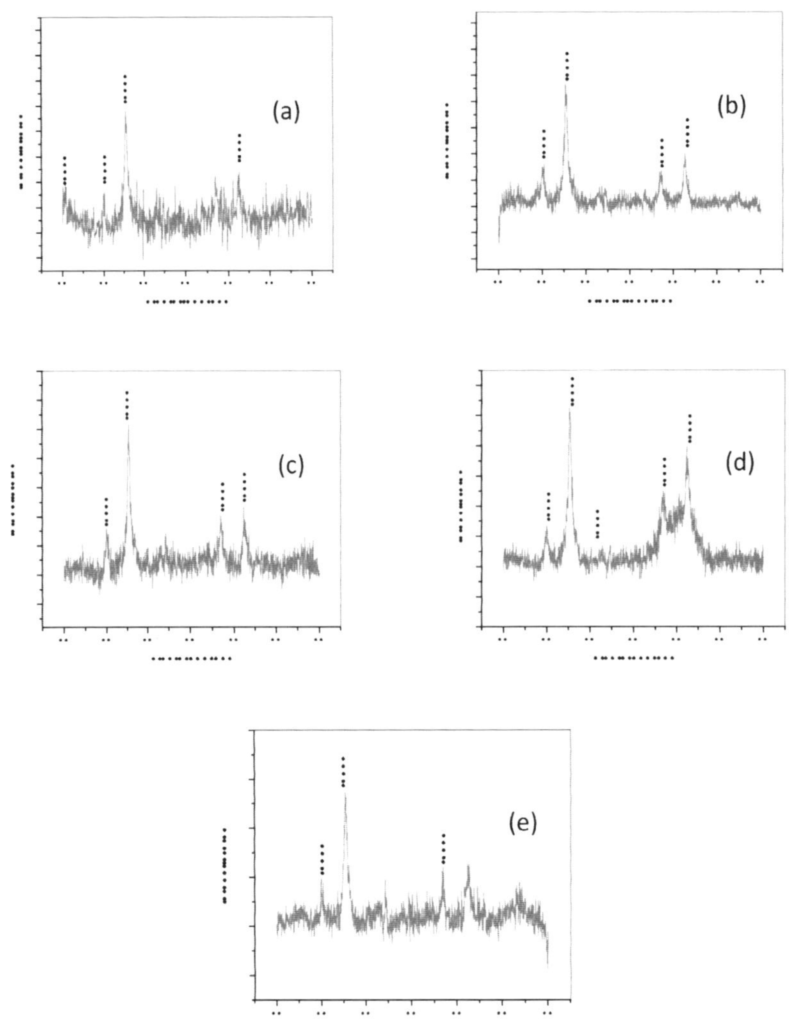

Figure 1. XRD pattern of Mn-Zn ferrite (a) as prepared, sintered at (b) 350 °C (c) 400 °C (d) 450°C (e) 500 °C

The prepared sample showed the spinel structure [8-9]. The sintering temperature was from 350 °C to 500 °C with the step difference of 50 °C. It was observed that by increasing the sintering temperature the lattice constant and the crystallite size of the sample were increased. The lattice constant varies from 8.38Å to 8.42Å. The maximum crystallite size of 19nm was observed at the sintering temperature of 450 °C.

3.2. Electrical properties

The number of grain boundaries would increase if the grain size is small. By increasing the sintering temperature the grain growth increased and the porosity decreased. Due to that the charge carriers faced less pores and the resistivity decreased. In Figure 2, the variation in dielectric constant (ε') as a function of temperature is shown.

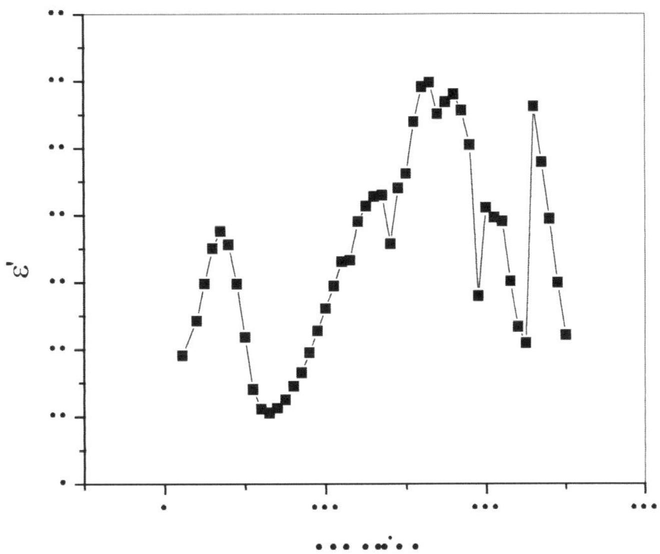

Figure 2. Variation in Dielectric constant (ε') as a function of temperature for a fixed frequency of 1 MHz.

The frequency was kept constant at 1MHz and the temperature ranged was from room to 500 °C. It was observed that the dielectric constant of the fully sintered sample increased by increasing temperature but after reaching at a certain value it decreased. The further increase in the value of the dielectric constant was observed at the Curie temperature. The Curie temperature for the prepared composition was 380K [10].The similar variation of the dielectric constant with respect to temperature was reported earlier [11].The variation of the dielectric constant due to temperature in the prepared composition is an important characteristic of a disordered perovskite structure with a diffuse phase transition. Due to the disordered arrangement of the cations on one or more crystallographic sites the microscopic heterogeneity in the composition created as a result of which different local Curie points generated [12-13].

In Figure 3, the variation in loss tangent (tan δ) is shown. It was observed that theloss tangent remained constant for the smaller value of temperature however, for the higher values of temperature it increased. This is due to the thermally activated hopping process. Dielectric properties of any material are strongly dependent on polarization. By increasing temperature interfacial polarization of the material increased which in return increased the value of loss tangent and the dielectric loss.

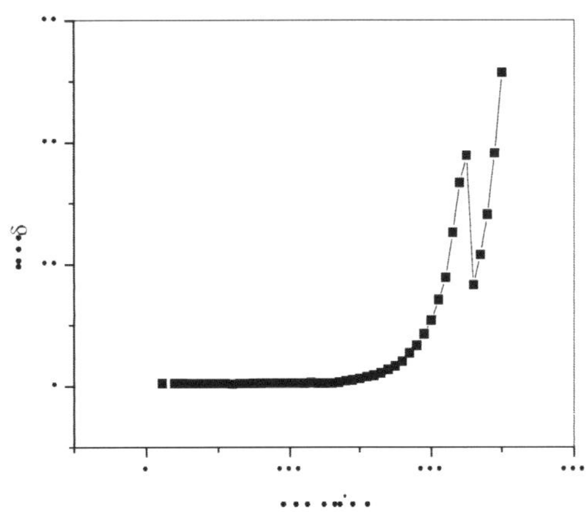

Figure 3. Loss tangent (tan δ) as a function of temperature for a fixed frequency of 1MHz

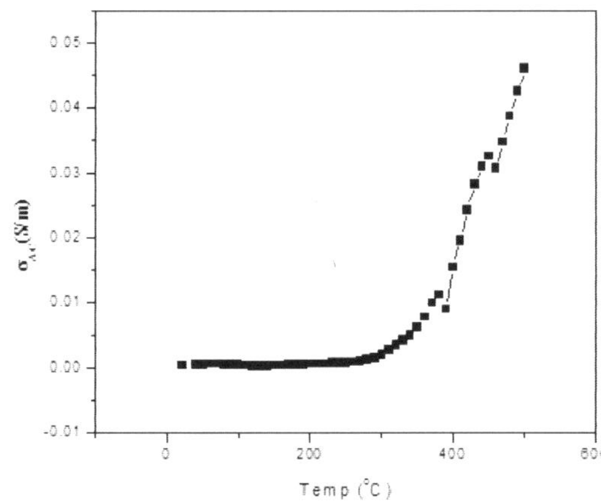

Figure 4. AC conductivity (σ_{AC}) as a function of temperature for a fixed frequency of 1 MHz

In Figure 4, AC conductivity (σ_{AC}) is shown as a function of temperature. The conductivity of a prepared material increased by increasing temperature. This is in accordance with the ion hopping mechanism. Temperature increased the mobility of the charge carriers and as a result conductivity also increased.

4. Conclusions

Phase pure $Mn_{0.5}Zn_{0.5}Fe_2O_4$ was successfully synthesized using co-precipitation method. Increase in the sintering temperature (350 °C to 500 °C with the step of 50 °C for 2 hours) affected the structural as well as the electrical properties. The XRD showed that the crystallite size were in nano meter range for both as prepared and the sintered samples confirming the selection of suitable temperature range for sintering. More variation in the ε' at higher temperatures showed the disordered perovskite structure of the prepared samples. The microscopic heterogeneity in the prepared composition was observed which results in the formation of local Curie points. The tanδ and the σ_{AC} remained constant for the smaller values of temperature however at higher values these increased, which is in accordance with thermally activated hopping process. The prepared composition can be usedin high frequency device applications.

5. References

[1] Elhiti M A1999 *J. Magn. Magn. Mater*192 305
[2] Tomar M S, Singh S P, Perez O P, Guzman R P, Calderon E, Ramos C R 2005 *J. Microelectronics,* 36475
[3] Arulmurugan R, Vaidyanathan G, Sendhilnathan S, Jeyadevan B 2006*J. Magn. Magn. ater.*298 83
[4] Taketomi S, Ozaki Y, Kawasaki K, Yuasa S, Miyajima H 1993*J. Magn. Magn. Mater.*1226
[5] Mergen A, Qureshi A 2009*J. Alloys Compd.*478741
[6] Barakat M M, Henaish M A, Olofa S A, Tawfik A 1991*J.Therm. Anal.*37 605
[7] Verma A, Goel T C, Mendiratta R G 2000 *Second International Conference on Processing Materials for Properties: TheMineral, Metal &Material Society* pp. 493-497.
[8] Nasir S, Anis-ur-Rehman M 2011*Phys. Scr.*84025603.
[9] Jacob B P, Thankachan S, Xavier S, Mohammed E M 2013*J. Alloys Compd.*578 314
[10] Lan N T, Hien T D, Duong N P, Truong D V 2008*J. Korean Physic. Soc.*521522
[11] Olofa S A 1994*J. Magn. Magn. Mater.*131103
[12] Yadav K L, Choudhary R N P 1994*Mater. Lett.*1961
[13] Bera S, Choudhary R N P 1995*Mater Lett.*22197

Single and double shock initiation modelling for high explosive materials in last three decades

T. Hussain[1] and Liu Yan[1]
Department of Physics and Applied Mathematics,
Pakistan Institute of Engineering and Applied Sciences, Islamabad, Pakistan
[1]State Key Laboratory of Explosion Science and Technology,
Beijing Institute of Technology, Beijing 100081, P.R China

E-mail: hushan@bit.edu.cn

Abstract. The explosives materials are normally in an energetically metastable state. These can undergo rapid chemical decomposition only if sufficient energy has first been added to get the process started. Such energy can be provided by shocks. To predict the response of these materials under impacts of shocks of different strengths and durations and at various conditions, mathematical models are used. During the last three decades, a lot of research has been carried out and several shock initiation models have been presented. The models can be divided into continuum based and physics based models. In this study the single and double shock initiation models presented in last three decades have been reviewed and the ranges of their application has been discussed.

1. Introduction

Keeping in view the high costs of experiments, the demand to develop the analytical and computational models for shock initiation of explosives materials to provide both an understanding and predictive capability has always been present. In last three decades, a large number of shock initiation models, also known as reactive burn models, have been developed. The development of a large number of initiation models points out, on one hand the importance and necessity of the modeling of shock initiation process, while on the other hand reflects the difficulty in modeling the process. These models make use of concept of hotpots, that take roots from the experimental findings that the initiation mechanism of both Homogeneous and heterogeneous explosives are entirely different [1-2]. Bowden and Yoffe [3] first introduced the concept of hot-spot generation during initiation of heterogeneous explosives. However only few experimental evidences [4-5] could be found. The shock initiation process is usually described as having two phases: (i) an ignition phase where 'hot-spots' are generated due to shock compression heats up the explosive at the discontinuities and the subsequent chemical reactions further raise the temperature of these localized heated regions or hot spots, and (ii) a growth phase where the 'hot-spots' grow and coalesce to consume the remainder of the explosive material. In spite of little knowledge regarding the evolution of 'hot-spots' behind a shock wave, Field in his two landmark papers [6-7] summarized ten possible mechanisms for 'hot-spot' generation. However in a typical shock initiation problem any one or more than one mechanism may be responsible for producing of hot spots, and no universal agreement exist among the research community about the mechanism(s) responsible to produce 'hot-spots'. Pore collapse [8], shear banding [9-12] and frictional heating [13] are the three relatively more plausible mechanisms

Content from this work may be used under the terms of the Creative Commons Attribution 3.0 licence. Any further distribution of this work must maintain attribution to the author(s) and the title of the work, journal citation and DOI.
Published under licence by IOP Publishing Ltd

[13] responsible for shock initiation. The reactive burn model must also model the effects of the factors significantly affecting the shock initiation and sensitivity of explosives such as initial density (initial porosity) [14-16], crystal or grain size distribution [17], crystal orientation, some microstructure effects [18-21] and initial temperature [22-23] of the explosive.

In the present paper we will review the models which are well known and widely used in shock initiation studies and have been key developments in the field. These models can be classified into two types: (i) continuum-based models, and (ii) physics-based or mesoscale models. There is yet another class of models which may be called hybrid models will also be reviewed.

2. Continuum-based reactive burn models

These models are phenomenological in nature. The reactive Euler equations are widely used for modeling the initiation of explosives. For engineering simulations the cell size is much greater than the scale of material heterogeneities. The limited resolution necessitates using a homogenized constitutive relation for the explosive. These models have been in use for calculations for practical engineering application because of their realistic requirement of the computational resources. Since the assumption of homogenization eliminates the temperature fluctuations at the discontinuities, burn models have to account for the effect of hot spots on the reaction rate.

2.1. Forest fire model

The Forest Fire model [24-26] assumes that the reaction rate depends only on the pressure. The pressure dependence is determined such that the lead reactive shock follows the buildup curve derived from the Pop plot [27]. In addition to the Pop plot, the Forest Fire rate requires the reactive shock Hugoniot. It is determined from the initial particle velocity and shock velocity measured in a series of wedge experiments. The reaction rate equation is given as

$$\frac{d\lambda}{dt} = (1-\lambda)G(P)$$

where λ is the reaction progress variable and t in above equation denotes time, $\lambda = 0$ corresponds to the unreacted explosive while λ=1 corresponds to the fully reacted explosive. $G(P)$ is assumed to be an exponential power series in local pressure variables and is of the form: $G(P) = \sum_{i=1}^{n} C_i P^i$, where C_i are determined by using the reactant's equation of state, reactive Hugoniot and equation of state of detonation products using square wave assumption.

2.2. Lee-tarver ignition and growth model

The Lee-Tarver model [28-29] uses two Jones-Wilkins-Lee (JWL) equations of state, one for the unreacted solid explosive and the other for reaction products, in the temperature dependent form:

$$P = Ae^{-R_1 V} + Be^{-R_2 V} + \omega C_v T / V$$

where P is pressure, V is relative volume, T is temperature, ω is the Grüneisen coefficient, C_v is the average heat capacity, and A, B, R_1 and R_2 are constants. The ignition and growth concept is represented by a reaction rate equation which was originally divided into an ignition term and a single growth term [29], but was later modified [30] to model the short pulse duration shock initiation experiments by the addition of an extra growth term. The reaction rate (RR) equation [30] is given as

$$\frac{d\lambda}{dt} = I(1-\lambda)^b(\rho/\rho_o - 1 - a)^x + G_1(1-\lambda)^c \lambda^d P^y + G_2(1-\lambda)^e \lambda^g P^z$$

where ρ is the current density, ρ_0 is the initial density and the parameter a is a critical compression that is used to inhibit ignition, and hence reaction, until a certain degree of compression has been reached. I, b, x, G_1, G_2, c, d, y, e, g and z are constants. The ignition term runs from $0 < \lambda < \lambda_{igmax}$, the first growth term from $0 < \lambda < \lambda_{G1max}$ while the second growth term works from $\lambda_{G2min} < \lambda < 1$. The form factors $(1-\lambda)^c \lambda^d$ and $(1-\lambda)^e \lambda^g$ are in accordance with the grain burning topology [31] for spherical grains burning from inside and outside.

This model [30] has been applied to many experimental shock initiation and detonation studies of solid explosives [16, 32-37] and propellants and embedded in several hydrocodes, making it the most popular reactive burn model in use today in spite of its short comings in some cases [38, 39]. The parameters of reaction rate equation for quite a large number of explosives have been determined [40] at different conditions. Various shock induced phenomena have successfully been predicted, including run-distance to detonation (Pop-Plot) data [29], short pulse shock initiation data [30], in-material pressure gauge measurements examining growth of reaction in shock initiated explosives [28], detonation propagation and failure [36] , and corner turning data [28]. Lee-Tarver model has also been used to explore the shock initiation criteria of explosives subject to the flyer impact [41].

Murphy [42] proposed an improved reaction rate equation with less parameters, making it easier to parameterize and optimize. The main problem with the model is that the parameters are empirical fit to some particular experimental data. Thus, there are doubts over the ability of the model to predict a wide range of phenomena with a single set of parameters, or to predict outside its fitting regime.

2.3. Johnson-Tang-Forest model

The Johnson-Tang-Forest (JTF) model [43] is a kind of ignition and growth model elaborating the contribution of the 'hot-spot' ignition phase in describing the shock initiation of heterogeneous explosives. It introduces internal degree of freedom for mass fraction of hot spots f, the incident shock pressure P_s and the hot spot temperature T_{hs}. The average burn rate if given by

$$\frac{d\lambda}{dt} = \mu\frac{df}{dt} + [1-\lambda-\mu(1-f)]\left(\frac{f-f_e}{1-f_e}\right) + G(P,P_s)$$

where μ is the mass fraction of explosive turned into hot spots, f_o is the threshold value for mass fraction of hot spots f and $G(P, P_s)$ is an empirical function which is constrained to approach the Forest Fire rate at high pressures, and is given as

$$G(P,P_s) = G_0(P)\left[1+\frac{F(P_s)}{G_0(P_s)}\right]$$

where $F(P_s)$ is the Forest Fire rate and G_o is a fitting parameter. In this model the ignition process includes average 'hot-spot' temperature as an explicit parameter. The ignition phase is based upon a thermal explosion where the induction time is a function of average 'hot-spot' temperature and the growth phase is represented by a polynomial in pressure that is similar to the Forest Fire reaction rate.

2.4. CREST reactive burn model

The CREST reactive burn model [44] is an entropy-dependent reactive burn model based on the analysis [45] of high quality, in-material gauge results of velocity histories in shock to detonation transition experiments carried out by Gustavsen et al. [46] for two HMX based explosives and one TATB based explosive. CREST uses a finite strain form of equation of state [47] for the unreacted explosive, JWL equation of state for the reaction products and isentropic solid equation of state [48] model is used for the reacting mixture. Pressure equilibrium assumption is used for the solid and the gaseous components during the reaction. The unreacted explosive equation of state is porosity dependent. The snow plough porosity model [47] is used [49] to account for the effect of porosity and initial temperature [50] on the solid unreacted explosive response. The reaction model utilizes two reaction rates: (i) a fast reaction rate $d\lambda_1/dt$ which encompasses all primary sources of reactions such hot spot, and (ii) a slower reaction rate $d\lambda_2/dt$ corresponding to the bulk burn of the explosive. The overall reaction rate $d\lambda/dt$ is the weighted sum of the two rates, as given by

$$\frac{d\lambda}{dt} = m_1 \frac{d\lambda_1}{dt} + m_2 \frac{d\lambda_2}{dt}$$

where m_1 and m_2 are the weighting factors. The description of the individual reaction rate can be seen in reference [44, 51]. CREST model has been successfully used in predicting the effect of porosity [52], initial temperature [50, 53] on shock sensitivity of explosives effects, corner turning experiments at different initial temperatures [53, 54], desensitization phenomena [53] and failure diameters [55]. Since CREST does not explicitly model the microstructure of explosives, it cannot predict the effect of particle size on shock sensitivity of the explosives [55].

2.5. Wescott-Stewart-Davis (WSD) model

The WSD model [56] utilizes a wide-ranging equation of state and a reaction rate law. The equation of state is calibrated to fit the shock and product Hugoniot curves that fit experimental data. The reaction law model is calibrated to fit detonation shock dynamics [57] relations (D_n-κ) and shock initiation data. The reaction rate similar to Lee-Tarver model is given by

$$R = R_I S_I(\lambda) + R_G S_G(\lambda) + [1 - S_G(\lambda)] R_B$$

where the subscripts I and G refer to the ignition and growth terms respectively. The details of the variables can be found in reference [56]. The reaction rate equation differs from Lee-Tarver model in that there are two different pressure sensitivities in the growth term, one for shock initiation and another for full detonation, thus allowing the matching of a wide variety of experiments with one set of parameters. This also allows the simulation of problems with both shock initiation and highly curved detonation waves. The application of this model to detonation diffraction and detonation-material interaction problems was encouraging.

3. Physics-based shock initiation models

Physics-based or mesoscale models for shock initiation of explosives can be classified into mainly two categories: (i) statistical 'hot-spot' models, and (ii) explicit 'hot-spot' models where some physical phenomena are used to model the hot spots.

3.1. The statistical 'hot-spot' models

Because of random distribution in sizes an orientation of the constituent particles of a typical heterogeneous explosive, the exact position and mechanism of hot spots formation is hard to predict. Even the so called explicit hot spot models elaborating some specific mechanism such as pore collapse for shock initiation take and average distribution of hot spots. The statistical hot spot models such as in reference [58] rely on an assumption of the size and number density of 'hot-spots' in shocked compressed material as the starting point for their calculations. Such information is usually obtained from the results of large scale mesoscale simulations of an ensemble of the explosive. The microstructure of the explosive consisting of the pore size distribution, grain size distribution, binder, and other solid inclusions is given as input to large scale, direct numerical simulations codes [59-61] and the size and distribution of 'hot-spots' as the initial conditions for a statistical hot spot model calculation. However, the computational resources requirement is pretty high to perform the direct numerical simulations of typical problems [62]. Moreover the task of extracting data from the direct numerical simulation codes to input the statistical hot spot modelling codes is complex. In the model of Hamates et al. [63], rather than keeping track of distributed 'hot-spots' (as in reference [58]), an aggregated 'hot-spot' concept is used representing the overall contribution of the distributed 'hot-spots'. An approach [64] to computationally predict and quantify the stochasticity of the ignition process in polymer-bonded explosives (PBXs) under impact loading is a recent development in statistical modeling of ignition.

3.2. Mechanism-based explicit hot spot models

Most of the physical based models take one particular mechanism for shock initiation. To date, no theory of hot-spots has attempted to cover all of the different possible 'hot-spot' phenomena, and most of the models have modeled 'hot-spot' ignition from this perspective. Numerous experimental studies e.g. [15] as well as experimental results e.g. [16] have shown that increasing porosity can make an explosive material more sensitive to shock initiation. This emphasizes the idea that the pores are prime candidates as potential 'hot-spot' sites. A large number of models based on pore collapse theories have been presented during last three decades. The new models and their applications reported are AMORC [65], the Cook-Haskin's CHARM model [66], the model of Bennett [67], the model of Khasainov et al. [68], Belmas et al. [65], Frey [69], Kang et al. [70], Bonnett and Butler [71], Massoni et al. [72], Whitworth[73], Bouma et al. [74], Brundage et al. [75], Springer et al. [76], Cheng et al.[77] and recently Akiki et al. [78]. Khasainov *et al.* [68] were the pioneer to build a viscoplastic pore collapse model and apply it to 'hot-spot' formation in explosives. The model of Khasainov [68] has since been implemented into the Autodyn hydrocode, and the two-dimensional Eulerian code EDEN [79]. Nonetheless, there are some models which simultaneously model more than one phenomenon for shock initiation of explosives. Example is the model by Shaw et al. [80] and Howe et al. [81].

4. Hybrid models

This class of models which may be called hybrid models, because in such models the ignition is modelled by some mesoscopic process such as pore collapse while the growth is modelled by using some empirical relations similar to continuum based models. Two of them are discussed below.

4.1. Kim's model

Kim's model [82, 83] is a kind of an ignition and growth model where the elastic viscoplastic pore collapse mechanism similar to Carroll and Holt [84] is assumed to be responsible for the ignition of explosives. After ignition, the growth of reaction is modeled by using the empirical relations somewhat similar to those of Lee-Tarver's model. The reaction rate is given as

$$\frac{d\lambda}{dt} = \left(\frac{d\lambda}{dt}\right)_{ig} + \frac{F}{r_0}\lambda^{2/3}P^{y} + \frac{G}{r_0}P^{z}(1-\lambda)^{2/3} \quad \frac{d\lambda}{dt} = \left(\frac{d\lambda}{dt}\right)_{ignition} + A\frac{p^{y}\lambda^{\frac{2}{3}}}{r_0} + B\frac{p^{z}(1-\lambda)^{\frac{2}{3}}}{r_0}$$

where r_o is average particle radius in micrometers, F, G, y and z are growth constants. Kim [82] used his model to study the SDT experiments for a specific particle size formulation of only one explosive (PBX9404) to achieve good agreement.

4.2. DZK model

Duan's ignition model (later named as DZK hot-spot model [85]) is also a kind of ignition and growth model, where the ignition term is based on the collapsing of the pores while the growth terms are based on the burning topology of pores. It has been implemented in DYNA2D software [86] and has been used to predict the effects of the particle size [87], the porosity of the explosive [88], the strength and content of the binder in a PBX explosive [88] and recently the effect of initial temperature on the shock initiation [85]. The DZK reaction rate equation is a sum of three terms: i) the ignition term by DZK hot spot model [89], ii) the slow growth term suggested by Kim [83], and iii) the fast growth term proposed by Zhang [90], and is of the form [89]:

$$\frac{d\lambda}{dt} = \left(\frac{d\lambda}{dt}\right)_{ignition} + \frac{3\lambda^{2/3}}{r_0} aP^{n} + GP^{z_1}(1-\lambda)^{x},$$

where the empirical parameters a, G, n, z_1 and x are determined by experiments.

5. Double shock initiation and desensitization phenomena

The models described is previous sections usually address single shock initiation. The double shock initiation phenomena is little more complex. The most interesting phenomena arising due to double or multiple shocking of an explosive is desensitization. It is now an established fact that passage of a weak shock desensitizes the explosive for the succeeding stronger shock. This phenomenon is either a desirable phenomenon for safety scenarios or, on the other hand, a highly undesirable phenomenon for design of explosive applications where the performance and yield of the device is of primary concern. A number of experimental findings [38-39, 91-98] have been reported confirming that the preshocks cause the desensitization effects in heterogeneous explosives. The SC jet initiation tests [99-105] for heterogeneous explosives also indicate the occurrence of desensitization phenomena.

5.1. Modeling desensitization using continuum based models

Unfortunately, most of the phenomenological models are unable to simulate the situations involving the preshock desensitization. Therefore separate desensitization procedures have to be annexed with them to enable such models to account for the desensitization effects. For example, the JTF model [43] accounts for the preshock desensitization phenomena by restricting creation of hot spots to the first shock only, and allowing a second shock to provide only adiabatic heating. Wescott-Stewart-Davis (WSD) model [56] has been improved [106] to a temperature dependent form WSD (T) for such calculations. Mader [25] used multiple shock Forest-Fire model to simulate the shock desensitization experiments [105]. However a reactive burn model CREST [44] that uses entropy-dependent reaction rates, has been successful [53] in simulating the persistent dead zones in corner turning experiments [107] on PBX 9502 explosive. However the model is still being improved at Atomic Weapon Establishment (AWE) of UK [51].

The standard Lee-Tarver model [29] cannot reproduce the desensitization effects. Several attempts have been made to modify it. Long ago, Tarver [38] added a condition in the reaction rate equation to account for desensitization of LX-17 by reflected shocks. The condition was that if a critical range of pressures is applied to the explosive, the reaction rate was forced to zero. The calculated growth histories using the modified reaction rate equation agreed well for the reflected shock experiments, except for few exceptions. Whitworth [108] described a simple model to predict shock desensitization effect by modifying the reaction rate equation by making it dependent on the preshock pressure P_{shk}. In place of local pressure P in reaction rate equation, $P=min\,(P, P_{shk})$ was used to find the reaction rate. A qualitative agreement was reported for the calculations of Mulford's precursor shock experiment [92] using the modified reaction rate equation. Whitworth also applied his modified reaction rate equation for detonation quenching experiments. However, contrary to the experimental results [109], the rate of quenching increased with decreasing pressure of the precursor shock.

In order to model the qualitative features of the corner-turning failure in LX-17, Sours [107] presented a failure model that cut off the reaction rate below certain detonation velocities. A single reaction rate equation was used which was a stripped down version of Lee-Tarver model, and the coefficients of this rate equation were different in different regions of the explosive. In a more recent approach, Sours [110] later refined his model by employing a single reaction rate format which varied for different pressure regimes in explosive to simultaneously model dead zone, failure region and the detonation region in corner turning experiments. This model could reproduce various experimental features including detonation failure region in corner turning experiments of LX-17.

Oliveira [111] proposed a modification in Lee-Tarver model by introducing the desensitization rate S as $S=MP(1-\varphi)(\varphi+\varepsilon)$, where φ is the degree of desensitization, which equals 1for virgin state while 0 for completely desensitized explosive. M is rate constant and ε is a small positive number to ensure some desensitization at $\varphi=0$ when $P>0$. Thus the Ignition term in reaction rate equation becomes as $R_I=I(1-\lambda)^b(\eta-1-a(\varphi))^x$, where the modified critical compression $a(\varphi)$ is defined as $a(\varphi)=a_0(1-\varphi)+a_1(\varphi)$, where $1+a_0$ is critical compression for ignition of the virgin material and $1+a_1$ is the same for the completely desensitized material. The second change proposed [111] was to change the switching limit of the growth term from 0 to $\lambda_{G1min}(\varphi)$, defined by $\lambda_{G1min}(\varphi)=\lambda_c\,\varphi$, where $\lambda_c > 0$.

This modification requires four additional empirical constants to enable the ignition and growth model to account for the desensitization effect. Using the augmented model, Oliveira [111] qualitatively reproduced the main features of the corner turning experiments including dead zones. It has been used by Tarver et al. [37] for simulation various experiments [112] involving desensitization. The form factor $(1-\varphi)\,\varphi$ for desensitization rate equation seems similar to the form factor $(1-\lambda)^c\,\lambda^d$ of the growth terms in original Lee-Tarver reaction rate equation, with λ replaced by φ. However we find no logic behind this similarity. Moreover the form factor also implies that the desensitization rate is maximum when the material is 50% desensitized, again without any experimental evidence. To improve the Oliveira's model, the authors have developed a modification in Lee-Tarver model and successfully applied [113] to simulate the shaped charge jet initiation of covered explosives.

5.2. Modeling desensitization using physics based models

The pores in explosives are considered as the potential hot spot sites. The preshock compresses some of the pores creating the hotspots while most of the remaining unreacted pores are squeezed, thus making it almost impossible for the second (stronger) shock to find additional potential hot spot sites. Unlike for the phenomenological reactive flow models discussed in the previous section the mesoscopic reactive flow models based on pore collapse are ideal candidates to simulate the situations involving the preshock desensitization of the solid explosives. Since the desensitization is believed to occur due to closure of pores, the reactive flow models based on pore collapse mechanisms should be able to account for the desensitization phenomena induced by preshocking of the heterogeneous explosives, without any special treatment such as annexation of a separate desensitization model.

During the preshocking process, the pores are either fully or partially collapsed depending upon the preshock pressure and the burn fraction is very low to lead to the growth of reactions. Obviously it can be thought that once a pore is closed/shrunk by the preshock, the preceding stronger shock will hardly find any pore to compress, inhibiting the ignition and growth. The explosive will behave as an inert material until the preceding stronger shock overtakes the weaker shock, after that explosive could be ignited depending upon the shock strength. Hence it is obvious that such models should be able to simulate multiple shock experiments, where desensitization effects play an important role. The AMORC model [65] has been shown to be able to describe the phenomenon of desensitization by preshocking in HMX, TATB, and HMX/TATB compositions [114]. In an unpublished work, the authors have demonstrated the use of the DZK model for simulation of desensitization experiments.

6. References

[1] Campbell A W, Davis W C and Travis J R, *Phys. Fluids (1958-1988)*, 1961, 4(4), 498-510.
[2] Campbell A W, Davis W C, Ramsay J B and Travis J R, *Phys. Fluids (1958-1988)*, 1961, 4(4), 511-21
[3] Bowden F P, Yoffe A D and Hudson G E, *Am. J. Phys.*, 1952, 20(4), 250-1
[4] Eden G, Belcher R A, Andrew M I and Marlow W R. *Proc., int. Symp. Detonation, 9th.*, Portland, Oregon, USA, 1989, p 1253–9
[5] Bourne N K and Field J E, *J. Fluid Mech.*, 1992, 244, 225- 40
[6] Field J E, Bourne N K, Palmer S J P, Walley S M, Sharma J and Beard B C, *Philosophical Transactions: Physical Sciences and Engineering*, 1992, 339(1654), 269-83
[7] Field J E, *Acc. Chem. Res.*, 1992, 25(11), 489-96
[8] Austin R A, Barton N R, Howard W M and Fried L E, *J. Phys.: Conf. Ser.*, 2014, 500(5), 052002
[9] Winter R E and Field J E, *Procedings A*, 1975, 343(1634), 399-413
[10] Williamson D M, Siviour C R, Proud W G, Palmer S J P, Govier R, Ellis K, Blackwell P and Leppard C, *J. Phys. D: Appl. Phys.*, 2008, 41(8), 085404
[11] Balzer J E, Proud W G, Walley S M and Field J E, *Combust. Flame*, 2003, 135(4), 547-55

[12] Balzer J E, Field J E, Gifford M J, Proud W G and Walley S M, *Combust. Flame*, 2002, 130(4), 298-306
[13] Chaudhri M M. *Proc., int. Symp. Detonation, 9th.*, Portland, Oregon, USA, 1989, p 857–67.
[14] Stirpe D, Johnson J O and Wackerle J, *J. Appl. Phys.*, 1970, 41(9), 3884-93
[15] Gustavsen R L, Sheffield S A, Alcon R R and H L G, Shock Initiation of New and Aged PBX 9501 Measured With Embedded Electromagnetic Particle Velocity Gauges, 1999
[16] Frank G, Kevin S V and Craig M T, *J. Phys.: Conf. Ser.*, 2014, 500(5), 052048
[17] Moulard H. *Proc., int. Symp. Detonation, 9th.*, Portland, Oregon, USA, 1989, p 18-24
[18] Conley P A, Benson D J and Howe P M. *Proc., int. Symp. Detonation, 11th.*, snowmas, CO, 1998, p 768–80
[19] Setchell R E, *Combust. Flame*, 1984, 56(3), 343-5
[20] Price D, Critical Parameters for Detonation Propagation and Initiation of Solid Explosives, 1981
[21] Eadie J. *Proc., int. Symp. Detonation, 4th.*, 1965, p 399-403
[22] Urtiew P A and Tarver C M, *Combustion, Explosion and Shock Waves*, 2005, 41(6), 766-76
[23] Hill L G, Bdzil J B, Davis W C and Critchfield R R. *Proc., int. Symp. Detonation, 13th.*, Norfolk, VG, 2006, p 331-41
[24] Mader C L and Forest C A, Two-dimensional homogeneous and heterogeneous detonation wave propagation, 1976
[25] Mader C L, *Numerical Modeling of Explosives and Propellants*. Third ed., CRC Press, Taylor and Francis, 2008
[26] Menikoff R and Shaw M S, *Combust. Theory Modell.*, 2008, 12(3), 569-604
[27] Ramsay J B and Popolato A. *Proc., int. Symp. Detonation, 4th.*, 1965, p 233–8
[28] Tarver C M and Hallquist J O. *Proc., int. Symp. Detonation, 7th.*, MD, USA, 1981, p 488-97
[29] Lee E L and Tarver C M, *Phys. Fluids (1958-1988)*, 1980, 23(12), 2362-72
[30] Tarver C M, Hallquist J O and Erickson L M. *Proc., int. Symp. Detonation, 8th.*, Albuquerque, NM, 1985, p 951–61
[31] Wackerle J and Anderson A B, *Chapter XIII : 5 - Burning topology in the shock-induced reaction of heterogeneous explosives, in Shock Waves Condens. Matter, Proc. Am. Phys. Soc. Top. Conf.* , J.R.A.A.G.K. Straub, Editor Elsevier, Amsterdam, 1984, pp. 601-4
[32] Murphy M J, Lee E L, Weston A M and Williams A E. *Proc., int. Symp. Detonation, 10th.*, Boston, MA, 1993, p 963-70
[33] Tarver C M and McGuire E M. 2002, p
[34] Tang Z P and Wang W W, *AIP Conf. Proc*, 2002, 620(1), 679-84
[35] Tarver C M, *Propellants, Explos., Pyrotech.*, 2005, 30(2), 109-17
[36] Tarver C M and Chidester S K, *AIP Conf. Proc*, 2007, 955(1), 429-32
[37] Tarver C M, *J. Physical Chem. A*, 2010, 114(8), 2727-36
[38] Tarver C M, Cook T M, Urtiew P A and Tao W C. *Proc., int. Symp. Detonation, 10th.*, Boston, Massachusetts, 1993, p 696-703
[39] Winter R E, Sorber S S, Salisbury D A, Taylor P, Gustavsen R, Sheffield S and Alcon R, *Shock Waves*, 2006, 15(2), 89-101
[40] Souers P C and Haselman L C J, Detonation equation of state at LLNL, 1993, *UCRLID-116113*, 1994
[41] Ceder R. *Proc., Symp. Ballistics, 28th.*, Atlanta, GA, 2014, p 150-60
[42] Murphy M J. *Proc., int. Symp. Detonation, 14th.*, Coeur d'Alene, Idaho, 2010, p 1072-8
[43] Johnson J N, Tang P K and Forest C A, *J. Appl. Phys.*, 1985, 57(9), 4323-34
[44] Handeley. *Proc., int. Symp. Detonation, 13th.*, Norfolk, VG, 2006, p 864-73
[45] James H R and Lambourn B D, *J. Appl. Phys.*, 2006, 100(8), 084906
[46] Gustavsen R L, Sheffield S A and Alcon R R, *AIP Conf. Proc*, 2004, 706(1), 973-6
[47] Lambourn B D, Whitworth N J, Handley C A and James H R, *AIP Conf. Proc*, 2007, 955(1), 137-40

[48] Cowperthwaite M. *Proc., int. Symp. Detonation, 7th.*, 1981, p 498–505

[49] Lambourn B D, *AIP Conf. Proc*, 2006, 845(1), 165-8

[50] Whitworth N J and Lambourn B D, *AIP Conf. Proc*, 2009, 1195(1), 458-61

[51] Lambourn B D and James H R, *AIP Conf. Proc*, 2012, 1426(1), 591-4

[52] Lambourn B D, Whitworth N J, Handley C A and James H R, *AIP Conf. Proc*, 2009, 1195(1), 53-6

[53] Whitworth N J, *J. Phys.: Conf. Ser.*, 2014, 500(5), 052050

[54] Lambourn B D, Handley C A, James H R and Whitworth N J. *Proc., int. Symp. Detonation, 14th.*, Coeur d'Alene, Idaho, 2010, p

[55] Whitworth N, *Mathematical and Numerical Modelling of Shock Initiation in Heterogeneous Solid Explosives*. PhD Thesis, Cranfield university, 2008

[56] Wescott B L, Stewart D S and Davis W C, *J. Appl. Phys.*, 2005, 98(5), 053514

[57] Rofi S A and Fenglei H, *Propellants, Explos., Pyrotech.*, 2008, 33(4), 311-5

[58] III A L N and Tarver C M. *Proc., int. Symp. Detonation, 12th.*, San Diego, CA, 2002, p 489–96

[59] Baer M R, Kipp M E and Swol F v. *Proc., int. Symp. Detonation, 11th.*, Snowmass, Colorado, 1998, p 788-800

[60] Baer M R and Trott W M. *Proc., int. Symp. Detonation, 12th.*, San Diego, CA, 2002, p 939–50

[61] Baer M R, *AIP Conf. Proc*, 2000, 505(1), 27-34

[62] Bardenhagen S G, KMRoessig, Byutner O, Guilkey J E, Bedrov D and Smith G D. *Proc., int. Symp. Detonation, 12th.*, San Diego, CA, 2002, p 497–503

[63] Hamate Y, Lu X R and Horie Y. *Proc., int. Symp. Detonation, 13th.*, Norfolk, Virginia, 2006, p 1270–5

[64] Barua A, Kim S, Horie Y and Zhou M, *J. Appl. Phys.*, 2013, 113(18),

[65] R. Belmas, Plotard J P and Bianchi C. *Proc., int. Symp. Detonation, 10th.*, Boston, MA, 1993, p 724–30

[66] Cook M D, Haskins P J and Stennett. C. *Proc., int. Symp. Detonation, 11th.*, snowmas, CO, 1998, p 589–98

[67] Bennett L S. *Proc., int. Symp. Detonation, 11th.*, snowmas, CO, 1998, p 612–20

[68] Khasainov B A, Borisov A A, Ermolaev B S and Korotkov A I. *Proc., int. Symp. Detonation, 7th.*, 1981, p 435–47

[69] Frey R B. *Proc., int. Symp. Detonation, 8th.*, 1985, p 68-80

[70] Kang J, Butler P B and Baer M R, *Combust. Flame*, 1992, 89(2), 117-39

[71] Bonnett D L and Butler P B, *J. Propul. Power*, 1996, 12(4), 680-90

[72] Massoni J, Saurel R, Baudin G and Demol G, *Phys. Fluids (1994-present)*, 1999, 11(3), 710-36

[73] Whitworth N J, *AIP Conf. Proc*, 2002, 620(1), 991-4

[74] Bouma R, Verbeek H J, Gruselle B and Steen. A C V d. 1997, p 139–52

[75] Brundage A L, Wixom R R, Tappan A S and Long G T, *AIP Conf. Proc*, 2009, 1195(1), 315-8

[76] Springer H K, Tarver C M, Reaugh J E and May C M, *J. Phys.: Conf. Ser.*, 2014, 500(5), 052041

[77] Cheng L R, Chen R, Shi H J and Lu F Y, *Shock and Vibration*, 2014, 2014(Article ID 972414), 8

[78] Akiki M and Menon S, *Combust. Flame*, 2015(0)

[79] Longbottom A, Boulton D and Milne A, Viscoplastic hot-spot ignition and growth model, 2002

[80] M. Sam Shaw T- and Ralph Menikoff T-. *Proc., int. Symp. Detonation, 14th.*, Coeur d'Alene, Idaho, 2010, p 1487-96

[81] Howe P M and Benson D J, *AIP Conf. Proc*, 2004, 706(1), 343-6

[82] Kim K. *Shock waves in condensed matter*, 1987, p

[83] Kim K. *Proc., int. Symp. Detonation, 9th.*, Portland, Oregon, USA, 1989, p 593–603

[84] Carroll M M and Holt A C, *J. Appl. Phys.*, 1972, 43(4), 1626-36
[85] Duan Z-P, Liu Y-R, Zhang Z-Y, Ou Z-C and Huang F-L, *Int. J. Nonlinear Sci. Numer. Simul.*, 2014, 15(5), 299
[86] Whirley R G and Engelmann B E, *DYNA2D: A Non-linear, Explicit, Two-dimensional Finite Element Code for Solid Mechanics - User Manual.* California University of California Press, 1992
[87] Wen L-J, Duan Z-P, Zhang L-S, Zhang Z-Y, Ou Z-C and Huang F-L, *Int. J. Nonlinear Sci. Numer. Simul.*, 2012, 13(2), 189
[88] Wen L J, *Research on mesoscopic reaction rate model of shock initiation of PBX.* PhD Thesis, Beijing, Beijing Institute of Technology, 2011
[89] Duan Z-P, Wen L-J, Liu Y, Ou Z-C, Huang F-L and Zhang Z-Y, *Int. J. Nonlinear Sci. Numer. Simul.*, 2010, 11(Supplement), 19-23
[90] Zhang Z Y, *Explosion and Shock Waves*, 1999(19), 360–4
[91] Setchell R E, *Combust. Flame*, 1983, 54(1–3), 171-82
[92] Mulford R N, Sheffield S A and Alcon R R, *AIP Conf. Proc*, 1994, 309(1), 1405-8
[93] Mulford R N and Swift D C, *AIP Conf. Proc*, 2000, 505(1), 895-8
[94] Bordzilovskii S A and Karakhanov S M, *Combust., Explos. Shock Waves*, 1995, 31(2), 227-35
[95] Bat'kov Y V, Glushak B L and Novikov S A, *Combust., Explos. Shock Waves*, 1995, 31(4), 482-5
[96] Gustavsen R L, Sheffield S A, Alcon R R, Winter R E, Taylor P and Salisbury D A, *AIP Conf. Proc*, 2002, 620(1), 999-1002
[97] Balagansky I A, Glumov A I, Samsonov A V, Matrosov A D and Stadnichenko I A, *Int. J. Mod. Phys. B*, 2008, 22(09n11), 1305-10
[98] Vandersall K S, Garcia F, Tarver C M and Fried L E. *Proc., int. Symp. Detonation, 15th.*, San Francisco, CA, 2014, p
[99] Held M, *Propellants, Explos., Pyrotech.*, 2004, 29(5), 267-73
[100] Arnold W and Rottenkolber E, *Procedia Engineering*, 2013, 58(0), 184-93
[101] Chick M C and Hatt D J, *Propellants, Explos., Pyrotech.*, 1983, 8(4), 121-6
[102] Held M. *Proc., int. Symp. Detonation, 9th.*, Portland, Oregon, USA, 1989, p 1416-26
[103] Held M, *Propellants, Explos., Pyrotech.*, 2001, 26(1), 38-42
[104] Held M, *Propellants, Explos., Pyrotech.*, 2002, 27(2), 88-93
[105] Mader C L, *J. Energ. Mater.*, 2014, 32, 80-7
[106] Desbiens N, Matignon C and Sorin R, *J. Phys.: Conf. Ser.*, 2014, 500(15), 152004
[107] Souers P C, Andreski H G, Cook C F, Garza R, Pastrone R, Phillips D, Roeske F, Vitello P and Molitoris J D, *Propellants, Explos., Pyrotech.*, 2004, 29(6), 359-67
[108] Whitworth N J and Maw J R, *AIP Conf. Proc*, 1996, 370(1), 425-8
[109] Campbell A W and Travis J R. *Proc., int. Symp. Detonation, 8th.*, Albuquerque, New Mexico, 1985, p 1057-68
[110] Souers P C, Andreski H G, Batteux J, Bratton B, Cabacungan C, Cook C F, Fletcher S, Garza R, Grimsley D, Handly J, Hernandez A, McMaster P, Molitoris J D, Palmer R, Prindiville J, Rodriguez J, Schneberk D, Wong B and Vitello P, *Propellants, Explos., Pyrotech.*, 2006, 31(2), 89-97
[111] DeOliveira G, Kapila A, Schwendeman D, Bdzil J, Henshaw W and Tarver C. *Proc., int. Symp. Detonation, 13th.*, Norfolk, Virginia, 2006, p 13-23
[112] Tarver C M, *J. Energ. Mater.*, 2012, 30(3), 220-51
[113] Hussain T, Tanver A, liu y and Huang F, *J. Energ. Mater.*, 2015
[114] Plotard J P, Belmas R, Nicollet M and Leroy M. *Proc., int. Symp. Detonation, 10th.*, Boston, MA, 1993, p 507–14

An investigation of preload relaxation behaviour of three zinc-aluminum alloys

A. A. Mir
Division of Engineering, Grimsby Institute of Further and Higher Education, Grimsby DN34 5BQ, England, UK

E-mail: mirlaa@gmail.com

Abstract. Zinc alloy castings are usually assembled together or mounted by screwed steel fasteners, and are tightened to a predetermined torque to develop the required tensile preload in the fastener. Due to relaxation processes in the castings, creep may cause a partial preload loss at an elevated temperature. The equipment used for load relaxation tests consists of a load-monitoring device, an oil bath, and a data-acquisition system. A load cell monitoring device is used to monitor the load loss in an ISO-metric M6×1 steel screw set into sand castings made from alloys No. 3, No. 5 and No. 2 and tightened to produce an initial preload of 6 kN. The castings were held at constant temperature in the range 80 – 120°C in an oil bath. The oil bath maintains the desired test temperature throughout the experiment. All tests were conducted for periods of up to 160 h. For all alloys, the initial load loss was high, decreasing gradually with time, but not ceasing. The load loss increased rapidly with test temperature, and almost all of the relaxation curves approximated to a logarithmic decay of load with time. Alloy No. 2 had the best resistance to load loss, with No. 5 next and No. 3 worst at all temperatures. The lower resistance to relaxation of alloy No. 3 was mainly due to the lower relaxation strength of copper-free primary dendrites, whereas in alloys No. 5 and No. 2, the higher copper contents contribute greatly to their relaxation strength in the form of second-phase particles.

1. Introduction

Load relaxation is the time and temperature dependent decrease of load in a material due to conversion of elastic into inelastic (plastic) strain [1]. During this process, the total strain is kept constant, while the initial applied load decreases with time [2]. It has been shown that load relaxation is a problem in rivet and screw joints, particularly in automotive components where the temperature is usually high. It is also considered important in other stressed components, such as wheel rims and consoles for spring brackets in cars that are used in a warm climate [3].

Zinc alloy castings are widely used for automotive components due to their excellent mechanical properties, near-net shape capability and over-all low manufacturing costs. In all these components, fasteners are used to fix them to other components and structures, or join individual parts together. The fasteners (bolts or screws) with normal or self-tapping thread forms are set in pre-threaded or plain holes [4].

Content from this work may be used under the terms of the Creative Commons Attribution 3.0 licence. Any further distribution of this work must maintain attribution to the author(s) and the title of the work, journal citation and DOI.
Published under licence by IOP Publishing Ltd

Tightening the fasteners to a predetermined torque produces a tensile preload in the fastener. A small elastic displacement of the clamped parts does not increase the preload of the fastener appreciably because the fasteners are much less stiff than the bosses or flanges that they clamp together. Therefore, torques may be applied to fasteners safely and the initial normal stress is close to the yield stress for the material [5]. It is therefore possible to induce a high preload in the fastener such that the components are joined so tightly together that their relative movement is stopped and the whole structure is integrated for the projected life of the assembly. A study of the stress distributions in steel fasteners set into steel nuts has revealed [6] that most of the loading is concentrated on the first few threads of engagement, and also the distribution of the stress depends critically on the exact dimensions of the threaded parts.

Torque-relaxation has been observed in assemblies subjected to high service temperatures and creep effects may become significant. In automotive applications, the initial preload is lost progressively because vibration may cause gradual rotation of fasteners in their threaded holes. Both effects are detected by an apparent slackening (loss of torque) of the bolt or screw, which may be temporarily rectified by re-tightening. In steel assemblies, this problem has been found to be due to creep elongation of the fastener causing partial loss of the essential preload [7].

Due to the high melting temperature of steel, the creep process in steel fasteners is insignificant at temperatures less than about 350°C. On the other hand, creep may occur at near-ambient temperatures for zinc-rich alloys [8]. Thus for steel fasteners used in zinc alloy castings at a service temperature of more than 0.4 T_m, where T_m is the melting temperature of zinc alloys, the contribution of fastener creep is negligible, and any preload relaxation can be attributed to plastic deformation in the casting. During this process, creep allows a gradual displacement of the fastener relative to the casting, so that the preload is reduced.

Although no theoretical studies of this type of creep process are known, recently some preliminary experimental investigations [4-5] showed that for steel screws in a range of commercial zinc alloy pressure-die castings, preload relaxation occurred rapidly at temperatures below 100°C.

Since these preliminary tests were undertaken on zinc alloy pressure-die castings at temperatures up to 80°C, there was therefore a need to investigate the load relaxation process in these zinc alloys at higher temperatures with different casting process i.e. sand casting. Another objective of this study was to compare the load relaxation behaviour of the commercial zinc alloys under similar testing conditions.

The load-relaxation data has little further use un-less some mathematical relationships are used to correlate it satisfactorily and allowing a quantitative analysis of the relaxation behaviour of alloys under different testing conditions. For this purpose, the relationship derived by Murphy [9] has been used to analyse the load-relaxation data of these alloys.

2. Experimental work

1.1. Chemical analysis of alloys

The chemical composition of the alloys was determined by atomic-absorption spectroscopy and the results are given in Table 1.

The test specimens used for load relaxation tests were made from the sand castings of zinc alloys No3, No5 and No2. Alloys No3 and No5 were provided by Britannia Alloys and Chemical Ltd UK in the form of ingots with guaranteed composition while alloy No2 was prepared at the foundry of the University.

These high-zinc alloys No3, No5 and No2 represent a series in which the copper content increases from 0 to 0.85 to 2.8%, respectively whereas aluminium content is constant for all three alloys. The specimens were of cylindrical shape, having the following dimensions: length, 30 mm; diameter, 13 mm and bore (threaded), 6 mm. These specimens for all tests were produced on a lathe in the Manufacturing and Production Engineering Laboratory of the University. During the production of the test-pieces, the machining operation was carefully controlled so as to reduce the surface finish variations to a minimum.

Table 1. Chemical composition of experimental alloys

	Al	Cu	Mg	Zn
No3	4.23	0.008	0.05	balance
No5	3.96	0.84	0.04	balance
No2	4.21	2.79	0.03	balance

1.2. Testing equipment

The equipment used for testing consists of a load-measuring device, an oil bath and a data-acquisition system. The load-measuring device was used for the continuous monitoring of load in the commercial fasteners used. It consisted of a short tension rod to which the head of the fastener was attached, which reacted to the tensile stress in the fastener through a compression load cell. When the casting was screwed onto the fastener, it simulated the effect of inserting a screw into a hole. The load cell was used to measure the force exerted on the screw thread of a casting. It was a small, high capacity load cell, having a range of 0.15-13600 kg and its operating temperature range was -54 to 121°C. The oil-bath was used to heat the specimen and maintain the same temperature throughout the test. The maximum operating temperature of bath was 150°C, and had a temperature controlling accuracy of ±0.1°C. At one time, up to seven specimens could be placed in the oil bath.

The data-acquisition system was used to record the results of tests in the form of retained load(N) versus the test time (Minutes).Variable speed logging was a useful feature provided by the system. This allowed programming the data logging speed to suit the application, i.e. at the initial stage of the tests, more data readings were measured and recorded than the later stages of the test. The time interval between two readings at the start of the tests was 5 and 10 seconds, whereas in the last stage of the tests, this interval was kept as one hour.

An M 6x1 (ISO standard) threaded screw was locked into the tension rod, and the load monitor assembled. The test piece was then threaded onto the screw up to 16mm depth, which was the desired engagement of screw. The entire monitoring assembly was dipped into the hot oil at constant required test temperature, and left for about 2-3 hours to equilibrate. To start the experiment, the tension rod and screw were loaded by turning the fine-pitch nut at the top of the rod. This process was continued until the initial load of approximately 6000 N which was recorded by the load cell, could be achieved. This whole process completed in a very short time, i.e. 15 ~ 20 s. The data-acquisition (logging) system was started as tightening commenced. The preload loss was then continuously monitored by the load cell for a period of up to about 155 hours, after which the test was terminated. Up to four tests were carried out simultaneously, and the test temperature was kept constant with a variation of ± 0.5°C.The results of experiments were recorded by data-acquisition system, and displayed on a computer in the form of time (Minutes) versus load (N) relaxed.

2. Results and discussion

All load relaxation tests were carried out in duplicate, and if results were not satisfactory, i.e. appreciable scatter was found in the results, some additional tests were also performed at those particular test conditions. The average values of the results were calculated and the mean values taken to represent the load relaxation behaviour of these zinc-based alloys.

The retained load (N) was measured continuously and plotted as a function of test time (minutes) for all three alloys at 80, 100 and 120°C, the results are shown in Figure 1 at 100°C. In all cases, the initial load loss was high, decreasing gradually with time but no ceasing. The amount of load loss increased rapidly with test temperature and almost all of the curves approximated to a logarithmic decay of load with time although there was some variation in the initial part of the relaxation curves. This variation was probably due to the effect of localised shear creep of the threads in the castings, which caused a redistribution of the loading. The results indicated clear differences in the comparative resistance to load loss of these alloys. It was revealed that alloy No 2 had the best resistance to load loss, with alloy No5 next and No3 worst at all temperatures as shown in Figure 1. Graphs were drawn to show the variation of retained load (N) with reciprocal temperature at 50, 100 and 150 h for all

IOP Conf. Series: Materials Science and Engineering **146** (2016) 012042 doi:10.1088/1757-899X/146/1/012042

alloys; an example is shown for 100h, Figure 2. From these plots, the effect of short and long-term load relaxation tests can be observed. The graphs at 50 h time showed the short-term effects of load relaxation while 150h time graphs demonstrated the long-term effect. These plots showed that alloy No 2 was the best resistant to relaxation while No 5 was better than No 3 at both short and long time periods. To show the effect of copper content on load relaxation behaviour of alloys, the variation of the retained load (N) against the copper content (%) at different temperatures was plotted at 50, 100 and 150 hours, as shown in Figure 3 for 100 h. Alloy No3 was much inferior in load relaxation strength than alloys No5 and No2 at all temperatures, on the other hand alloy No2 was better than No5, but difference was greater at 100°C as compared to other temperatures. This indicated that for these alloys, the resistance to load relaxation increased with increasing copper content.

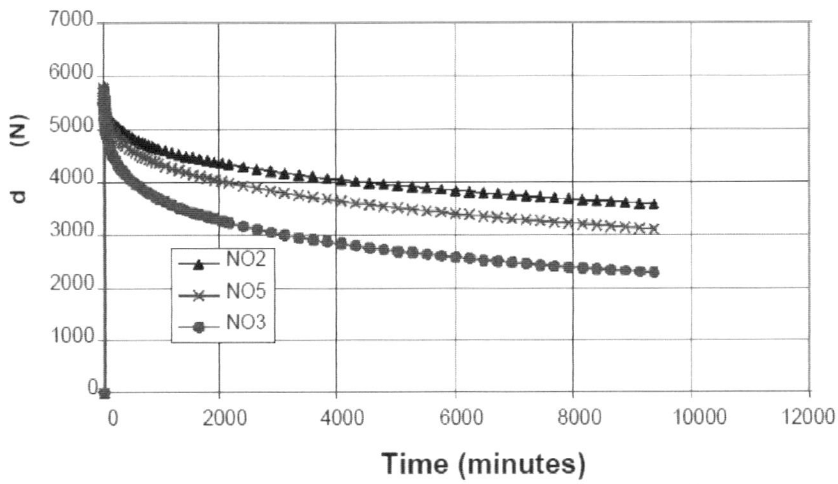

Figure 1. Retained load vs test time for alloys at 100°C

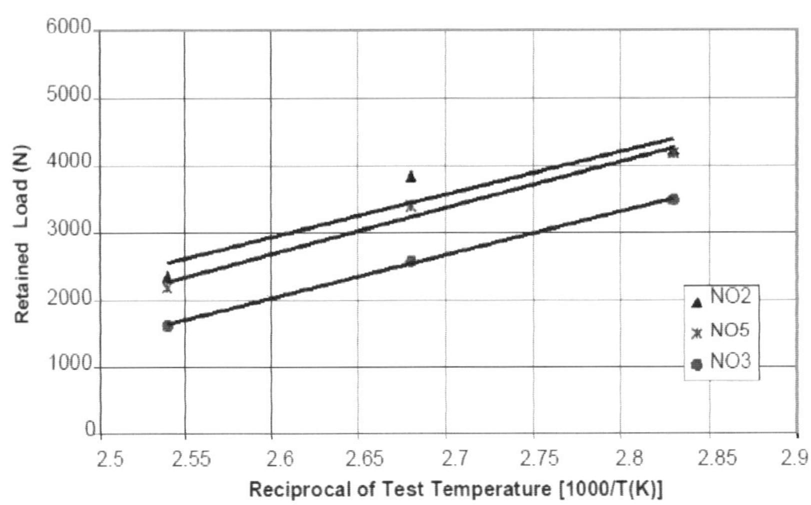

Figure 2. Retained load vs reciprocal temperature for duration of 100 h

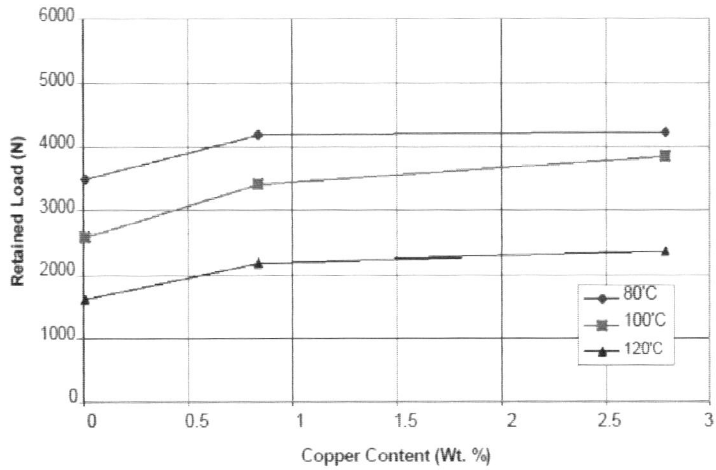

Figure 3. The variation of the retained load against the copper content (%) at different temperatures for test time of 100 ht

Load relaxation data for the sand-cast zinc alloys during the current investigation was obtained from a limited number of tests of relatively short duration. This data has little further use unless some mathematical relationships are used to correlate it satisfactorily and allowing a quantitative analysis of the relaxation behaviour of alloys under different testing conditions. For this purpose, the following relationship derived by Murphy [9] has been used to analyse the load-relaxation data:

$$\ln L = \alpha \, [Q/RT - \ln B - \ln t] \quad (1)$$

Where L is the load after time t, Q is an activation energy for load relaxation, R is the universal gas constant, and α and B are constants. Using the experimental data, the mean values of constants α, ln B and Q were derived and finally the retained loads (N) were calculated for alloys by using the general equation for load relaxation (Eqn. 1) at 50°C, 80°C, 100°C and 120°C after 30,000 h of service time, as shown in Table 2. The calculated values of retained loads (N) showed that alloy No2 was the best resistant to relaxation, then No5 and No3 was the worst at all temperatures.

Table 2. Chemical composition of experimental alloys

Alloy	50°C	80°C	100°C	120°C
No3	1915	974	659	464
No5	2842	1595	1144	848
No2	3214	1934	1442	1108

Load relaxation is a phenomenon that is closely related to creep process and follows the same mechanisms as creep. It is therefore logical to analyse the results of load relaxation in the light of general creep theories.

It has already been observed that the rate-controlling mechanism operating in compressive creep of zinc alloys No3, No5 and No2 was dislocation climb [10]. The values of the stress exponents are within the range of 3.0 – 5.5, indicating Weertman's dislocation climb as the most likely mechanism [11]. The climb-controlled creep model was proposed by Weertman [12] and has reasonable agreement with some previous experimental results during creep of alloys [12-13]. The current load relaxation results may therefore be analysed considering the dislocation climb as a rate-controlling process. From the analysis of load relaxation results and structure of the experimental alloys, it may be concluded that the higher copper content in alloys No5 and No2 contribute greatly to their relaxation strength in the form of second-phase (ε) particles [10]. These second-phase particles play an important role in preventing dislocation movement and therefore reducing the relaxation rate as observed for creep experiments of these alloys. The lower resistance to load relaxation of alloy No3 was mainly due to the lower relaxation strength of copper-free primary h dendrites, having greater volume than the eutectic in the microstructure.

Copper also influences the precipitation properties of alloys by increasing the precipitate volume and thus gives a certain strengthening to alloys No5 and No2 showed much better relaxation strength than alloy No3, based on precipitation hardening due to the presence of small copper-rich e-phase particles. Another well-known effect of copper additions to these alloys is a reduced tendency to inter-crystalline weakness due to copper-stimulated precipitation within the grains.

Conclusions

I. The preload relaxation equation: $\ln L = \alpha\ [Q/\ RT - \ln B - \ln t]$ could be used to correlate the experimental data of the zinc-rich alloys
II. The retained loads determined from the experimental data showed that alloy No2 was the mostresistant to relaxation, then No5 and No3 was the worst at all temperatures
III. The resistance to load loss increased with increasing copper contents under all testing conditions
IV. The lower resistance to preload relaxation of alloy No3 was mainly due to the lower relaxation strength of copper-free primary η dendrites whereas in alloys No5 and No2, the higher copper contents contribute greatly to their relaxation strength in the form of second-phase (ε) particles
V. Dislocation climb was considered as the rate-controlling mechanism for load relaxation in these zinc-rich alloys
VI. The values of power law constants, i.e., α and B generally increased with the increase of temperature. The mean values of α showed a decrease with the increase of copper content
VII. The values of the activation energy Q were 95, 103 and 110 kJ/mole for alloys No3, No5 and No2, respectively, indicating an increase with the increase of copper content

3. References

[1] Fox A, (Ed.) 1979(May 24-25) Stress Relaxation Testing, *ASTM Special Technical Publication* 676, Kansas City, MO
[2] Garofalo F 1965*Fundamentals of Creep and Creep-Rupture in Metals*, MacMillan, New York
[3] Solberg J K and Thon H 1985*Materials Science and Engineering*75105-116
[4] Murphy S and Goodwin F E 1996(Feb. 26-29) *International Congress and Exposition, (SAE Technical Paper Series)* (Detroit, Michigan) p 1-10
[5] Murphy S and Haines C W 1994 *Proceedings of Advances in Science, Technology and Applications of Zn-AlAlloys*(Mexico) p 3-7
[6] Sopwith D G 1948*Proc. Inst. Mechanical Engineering*159 (London) p 373
[7] Markovets M P 1970*Thermal Engineering*17 77
[8] Goodwin F E and Ponikvar A L (Ed.) 1989 Engineering Properties of Zinc Alloys. *Int. Lead Zinc Research Organisation Inc.*(New York) 3rd Edition p 41-47, 64-66
[9] Murphy S 1996Torque Relaxation in Zinc Die castings, Project Report, Project ZM 414(UK)
[10] Mir A A 1998*The Creep Properties of a Series of Zinc-Rich Zinc-Aluminium Alloys* PhD Thesis, Aston University (UK)
[11] Bird J E, Mukherjee A K and Dorn J E 1969 Proc. *An Int. Conf. on Quantitative Relation between Properties and Microstructure* Editors: Brandon D G and Rosen A (Israel Universities Press, Jerusalem)
[12] Weertman J 1957*J Applied Physics* 28 362-364
[13] Lagneborg R 1972Dislocation Mechanisms in Creep *Int. Metall. Rev*171

IOP Conf. Series: Materials Science and Engineering **146** (2016) 012043 doi:10.1088/1757-899X/146/1/012043

Sn-Pb and lead free solders containing active carbon particles

S. Talas, B. Gökçe and M. Çakmakkaya
Afyon Kocatepe University, Faculty of Technology, Department of Metallurgical and Materials Engineering, Afyonkarahisar, 03200, Turkey

E-mail: stalas@aku.edu.tr

Abstract. Upon the legislations issued by the governmental agencies, many companies are in effort of using lead free solders for their electronic products. Many researchers have also focused on lead free solders and determined their physical properties to the merit of their desired strength and conductivity which turns out to be a potentially advantageous after all. The addition of nano particles into the solder alloys has been attempted to investigate the property change caused by such addition from which a main outcome was a limited improved mechanical and physical properties such as lowering the melting temperature. In this study, the addition of nano active carbon particles to Pb-Sn and Pb-free solder alloys were made and characterization studies were conducted to determine their basic properties such as electrical conductivity, microstructural study and also phase transformations. The results indicate that the addition of active carbon particles brings about a change in thermal properties more markedly than other properties with respect to the amount of addition.

1. Introduction

The activated carbon is produced from organic material such as food extracts, for places where high surface area for filtration of contaminants from air and water. Activated carbon is extremely porous substance and hence can be used for adsorption of relatively low-molecular-weight organic compounds such as phenols [1-2].

The environmental concern over the toxicity of Pb has led the companies to abide by strict regulations of government agencies and the research on lead-based solders is promoted through regulations [3]. Various research agencies and companies have produced enormous data on binary alloys as candidates for replacement of the traditional Sn-Pb solder alloy. Properties such as wettability, electrical and thermal conductivities and the amount of alloying additions, mechanical properties such as ductility, creep resistance, thermal fatigue and also manufacturability are desirable main characteristics for mainstream solder alloys. Solders alloys of Sn-Ag and Sn-Cu binary system and ternary additions are favourable for their good mechanical behaviour [4-5]. The typical hypoeutectic Sn-X alloys have usually Sn-rich dendritic matrix and a eutectic mixture of a Sn-rich phase and intermetallic phases such as Cu_6Sn5 and Ag_3Sn [5-6]. Sn-Cu alloys as an alternative to Sn-Pb alloys are cheaper to produce with conventional methods and have a eutectic reaction at 227°C situated between intermetallic Cu_6Sn_5 phase and the Sn-rich phase and the addition of some alloying elements, such as Ni, Ag and Zn have been the subject of study for the improvement of their certain properties [6-8]. These intermetallic phases accelerate the corrosive deterioration due to galvanic corrosion mechanisms. The dissolution of tin is main breakdown mechanism for the failure [9]. There are very few studies regarding the addition of carbon derivatives into solders which may potentially prevent such failure and improve the properties [10-12]. The addition of carbon black was found to

Content from this work may be used under the terms of the Creative Commons Attribution 3.0 licence. Any further distribution of this work must maintain attribution to the author(s) and the title of the work, journal citation and DOI.
Published under licence by IOP Publishing Ltd

improve the mechanical properties of Sn-3.5Ag alloy due to uniform distribution of the Ag_3Sn particle and help inhibit the growth of Ag_3Sn IMC layer by the adsorption of carbon black particles at the phase boundary in the IMC layer[10]. There was no change associated with melting temperature. The addition of carbon nanotubes, however, caused a change in melting temperatures and improved the mechanical properties and did not cause a substantial change in electrical resistance, too[12].

In this study, the additions of nanosized active carbon (AC) particles in the lead containing (Sn40Pb60) and lead free solder alloys (SnCu3 and SnCu0.3) have been made and characterization of alloys was conducted to determine their basic properties.

2. Experimental procedure

The activated carbon was produced from orange peel in a high temperature reactor made of stainless steel which can be pressurized upto 15 Bar and can be operated at 300°C. The food extracts such as orange peel, hazelnut shells have been used to produce active carbon powders. The reactor has a steam disposal mechanism to let the unwanted gasses and water vapour come out of the reactor. The heating process was conducted for 2 hrs at a maximum pressure of approximately 5 bar. The extract was then dried over a night at a temperature of 90°C in N_2 containing autoclave. Following the manufacturing of activated carbon, BET (Micromeritics Gemini 2360) was used to measure the surface area and average size of powders and it was shown that the average particle size was approximately 146±58 nm. Following the manufacturing of activated carbon, BET (Micromeritics Gemini 2360) was used to measure the surface area and average size of powders and it was shown that the average particle size was approximately 146±58 nm.

Three types of solders have been used for the experiments: Sn40Pb60 (LK1X), SnCu3 (3% Cu) (LK2X), SnCu0.3 (0.3% Cu) (LK3X). The mixing process of active carbon into solders was made by remelting and casting of solders containing no addition (LKXA series), approx. 0.14 (LKXB series) and 0.3 wt% (LKXC series) of active carbon additions. At a temperature of 270°C, the casting was made into Al crucible which was heated to 45°C, then; the specimens were subjected to standard metallographic examination. SEM with EDX attachment was employed for microscopic investigation to reveal the elemental change within the matrix. No etching was carried out but BSE (Back Scattered Electron) mode was used for imaging during SEM analysis.

Figure 1. Remelting pot for casting of solder without and with active carbon additions

Figure 2. Resistance measurement jig for solder alloys

Hardness measurements were made using Shimadzu microhardness tester HMV01 with a load of 100 g using Vickers tip. The measurement was repeated more than five times and results were given in average of total measurements.

Electrical conductivity measurements were conducted by using GW Instek GOM 802 milliohmeter with a precision of 1 micro ohm. An apparatus was built, as seen in Figure 2, with a constant load of approximately 1.3 Kg applied by a spring. The measurements were made at room temperature using pure copper pins between which the inter distance length from pin to pin was 11.5 mm.
Melting point of alloys was measured using a electrically isolated thermocouple sensor with a diameter of 1mm by dipping the thermocouple into the melt. Data logger DI145 of DATAQ was used to capture the change in temperature during cooling after casting.

3. Results and discussion

3.1. Hardness and Electrical conductivity

In table 1, hardness of each solder alloy were measured and the results showed that there is slight increase in hardness with respect to addition of carbon particles, which is possibly result of matrix hardening effect from particles as well as the cooling rate associated with casting process. Since there is no difference in casting method, it is difficult to determine the reason for hardening based on casting effect. However, it is possible that the use of Al pipe to direct the melt to the mould may have helped to the introduction of Alumina into the alloys, contributing to the hardness of the matrix. The hardness values are usually similar to each other with very little difference but this not true for the series of LK2X which contain 3% Cu.

Table 1. Hardness values of solders in $HV_{0.1}$ (±%2.2)

LK 1A	LK 1B	LK 1C	LK 2A	LK 2B	LK 2C	LK 3A	LK 3B	LK 3C
12.5	13.1	14.4	11.2	11.7	12.8	9.3	10.8	11.3

The experimental results of electrical conductivity measurements are presented in table 2. The results revealed that the electrical conductivity of the bulk composite solders were comparable to that of the unreinforced Sn-Cu solder and were also better than that of conventional Sn40Pb60 solder. In fact, in all series the addition of AC (Active Carbon) resulted in changes in the measurement values especially in Sn40Pb60 series which caused serious changes hardness and resistivity/conductivity values. The disturbance of normal motions of electrons is primarily attributed to lattice scattering and impurity scattering which reduces the mean free path of electron motion within the matrix phase. A reduction in electron mobility and hence an increase in the resistivity value are inevitable for composite solders[12]. The total volume fraction of the electron scattering will result in an increase in the materials resistivity value, however, the electron contribution from other sources will result in a positive change in resistivity.

Table 2. Electrical conductivity coefficient of solder alloys (μohm.m)

LK 1A	LK 1B	LK 1C	LK 2A	LK 2B	LK 2C	LK 3A	LK 3B	LK 3C
0.179	0.165	0.136	0.122	0.110	0.116	0.118	0.113	0.101

Table 3 shows the approximate melting temperatures of each solder studied, measured during cooling with dipped thermocouple. In general, the melting temperatures of solders without AC additions are similar to their original manufacturer's data, however, the additions of AC has caused slight changes in the melting point. The mechanism for such improvement may be due to the presence of AC particles by inhibiting the formation of inter-metallic compounds, which is not studied in this work. It is also possible that inhibition the IMC formation may be due to a layer of AC at the interface of Sn rich and other phases acting as a diffusion barrier.

Table 3. Approximate transformation temperatures of solders studied

LK 1A	LK 1B	LK 1C	LK 2A	LK 2B	LK 2C	LK 3A	LK 3B	LK 3C
186.6	185.8	186.2	226.3	225.2	224.4	228.6	228.3	227.8

3.2 Microscopy

3.2.1. Leaded solders

Electron microscopy results were obtained from specimens with and without additions of nanoactive carbon particles are given in Figures 3-5 below. The results indicate that, in Sn/Pb standard alloy, particles are embedded in the softer phase of Pb, which is in light grey colour in Figure 3. All alloys in this series possess a matrix microstructure of primary Sn dendrites with inter-dendriticly formed Sn-Pb eutectic, which is partly deteriorated during casting processing. The processing route to cast may influence the shape and size of the primary Sn phase within the microstructure. This has been well documented by Fan (9) and Ji and Fan (13). Since there is no additional thermal or mechanical processing on liquid/solid phase during or after the casting, which may affect grain sizes detrimentally (14), the grain sizes appear to be similar in all specimens. This is also dependent on the heat removal process at the interface of alloy and Al mold which was heated upto 35 degrees centigrade prior to casting to prevent excessive fast cooling. Sn phase seperation is ensured through this process but two phase interdendritic phase was not observed possibly due to unetched surface.

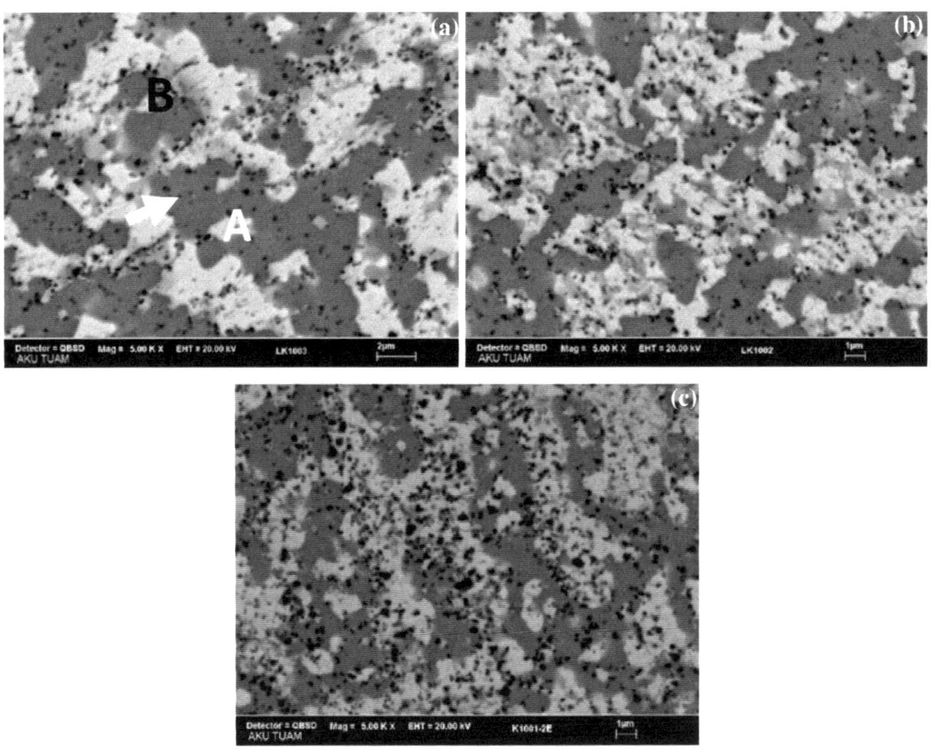

Figure 3. a) Sn/Pb alloy without addition (LK 1A), b and c) Active carbon additions were made to Sn/Pb solder alloys at the amounts of 0.14 (LK 1B) and 0.30 wt% (LK 1C), respectively. (A indicates Pb rich and B indicates Tin rich phase. The arrow in Figure 3a and Figure 3c indicate particles and the dendrite arm, respectively.)

IOP Conf. Series: Materials Science and Engineering **146** (2016) 012043 doi:10.1088/1757-899X/146/1/012043

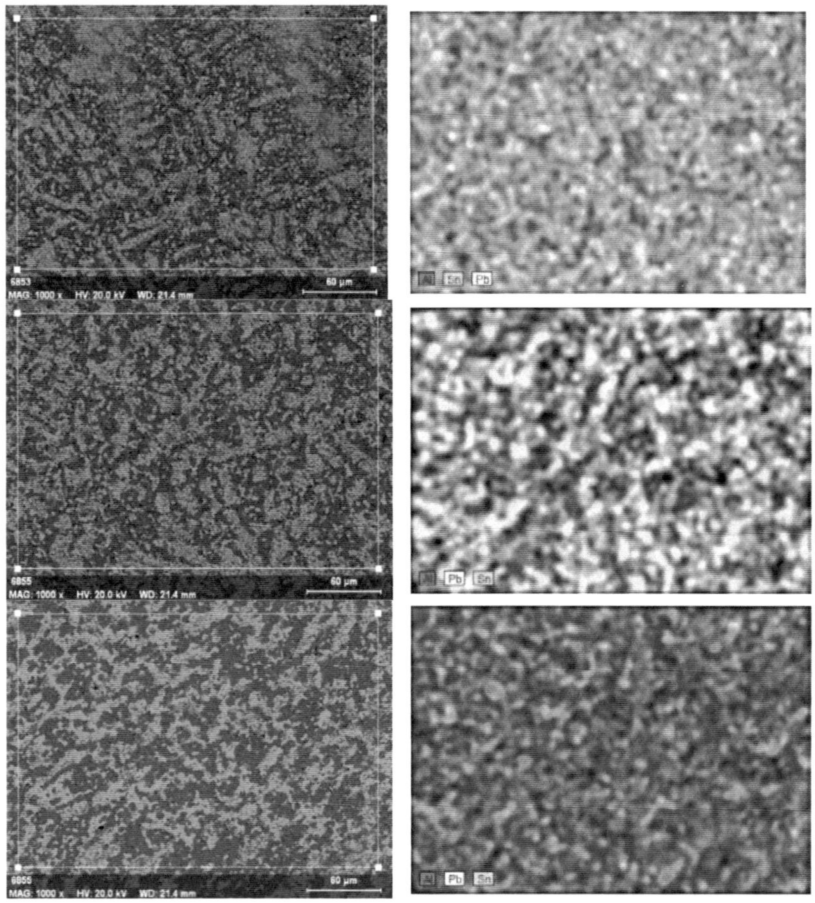

a)
b)
c)

Figure 4. Effects of active carbon additions on the microstructural development of Sn/Pb alloys; a) no addition, b) 0.14 wt% and c) 0.3wt% AC additions with EDX mapping

SEM images in Figure 4 show that, as the amount of active carbon increases, the phase separation becomes more distinctively different by the selective separation of Tin into Tin rich regions, leaving dendrites morphologically deformed and poor in Tin. At high concentration of AC, it is more obvious to discern that the phase separation and the breakdown of dendrites may be associated with high surface area of AC particles which increases the surface energy of second phase regions in liquid state with instability of interface to dissociate in ample time.

3.2.2. Lead free solders

The unetched EBSD image of SnCu3 (Figure 5) and SnCu0.3 (Figure 6) alloy microstructures are shown with respect to the amounts of additions. A relatively well distributed carbon particles are random. The mixing of carbon particles are relatively successful, however, the coagulation of these particles are possible as seen in Figures 5 and 6. SEM image in EBSD mode does not show significant compositional difference due possibly to single phase of Sn matrix, although, hypothetic Sn-Cu alloy has a eutectic point. The inhibition of second phase may well be due to the presence of AC in Figure 5b and Figures 6b and c. Figure 5a and Figure 6 a do not contain any AC, which indicates that the amount of Cu is so small that it is not detectable by EDX sensor. Considering the pattern of particles, it can be said that particles follow the solidification fronts of grains and trapped in the grain boundaries as in Figure 5a. As the amount of AC increased, it has become more complex and indiscernable.

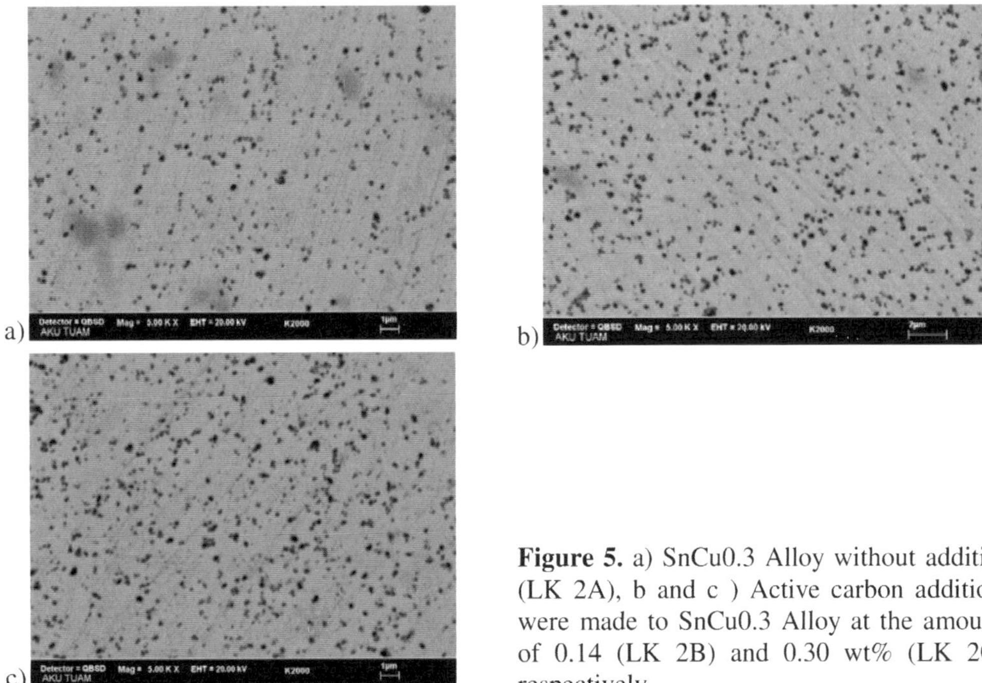

Figure 5. a) SnCu0.3 Alloy without addition (LK 2A), b and c) Active carbon additions were made to SnCu0.3 Alloy at the amounts of 0.14 (LK 2B) and 0.30 wt% (LK 2C), respectively

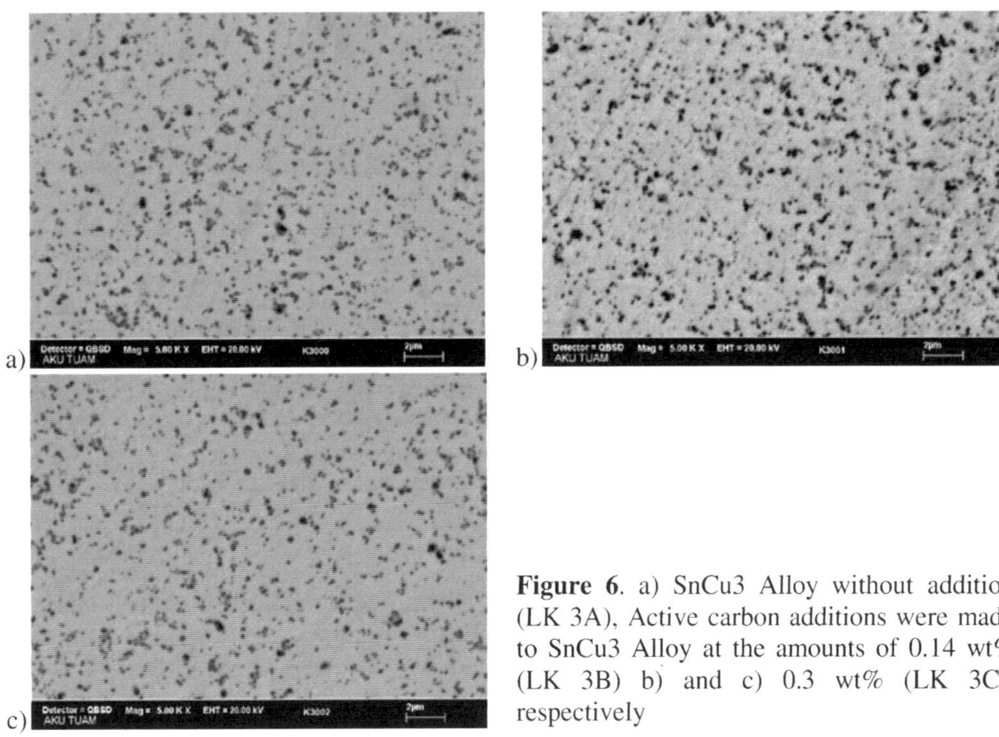

Figure 6. a) SnCu3 Alloy without addition (LK 3A), Active carbon additions were made to SnCu3 Alloy at the amounts of 0.14 wt% (LK 3B) b) and c) 0.3 wt% (LK 3C), respectively

4. Conclusions

Leaded solders were micro structurally affected by active carbon additions which possibly promoted phase separation at microscopic level. A possible mechanism for the phase separation may be due to the difference in surface energy effect of active carbon particles on the phases, resulting in the increase of their surface energies. Unleaded solders did not show any differentiation in their microscopic properties because of their single phase matrix and the additions did not cause any phase separation. However, the resistivity values may be positively affected by the addition of active carbon particles.

5. Acknowledgments

This study is financially supported by Afyon Kocatepe University, Scientific Research Project Commissions (Project Number: 15.FEN.BIL.43). Authors are also grateful to Prof. Selçuk Aktürk of Muğla University for providing active carbon.

6. References

[1] Dąbrowski A, Podkościelny P, Hubicki Z and Barczak M 2005 *Chemosphere* 58 1049
[2] Dubinin M M 1989 *Pure Appl. Chem.* 61 1841
[3] European Union, Directive 2002/95/EC of the European Parliament and of the Council of 27 January 2003 on the restriction of the use of certain hazardous substances in electrical and electronic equipment, Official J. Eur. Union 2003, 46, 19-23
[4] K. Kanlayasiri K and Ariga T 2010 *J. Alloys Compd.* 504 15
[5] Lin F, Bi W, Ju G, Wang W and Wei X 2011 *J. Alloy Compd.* 509 6666
[6] Deghaid P P, Eduardo S J and Garcia A 2013*Mater. Des.* 45 377
[7] Hammad A E, 2013 *Mater. Des.* 50 108–116
[8] El-Daly A A and El-Taher A M 2013 *Mater. Des.* 47 607
[9] Z. Fan, 2002 *Inter. Mater. Rev.* 47 49
[10] Son P V, Fujitsuka A and Ohshima K 2012 *Journal of Electronic Materials* 41 1893
[11] Kumar K M, Kripesh V and Tay A A O 2008 *J. Alloys Compd.* 455 148
[12] Nai S M L, Wei J and Gupta M 2008 *Journal of Electronic Materials* 37 515
[13] Ji S and Fan Z 2002 *Metall. Mater. Trans. A* 33 3511
[14] Kim Y K and Qin RS 2010 *Materials Science Forum* 649 415

Hydrothermal synthesis and characterization of fluorine & manganese co-doped PZT based cuboidal shaped powder

H. Nawaz, M. Shuaib, M. Saleem, A. Rauf and A. Aleem
Institute of Industrial Control Systems, P.O. Box 1398, Rawalpindi, Pakistan

E-mail: rhn.qau@gmail.com

Abstract. Cuboidal shaped PZT powder particles based composition $Pb_{0.89}(Ba, Sr)_{0.11}(Zr_{0.52}Ti_{0.48})O_3$ co- doped with 1 mol% manganese and 2 mol% fluorine was prepared through hydrothermal route. 200-250nm size cuboidal particles were observed under FE-SEM. XRD technique revealed that the perovskite type ceramic structure has a dominant rhombohedral phase. The resultant powder particles were then spray dried, uniaxially pressed and sintered at different temperatures to achieve maximum theoretical density. 98% density was obtained in the pellets at a sintering temperature of 1190°C with an average grain size of 1-3um. The electrical properties of sintered samples were also measured before and after poling to evaluate the effect of dopants on piezoelectric properties.

1. Introduction

The composition of piezoelectric materials can be tailored to achieve properties suited for many applications such as medical imaging, dielectrics, thermal detectors, microwave resonators and transducers [1]. Out of these piezoelectrics, perovskite type, lead zirconate titanate (PZT) are of particular interest. Although pure PZT ceramics are not often used, their electrical properties can be modified by the addition of dopants. The dopants are classified into two classes i.e., donors and acceptors. The donor dopants induce soft piezo character i.e., increase dielectric constant & charge coefficient d_{33}. The excess of positive charge in soft PZT is compensated by lead vacancies. Whereas the acceptor dopants give rise to hard PZTs i.e., lower dielectric constant and d_{33}. These type of dopants induce oxygen vacancies in the system [2]. The effect of co-doping on the properties of Piezoceramics has also been studied extensively. Guiffard et al. have reported that the Mn-F co-doped PZT, prepared through co-precipitation method, have high mechanical quality factor Q_m and better electromechanical & dielectric characteristics [3].

The PZT ceramic powders can be prepared by mechanical alloying, co-precipitation and hydrothermal routs. However, the powder particles prepared via hydrothermal route is advantageous in many prospects including chemical homogeneity and lower sintering temperature [4]. As a result lead volatilization is minimized during sintering process and sintered product is expected to have uniform chemical composition.

According to best of our knowledge, (Mn,F) co-doped PZTs have not yet been synthesized through hydrothermal route. This study focuses on the structural and electrical characterization of the hydrothermally synthesized composition $[Pb_{0.89}(Ba, Sr)_{0.11}(Zr_{0.52}Ti_{0.48})O_3]$ co-doped with 1 mol% manganese and 2 mol% fluorine.

Content from this work may be used under the terms of the Creative Commons Attribution 3.0 licence. Any further distribution of this work must maintain attribution to the author(s) and the title of the work, journal citation and DOI.
Published under licence by IOP Publishing Ltd

2. Materials

The materials used in this study includes Iso-Propanol (99.5%, BDH), Lead Nitrate (99%, Fisher), Titanium (IV) isopropoxide (99.5%, Fluka), Zirconium (IV) n-propoxide (70%, Alfa Aesar), Barium Acetate (99%, Riedel-de Haen), Strontium Acetate (99%, Alfa Aesar), Lead Fluoride (99.5%, Alfa Aesar), Manganese Acetate (99%, Acros organics), Poly Vinyl Alcohol (Merck), DOLOPIX (PC 67, ZS Chimmer & Schwarz), Acetic Acid (Sigma Aldrich) and KOH (Merck)

3. Experimental

A solution of Zirconium n-propoxide and Titanium isopropoxide was prepared in Iso-Propanol. The solution was co-precipitated in water and filtered. The filtrate was mixed with the stoichiometric amounts of Lead Nitrate, Barium Acetate, Strontium Acetate, Lead Fluoride and Manganese Acetate. Distilled water was then added upto $3/4^{th}$ of the vessel and the pH of the system was raised to 14 using KOH pellets. The prepared solution was loaded into the chamber of R-201 type reactor system and a hydrothermal treatment was carried out at 260 °C under auto-generated pressure. After completion of reaction, the synthesized powder was filtered and washed twice with distilled water and once with dilute acetic acid to remove unreacted Lead Acetate and again with distilled water. Finally, the filtrate was dried in an oven at 100 °C.

The powder particles, thus obtained, were mixed with binder (PVA) and dispersant (DOLOPIX) and spray dried. The resulting powder was uniaxially pressed in the form of pellets and sintered at different temperatures to optimize the density.

4. Characterization

Size and morphology of the powder particles was characterized using a *Quanta-450* FEG-SEM technique. The structural characterization of the powder samples was carried out by *JEOL JDX9C* X-ray diffractometer using CuKα (λ = 1.54056 Å) radiations. The samples were scanned from 20° - 80° with a scan rate of 1° per second at a current and voltage ratings of 35mA and 30KV, respectively. The morphology of the sintered pellets was observed through JSM-5910 LV, JEOL Scanning Electron Microscopy (SEM), whereas the density was measured using Archimedes' Principle. For the electrical measurements, sintered samples were electroded with silver paste on both sides and fired at 700 °C. The samples were then poled with an applied field of 3KV/mm in a silicone oil bath at 120 °C. Electrical properties such as capacitance, piezoelectric coefficient (d_{33}) and mechanical quality factor (Q_m) were measured using *7600 Precision* LCR meter, *PM-300* Piezo Meter and TA-2000 respectively.

5. Results and discussion

5.1. X-ray diffraction analysis

Figure 1 shows the typical XRD pattern of provskite PZT based powder particles along with miller indices for each peak. Mn addition led to the formation of rhombohedral phase instead of tetragonal phase as depicted by (200) peak instead of (211) which is a typical behaviour of manganese doped PZT systems [5-6].

Possible explanation for phase change may be related to the ionic state of manganese (Mn^{+4}, Mn^{+2} and Mn^{+3}) and its location in the perovskite structure where it substitutes the B-site ion. Mn^{+2} assumes high spin states $t_{2g}^3 e_g^2$ (d5) and Mn^{+3} $t_{2g}^3 e_g^1$ (d4) as O^{-2} is a weak ligand. The presence of symmetrical spin state $t_{2g}^3 e_g^2$ suggests that d^4 electron state of Mn^{+3} ion makes it a Jahn–Teller ion along with the generation of Jahn–Teller distortion [7].

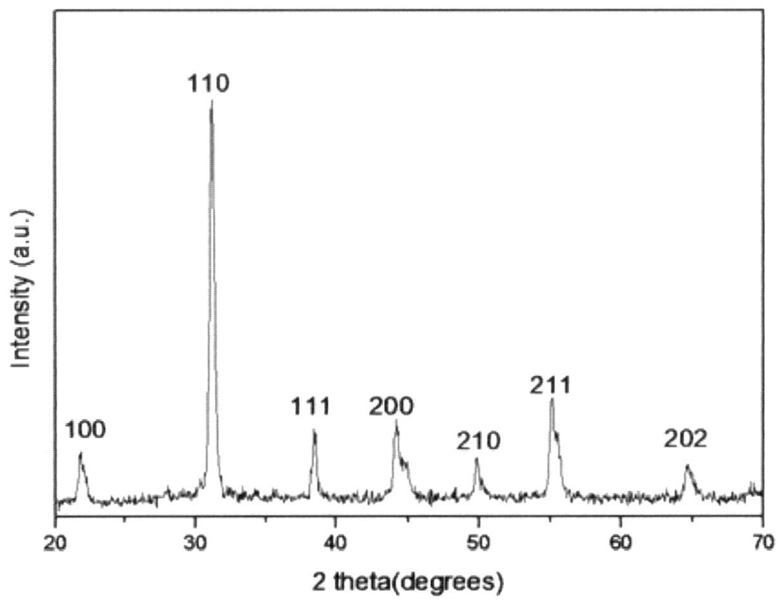

Figure 1. XRD pattern of $Pb_{0.89}$ (Ba, $Sr)_{0.11}(Zr_{0.52}Ti_{0.48})$ O_3 co-doped with 1 mol% manganese and 2 mol% fluorine powder particles

Splitting of orbitals from the Jahn-Teller effect for Mn^{+3} ion has been shown in Figure 2. The Figure shows that d_z^2 orbital is occupied by none of the d-electron which results in higher electron density in xy-plane as compared to z-direction. Thus the electrostatic attraction between Mn^{+3} and O^{-2} ligands experiences low shielding effect in z-direction as compared to that of xy-plane. This decreases the c/a ratio in the unit cell so rhombohedral phase is exhibited by the perovskite type structure with Mn addition.

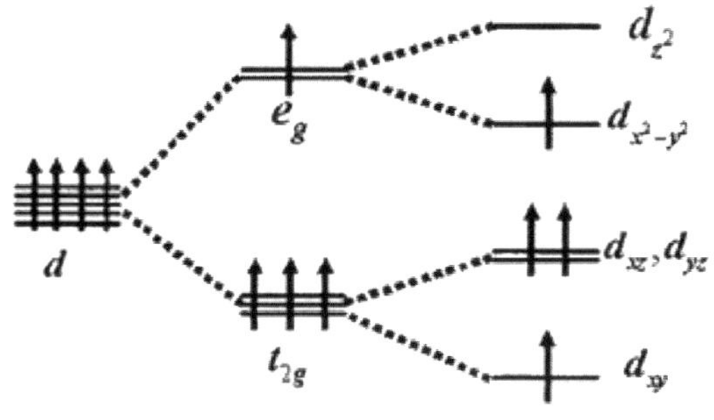

Figure 2. Splitting of orbitals of Mn+3 ion from the Jahn-Teller effect

5.2. *Morphological analysis*

The morphological analysis of the synthesized powder and sintered pellets was carried out using FE-SEM and SEM respectively (Figure 3). Figure 3(a) & (b) show the cuboidal geometry of powder particles having size of 200-250 nm size. The uniformity in the geometry is a characteristic feature of hydrothermal route. Such powder particles lead to a uniform grain growth during sintering [4] as shown in Figure 3 (c) & (d). Fully dense structure with 1-2um grain size was observed in the sintered pellets. Smaller grain size may be attributed to the hydrothermal processing which resulted fine powder particles. The grain size may also be correlated with the dopant addition. During sintering the build-up of Mn at grain boundaries restricts the grain growth [8] however, on the other hand the addition of Fluorine leads to PbO and/or PbF2-rich secondary phases causing an increase in the density along with grain growth promotion [9,10].

Figure 3. (a & b) FE-SEM micrographs of synthesized PZT powder showing cuboidal particles, (c & d) SEM images of fractured PZT pellets showing highly dense structure

5.3. Density measurement

The powder particles were pressed and sintered at different temperatures from 1150-1200 °C for 2 hours to optimize the density. Figure 4 shows the variation of density as a function of temperature. It can be seen from the graph that >98% of the theoretical density was achieved at 1190 °C.

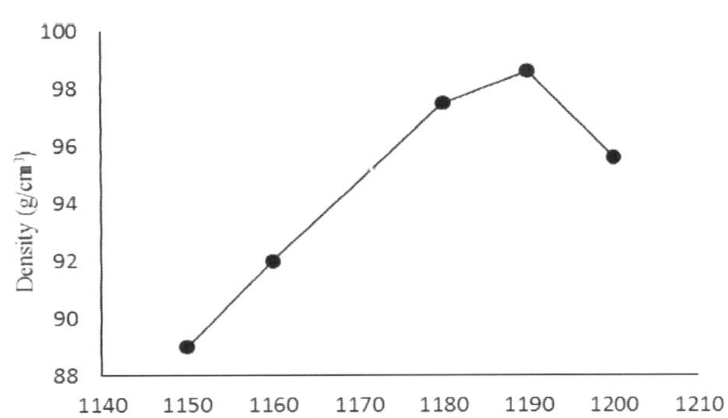

Figure 4. Influence of sintering temperature on densification at constant stay time (2 hours) and heating rate (5 °C/min)

5.4 Electrical properties

The electrical properties dielectric constant K_{33}^{T}, dielectric loss factor tanδ, piezoelectric coefficient d_{33}, the electromechanical coupling coefficient k_{33} and the mechanical quality factor Q_m, of the samples have been listed in Table 1. The properties listed in the table are in good agreement to the previous studies [3]. The piezoelectric activity has been improved to a great deal due the presence of dopants.

Table 1. Electrical properties of sintered samples before and after poling

Before Poling			After Poling					
K_{33}^{T}	$R\ (K\Omega)$	$tan\delta$	K_{33}^{T}	$R(K\Omega)$	$tan\delta$	$k_{33}\,(\%)$	Q_m	$d_{33}\,(pC/N)$
1150	4.5	0.026	1280	3.1	0.018	56	320	280

The combined effect of manganese and fluorine imparts unique properties in the system. Fluorine is expected to substitute for oxygen defects caused by the manganese doping, making possible to pole PZT system at 3KV/mm [11]. On the other hand, Mn addition lowers the saturation field of perovskite type ferroelectrics by decreasing the c/a ratio of crystal lattice and thus increases electromechanical coupling coefficient. The trivalent B site ion (Mn^{+3}) also forms dipoles with oxygen atoms which gives rise to "clamping effect" as a result the wall movement of ferroelectric domain is restricted [12]. This ultimately increases the mechanical quality factor.

6. Conclusion

The effect of co-doping of 1 mol% manganese and 2 mol% fluorine on hydrothermally prepared $Pb_{0.89}(Ba,\ Sr)_{0.11}(Zr_{0.52}Ti_{0.48})\ O_3$ was evaluated in the present study. Powder particles of 250-300 nm size with a cuboidal morphology were observed under FE-SEM. XRD analysis of powder particles confirmed the formation of rhombohedral phase in the PZT perovskite structure. It was found out that more than 98% theoretical density is achieved at a sintering temperature of 1190 °C, for the studied composition. The grain size of 1-2um was observed in the sintered pellets. Moreover the electrical properties such as the piezoelectric coefficient d_{33}, the electromechanical coupling coefficient K_{33} and the mechanical quality factor Q_m increase by the addition of dopants.

7. References

[1] Thomann H and Heywang W 1984 Annu. Rev. Mat. Sci. 14 27-47
[2] Berlincourt D 1992 J. Acoust. Soc. Am 91 3034-40
[3] Guiffard B, Boucher E, Lebrun L and Guyomar D 2007 Mat. Sci. and Engg. B 137 272-7
[4] Maria T. Christian C, Anne L and Bernard T 1999 J. Euro. Cer. Soc. 19 1023-6
[5] Ibrahim S, Hanna F and Marei L 2003 Egypt. J. Sol. 26 139-50
[6] Hou Y, Zhu M, Gao F, Wang H, Wang B, Yan H and Tian C 2004 J. Am. Ceram. Soc. 87 847-50
[7] Porter B and Harvey K 1972 Introduction to Physical Inorganic Chemistry: (Addison-Wesley)
[8] Izaki T, Haneda H, Watanabe A, Uchida Y, Tanaka J and Shirasaki S 1992 Jpn. J. Appl. Phys. 31 3045
[9] hiang S, Nishioka M, Fulrath R and Pask 1981 J. Ceram. Bull. 60 484
[10] Ohtaka O, Von DerM¨uhll R and Ravez J 1995 J. Am. Ceram. Soc. 78 805
[11] Eyraud L, Eyraud P, Audiger D, and Claudel R 1995 Ferroelectric. 175 241−50
[12] Kamiya T, Suzuki T, Tsurumi T and Daimon M 1992 *Jpn. J. Appl. Phys.* 31 3058-60

An integrated experimental and computational approach to material selection for sound proof thermally insulted enclosure of a power generation system

R. Waheed, W. Tarar and H. A. Saeed
Department of Mechanical Engineering, College of Electrical and Mechanical Engineering, National University of Sciences and Technology, Islamabad, 44000, Pakistan

E-mail: rehanwaheed107@hotmail.com, wasimakram@ceme.nust.edu.pk hasan.saeed@ceme.nust.edu.pk

Abstract. Sound proof canopies for diesel power generators are fabricated with a layer of sound absorbing material applied to all the inner walls. The physical properties of the majority of commercially available sound proofing materials reveal that a material with high sound absorption coefficient has very low thermal conductivity. Consequently a good sound absorbing material is also a good heat insulator. In this research it has been found through various experiments that ordinary sound proofing materials tend to rise the inside temperature of sound proof enclosure in certain turbo engines by capturing the heat produced by engine and not allowing it to be transferred to atmosphere. The same phenomenon is studied by creating a finite element model of the sound proof enclosure and performing a steady state and transient thermal analysis. The prospects of using aluminium foam as sound proofing material has been studied and it is found that inside temperature of sound proof enclosure can be cut down to safe working temperature of power generator engine without compromise on sound proofing.

1. Introduction

Power generation systems are widely used in industry, defence and domestic applications. A power generation system consists of an engine of required power, power transmission system in the form of a gear box or pulley drives and a generator for producing output power. An enormous amount of heat and sound are generated during power generation process and need to be controlled. This paper presents a solution to engine overheating problem inside a sound proof enclosure through an integrated experimental and computation approach. Thermal insulation of the cabin is also required in order to prevent the heat from entering into a crew compartment. The thermal insulation further complicates the problem as the enclosure temperature continues to rise during the process making the ambient air available to the engine radiator hotter with the time. The sound proofing requirement limits the required air intake for the radiator resulting in engine overheating problem during the continuous operation. An optimum solution is developed providing sound suppression to required levels and ensuring smooth operation of engine without overheating. Various materials including PU, Aluminum foam and Glass wool are evaluated for cumulative required performance for thermal as well as sound proofing problem. Analytical methods, transient thermal analysis and experiments are used for obtaining the final solution and validation. A complete procedure is proposed for obtaining solution

Content from this work may be used under the terms of the Creative Commons Attribution 3.0 licence. Any further distribution of this work must maintain attribution to the author(s) and the title of the work, journal citation and DOI.
Published under licence by IOP Publishing Ltd

for this problem of multi physics nature. After detailed analysis and comparison aluminium foam is suggested as preeminent material for such type of application.

2. Literature review

Sound proofing of diesel generators has been a topic of interest for researchers in the past. Pish et al [1] developed a sound proof enclosure for a 0.3KW military purpose small engine. There are many engineering applications in which both sound proofing and thermal insulation needs to be treated simultaneously one such application is the automotive exhaust system. The exhaust pipe of automotive exhaust system passes through intricate spaces beneath the chassis. Berbner et al [2] developed a sound proof thermal shield to protect heat and sound from entering the automotive inner environment. Previously the sound thermal shield of exhaust system was made from two aluminum foil layers with a rock wool or glass wool layer sandwiched in between. Berbner et al used a special knit aluminum wire pressed in two layers of perforated aluminum foil. The added advantage of their design was that it can be reinstalled multiple times during repair works. A similar attempt has been made by Muller et al [3]. They use a perforated aluminum support layer that is placed such that it does not come in contact with the automotive body thus preventing the heat from exhaust system to be transferred to automotive main frame. Use of aluminum foam as sound proofing material in automobile has also been studied. Ilgaz Akseli [4] in his research work has used aluminum foam under the front hood of different automobiles and analyzed its effectiveness in terms of thermal conductivity and sound absorption coefficient against other commercially available sound proofing materials. Jorge P. Arenas and Malcom J. Crocker [5] have discussed in detail the latest trends in use of porous sound absorption material. They concluded that with the use of piezoelectric actuators a hybrid smart foam can be produced which utilizes both active and passive noise control techniques and covers a broad spectrum of sound waves. Jorge et al also concluded that new sound proofing materials are lighter, safer, environment friendly and can be recycled. Francois Xavier et al [6] in their research have developed a composite porous sound absorbing material by filling the perforations of a sound absorbing material with another porous material. One of the major disadvantages of fibrous sound proofing materials developed in early 1960 was their handling. These fibrous materials cause severe irritation when touched by a person also their fibers pollute the air in form of fine particles during installation. The environmental effect of various sound absorbing materials has been discussed by Francesco et al [7]. Another disadvantage of using fibrous material is the resistance to flow of air especially when the sound proofing material is applied to an area where air flow rate is of prime importance. Massimo et al [8] developed an empirical relation for the friction on polyester fiber materials. In this research work a comparison has been made between glass wool and aluminum foam as sound absorption material.

3. Problem statement

In this research work a diesel generator with specifications given in Table 1 have been selected.

Table 1. Diesel Generator general specifications

Generator Data	
Engine rated power at 2800 rpm	42 KW
Engine rated power at 2500 rpm	37 KW
Electrical power produced by Generator	16 KW
Oil cooler temperature	$<75^{o}C$
Engine exhaust temperature	$<160^{o}C$
Required Cfm	4200 l/m

The above generator is installed in a sound proofed cabin. The construction of cabin is such that the cabin frame structure is made up of steel box sections. The outer layer is of GI sheet 22 gauge (0.8mm thick). The inner lining is made up of glass wool bats packed in perforated steel sheet. The pockets between main frame pipe sections are filled up with polyurethane foam. A schematic view of the sound proof cabin is shown in Figure 1.

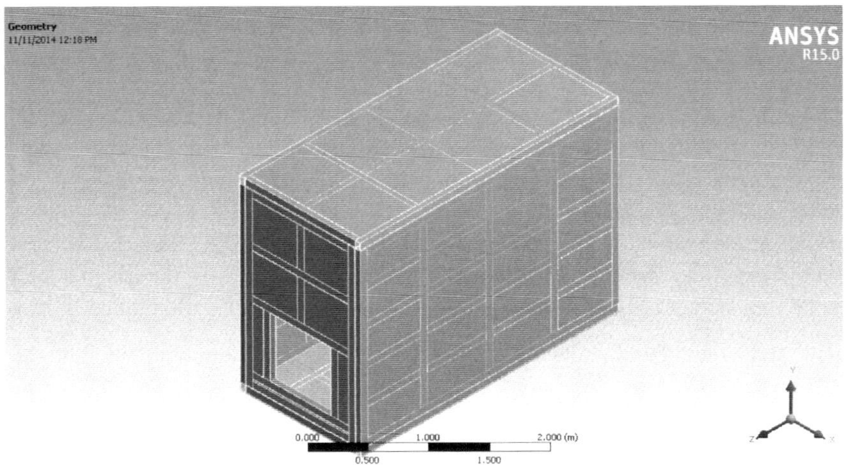

Figure 1. Sound proof cabin

The physical properties of different materials used in this research are given in Table 2.

Table 2. Physical Properties of Materials

Material	Density (Kg/m^3)	Specific heat (J/Kg.K)	Thermal Conductivity (W/m.°K)
Steel	7850	434	60.5
PU	30	1500	0.03
Glass Wool	40	1400	0.02
Al Foam	700	890	30.0
Air	1.16	1007	0.026

To collect acoustic data, sound producing equipment needs to be tested in a perfect sound proof environment preferably anechoic chamber, since the sound produced by generator has a substantial difference with the ambient noise level of an open environment (≈60 db) away from traffic and machine/equipment noise therefore the first test of generator is performed in open space. The sound pressure level (SPL) produced by the generator was 105 db. The second test was performed by placing the generator inside the sound proof cabin. It was observed that majority of sound is coming from the engine radiator side. The reason behind is that the air flows along the length of engine as it is sucked up by radiator fan and blown out of radiator. The sound proof hood installed in front of the radiator attenuates this sound and bring it down to 80 db. It was also observed that engine gets over heated after an operation of 2 hours. After various experiments and analysis of data the root cause of engine heat up issue was found out to be reduced air at inlet and out let of sound proof cabin. This observation was inferred by measuring cfm of air at inlet and exhaust side of cabin. After removing the sound proof hood at exhaust side the engine heat up problem can be resolved but it takes up the sound pressure level to alarming limits of 95db

The main objectives of this research work was to operate the engine at its intended time stretch of 8hrs under allowable temperature limits and maintaining the sound pressure level <80db. To resolve this multi constrained problem the strategy adopted has been explained in next section.

IOP Conf. Series: Materials Science and Engineering **146** (2016) 012045 doi:10.1088/1757-899X/146/1/012045

4. Methodology

First of all data has been gathered through a series of experiments to identify all problems. It has been found through the experiments that there are five main contributors to the sound attenuation of generator set.

I. Sound proof material applied to the inner walls of cabin (absorbs engine sound as it strikes the walls and restrict it to be transferred outside)
II. Exhaust system of engine (takes heat and exhaust gases out of the cabin along with engine noise).
III. Rubber mounts to install engine on the floor of cabin (controls engine rattling and vibrations transferred to the floor of cabin)
IV. Strainers at inlet of cabin (to reduce the noise of air flowing in)
V. Sound proof hood at cabin outlet (the hood is designed such that all the air going out of the cabin must strike some sound absorbing material on the way thus attenuating sound).

The contribution of above five aspects in attenuating generator sound level has been calculated by applying them in isolation with the others e.g. to calculate the effectiveness of sound proof hood, the hood is removed and engine is operated with air flowing free out of the cabin. The amount of contribution is shown in Table 3 in terms of sound pressure level.

Table 3. Parameters contributing to sound attenuation

Modifications in Generator and Cabin	Difference in Attenuation of Sound Pressure Level (db)	Percentage of contribution
Sound proof material at inner walls of cabin	4 db	16 %
Modification in Engine Exhaust System	2 db	8 %
Modified Rubber mounts	3 db	12 %
Strainers at cabin inlet	4 db	16 %
Sound proof hood at cabin outlet	12 db	48 %

As mentioned earlier the engine heat up problem is due to the insufficient amount of air entering and leaving the cabin. The air flow level can be maintained by removing the sound proof hood. From Table 3 it can be seen that by removing the sound proof hood required sound attenuation level cannot be achieved as it is contributing to 48% of the aver all noise control. To further strengthen this argument a steady state and transient thermal analysis of generator cabin has been performed. These analyses revealed that the glass wool material present on the inner walls of cabin is the cause of heat accumulation inside the cabin. When the hot air coming out of engine radiator is restricted it transfers heat to air surrounding generator through conduction and convection. This heat starts to accumulate inside the cabin as inner walls have a layer of glass wool which is thermal insulator. A steady state and transient thermal analyses are further performed by removing the inner glass wool layer. The results were better than the previous. In this case sound attenuation is further compromised by 12%. As case study aluminium foam is selected for sound proofing of inner walls of cabin. The thermal conductivity of aluminium foam is very high compared to glass wool (can be seen from Table 2). Aluminium foam can transfer heat quickly in addition to restricting the sound waves from crossing the cabin boundaries. The results with aluminium foam as sound proof material were even better.

A further investigation has been made by performing all the above analysis in two sets of ambient air condition. One is at ambient temperature of 30°C while the other is at ambient temperature of 50oC keeping in view the local extreme weather conditions. The reason behind these analyses was to check the heat transferring rate of aluminium foam in hot weather conditions. The steady state and transient thermal analysis performed in this research are further elaborated in the preceding section.

A. Steady state and transient thermal analysis with insulation
B. Steady state and transient thermal analysis without insulation
C. Steady state and transient thermal analysis with Aluminium foam as sound proof material

5. Steady state and transient thermal analysis

To check the behaviour of sound proof cabin under various heat loads and thermal insulation both steady state and transient thermal analysis were performed. For transient analysis a time period of two hours was selected to check the operation of generator set in sound proof cabin with convective boundary conditions. The external heat load which is the ambient temperature was taken as 30°C and 50°C. The heat produced by the engine during operation was recorded through temperature sensors. The engine is modelled in the analysis as heat producing body. Engine is cooled continuously by air flowing on its surface by forced convection. The convection is taken as 5 W/m^2.°C. In heat producing bodies there is always a heat sink designed to allow a passage for heat flow. The thermal insulation on walls of the cabin was a hindrance in the path of heat flow as can be seen in the results given in Table 4 below.

Table 4. Steady state and transient thermal analysis

Analysis	Ambient Temperature 30°C		Ambient Temperature 50°C	
	With insulation	Without Insulation	With insulation	Without Insulation
Steady state	82.31°C	47.69 °C	85.265 °C	63.43 °C
Transient	79.87 °C	47.646 °C	83.621 °C	62.46 °C

The temperature is taken on all the walls roof and floor. In Table 4 the temperature of front wall near to the engine radiator is given. Maximum temperature is observed in this area.

There is a slight difference between values of temperature in steady state and transient analysis however the difference in temperature is large when analysis is done without insulation. At 30°C the difference in temperature is 32°C considering transient analysis. At 50°C the temperature difference is reduced to 21°C. The temperature values without insulation were a clear indication that temperature inside sound proof cabin can be further reduced by introducing a material that should act as a heat sink. For this purpose aluminium foam was selected, which can act both as sound absorber and heat sink. Another analysis was run with layer of aluminium foam on the walls and roof of sound proof cabin in placement of glass wool.

6. Comparison of results

The results of aluminium foam are compared with cabin walls with glass wool as sound proof material at 50°C ambient temperature.

Figure 2. a) Temperature inside cabin with aluminium foam

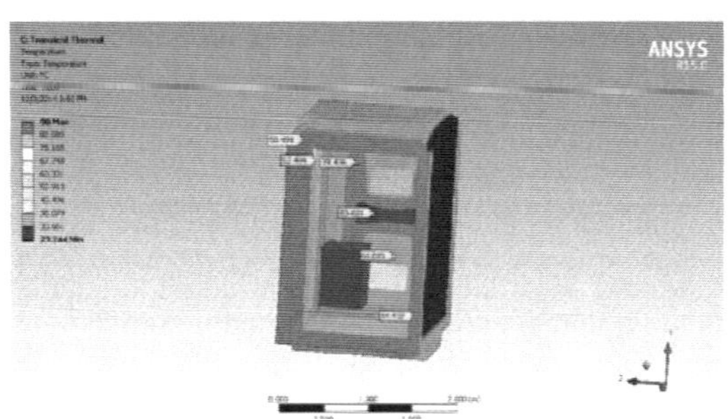

Figure 2. b) Temperature inside cabin with glass wool

It can be seen from Figure 2a and Figure 2b that temperature inside sound proof cabin has been reduced by the use of aluminium foam. As the properties of aluminium foam are slightly different from pure aluminium therefore the advantage of high thermal conductivity of pure aluminium cannot be gained however temperature difference of 21°C is still good enough to prove aluminium foam as suitable material for sound proof cabins of diesel generators.

7. Conclusion

A steady state and transient thermal analysis of sound proof cabin for diesel generator set has been performed with thermal insulator sound proof material, without sound proof material and with aluminium foam as sound proof material. It is concluded that glass wool has raised the cabin temperature beyond maximum allowable limits for safe working of diesel generator. By using aluminium foam the cabin temperature has been reduced substantially. A total removal of sound absorbing material is helpful in reducing temperature but sound absorption is compromised. It can be further concluded from this research that temperature effects inside a sound proof cabin are more profound at low ambient temperature than at higher temperature.

8. Future work

Air entering and leaving a sound proof cabin is major source of noise. As the air coming out of cabin is the only sound carrier, sound reduction can be achieved by designing sound proof hood at cabin exhaust point in front of the engine radiator. Care must be taken in designing of the sound proof hood as the reduction in volume of air leaving the cabin can elevate inside temperature of cabin. Aluminium foam can also be used in sound proof hoods as it does not deteriorate easily and has very low surface friction

9. References

[1] Pish, R. and Hull, R., "Noise Reduction Techniques as They Apply to Engine-Generator Design and Treatment," SAE Technical Paper 690755, 1969
[2] Berbner et al Sound Proof Thermal Shield US Patent No:7445084132. November 4, 2008
[3] Muller et al Sound absorbing heat shield with perforate support layer US Patent No. 5196253A March 23,1993
[4] Ilgaz Askeli "The application of aluminum foam for the heat and noise reduction in automobiles. Dissertation MS. Mechanical Engineering Izmir Institute of technology June 2005.
[5] Jorge P. Arenas and Melcom J Crocker "Recent trends in porous sound absorption materials" Journal of sound and vibration July 2010
[6] Francois-Xavier Becot, Luc Jaouen and Frank S guard "Noise control strategies using composite porous materials-Simulations and experimental validations on plate cavity system"Journal of noise control engineering September-October 2011
[7] Francesco Asdrubali "Survey on the acoustical properties of new sustainable materials for noise control. Euro Noise Conference University of Perugia Italy 30th May- 1st June 2006)
[8] Massimo Garai, Francesco Pompoli "A simple empirical model of polyester fiber materials for acoustical applications. Journal of applied acoustics 26th July 2005

IOP Conf. Series: Materials Science and Engineering **146** (2016) 012046 doi:10.1088/1757-899X/146/1/012046

Investigation of mechanical properties based on grain growth and microstructure evolution of alumina ceramics during two step sintering process

U. A. Khan, A. Hussain, M. Shah, M. Shuaib[1] and F. Qayyum
Department of Mechanical Engineering, University of Engineering and Technology, Taxila 47050 Pakistan
[1]Ibn-e-Seena Institute of Technology, Sector H-11, Islamabad, Pakistan

E-mail: umar.ali.2289@gmail.com

Abstract. Alumina ceramics having small grain size and high density yield good mechanical properties, which are required in most mechanical applications. Two Step Sintering (TSS) is used to develop dense alumina ceramics. In this research work the effect of sintering temperatures on microstructure and density of the alumina specimens developed by using TSS has been investigated. It has been observed that TSS is more efficient in controlling grain growth and increasing the density as compared to One Step Sintering (OSS) of alumina. Scanning electron micrographs of sintered alumina specimens have been compared. It has been observed that TSS proves to be a better technique for increasing density and controlling grain growth of alumina ceramics than OSS. More relative density, hardness, fracture toughness and small grain size was achieved by using TSS over OSS technique.

1. Introduction

Alumina ceramics are well known and widely used as engineering material due to their good mechanical properties like hardness, wear resistance, and structural stability at elevated temperature. They are widely used for making good quality abrasive tools, seal rings, thread guides, cutting tools etc. It has been continuously found by different researchers that alumina ceramics having small grains and fine microstructure show improved strength [1], hardness [2] and wear resistance [3].

Sintering process involves coalesce of powder particles into dense object under action of heat in which accelerated grain growth occurs at final stage of sintering [4]. Since the invention of process many researcher have devised several sintering techniques to attain required density and grain size which eventually helps in achieving required mechanical characteristics. The techniques involved are Pressureless sintering and Pressure assisted sintering. Pressureless sintering involves sintering of powder compact without applying pressure during heating that is economical and produces near net shaped parts [5-7]. High temperature is required to achieve good densification which also increases grain size. Pressure assisted sintering utilize pressure on powder compact during heating which is important for manufacturing dense parts of ceramics quickly [4]. Spark Plasma sintering [2, 8, 9], hot pressing sintering [3, 10], and hot Isostatic pressing [1, 11]are a few well know pressure assisted sitering techniques, these techniques help in reduction of grain growth speed while achieving high density.

Two step sintering (TSS) is a Pressureless sintering technique which was developed by Chen and Wang [12] in year 2000 to reduce the grain growth in yttria ceramic in the final stage of sintering.

Content from this work may be used under the terms of the Creative Commons Attribution 3.0 licence. Any further distribution of this work must maintain attribution to the author(s) and the title of the work, journal citation and DOI.
Published under licence by IOP Publishing Ltd

Later on many researchers have successfully used TSS to increase ceramic density and to control grain size during sintering of Yttria [12, 13], Barium Titanate [14], Zinc Oxide [15, 16], Alumina [15, 17, 18] and Silicon carbide [11] specimens.

In current research alumina ceramic was manufactured by using OSS and TSS techniques, the effects of sintering temperature on microstructure and density of the specimens were observed. Mechanical properties on the basis of grain growth and microstructure evolution were presented. The etched specimens were observed under TESCAN Field Emission Scanning Electron Microscope (FESEM) to determine their grain sizes and the results have been presented.

2. Experimentation

Alumina powder CT 3000 (Almatis) designated as AP-1 and alumina powder (Nanophase Technology) designated as AP-2were used for sintering of specimens. Powder specifications are mentioned in Table 1.

Table 1. Alumina powder specifications

Alumina Powder	Purity	Particle Size	Specific Surface Area (m^2/g)
AP1	99.7 %	600 nm	7
AP2	99.9%	47 nm	35

Alumina green compacts were made in a custom die (having 28mm internal diameter) mounted on Material Testing System MTS 810 which is shown in Figure 1 having force capacity of 100 kN. Uniaxial pressure of 105 MPa was applied to successfully develop green compacts.

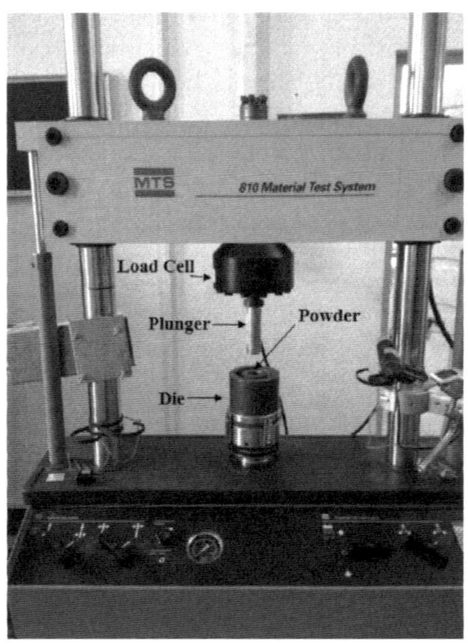

Figure 1. Test setup for uniaxial pressing

The green density is measured by volumetric method. S-1, S-3 specimens were formed by Alumina Powder AP-1 (600 nm) and S-2, S-4 specimens by alumina powder AP-2 (47nm).The green density and specimens dimensions are shown in Table 2.

Table 2. Green Density of AP -1 & AP-2 Specimens pressed at 105 MPa

Specimen	Mass (g)	Mass of Al_2O_3 (g)	Diameter (mm)	Thickness (mm)	Green Density (g/cm^3)	Relative Density (%)
S-1	5	4.37	28	3.40	2.08	52.1
S-3	5	4.34	28	3.40	2.07	52
S-2	3	2.28	28	2.9	1.27	31.8
S-4	3	2.16	28	2.85	1.23	30.8

During OSS green compacts were heated to1450°C for soaking time of 6 hour while during TSS the green compact to 1450°C then suddenly cooled down to 1350°C for soaking time of 6 hour shown in Figure 2. Electric furnace with SiC heating elements (Vecstar) of heating chamber size 5.9×7×13.9 (inch) was used for sintering of green compacts.

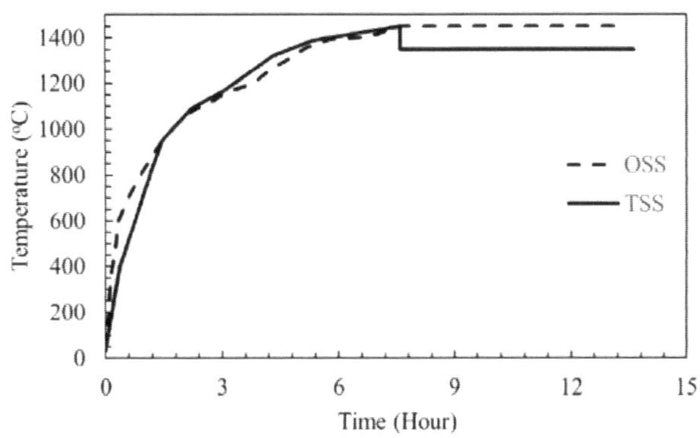

Figure 2. OSS and TSS sintering Curves

Densification of sintered alumina specimens was measured by Archimedes technique using MD-300s balance in which specimen values are measured by immersing them in distilled water. The measurements are presented in Table 3.

Table 3. Bulk density of AP-1 & AP-2 sintered specimens

Specimen	Diameter (mm)	Thickness (mm)	Bulk Density (g/cm^3)	Relative Density (%)
S-1	24	3.22	3.23	81
S-3	23.30	3.07	3.61	90.5
S-2	20.80	2.38	3.56	89.3
S-4	19.72	2.29	3.66	91.8

As discussed previously that microstructure of alumina ceramics play an important role in defining mechanical properties. To reveal the microstructure of the specimens and prepare them for microscopy, the specimens were ground using 400, 800, 1200 and 2000 grade emery papers consecutively. The specimens were polished with 1 μm diamond paste with the help of a buffing wheel. The specimens were thermally etched at 1250°C for 30 min in electric furnace and observed under TESCAN Field Emission Scanning Electron Microscope (FESEM).

Vickers hardness (H_v) measurements were conducted by a fully calibrated Vickers hardness tester (Time Instrument). Indentations were conducted on polished faces of the specimens with a diamond indenter using 10 kg load for 15 sec. Three measurements were taken and average result are reported using Eq. (1)[19].

$$H_V = 1.854\left(\frac{P}{d^2}\right) \quad (1)$$

Where P is applied load in Kgf and d (mm) is mean value of diagonal length.
Fracture toughness (K_c) was computed by following Eq. (2) [20].

$$K_C = 0.0726\left(\frac{P}{c^{\frac{3}{2}}}\right) \quad (2)$$

Where P is applied load in Kgf and c (mm) is the mean crack length of indented marks.

3. Results and discussion

Green compacts were manufactured on MTS 810 system, due to its fully calibrated load cells and LVDT's we were able to record detailed compactness curves of alumina powder which are presented in Figure 3. Compactness curves shows that AP-2powder was compressed more than AP-1 during uniaxial pressing which is due to fine and small particle size of AP-2 powder. The finer alumina powder gives greater contact points of particles which form cold welds in green compact due to which powder compactness increases[21].

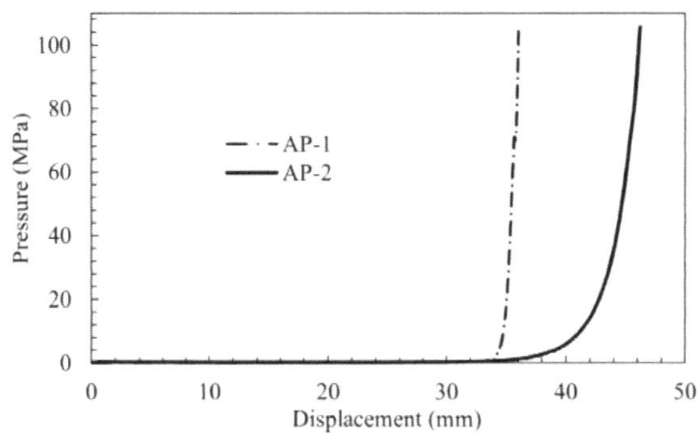

Figure 3. Pressure vs displacement curve of alumina powders during uniaxial pressing

The results obtained after OSS and TSS are summarized in Table 4. The results showed that alumina powder AP-2 shrinked more than AP-1 because of fine particles, high specific surface area of the AP-2 powder and greater compactness during uniaxial pressing. Fine particles haves horter diffusion distances and larger area of diffusion. High surface area of powder delivers higher driving force and stimulates densification leading to decrease of interfacial energy [22].

Also TSS has increased specimens relative density of AP-1 specimens by 9.5 % and AP-2specimens by 2.5% as compared to OSS specimens. The microstructure of alumina specimens obtained by one step sintering and two step sintering are shown in Figure 4 and 5. It was observed that two step sintered specimens have small grains as compared to one step sintered and both contain inter granular porosity. During sintering of ceramics open pores stick grain boundaries and block grain boundary motion at intermediate stage but when open pores close, grain growth starts to occurs[23]. Chen and Wang developed two step sintering, to increase density without grain growth, grain boundary diffusion must be active and grain boundary migration must be frozen [12]. They found that grain boundary migration require higher activation energy than grain boundary diffusion so they decreased temperature in second step (T_2) in order to freeze grain boundary migration.

It can be seen from SEM micrographs that TSS specimens S-3 and S-4 have grain sizes of 1.28 μm and 0.76 μm which is smaller than the grain size of OSS specimens.

Table 4. Green and Bulk density of sintered specimens

Specimen	Green Density (g/cm^3)	Bulk Density (g/cm^3)	Densification (%)
S-1	2.08	3.23	35.6
S-3	2.07	3.61	42.6
S-2	1.27	3.56	64.3
S-4	1.23	3.66	66.4

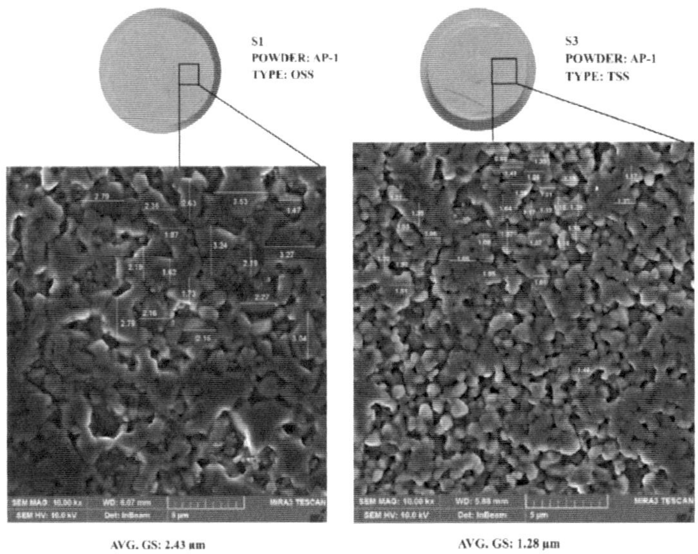

(a) (b)

Figure 4. SEM micrographs of AP-1 specimens (a) S-1 OSS (b) S-3 TSS

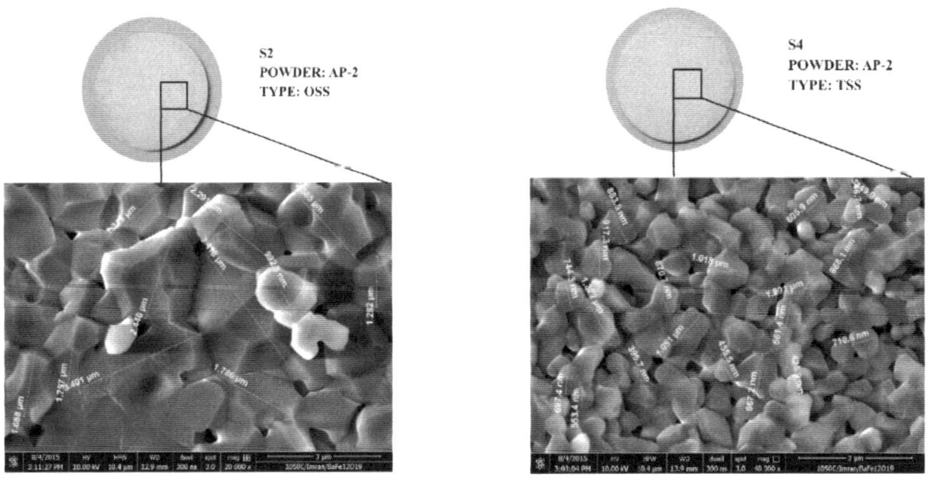

(a) (b)

Figure 5. SEM micrographs of AP-2 specimens (a) S-2 OSS (b) S-4 TSS

The variation of relative density and grain size of alumina ceramic specimens sintered via one step and two step sintering are shown in Figure 6. One step sintered specimens S-1 and S-2 have large grain size because the decrease in total interfacial energy arises via densification and particle coarsening, smaller particles merge with the large ones producing larger grains[24]. While Two step sintered specimens S-3 and S-4have small grain size because specimens heated to high temperature to decrease pores that are unstable against shrinkage and cooled at lower temperature to attain densification via suppression of grain boundary migration [12].

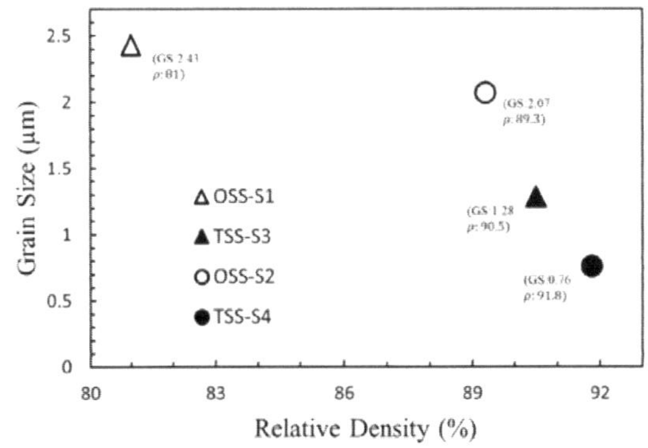

Figure 6. Grain size versus relative density of alumina specimens sintered by single and two step sintering

Mechanical properties of OSS and TSS specimens were calculated by using Eq. (1) [19]. and Eq. (2) [20]. are shown in Table 5.As gain size has strong influence on mechanical properties [1]because fine grain size of a microstructure contain high density of grain boundaries that restrain deformation of specimen in loading.

Table 5. Mechanical Properties of sintered alumina specimens

Specimens Sintered	Hardness (GPa)	Fracture Toughness ($MPa.m^{1/2}$)
S-1	12.41	1.53
S-3	13.27	1.87
S-2	13.64	2.16
S-4	14.45	2.30

It is found that TSS has controlled alumina grain growth and increased density, hardness and fracture toughness.

4. Conclusion

I. Hence it is evident from research that alumina ceramics developed by using fine powder and TSS technique yield better ceramics of smaller grain size, increase hardness and toughness value.

II. Two step sintering has increased density and controlled grain size of AP-1(600 nm) and AP-2 (47 nm) ceramics.

III. Alumina AP-2 (47 nm) showed greater sinter ability due to small powder particle size, high compatibility and specific surface area of the powder.

5. Acknowledgements

I. The authors are grateful:

II. For financial support from Board of Post Graduate studies, University of Engineering and Technology Taxila, Pakistan

III. For specimen characterization support, Institute of Space Technology and Ibn-e-Seena Institute of Technology Islamabad.

6. References

[1] O, Y.T., et al., *Effect of grain size on transmittance and mechanical strength of sintered alumina.* Materials Science and Engineering: A, 2004. 374(1–2): p. 191-195
[2] Santanach, J.G., et al., *Spark plasma sintering of alumina: Study of parameters, formal sintering analysis and hypotheses on the mechanism(s) involved in densification and grain growth.* Acta Materialia, 2011. 59(4): p. 1400-1408
[3] Felten, E.J., *Hot-Pressing of Alumina Powders at Low Temperatures.* Journal of the American Ceramic Society, 1961. 44(8): p. 381-385
[4] M.N.Rahaman, *Ceramic Processing and Sintering.* 2nd ed. 2003, New York: Marcel Dekker.
[5] Somiya, S., *Handbook of Advanced Ceramics: Materials, Applications, Processing, and Properties.* 2013: Academic Press
[6] Eskandari, A., et al., *Effect of high energy ball milling on compressibility and sintering behavior of alumina nanoparticles.* Ceramics International, 2012. 38(4): p. 2627-2632
[7] Cutler, I.B., et al., *Sintering of Alumina at Temperatures of 1400°C. and Below.* Journal of the American Ceramic Society, 1957. 40(4): p. 134-139
[8] Kim, B.-N., et al., *Spark plasma sintering of transparent alumina.* Scripta Materialia, 2007. 57(7): p. 607-610
[9] Shen, Z., et al., *Spark plasma sintering of alumina.* Journal of the American Ceramic Society, 2002. 85(8): p. 1921-1927
[10] Chang, S., et al., *Hot-Pressing of Nano-Size Alumina Powder and the Resulting Mechanical Properties.* International Journal of Applied Ceramic Technology, 2004. 1(2): p. 172-179
[11] Besson, J. and M. Abouaf, *Grain growth enhancement in alumina during hot isostatic pressing.* Acta Metallurgica et Materialia, 1991. 39(10): p. 2225-2234
[12] Chen, I.-W. and X.-H. Wang, *Sintering dense nanocrystalline ceramics without final-stage grain growth.* Nature, 2000. 404(6774): p. 168-171
[13] Wang, X.H., P.L. Chen, and I.W. Chen, *Two-Step Sintering of Ceramics with Constant Grain-Size, I. Y2O3.* Journal of the American Ceramic society, 2006. 89(2): p. 431-437
[14] Polotai, A., et al., *A novel approach to sintering nanocrystalline barium titanate ceramics.* Journal of the American Ceramic Society, 2005. 88(11): p. 3008-3012
[15] Maca, K., V. Pouchly, and P. Zalud, *Two-Step Sintering of oxide ceramics with various crystal structures.* Journal of the European Ceramic Society, 2010. 30(2): p. 583-589
[16] Mazaheri, M., A. Zahedi, and S. Sadrnezhaad, *Two-step sintering of nanocrystalline ZnO compacts: Effect of temperature on densification and grain growth.* Journal of the American Ceramic Society, 2008. 91(1): p. 56-63
[17] Kim, H.S., S.T. Oh, and Y.D. Kim, *Effects of the two-step sintering process on the optical transmittance and mechanical strength of polycrystalline alumina ceramics.* Ceramics International, 2014. 40(9, Part A): p. 14471-14475.
[18] Wang, C.-J., C.-Y. Huang, and Y.-C. Wu, *Two-step sintering of fine alumina–zirconia ceramics.* Ceramics International, 2009. 35(4): p. 1467-1472
[19] *ASTM Standard Designation C 1327-03. Standard test method for vickers indentation hardness of advanced ceramics.* ASTM USA :International
[20] Sergejev, F. and M. Antonov, *Comparative study on indentation fracture toughness measurements of cemented carbides.* Proc. Estonian Acad. Sci. Eng, 2006. 12(4): p. 388-398
[21] Datta, B.K., *Powder Metallurgy: An Advanced Technique of Processing Engineering Materials.* 2014: PHI Learning Pvt. Ltd
[22] Bowen, P. and C. Carry, *From powders to sintered pieces: forming, transformations and sintering of nanostructured ceramic oxides.* Powder Technology, 2002. 128(2–3): p. 248-255
[23] Li, J. and Y. Ye, *Densification and Grain Growth of Al2O3 Nanoceramics During Pressureless Sintering.* Journal of the American Ceramic Society, 2006. 89(1): p. 139-143
[24] Mayo., M., in *Int Mater Rev.* 1996. p. 85-115

Evaluation of performance parameters of indigenously developed roots pumping system

M. Maqsood[1], A. Usman[1], M. F. Bodla[1,2], and J. Ali[1,3]
[1]National Institute of Vacuum Science and Technology,
Shahdara Road, NCP-Complex, Islamabad, Pakistan
[2]Department of Physics, Allama Iqbal Open University, Islamabad, Pakistan
[3]Hazara University, Mansehra - Pakistan

E-mail: maqsood@ninvast.edu.pk

Abstract. Roots pumping systems are widely used in industries to generate vacuum with high pumping speed. In the present work, the performance parameters of indigenously developed Roots pumping system have been studied. The performance parameters being studied are the ultimate pressure, working temperature, compression ratio and pumping speed. Ultimate pressure of the Roots pump after continuous running of eight hours is found to be 1.1×10^{-3} mbar. The most important parameter of the roots pump is the zero-gas flow compression Ratio (K_0) which is found to be 18 for the pumping system under study. Efficiency of Roots pump is found to be 76% which is in good agreement as reported in the literature.

1. Introduction

Roots pumping systems have a wide range of applications. The roots pumping systems are very useful in evacuating large volumes in shorter periods of time due to very high pumping speed. This property make roots pumping systems, a preferable choice in coating and semi-conductor industry, chemical and process technology and research and development setups. Roots pumps, when combined with suitable backing pumps, make them adaptable for different applications.

The performance of the vacuum pumps can be evaluated from its ultimate pressure, operating temperature, compression ratio, pumping speed etc. In 1963, American Vacuum Society (AVS) issued tentative standard for measuring pumping speed of the vacuum pump, which was revised and published in 1971. In 1987, M. H. Hablanian recommended some practices for the pumping speed measurement of the vacuum pumps by incorporating newly established pumping methods and improvements in the high vacuum techniques. He explained the specifications of the experimental setups in detail [1]. B. R. F. Kendall recommended the procedures for measuring the pumping speed of the positive displacement mechanical vacuum pumps in 1989 [2]. In 1996, W. Jitschin*et.al.* measured the pumping speed of the rough vacuum pumps with stationary method and intermittent pump down method [3].

In the present work, the indigenously developed roots pumping system was tested. The roots pump is backed by a dual stage rotary pump. Its ultimate pressure and working temperatures are worked out for 7 hours of operation. The compression ratio and pumping speed of the roots pump is studied as a function of outlet pressure.

Content from this work may be used under the terms of the Creative Commons Attribution 3.0 licence. Any further distribution of this work must maintain attribution to the author(s) and the title of the work, journal citation and DOI.
Published under licence by IOP Publishing Ltd

2. Ultimate pressure and working temperature

The roots pumping system was tested for its ultimate pressure and working temperature before determining other performance parameters. The system was continuously operated for seven hours. The ultimate pressure of the roots pump was found to be 1.1×10^{-3} mbar with the maximum operating temperature of 37°C.

3. Compression ratio

An important characteristic of the roots pump is the zero-load compression ratio K_0. It is the ratio of the pump outlet pressure to the pump inlet pressure [4].

$$K_0 = \frac{P_o}{P_i} \tag{1}$$

For roots pump, compression ratio is worked out without the admittance of gas flow to the pump inlet [5]. It is measured by the experimental setup shown in the Figure 1.

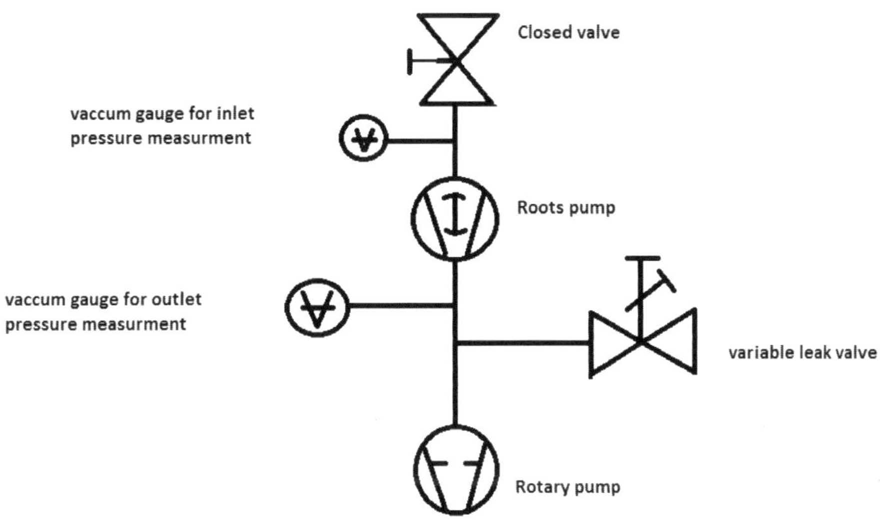

Figure 1. Experimental setup for determining compression ratio

The roots pump inlet is blanked off with a vacuum gauge and a rotary pump is attached to its outlet as backing pump. A low vacuum gauge is installed in the baking line to monitor the backing pressure of the roots pump. The air is admitted to the backing line via variable leak valve to adjust the backing pressure of the roots pump by throttling the rotary pump. This arrangement makes it possible to calculate the compression ratio over a wide range.

The observations were taken for the pressure range of 5×10^{-2} mbar to 118 mbar. This was done by the necessary throttling the rotary pump via the variable leak valve. The compression thus obtained, shown in Figure 2, has the conventional trend as reported in literature [6].

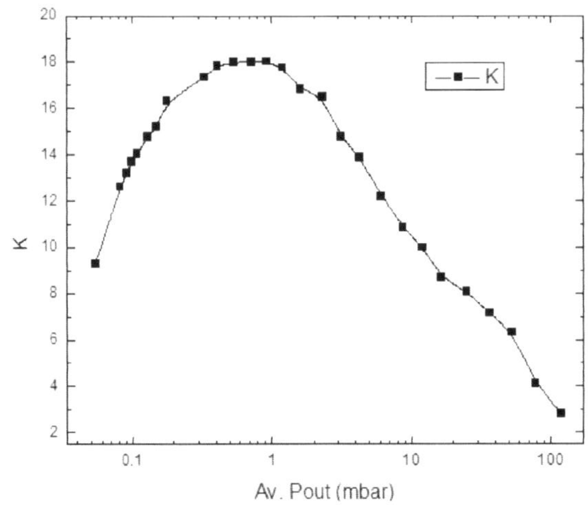

Figure 2. Behavior of compression ratio with outlet pressure

The maximum compression ratio of the roots pump is found to be 18 at the outlet pressure of 1 mbar. The compression ratio gradually decreases to about 3 upon the gradual increase of outlet pressure to 118mbar. Compression ratio also decreases to about 9 at the pressure of 5×10^{-2} mbar. The decrease of compression ratio values below 10 at 10 mbar and 10^{-2}mbar is due to the dominance of the back flow at higher pressure side and increase gas load due to desorption of gas molecules from rotors at lower pressure side [6].

4. Pumping speed

Pumping speed is the most important parameter of any vacuum pump. In ISO 1608/1, the pumping speed is defined as the volume of gas which flows through the pump inlet in unit time under ideal conditions. For experimental purposes, the volume rate of flow of a vacuum pump is taken to be the quotient of the throughput of that gas and the equilibrium pressure at a specified position in a test dome under specified conditions [7].

$$S = \frac{Q}{P} \tag{2}$$

The units most commonly used for the pumping speed are cubic meter per hour (m^3/h), liter per second (l/s) and cubic feet per minute (cfm).
The theoretical pumping speed of the roots pump is:

$$S_{th} = 4\times n\times V' \tag{3}$$

Where n is the rotational speed of the rotors and V' is the volume of gas isolated from the inlet [8]. Therefore, at high rotational frequencies, high pumping speeds can be achieved with small roots pump. For the pump under study, the theoretical pumping speed is found to be $271m^3/h$.
For the practical pumping speed of the roots pump at a particular backing pressure P_2 can be calculated by:

$$S = \frac{S_b S_{th} K_0}{S_{th} + S_b K_0} \tag{4}$$

Where, S_b is the pumping speed of the backing pump at the backing pressure P_2, S_{th} is the theoretical pumping speed of the roots pump and K_0 is the zero flow compression ratio of the roots pump at the backing pressure of P_2 [6].
The behavior of the pumping speed of the roots pump as a function of the outlet pressure is shown in Figure 3.

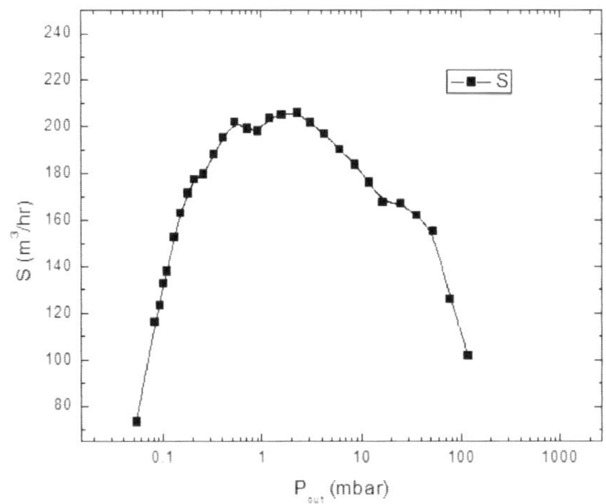

Figure 3. Behavior of pumping speed with outlet pressure

The maximum pumping speed of the roots pump is found to be 205m^3/hr at the outlet pressure of about 2 mbar. The pumping speed of the roots pump is found to be higher in the pressure range of 10^{-1} to 10 mbar. Beyond this pressure range, the pumping speed of the roots pump drops. The maximum efficiency of the roots pump is found to be 75% at 2mbar.

5. Effective pumping speed

The Effective Pumping Speed of the Roots pump is the product of Efficiency and theoretical pumping speed of Roots pump.

$$S_{eff}=\eta S_{th} \quad (5)$$

Where 'η' is Efficiency of Roots pump and $S_{th}$is theoretical pumping speed of Roots pump [5] . The Effective pumping speed is found to be 203m^3/h.

6. Gradation of pumping speed between fore pump and roots pump

There are two main aspects to determine the Gradation of the pumping speed from the Roots pump to the backing pump.

I. High volumetric efficiency required.
II. The maximum tolerable pressure difference of the Roots pump shall not be exceeded [8].

The Staging Ratio or Gradation of Roots pump is the ratio of actual pumping speed and pumping speed of backingpump. The staging ratio between roots pumping speed and backing pump is 1:5.7.

7. Max. pressure difference

Roots pump generate heat, when pumping gas at high pressure. Heat causes the rotors to expand; if unchecked, rotors expansion could destroy the pump. To avoid this heating effect, a maximum pressure difference between the inlet and outlet pressure of a Roots pump is specified.

$$\Delta P_{max}=P_{out}-P_i \quad (6)$$

for a short time without harm to the pump.[9]

8. The volumetric efficiency of roots pump

The formula for the Volumetric Efficiency of Roots pump is

$$E_v = \frac{K_o}{K_o + K_{th}} \tag{7}$$

Where K_0 is the zero flow compression ratio of the roots pump, K_{th} is ratio of theoretical pumping speed of Roots pump and pumping speed of backing pump [10]. The Volumetric Efficiency is found to be 75% which is good agreement with Mechanical Efficiency.

9. Conclusions

The performance of indigenously developed roots pumping system was tested at NINVAST in detail. Its ultimate pressure, operating temperature, compression ratio and pumping speed are some of the performance parameters being studied thoroughly. The roots pump with the ultimate pressure of 1.1×10^{-3} mbar, was found to be efficient in the outlet pressure range of 10^{-1}mbar to 1 mbar. It was found to be 75% efficient at the outlet pressure of 2mbar with 205m^3/hr pumping speed. Hence, it is found to be good to evacuate the system with high pumping speed.

10. References

[1] M. H. Hablanian, J. Vac. Sci. Tech. A, Vol. 5, No. 4, July/Aug, 1987
[2] B. R. F. Kendall, J. Vac. Sci. Tech. A, Vol. 7, No. 3, May/Jun, 1989
[3] W. Jitschin, K. H. Bernhardt, R. Lachenmann, P. Bickert and F. J. Eckle, Vacuum, Vol. 47, No. 6-8, 505 – 509, 1996
[4] D. J. Hucknall and A. Morris, Vacuum Technology: Calculations in Chemistry, Royal Society of Chemistry, 01-Jan-2003
[5] D. J. Hucknall and A. Morris,Vacuum Technology: Calculations in Chemistry, Royal Society of Chemistry, 01-Jan-2003
[6] Austin Chambers, Modern Vacuum Physics
[7] ISO 1608/1 - 1980
[8] K. Jousten, Handbook of Vacuum Technology
[9] M. H. Hablan, User Guide To Vacuum Technology
[10] Dorothy M. Hoffam, Handbook of Vacuum Science and Technology

Spectro-photometric determinations of Mn, Fe and Cu in aluminum master alloys

A. Rehan, A. Naveed, A. Shan[1], M. Afzal, J. Saleem, and M. A. Noshad
Soan Enterprises, P. O. Box 2723, Islamabad, Pakistan
[1]Department of Physics, University of Engineering and Technology, Lahore, Pakistan

E-mail: rehanchatta@gmail.com, chathashan@gmail.com

Abstract. Highly reliable, fast and cost effective Spectro-photometric methods have been developed for the determination of Mn, Fe & Cu in aluminum master alloys, based on the development of calibration curves being prepared via laboratory standards. The calibration curves are designed so as to induce maximum sensitivity and minimum instrumental error (Mn 0.1mg/100ml-2mg/100ml, Fe 0.01mg/100ml-0.2mg/100ml and Cu 2mg/100ml-10mg/ 100ml). The developed Spectro-photometric methods produce accurate results while analyzing Mn, Fe and Cu in certified reference materials. Particularly, these methods are suitable for all types of Al-Mn, Al-Fe and Al-Cu master alloys (5%, 10%, 50% etc. master alloys).Moreover, the sampling practices suggested herein include a reasonable amount of analytical sample, which truly represent the whole lot of a particular master alloy. Successive dilution technique was utilized to meet the calibration curve range. Furthermore, the workout methods were also found suitable for the analysis of said elements in ordinary aluminum alloys. However, it was observed that Cush owed a considerable interference with Fe, the later one may not be accurately measured in the presence of Cu greater than 0.01 %.

1. Introduction

Aluminum master alloy is the aluminum combined with one or two other elements (e.g., Al-Cu 50 % master alloy) and is used as a CHARGE for alloying during the melting of wrought aluminum alloys. Normally aluminum master alloys are produced for such elements that are the alloying elements of Aluminum wrought alloys. Commonly used Aluminum master alloys are Al-Mn, Al- Si, Al-Ni, Al-Cu, Al-Mg, Al-V, Al-Zn, Al-Zr, Al-Ti, Al-Cr and Al-Fe and are available in 5 or 7 Kg ingots. The use of Aluminum master alloys offers distinct advantages over the addition of individual element during alloying of aluminum wrought alloys. Three major advantages are: (a) these can be calculated to become part of the initial cold charge i.e. melted with the metal at low temperature. This rapid dissolution results in a significant reduction in dross formation, minimal gas pick up and extended furnace life (b) significantly reduces the amount of energy required and shortens the length of melt time (c) gives higher accuracy to meet specified composition limits [1-2].

The above mentioned benefits can only be attained by knowing the true composition of a particular master alloy [3]. Instantly, master alloy manufacturer provides a broad range of chemical composition for a single lot of particular master alloy, for example, in Al-Cu 50 % master alloy, Cuelement may varies from 47 to 52% or else as affirmed by the manufacturer. On the other hand for charge calculation one should know the nearest possible true composition of a particular master alloy so that melt could be adjusted as quickly as possible to save time, energy, and material. Also, chemical

Content from this work may be used under the terms of the Creative Commons Attribution 3.0 licence. Any further distribution of this work must maintain attribution to the author(s) and the title of the work, journal citation and DOI.
Published under licence by IOP Publishing Ltd

composition analysis of master alloys is indispensable for the strict quality control of master alloys production. Spark optical emission spectroscopy (OES) is a quick and reliable tool for analysis of aluminum alloys but it does not work for master alloys because of high percentage of alloy elements (e.g., Al-Cu 50% master alloy)[6]. Other techniques such as: Gravimetric technique, Atomic Absorption Spectrometry (AAS), Atomic Emission Spectrometry (AES), and X-ray fluorescence spectroscopic (XRF) technique are also not valid for such a high concentrations [4-8].

Previously Naveed et al [3] reported a method in which Aluminum master alloys were re-melted along with Aluminum ingots of 99.9% purity in an electrical crucible furnace and subsequently analyzed by spark optical emission spectrometer (OES). However, the technique was costly. Also, the obtained analysis results were significantly dependent upon recovery of a certain element in a particular melting process.

Spectro-photometry is a technique that uses the absorbance of light by an analyte (the substance to be analyzed) at a certain wavelength to determine the analyte concentration. Spectro-photometric analysis is based on development of calibration curve: several standards containing exactly known concentrations of the analyte are introduced into instrument, and the instrumental response is recorded. Ordinarily this response is corrected for instrument output obtained with a blank. Ideally, the blank contains all of the components of the original sample except for the analyte. The resulting data are then plotted to give a graph of corrected instrument response versus analyte concentration, and this graph in turn can be used to find the concentration of an unknown [9-10].

In this contribution we presented a cost effective and reliable spectro-photometric method for the composition analysis of aluminum master alloys. Analysis of Al-Mn, Al-Fe, & Al-Cu master alloys using the calibration curve of Mn, Fe, & Cu are reported. Furthermore, the designed methods are verified by analyzing certified reference materials (CRM) containing Mn, Fe, & Cu against their calibration curves.

2. Experimental methods

2.1. Development of calibration curve of Mn

A calibration curve ranging from 0.1mg/100ml to 2mg/100mlMn was obtainedaccording to ASTM standard [11].This was done by transferring 1.000 g of pure Mn metal powder(weighed accurately to 3rd decimal) to a 500-ml beaker and dissolved in 100 ml of H_2SO_4 (1+9 v/v). The solution was then boiled for 5 minutes and transferred to 1 liter volumetric flask (when cool down), diluted to the volumeand mixed. This 100mg/100ml solution was named as Mn standard solution-A.

From the Mn standard solution-A transferred 100 ml in 1 liter volumetric flask, diluted to the volume and mixed. This 10mg/100ml solution was named as Mn standard solution-B.

Calibration solutions were prepared by transferring0 (reagent blank), 1.0, 2.0, 5.0, 10.0, 15.0, and 20.0 ml of Mn Standard Solution B (1 ml = 0.1mg Mn (10mg/100ml)) to 100-mlvolumetric flasks. To each 15 ml of H_2SO_4 (1+1 v/v), 10 ml of HNO_3, and 5 ml of H_3PO_4 were added and Diluted to about 70 ml. For color development, the solutions were heated to boiling in a water bath and 20 ml of KIO_4 (25 g/L) solution was added in each. The boiling process was continued for an hour until the characteristic color developed. The solution was then cooled to room temperature and diluted to volume with potassium iodated water (water free from reducing agents). Photometric measurements were made at approximately 525 nm against the reagent blank.

2.2. Development of calibration curve of Fe

A calibration curve ranging from 0.02mg/100ml to 0.2mg/100mlfor Fe was obtained according to ASTM standard [11]. This was done by transferring 0.1000 g of pure Fe metal powder (weighed accurately to 3rd decimal) to a 500-ml beaker and dissolved in 50 ml of H_2O, 25 ml HCL, 1ml HNO_3. The solution was then heated to dissolve Fe and transferred to 1 liter volumetric flask (when cool down), diluted to the volume and mixed. This 10mg/100ml solution was named as Fe standard solution-A.

From the Fe standard solution-A transferred 100 ml in 500 ml volumetric flask, added 10 ml HCL, diluted to the volume and mixed. This 2mg/100ml solution was named as Fe standard solution-B. Calibration solutions were prepared by transferring 0 (reagent Blank), 1.0, 2.0, 4.0, 7.0, 10.0ml of Fe Solution B (1 ml = 0.02 mg Fe) to five 100-ml volumetric flasks. Diluted to 60 ml and for color development : added 8ml$NH_2OH{\cdot}HCl$ (100g/L) solution, 10ml Sodium acetate-acid (272g/L) buffer solution and 10ml of1,10-phenanthroline(2g/L) solution, diluted to volume, mixed and allowed to stand for 5 min. Photometric measurements were made at approximately 510 nm against the reagent blank.

2.3. Development of calibration curve of Cu

0.1g of pure Cu metal powder (weighed accurately to 3rd decimal) was dissolved in a 5ml of nitric acid and diluted to 100 ml in a volumetric flask (100mg/100mlstock Cu solution).

2ml, 4ml, 6ml, 8ml, and 10mlof the stock Cu solution were transferred in five different volumetric flasks of100ml. After that the ammonia solution was gradually added in each volumetric flask, resulting in a change of color(the color turns to dark blue). Adding ammonia solution was continued until a stabilized color was achieved (i.e., no further color change occurs by adding more ammonia solution). Additional 3mlof ammonia solution was added in each and diluted to the volume. Photometric measurements were made at approximately 610 nm against distilled water as a reagent blank.

2.4. Determination of Mn, Fe, and Cu in aluminum master alloys

After the development of calibration curves, some test runs were conducted to characterize the calibration curves. This was done by testing Mn, Fe and Cu in certified reference materials (CRM's). Sample preparation was started with the chipsformation ofCRM's using lathe machine. Similarly, master alloy samples were collected in chips form by drilling several master alloy ingots of a single lot (e.g., Al-Mn 10 % master alloy lot) and mixed them.The test and master alloy solutions were prepared in the same way as that of calibration solutions but using successive dilutions to meet with the ranges of Mn, Fe and Cu calibration curves.

3. Results and discussion

Calibration standards for Mn, Fe and Cu were prepared and their response in the visible region of electromagnetic spectrum was recorded. Figures 1-3 represent the corresponding calibration curves.

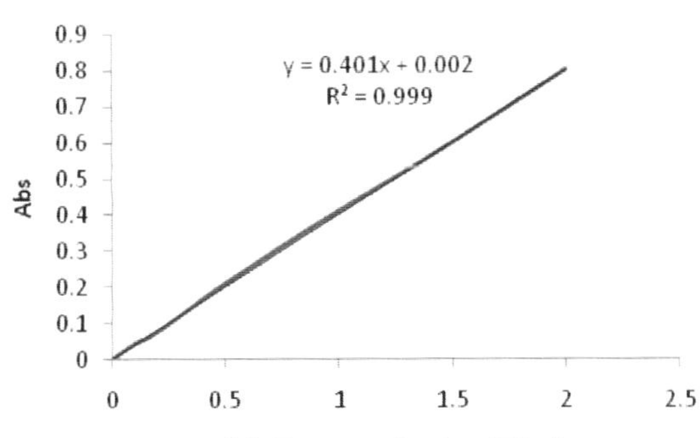

Figure 1. Calibration curve of Mn. Photometric measurements were made at 525 nm

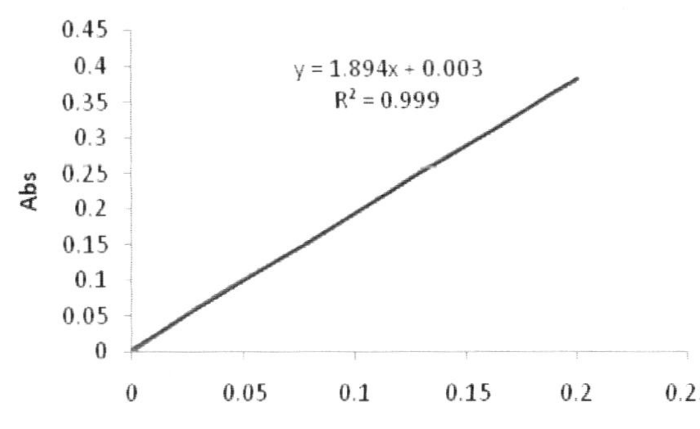

Figure 2. Calibration curve of Fe. Photometric measurements were made at 510 nm

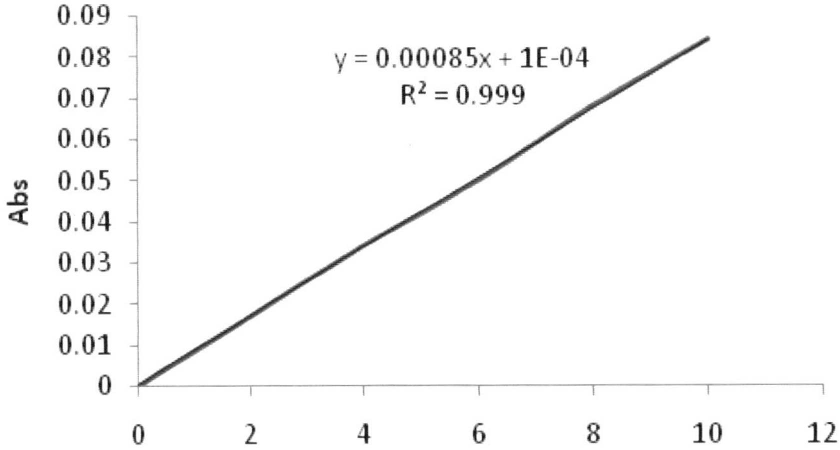

Figure 3. Calibration curve of Cu. Photometric measurements were made at 610 nm

At the outset the test run analysis is carried out in order to characterize the calibration curves. This is done by testing Mn (wt. %) in Certified Reference Materials (CRM's) against the calibration curve prepared for Mn. The photometric calculations and analysis results of Mn are shown in Table 1. It may be seen from Table 1 that the results of photometric analysis of Mn in CRM's (49712b-4 & A333.O AE) are in well agreement with certified values. Also the spectro-photometric results of Mn matches good with spark OES results. These results standardize the characteristic calibration curve of Mn. Table-1 also enlists the spectro-photometric calculations and analysis results of 5%, 10% and 20% Al-Mn master alloys. The Mn % is within the prescribed range for all types of master alloys.

Table 1. Spectro-photometric data for analysis of Mn

Sample Identity	**Known Value, Mn Wt.%**	**Sample weight/Dilution for photometry**	**Concentration found by photometry (mg/100ml)**	**Result, Mn Wt.%**
Alcan International Ltd. CRM ALC-A 333.0-AE	0.32	0.1g/100ml	0.333	0.333
Bureau of Analysed Samples (BAS) Ltd. CRM NCS HS 49712b-4	1.93	0.05g/100ml	0.967	1.934
AA 2024	0.601	0.1g/100ml	0.609	0.609
Mn 10% MA	9-11	0.1g/20x100ml	0.507	10.14
Mn 20% MA	18-22	0.1g/20x100ml	1.024	20.48

Where, CRM=certified reference material, MA=master alloy, x=multiplication

The photometric calculations and analysis results of Fe are represented in Table 2. It may be seen from Table 2 that the results of photometric analysis of Fe in CRM (49712b-4) are close tothe certified value. However for CRM (A333.O AE) there is a large deviation in spectro-photometric analysis results as compared to certified value. This happens because CRM (A333.O AE) contains high % of

Cu (3.5%) that interferes with Fe [11]. Also the spectro-photometric results of Fe for AA1050 are closed to OES value because of low % of Cu. These results standardize the characteristic calibration curve of Fe and express that the Fe can only be determined accurately in the absence of Cu. The spectro-photometric calculations and analysis results for 10% Al-Fe master alloy are also shown in Table 2. The Fe % is within the prescribed range.

Table 2. Spectro-photometric data for analysis of Fe

Sample Identity	known value, Fe Wt.%	Sample weight /Dilution for photometry	Concentration found by photometry (mg/100ml)	Result, Fe Wt.%
Bureau of Analysed Samples (BAS) Ltd. CRM NCS HS 49712b-4	0.242	0.1g/2x100ml	0.111	0.222(Cu=0.057)
Alcan International Ltd. CRM ALC-A 333.0-AE	0.78	0.1g/5x100ml	0.051	0.255*
AA 1050	0.37	0.1g/10x100ml	0.033	0.33 (Cu= 0.02)
Fe 10% MA	9-11	0.1g/100x100ml	0.101	10.1

Where, CRM=certified reference material, MA=master alloy, *low reading due to Cu (3.53%) interferes, x=multiplication

First the Cu (wt. %) is determined in CRM against the calibration curve prepared for Cu. The photometric calculations and analysis results of Cu are shown in Table 3. It may be seen from Table 3 that the results of photometric analysis of Cu in CRM (A333.O AE) are just close to the certified value. Also the spectro-photometric results of Cu for AA2219 do not match well with OES results.This variation as compared to the Certified or OES value is due to the reason that the absorption coefficient for copper tetra amine complex is very small (Figure 3. calibration curve of Cu). The spectro-photometric calculations and analysis results for 10% and 50% Al-Cu master alloys are also shown in Table 3. The Cu % is within the prescribed range but lies away from central value. Again this is because the absorption coefficient for copper tetra amine complex is very small which probably increases the instrumental error.

Table 3. Spectro-photometric data for analysis of Cu

Sample Identity	Known value, Cu Wt.%	Sample weight /Dilution for photometry	Concentration found by photometry (mg/100ml)	Result, Cu Wt.%
Alcan International Ltd. CRM ALC-A 333.0-AE	3.53	0.5g/5x100ml	3.3	3.3
AA 2219	6	0.5g/5x100ml	6.23	6.23
Cu 10% MA	9-11	0.5g/5x100ml	9.5	9.5
Cu 50% MA	47.5-52.5	0.1g/5x100ml	9.42	47.1

Where, CRM=certified reference material, MA=master alloy, x=multiplication

4. Conclusions

Chemical compositional analysis of Al-Mn, Al-Fe, and Al-Cu master alloys are necessary to adjust the melt chemistry quickly. Also, it is indispensable for the strict quality control of master alloys production. The presented spectro-photometric methods are reliable and cost effective. Particularly, these spectro-photometric methods cover all the ranges for Mn, Fe and Cu determinations in aluminum master alloys. Also, the sampling practices (chips formation and mixing)suggested here easily include a large number of master alloy ingots, whose average results are the true representatives of the whole lot.

5. Acknowledgement

The authors are deeply indebted to Mr. Muhammad Waseem and Mr. Muhammad Awais for their assistance during the preparation of manuscript and proof reading.

6. References

[1] Totten G E and Mackenzie D S 2003 Handbook of Aluminum: Alloy production and materials manufacturing(Marcel Dekker Inc. New York)2
[2] Moldovan P, Popescu G, Popescu C A and Apostolescu I2007*Advanced Materials Res.*23 295.
[3] Naveed A 2010 Pak. J. Anal. Environ. Chem. 11 2 51-55
[4] ASTM E60.98 1998 Standard practice for analysis of metals, Ores and related materials by molecular absorption spectrometry
[5] ASTM E1452.92 1996 Standard practice for preparation of calibration solutions forSpectro-photometricanalysis
[6] ASTM E1282.98 1998 Standard guide for specifying the chemical compositions and selecting sampling practices and quantitative analysis methods for metals
[7] Ingle J D and Crouch S R 1988 *Spectro chemical Analysis* (Prentice Hall International, Inc. New Jersey) 257
[8] Jenkins R, Gould R W and Gedcke D1995*Quantitative X-Ray Spectrometry* (Marcel DekkerInc. New York)
[9] Grieken R E V and Markowicz A A 2002*Handbook of X-Ray Spectrometry*(Marcel Dekker Inc. New York)
[10] ASTM E169.99 1999 General Techniques of Ultraviolet-Visible Quantitative Analysis.
[11] Jeffery G H, Bassett J, Mendham J, Denney R C 1989 *quantitative chemical analysis*(Longman Group UK Limited)
[12] ASTME34.94 2002 Standard test methods for chemical analysis of aluminum and aluminum-base alloys

Numerical simulation of armor capability of Al_2O_3 and SiC armor tiles

T. Rashid, M. A. Aleem, S. Akbar[1], A. Rauf[2] and M. Shuaib[2]
PIEAS, P.O. Nilore, Islamabad, Pakistan
[1]Quaid-e-Azam University, Islamabad, Pakistan
[2]Institute of Industrial Control Systems, P. O. Box 1398, Rawalpindi, Pakistan

E-mail: adeel713@gmail.com

Abstract. Alumina and Silicon Carbide armor plates have been tested numerically against 7.62x51 (mm x mm) armor piercing (AP) projectiles. A 2-D problem with axial symmetry has been designedand the simulations were carried out using commercial software ANSYS AUTODYN. Experiments were modeled for Alumina (99.5%), Alumina (99.7%) and SiC with a range of tile thicknesses (5, 10, 15 and 20 mm). The projectile was chosen as 7.62 x 51AP bullet (initial velocity 810 m/sec)with two different core materials Steel 4340 and WC, however, casing material was copper for both cores. SiC showed better defense against AP bullet as compared to Al_2O_3. The residual velocity and momentum of the bullet were found to decrease with increasing tile thickness. SiC tiles with thickness 15mm and 20 mm successfully sustained penetration against steel 4340 and WC core bullets, respectively. However none of the Alumina targets succeeded in stopping the bullet.

1. Introduction

Protection of soldiers had been under focus since ancient times. With the modernization of arms, demand for rise in protection levels against potential threats has also been increasing. For designing of armor vestvariousparameters are taken into account i.e. threat level, durability, weight and cost etc. Ceramic armor's are preferred over metals mainly for their low weight and increased durability.To design armor, hit and trial procedure is not viable for being costly and labor intensive.

Hazell et al. studied the penetration and failure mechanism of Silicon carbide tiles for 7.62AP projectiles using hydrocode Autodyne-2D. Critical tile area for optimized performance of square SiC tile has been worked out as 70 mm x 70 mm.M. V. Silva et al. [2]studied mechanical properties of different Al_2O_3 compositions (92, 96 and 99 %). Al_2O_3 (92 % and 99 %) showed higher Vickers hardness and bend strength as compared to $Al_2O_3$96%. However, all three compositions did not allow penetration against AP-projectile (865 m/s). Hosein Kia et al. [3] presented a comparison of theoretical and numerical results for effect of projectile density on its penetration capability into alumina armor. Ballistic limit velocity (BLV) of the alumina armor was calculated using LS-Dyna software. Results showed that increase in projectile density decreases BLV. Chetan Swaroop [4]studied the use of Smooth particle hydrodynamics (SPH) solver of Autodyne-2D to simulate the performance of Al_2O_3 (99.7%) by Lead Round Nose projectile (LRN).The main result was as empirical relation between impact velocity and corresponding residual velocity for 6 mm thick ceramic plate (Al2O3-99.7%) with linear fit.R. Azarafza et al. [5] studied the impact behviour of Alumina/Nano-SiC ceramic composite using both computational and experimental techniques.

Content from this work may be used under the terms of the Creative Commons Attribution 3.0 licence. Any further distribution of this work must maintain attribution to the author(s) and the title of the work, journal citation and DOI.
Published under licence by IOP Publishing Ltd

Composites with 10 % SiC doped Alumina showed improved ballistic behavior as well as low density as compared to pure Alumina.Z. Fawaz et al. [6] simulated the energy loses and stress distributions in ceramic-composite armor when subjected to normal and oblique 7.62 AP projectiles. Inter laminar stresses at the ceramic composite interface were found smaller for oblique impacts as compared to the normal impacts. Projectile erosion for oblique impacts was however greater than those for normal impacts. Daniel Bürger et al. [7] Simulations in ABAQUS were carried out for testing ceramic/fiber reinforced composite armors against 7.62 x 51 AP projectile impact. Results predicted a partial penetration at target speed 675 m/s and complete penetration at 765 m/s.

In the present work, AUTODYN-2D is used to evaluate and compare the impact resistance capabilities of three different ceramic materials against armor piercing bullets of two different types. Impact bearing capability of four different thicknesses of Alumina (99.7%), Alumina (99.5%) and Silicon carbide plates was simulated against armor piercing projectiles with steel4340 and tungsten carbide cores.

2. Materials and methods

2.1 Autodyn-2D

Autodyn-2Dis a module of commercial software ANSYS and is used for the non-linear analysis of high velocity impact problems, high and low distortion problems and for projectile and target interaction related problems [8]_ENREF_8. The 2D model was designed in the Preprocessor (ANSYS DESIGN MODELER). Geometry was then imported to AUTODYN-2D (Solver & Post Processor) for further analysis. General setup of the model is shown in Figure 1.

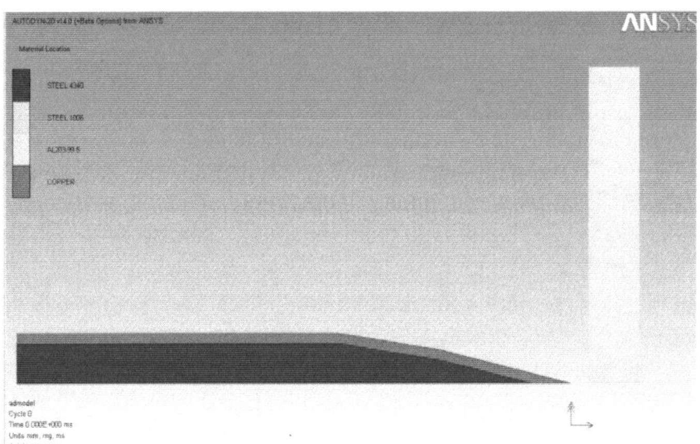

Figure 1. General setup of AP-bullet projectile and ceramic target

2.2 Projectile

In the present work, 7.62x51 mm NATO armor piercing (AP) bullet was used as projectile. The total length of the projectile is 51 mm and the maximum diameter is 7.61 mm. Projectile was integrated of hard material inner core (WC and Steel4340) and metal (copper) outer jacket/casing. Two different core materials, WC and Steel 4340 were used in simulation experiments. Thickness of the copper casing was taken 1 mm and diameter of the core was fixed as 6.61 mm, as shown in Figure 2. The initial velocity of 810m/sec was assigned to the bullet. The material model and properties for above mentioned materials were taken from AUTODYN Materials Library.

2.3 Target

Three different materials; Al_2O_3 (99.7%), Al_2O_3 (99.5%) and SiC were used as targets. Bullet penetration simulations were carried out for four different thicknesses; 5, 10, 15 and 20 mm of each of the three ceramic materials. Fixed boundary condition (or Support) was applied to the upper edge of the target. Height of each target was 30 mm as show in Figure 2.

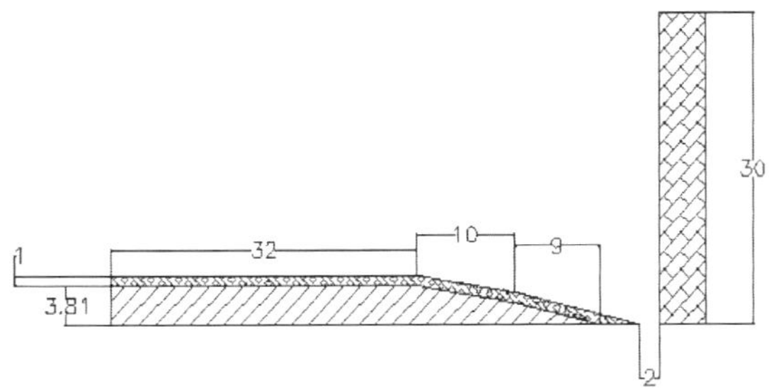

Figure 2. Mesh design for the bullet and the target

2.4 Meshing

In the AUTODYN-2D, quadrilateral meshing was introduced to both the bullet and the ceramic target. While size of quadrilateral element was fixed as 0.1mm. The number of elements and nodes for the casing and the target are given as

Table 1. Mesh parameters of Bullet and Target

	Bullet		**Target**			
	Casing	**Core**	**5mm**	**10mm**	**15mm**	**20mm**
No. of Elements	5886	16457	1962	2976	3362	3700
No. of Nodes	6449	16990	2090	2852	3240	3577
Skewness	0.843	0.843	0.00164	0.0029	0.0013	0.0029

2.5 Models used for materials

The material models and properties for the Steel 4340, Copper, Alumina 99.5, Alumina 99.7 and Silicon Carbide were taken from AUTODYN materials library_ENREF_8. The Johnson-Cook model was used for failure analysis of Steel4340 core and Copper jacket. At non zero strain rate the yield stress ($\bar{\sigma}$)depends on the equivalent plastic strain ($\bar{\varepsilon}^{pl}$), reference strain rate($\dot{\varepsilon}_0$) and softening temperature($\hat{\theta}^m$)given by [8]

$$\bar{\sigma} = \left[A + B(\bar{\varepsilon}^{pl})^n\right]\left[1 + C\ln(\dot{\bar{\varepsilon}}^{pl}/\dot{\varepsilon}_0)\right](1-\hat{\theta}^m) \tag{1}$$

Where *A, B, C, n and m* are material parameters.

$$\bar{\varepsilon}_f^{pl} = \left[d_1 + d_2\exp(d_3\frac{p}{q})\right]\left[1 + d_4\ln(\frac{\dot{\bar{\varepsilon}}^{pl}}{\dot{\varepsilon}_0})\right](1 + d_5\hat{\theta}) \tag{2}$$

Where q, p and d1 – d5 are Mises stress, mean stress and failure parameters, respectively. Values of d1 – d5 have been mentioned in Table 2.

Table 2. Failure parameters of Johnson-Cook Model for Steel 4340

D_1	D_2	D_3	D_4	D_5
0.05	3.44	2.12	0.002	0.61

The failure modelfor copper considers Failure strain of the value 0.4 and erosion geometric instantaneous strain is 2[8].

3. Results and discussion

3.1 Simulation of penetration

Figure 3 shows the penetration of steel 4340 core bullet in three different ceramic materials with varying thickness. It can be observed that the bullet penetrates almost un-hurt through 5 mm thick Alumina (99.5 and 99.7%), however its cone face in blunted in case of SiC (5mm) tile. Reduction in velocity and momentum of the bullet against all three tiles is almost negligible as evident from Fig. 5 and Fig. 7, respectively.

For 10 mm tile thickness, Alumina 99.5% is found to show better resistance as compared to Alumina 99.7%. For SiC 10 mm, the bullet (core steel 4340) has penetrated successfully but it is fragmented. Percentage reduction in velocity and momentum are less than 10, however SiC can be seen on the upper edge in Fig. 5 and Fig. 7.

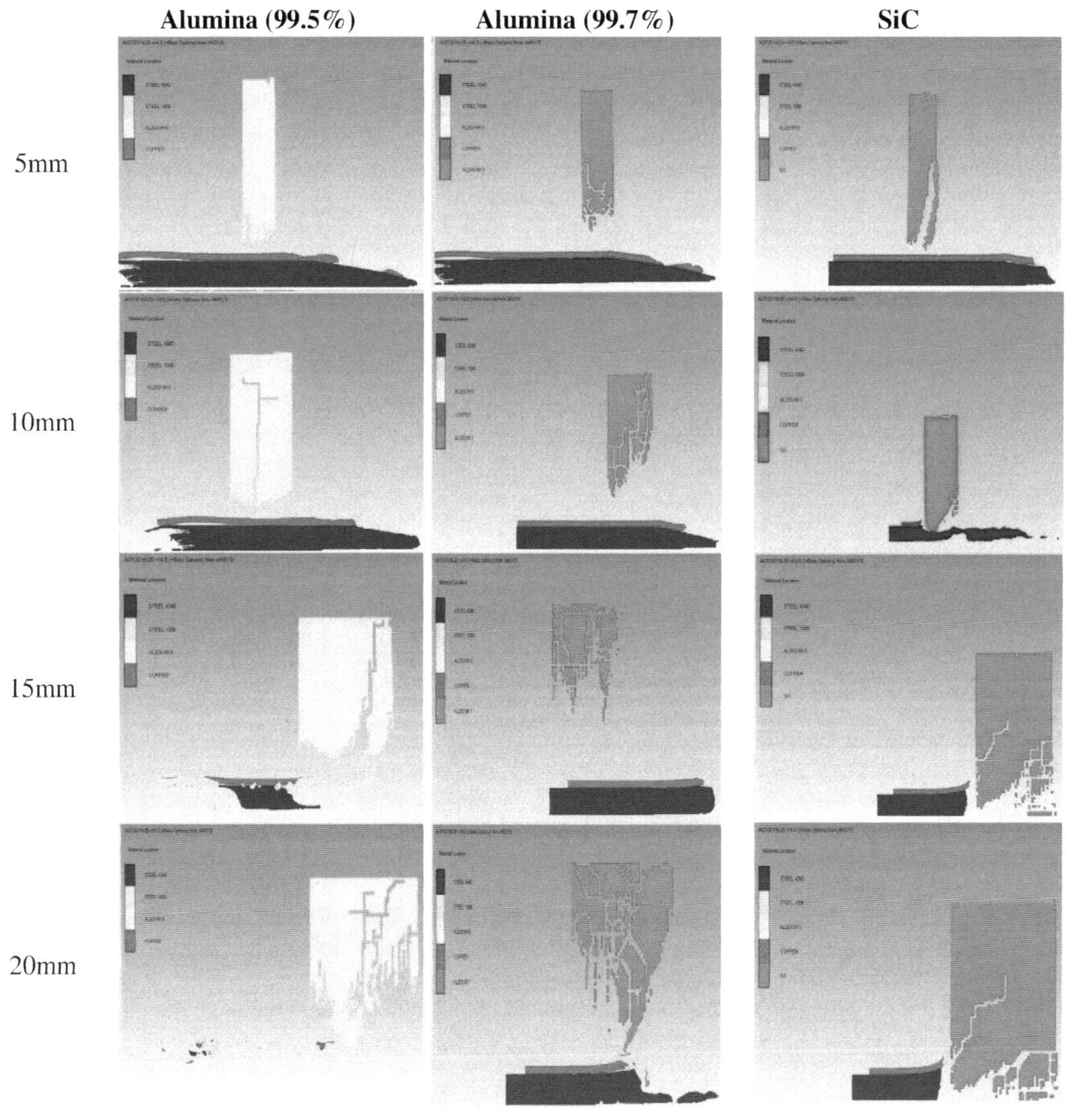

Figure 3. Penetration of 7.62 AP (Core: Steel4340) bullet in Alumina and SiC targets of thicknesses

For tile thickness 15mm, the bullet is found to penetrate the Alumina 99.7% target by scattering the tile into fragments. In case of Alumina 99.5% a single major crack can be observed in the tile and bullet is fragmented. SiC (15mm) tile has been able to survive the attack by blocking the bullet while developing few major cracks in itself. Reduction in velocity and momentum are almost identical for both Alumina-99.5% and Alumina-99.7%, which is around 20% of the initial. However, SiC has successfully absorbed the bullet energy while reducing the velocity and momentum of the bullet to zero.

By further increasing the tile thickness to 20 mm, it has been observed that Alumina 99.7% has failed to survive the attack of steel 4340 bullet, however Alumina 99.5% has been able to completely scatter the bullet into tiny fragments.

Figure 4 shows the penetration of WC core bullet in three different ceramic materials tiles with varying thickness. Complete penetration can be seen in all compositions with thickness 5mm, however minor damage to the copper casing is observed.

For 10 mm tile thickness, Alumina 99.5% imparted more damage to the bullet as compared to Alumina 99.7%. For SiC 10 mm tile, considerable damage can be observed as the bullet is divided into tiny fragments while penetrating through the tile.

In case of 15 mm thick tile, the WC core bullet has penetrated through all three materials. Alumina 99.7% presented the least resistance by only blunting the cone of the bullet. Alumina 99.5% posed comparatively better resistance as we see fragmented bullet however the tile also developed major cracks. Unlike Steel 4340 core bullet, the WC core bullet succeeded in completely penetrating the SiC 15mm tile however tearing itself into fragments.

For the 20mm tile, both Alumina 99.7 % and 99.5 % failed to survive the attack of the WC core bullet, however in this case as well SiC has been able to block the penetrator while developing a few major cracks in itself.

	Alumina (99.5%)	**Alumina (99.7%)**	**SiC**
5mm	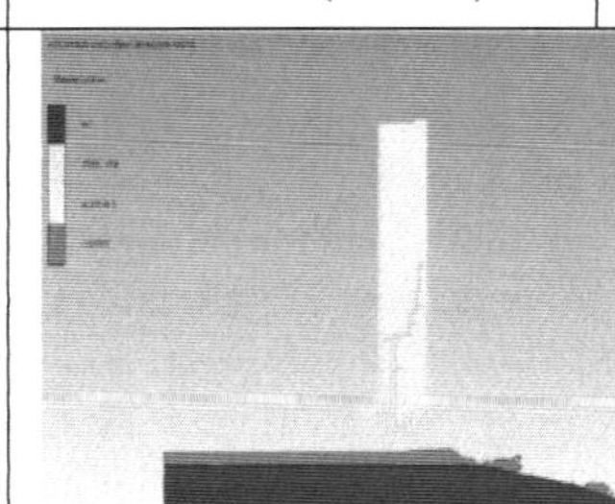	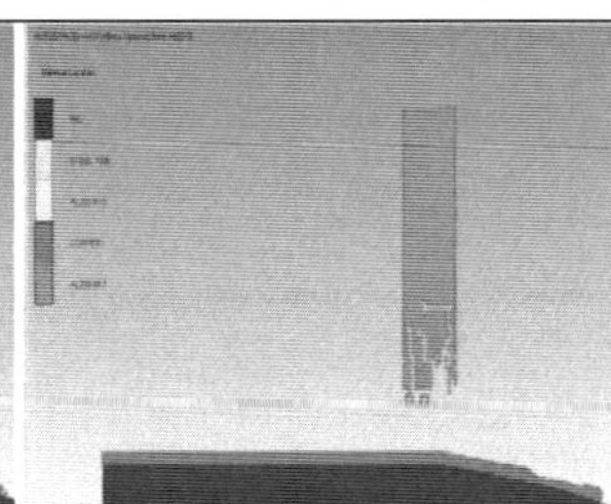	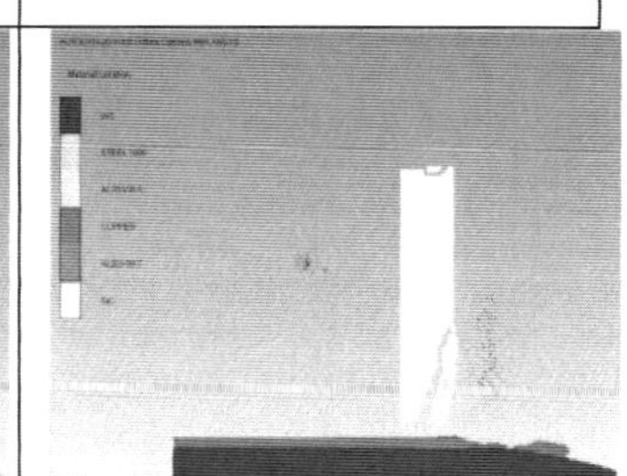
10mm	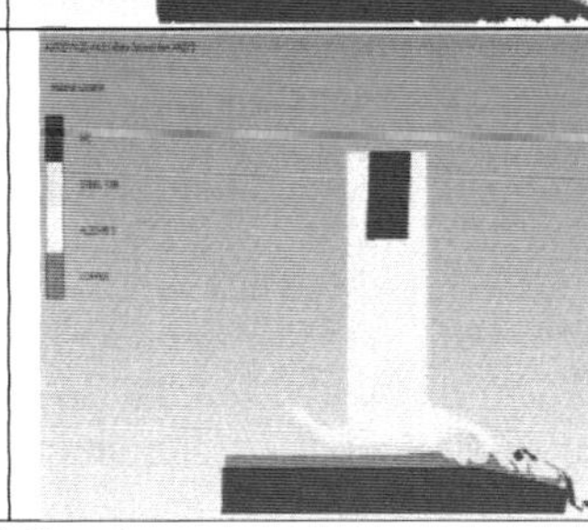	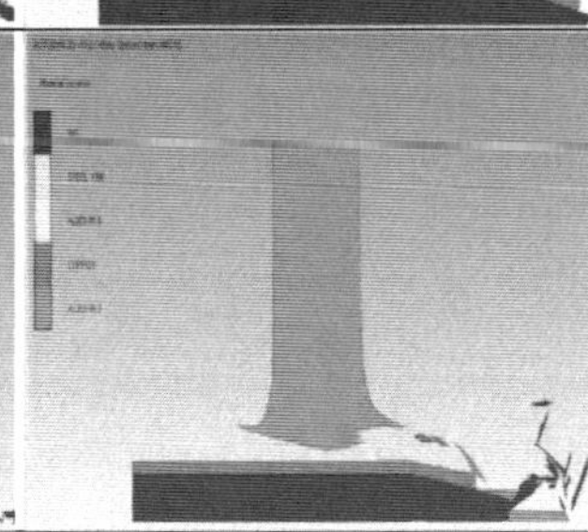	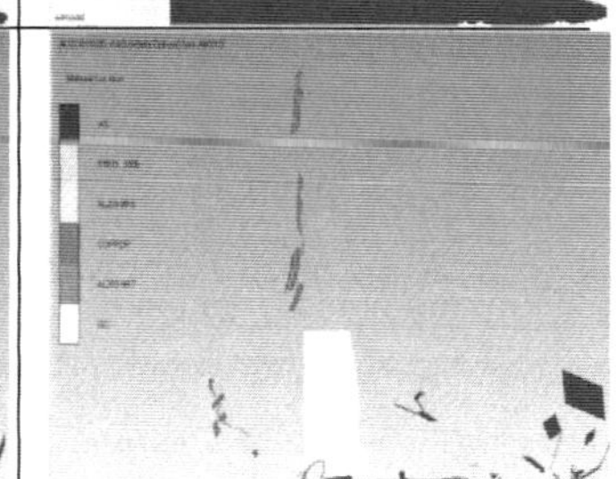

Figure 4. Continue

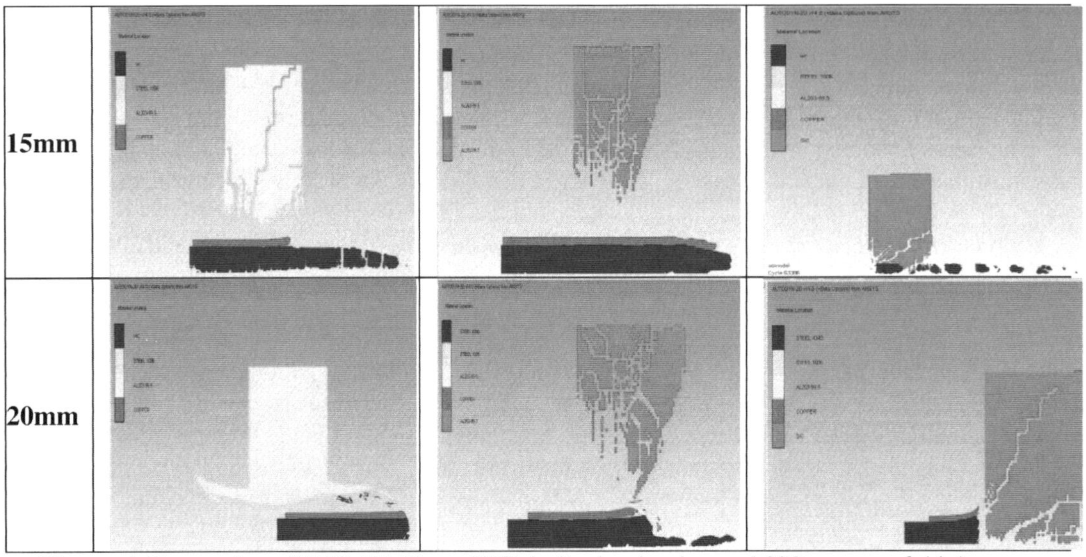

Figure 4. Penetration of 7.62 AP (Core: WC) bullet in alumina and SiC targets of thicknesses

3.2 Reduction in bullet velocity

Figure 5 and Figure 7 show residual velocity and residual momentum of the Steel4340 core bullet after striking the target. Results of residual velocity and momentum against Alumina (99.5% and 99.7%) targets showed almost same trend for the whole thickness range. In case of SiC tiles (5mm and 10mm) the residual velocity and momentum was comparable to that of Alumina (99.5% and 99.7%). However, 15mm and 20mm SiC tiles reduced the speed of the Steel4340 to dead level.

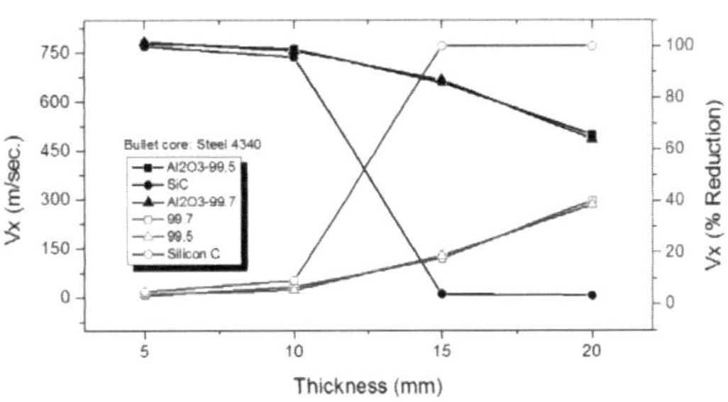

Figure 5. Reduction in velocity of Steel4340 core bullet against different thicknesses of alumina and SiC

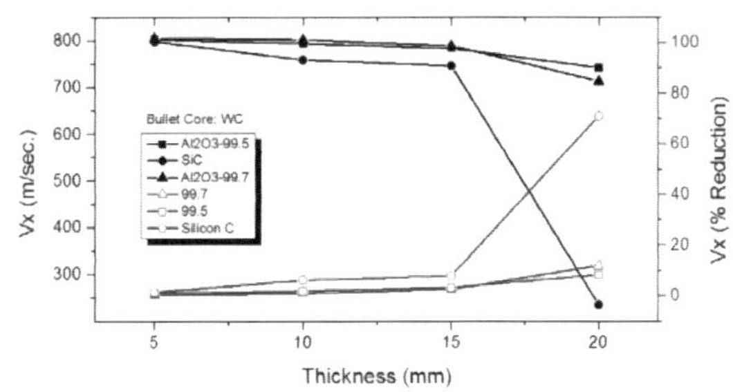

Figure 6: Reduction in velocity of WC core bullet against different thicknesses of alumina and SiC

Figure 6 and Figure 8 show residual velocity and residual momentum of the WC core bullet after striking Alumina and SiC targets. Results of residual velocity and momentum against Alumina (99.5% and 99.7%) targets showed almost same trend for the whole thickness range. In case of SiC tiles (5mm and 10mm) the residual velocity and momentum were comparable to that of Alumina (99.5% and 99.7%). However, 20mm SiC tiles reduced the speed of the Steel4340 core to dead level.

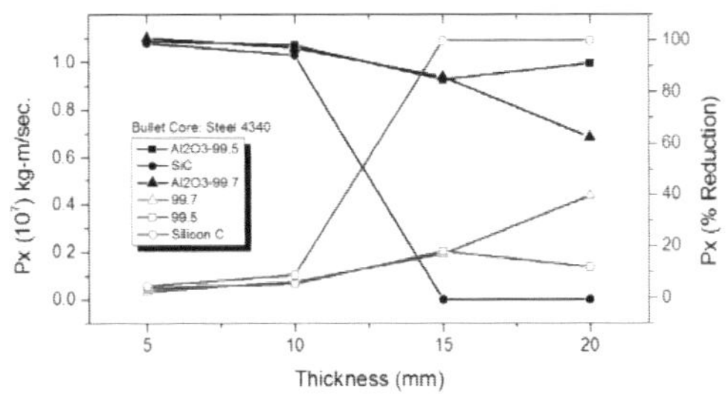

Figure 7. Reduction in momentum of Steel 4340 core bullet against different thicknesses of alumina and SiC

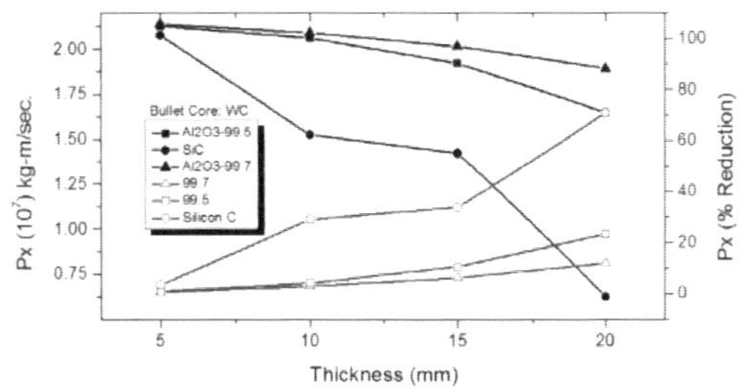

Figure 8. Reduction in momentum of WC core bullet against different thicknesses of alumina and SiC

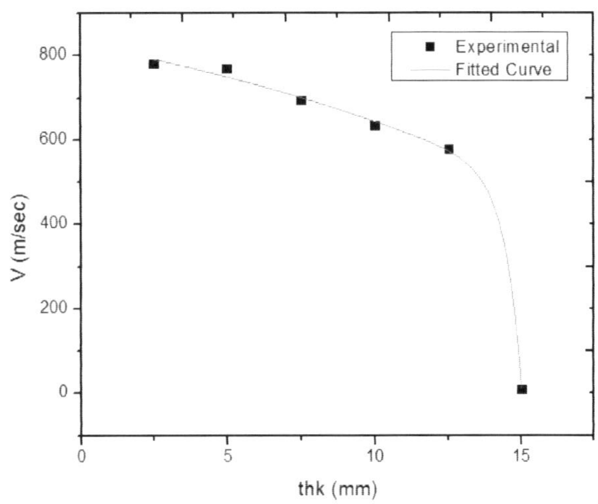

Figure 9. Curve fitting on residual velocity of sheet 4340 bullet against different SiC tile thicknesses

The relationship between residual velocity of steel 4340 bullet and SiC tile thickness is derived by curve fitting as shown in Figure 9.

$$V_r = V_0 + Ae^{-\frac{t}{\lambda_1}} + Be^{-\frac{t}{\lambda_2}}$$

Where V_0 is the initial velocity of the bullet, while A (-1.028E-05), B(-28.02) are amplitudes, while $\lambda_1(-0.8621)$ and $\lambda_2(-6.027)$ are decay constants or fitting constants.

4. Conclusions

Alumina and Silicon Carbide armor plates were subjected to high velocity (810 m/s) impact of 7.62x51 mm^2 AP projectiles. AUTODYN results show SiC plates is a better option for armor applications as compared to Alumina. SiC(15mm) stopped penetration effectively against Steel4340 core bullet, however, 20mm tile was found sufficient to stop WC core bullet. All thicknesses up to 20 mm failed to stop penetration against both steel 4340 and WC core bullets.

5. References

[1] C.J.R. Paul J Hazell*, Mauricio Moutinho, The design of mosaic armour: the influence of tile-size on the ballistic performance, Materials and Design, 29 (2007) 1497-1503

[2] D.S. M. V. Silva, H. A. Al-Qureshi, O. R. K. Montedo and D. Hotza, Alumina-Based Ceramics for Armor Application: Mechanical Characterization and Ballistic Testing, Journal of Ceramics, 2014 (2014) 06

[3] A.M.O. Hossein Kia, Analytical and Numerical Consideration of Projectile Density Effect on Its Penetration Ability in Alumina Armor, Journal ofMechanical Research and Application, 3 (2011) 21-28

[4] A.K.S.K.N.P. Chetan Swaroop, SPH Simulation of Ballistic Impact on Ceramic Plate, International Journal on Theoretical and Applied Research in Mechanical Engineering, 02 (2013) 05

[5] A.A.a.A.M. R. Azarafza, Impact Behavior of Ceramic-Metal Armour Composed of Al_2O_3-Nano SiC Composite, Int J Advanced Design and Manufacturing Technology, 5 (2012) 5

[6] W.Z. Z. Fawaz, K. Behdinan, Numerical simulation of normal and oblique ballistic impact on ceramic composite armours, Composite Structures, 63 (2004) 09

[7] M.V.D. Daniel Bürger*, Francisco C. L. de Melo and Sérgio F. M. de Almeida†, Ballistic impact simulation of an armour-piercing projectile on hybrid ceramic/fiber reinforced composite armours, 16th International Conference on Composite Structures ICCS 16, (2011) 2

[8] T. Borvik, Hopperstad, O. S., Berstad, T., and Langseth, M., Numerical simulation of plugging failure in ballistic penetration, International Journal of Solids and Structures, 38 (2001) 03

Microstructural study of as sprayed and heat treated Ni_3Al coatings deposited by air plasma spraying technique

K. Mehmood[1], M.A. Rafiq, A. Nusair Khan[1], M.M. Rauf[1]
University of Engineering and Technology, Lahore, Pakistan
[1]Institute of Industrial Control Systems, P. O. Box 1398, Rawalpindi, Pakistan

E-mail: engrkhalid80@yahoo.com, asifrafiq@uet.edu.pk, aamer.nusair@gmail.com
mudassar.79@hotmail.com

Abstract. Air plasma spraying system was utilized to deposit Ni_3Al coatings on AISI 321 steel samples. After plasma spraying the coatings were heat treated at different temperatures i.e. 500 °C to 800 °C for 10 to 100 hours. The characterization tools such as, X-Ray diffraction analysis, optical and scanning electron microscopy were used. By comparing the XRD scan data of as sprayed and heat treated coating, it was observed that the formation of NiO increases drastically with time and temperature. Due to the formation of NiO, hardness was also enhanced. The oxidation behavior was observed by using optical microscope and when it was studied that the oxidation was increasing with time and temperature. Further, the SEM tool was utilized to study the detail microstructural behavior such as shrinkage cavity and oxide particles. The other phases like alumina and spinel phases were determined by using Energy dispersive spectrometer method.

1. Introduction

The intermetallic compounds have been established speedily, in matching with other new materials. These materials have best applications to fulfil the requirement of industry such as power generation, coal-burning boilers, fluidized beds, gas turbines and combustion of heavy oils [1], synthetic fuels, and pulverized coal. To protect the components used in area, protective coatings play a vital role to protect the surface of hardware [2]. This complex validates decent structural properties such as low density, increase in yield strength with temperature, high strength in relation to the specific mass, oxidation resistance and excellent creep properties [3-7]. Nickel aluminde is comparatively less utilized, in industry, as compared to iron and chromium aluminde [8].

The coating techniques such as pack cementation, plasma spraying and chemical vapour deposition are utilized to deposit the metallic, ceramic, cermets and intermetallic powders on different substrates, but plasma spraying is showed to be the best technique due to its flexibility in depositing [9]. This technique can deposit thick coatings up to 1000 micrometers. More, this technique is useful for each size and shape of the substrate material.

The hot sections of components being used at high-temperature, oxidation is one of the problem in boiler tubes and gas turbines. Superalloys and Stainless steels are difficult to sustain both high-strength and oxidation at high temperature [10]. Thus, at high temperature oxidation behavior of aluminide coatings is important [11]. In this work, the as sprayed coatings were isothermal heat treated and oxidation behavior of Ni_3Al coating was studied.

Content from this work may be used under the terms of the Creative Commons Attribution 3.0 licence. Any further distribution of this work must maintain attribution to the author(s) and the title of the work, journal citation and DOI.
Published under licence by IOP Publishing Ltd

2. Experimental

2.1. Deposition of coating

Nickel aluminide intermetallic powder of irregular shaped (Figure 1) AMDRY-404, needing 45-125 μm size was used in these experiments. Air plasma spraying system of 80KW was utilized to deposit the coating on the stainless steel (AISI-321) substrates. The surface of the substrates were grit blasted with alumina particles before the coating. The grit blasting parameters are mentioned in the Table 1.The substrate surface roughness was ranged from 3 to 5 μm (Ra) after grit blasting. For plasma spraying 25 mm diameter samples were screwed in a steel fixture and the fixture was rotated at 120 rpm while the plasma torch move in to and fro manner along the rotation axis of the samples. The parameters used during spraying are mentioned in the Table 2. The spraying gun (80KW) having 7.5 mm anode internal diameter. The powder injector was fixed at 90° with respect to the spraying gun, whereas, the internal diameter of the powder injector was 1.5 mm.

Table 1. Parameters used in grit blasting

Parameters	
Air Pressure, psi	90
Angle between the substrate surface and nozzle, degree	90
Time, min	05
Grit size, mesh	46

Table 2. Parameters used in plasma spraying

Parameters	
Current, A	500
Voltage, V	70
Plasma gas (Ar) flow rate, liter/minute	75.5
Secondary gas (H_2) flow rate, liter/minute	4.72
Powder feed rate, g/min	113.6
Carrier gas (Ar) flow rate, liter/minute	16.5
Standoff distance, mm	125

The powder feed rate and the spray conditions were optimized by doing different experiments. After each experiment, the interface and the porosity of the coating was measured. The optimized spraying parameters showing minimum porosity level were finalized to spray the final coating. The temperature of the substrate was kept below 150°C by using compressed air at 100 MPa during plasma spraying.

2.2. Heat treatment of coating and characterization

The as sprayed samples were exposed at 500 to 800°C with 100°C increments, whereas, the holding time at each temperature was 10, 30, 60 and 100 hours. The heating rate was kept at 10°C per minute during heat treatment. As sprayed and heat treated samples were then characterized with the help of X-Ray Diffractrometer (XRD), Scanning Electron Microscope (SEM), optical microscope and hardness tester. Density of the coating was calculated by utilizing the Archimedes method. For this purpose the coating sample was delaminated from the surface of the substrate and then weighed in air and water.

For X-ray diffraction (XRD) analysis, the samples were scanned with JEOL JDX-99C, using Ni-filtered Cu-Kα radiation. The microstructures of the top and cross sectional coatings were studied using optical and scanning electron microscope, before and after heat treatment process. The chemical composition profiles and analysis of some important phases of as - sprayed and heat treated samples were estimated by Energy Dispersive Spectroscopy (EDS) technique.

3. Results and discussions

3.1. Metallography

To determine the thickness of coating, as sprayed sample was cross sectional mounted and it was revealed that about 250-270 micrometer thick coating was deposited on the steel substrates, Figure 2. The interface between the substrate and the coating was also studied and found satisfactory. It was observed that few porosity sites were also present in the coating, Figure 2. Further, un-molten particles were also present randomly within the coating, Figure 2. A thin grayish oxide film was observed between the splats, which was probably formed due to oxidation of powder, during the spraying process.

The detail microstructural study of the heat treated samples treated for 100 hours at different temperatures revealed that the oxidation of the coating increases with time and temperature. At low temperatures i.e. at 500°C for 100 hours, it was observed that the oxidation just started along the grain boundaries of splats as shown in the Figure 3. This coating can be utilized at 500°C up to 60 hours satisfactory and after treated for long time, it will dissociated in to new phase like NiO. After 600°C for 100 hours the oxidation of coating was gradually increased with time and temperature Figure 4. When the temperature was increased up to 700°C, it was observed that the oxidation of the coating was almost uniform throughout the coating Figure 5, however, at higher temperature as at 800°C this was more prominent at the top surface. This is because of the initial attack of oxygen which almost reached uniformly throughout the coating and formed a thin grayish layer around the splats, Figure 6.

It was noted that the oxygen was pre dominantly attacked on the top surface of the coating; this was more prominently visible in the samples heat treated at 800°C, Figure 6. Further, in the same samples, it was observed that the level of porosity also increased closed to the top surface of the samples, Figure 5. This is probably due to the formation of more oxides at the top surface which upon cutting and grinding broke away due to brittleness. , as heating timing is increased, more NiO oxides are formed on surfaces than smaller heating times.

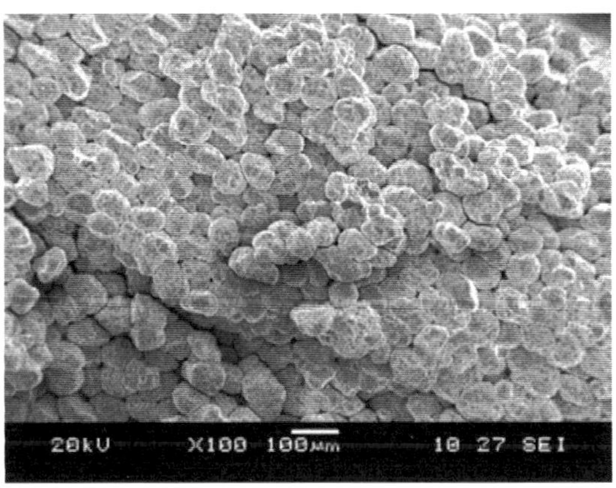

Figure 1. SEM image showing the morphology of spraying powder

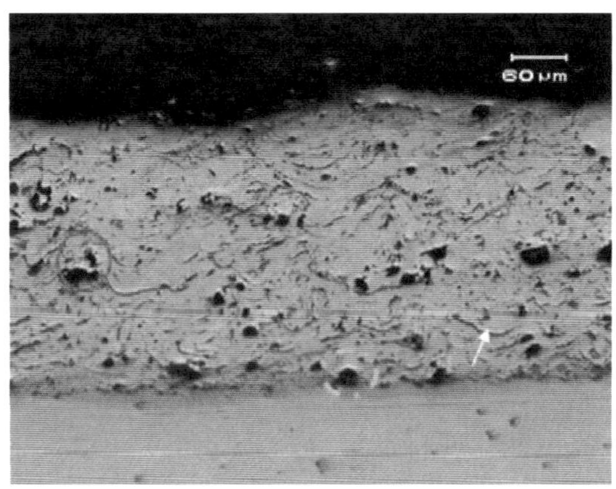

Figure 2. As sprayed nickel aluminide coating showing different features of the coating i.e. porosity (encirled), oxide layers (arrows) and un-melted particles (dotted box)

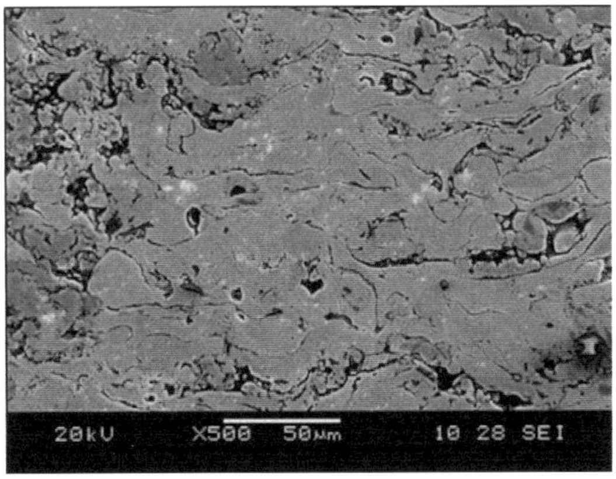

Figure 3. SEM micrograph showing the oxidation of samples treated for 100 hours at 500 °C temperature

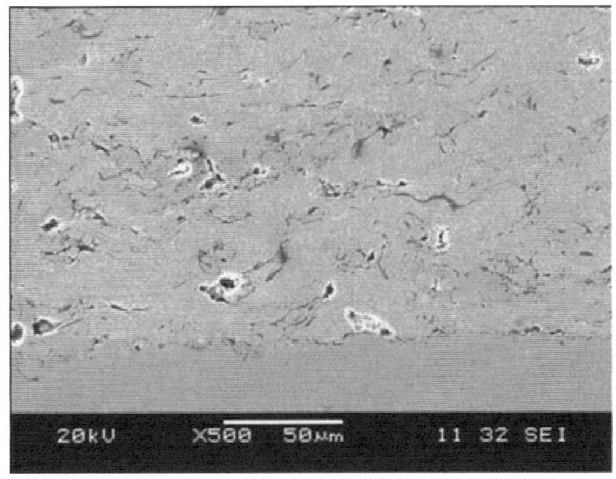

Figure 4. SEM micrograph showing the oxidation of samples treated for 100 hours at 600 °C temperature

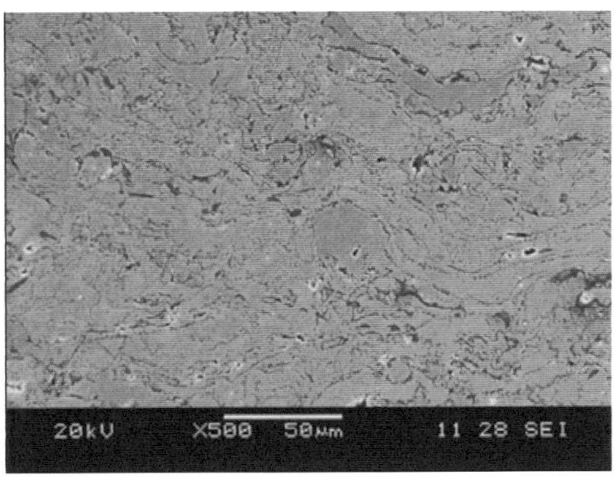

Figure 5. SEM micrograph showing the oxidation of samples treated for 100 hours at 700 °C temperature

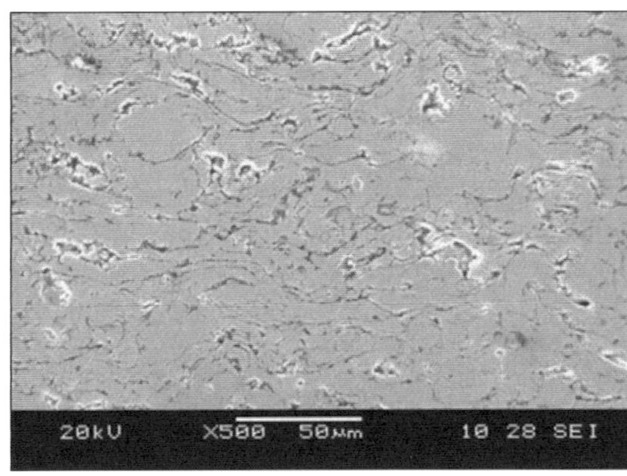

Figure 6. SEM micrograph showing the oxidation of samples treated for 100 hours at 800 °C temperature

3.2. Micro hardness testing

The as sprayed and heat treated samples were subjected to micro-hardness testing which was performed on all the samples. For this purpose the hardness was made on the cross section as well on the surface of the samples. All the hardness tests were performed on 100 gram load with 10 seconds dwell time. At least five to ten hardness values were taken from different portions of a sample. The results demonstrate that the hardness on the surface of the samples, continually increases with increasing temperature as shown in Figure 7. The increasing in hardness values are due to the formation of NiO phase. This nickel oxide phase is more dense and hard as compare to intermetallic phase. As heating time is increased, more hardness on surface is achieved than lesser heating time. The hardness data taken from the cross section of the samples, however, demonstrated a constant increase till 600°C and then a decreasing trend in the same as shown in Figure 8. The decrease in hardness values is maybe due to the cracking phenomenon observed at higher temperatures which caused relaxation of stresses. The decreasing hardness trend can be explained by the fact that the formation of NiO at the boundaries of the splats may weakened the overall structure, while the force was acting from the cross-section side.

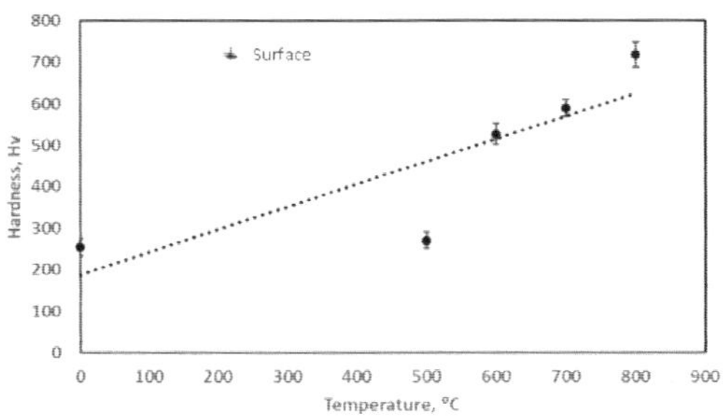

Figure 7. Effect of temperature on hardness of the coating, on surface, exposed for 100 hours

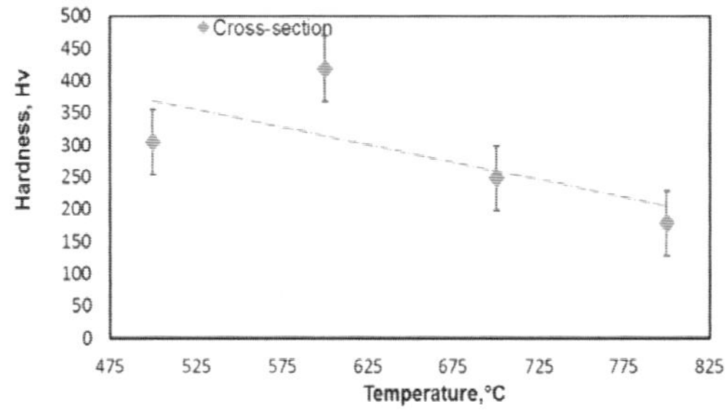

Figure 8. Effect of temperature on hardness of the coating, in cross section, exposed for 100 hours

3.3. Porosity measurement

After plasma spraying, density of the as sprayed coating was measured by taking the ideal density of Ni_3Al i.e. 7.29 g/cm^3 [12], and porosity was calculated from the measured value of the chipped off coating, which was 6.9 g/cm^3. It was detected that the porosity of the sprayed Aluminide was about 5.35%.

3.4. Phase analysis

For phase analysis of as-received powder for coating purpose it was revealed that Ni_3Al was predominately formed. XRD pattern of samples exposed at different temperature for different times are observed but here only two XRD scans of samples treated at 100 hours for 500°C to 800°C were discussed for comparison in Figures 9-10. X-Ray diffraction analysis demonstrates that small percentage of NiO formed after 100 hours exposure at 500°C. It was observed that by increasing the temperature above 500^{0}C, the formation of NiO became relatively faster and even noted after 10 hours exposure at 600^{0}C. It was noticed that the formation of NiO phase was linear at all the exposed temperatures (i.e. 500 and 800°C), Figures 9-10. However, it was observed that the slope of the curves was increased with increase in temperature.

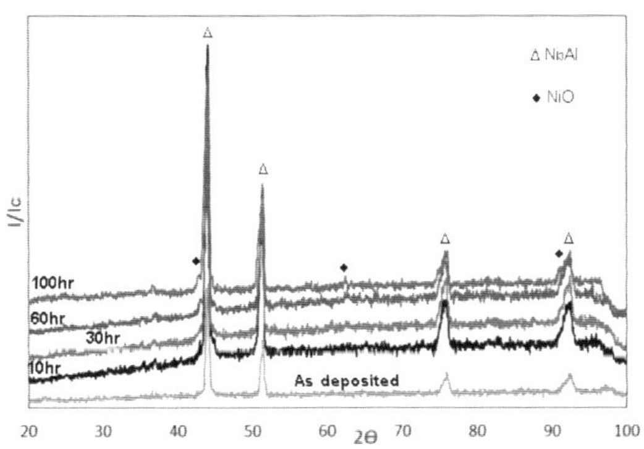

Figure 9. Comparison of XRD results of Ni_3Al coating treated at 500 °C for different times

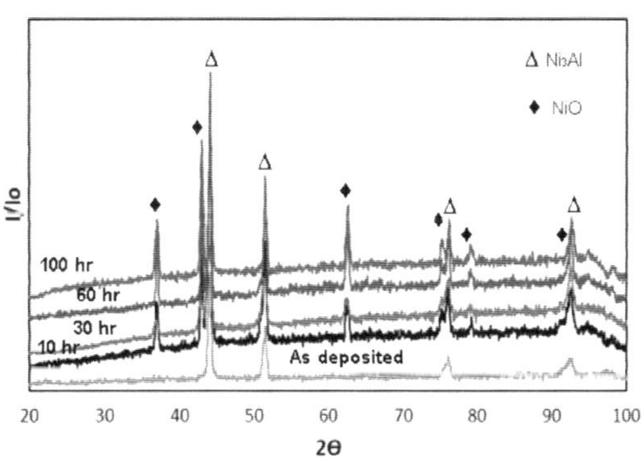

Figure 10. Comparison of XRD results of Ni_3Al coating treated at 800 °C for different times

4. Conclusion

Nickel aluminide coating when subjected to different temperature for different times revealed the formation of nickel oxide phase. This NiO phase frequently increases at different temperatures with time. It was also observed that due to the formation of NiO phase the hardness of the coating constantly increasing. The porosity level was also controlled by optimizing the spraying parameters. The percentage of other phases such as alumina and spinels were 1-2 % which were not detected by XRD scans due to less amount, only observed by EDS.

5. References

[1] Karin Gong, Zhifeng Zhou, P.W. Shum, Heli Luo, Zhiling Tian, Changhai Li, Tribological evaluation on Ni_3Al-based alloy and its composites under unlubricated wear condition, Wear 270 (2011) 195–203.

[2] S.B. Mishra, K. Chandra, S. Prakash, B. Venkataraman, Characterisation and erosion behaviour of a plasma sprayed Ni_3Al coating on a Fe-based superalloy, Materials Letters 59 (2005) 3694-3698

[3] G. Suuthoff, Mechanical Properties of Intermetallics at High Temperatures, Vortrag ASM Materials Week, Indianapolis, (1989)

[4] Wenming Wang, Bin Yang, Lingzhong Du, Weigang Zhang, Diffusion research between Ni3Al coating and titanium alloy produced by plasma spraying process, Applied Surface Science, 256 (2010) 3342–3345

[5] K. S. Kumar, C. T. Liu, Nickel and iron aluminide, Journal of Metals, 45 (1993) 28

[6] H. M. Saffarian, Q. Gan, R. Hadkar, G.W. Warren, Corrosion behavior of binary titanium aluminides intermetallics, Corrosion, 52 (1996) 626

[7] S. H. Kim, M. H. Oh, K. Kishida, T. Hirano, D.M. Wee, Deposition of NiAl for improvement of oxidation resistance of cold rolled Ni_3Al foils, Intermetallics 13 (2005) 129-136

[8] U. Malik, R. Ahmad, S. Ahmad, S. Ahmad, Prakt, High temperature oxidation behaviour of nickel aluminide coated mild steel, Metallogr. 29 (1992) 255

[9] Buta Singh Sidhu and S. Prakash, Degradation Behavior of Ni_3Al Plasma-Sprayed Boiler Tube Steels in an Energy Generation System, Journal of Materials Engineering and Performance, 14(3) (2005) 356-362

[10] Harpreet Singh, D. Puri, S. Prakash, Some studies on hot corrosion performance of plasma sprayed coatings on a Fe-based superalloy, Surface and Coatings Technology, 192 (2005) 27-38.

[11] P.S. Liu, K.M. Liang, S.R. Gu, High temperature oxidation behaviour of aluminide coatings on a cobalt base superalloy in air, Corrosion Science 43 (2001) 1217-1226

[12] Y. Tamarin, Protective coatings for turbine blades, ASM International, Materials Park, Ohio, USA, (2002) 56

Investigation of low stress rupture properties in Inconel-718 super alloy

T. Zaman, M. Farooque, S. A. Rizvi, I. Salam and M. Waseem
Institute of Industrial Control Systems, P. O. Box 1398, Rawalpindi, Pakistan

E-mail: ishmaelnn@gmail.com

Abstract. Inconel-718 is a Ni-Cr-Fe based super alloy. It is widely utilized in aircraft gas turbines, nuclear power systems, space vehicles and medical applications. Aim of the present study is to evaluate the effect of Ti and Nb content on high temperature stress rupture properties of Inconel718. OM, SEM and TEM were utilized for characterization of microstructure. Inconel718 is unique in that it forms large number of phases due to its composition and variety of heat treatments. γ''+ γ' precipitates and the effect of annealing on these precipitates have been studied using TEM. The main hardening phase was identified as metastable Ni3Nb (γ''). Other phases identified after annealing were secondary carbides (NbC) and stable acicular δ phase. Effect of γ'', δ, primary carbides and NbC on creep behavior was observed using OM and SEM. Higher Ti content(1.25 wt. %) resulted in poor creep properties due to large concentrations of primary carbides (TiC) at grain boundaries.

1. Introduction

Inconel718 is a nickel-based superalloy having good corrosion resistance and excellent mechanical properties at high temperatures. These alloys are particularly preferred in applications requiring good creep properties. These superalloys are extensively used as disc material in gas turbines and jet engines where service temperatures can be as high as 650°C [1-4].

Carbides play a very critical role in superalloys. Hence, studying the morphology, behavior and nature of these carbides is very important. Molybdenum, chromium, niobium, vanadium and titanium are carbide forming elements. MC, $M_{23}C_6$, M_6C and M_7C_3 type carbides are usually observed in superalloys [5]. Composition of alloy,heat treatment and deformation parameters dictate the morphology of these carbides. Inconel718 has an austenitic (γ) matrix. γ' and γ'' are the main strengthening phases of inconel718 superalloy. γ' is Ni_3Al while γ'' is Ni_3Nb. In full heat treated condition i.e. after solution treatment and double aging, the %volume fraction of γ'' and γ' is 20% and 5% respectively [6-7].

Xishan Xie et al studied the effect of titanium, niobium and aluminum on precipitation and strengthening behavior of 718 type superalloys [8]. They studied the characteristics of γ' and γ'' phase using scanning electron microscope and transmission electron microscope. They also determined the weight fraction of γ' + γ'' phases by electrolytic isolation followed by micro-chemical analysis. Moukrane Dehmas et al. utilized transmission electron microscope to study the high temperature precipitation of δ phase in inconel718 alloy [1]. They investigated a case where δ phase precipitated

Content from this work may be used under the terms of the Creative Commons Attribution 3.0 licence. Any further distribution of this work must maintain attribution to the author(s) and the title of the work, journal citation and DOI.
Published under licence by IOP Publishing Ltd

directly from the austenitic (γ) matrix. Presence of rotation-ordered domains in δ-plates was also confirmed.

Aim of this study was to investigate the influence of titanium content on high temperature stress rupture properties of Inconel718 superalloy. Secondly the effect of heat treatment on different precipitated phases (γ', γ'', δ, metallic carbides) was also studied using scanning transmission electron microscopy.

2. Experimental

Inco718 rod (diameter: 12mm) was utilized for this investigation. Chemical composition of the rod is given in Table 1. Chemical composition was obtained using Optical Emission Spectrometer (OES). Mechanical testing of the rod included: Hardness testing, tensile testing and creep testing.
Tensile testing was carried out using INSTRON 1195. For creep testing ATF-2330 creep tester was employed. Tests were conducted at 649°C at a stress level of 690 MPa.

Table 1. Chemical composition weight %

Elements	Ni	Cr	Fe	Nb	Mo	Ti	Al	(Al+Ti+Nb)at.%
ASTM B637-03	50-55	17-21	17-18	4.7-5.5	2.8-3.3	0.6-1.15	0.2-0.8	5.5
Composition wt.%	53	18.87	17.19	5.5	3.06	1.25	0.52	5.98

Samples were prepared before and after creep testing from cross-sectional and longitudinal regions. Samples were ground using180 to 1200 grit silicon carbide paper.1μm diamond paste was utilized for final polishing. Electrolytic etching of Inco718 samples was done with 10% oxalic acid. Average grain size was recorded using mean linear intercept method. Optical microscope OLYMPUS BX51 was used for studying polished and etched samples.

Fractured surfaces after high- temperature creep testing were observed using Scanning Electron Microscope JEOL JSM5910LV equipped with Energy Dispersive Spectrometer (EDS).
Scanning Transmission Electron Microscope JEOL-JEM-2000FXII was employed for study of precipitated phases in Inconel718. 3mm discs were manually punched out and ground using SiC paper, till final thickness of 0.1mm was achieved. These disc samples were then electrolytically prepared in Twin Jet Electropolisher using perchloric acid electrolyte (at 23V). Temperature was kept at approximately -15°C.

3. Results and discussion

3.1 Mechanical testing

Table 2 presents results of hardness and tensile testing. Yield strength of 1292MPa and ultimate tensile strength of 1512MPa was observed. %Elongation and % reduction in Area were recorded as 17.8% and 34.5% respectively. Hence, tensile properties of Inconel718 rod were well within the range of ASTM B637-03 standard.

Table 2. Mechanical properties: room temperature

Properties	Y.S (MPa)	UTS (MPa)	% Elong.	%Red. of Area	Hardness(HB)
ASTM B637-03	> 1034	> 1275	> 12	> 15	> 331 HB
Inconel718	1292	1512	17.8	34.5	434

Table 3 depicts results of creep testing done at 649°C (690MPa). Creep specimens sustained for total time of 80.3 hours before finally breaking. %Elongation observed in sample after 23 hours of testing was 1.07%. This %elongation is below the minimum required range of 4% (according to ASTM B637-03). A total %elongation of 4.24% was recorded in the sample after 80.3 hours. In order to investigate the root cause of low stress rupture value, chemical composition was observed which showed that the titanium content in rod was 1.25 weight %. This content is higher than the maximum permissible limit of 0.6-1.15(weight %). This abnormality in chemical composition formed the basis of investigation.

Table 3. Mechanical properties: high temperature

Properties	Temp.	Stress	Time Sustained hrs.	%E.(23 hrs.)	Total %E
ASTM B637-03	649°C	690 MPa	> 23	> 4%	--
Inconel718	649°C	690 MPa	80.3	1.07	4.24

(Al + Ti + Nb) atomic % is reported to be 5.5 atomic % for Inconel718 alloys [9]. However, in this study a higher (Al + Ti + Nb) atomic % of 5.98 atomic% was observed. Aluminum, titanium and niobium play key role in determining mechanical properties of Inconel718 alloy. These elements combine with carbon in order to give various metallic carbides. These metallic carbides are subdivided into: primary carbides (e.g. TiC) and secondary carbides (e.g. NbC).

3.2 Metallic carbides

Polished samples were studied under optical microscope in order to reveal metallic carbides. Metallography samples were prepared both before and after creep testing. Figure 1 shows TiC in polished Inconel718 alloy after creep testing. Average size of these primary carbides was recorded to be 5-6μm. these samples were then observed under Scanning Electron Microscope. SEM revealed a large quantity of titanium and niobium carbides located at the grain boundaries. Figure 2 shows primary carbides before creep testing (left) and after creep testing (right). It was observed that size of metallic carbides increased after creep testing. In terms of morphology, blocky/cube shaped, spherical and irregular carbides were mainly observed. The presence of primary carbides at grain boundaries effect high temperature stress rupture properties.

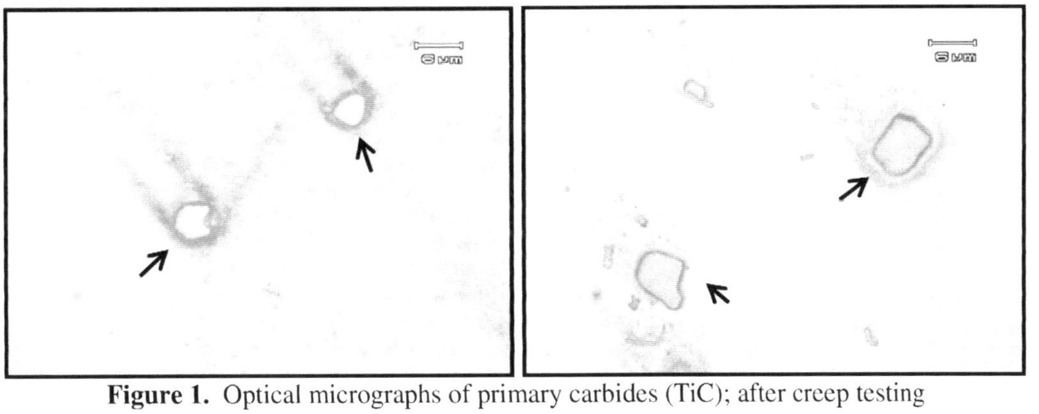

Figure 1. Optical micrographs of primary carbides (TiC); after creep testing

Figure 2. SEM micrographs depicting metallic carbides located at grain boundaries; before creep test (left) and after creep test (right)

3.3 Fractography

Figure 3 depicts Scanning Electron Microscope images of specimens fractured during high temperature creep testing. Microvoid coalescence phenomenon was dominant indicating ductile mode of failure. A large quantity of primary carbides was observed in the fractured surface. Presence of these carbides led to nucleation of primary micro-voids. Presence of large primary carbides of 5-6μm at grain boundaries deteriorated creep strength of inconel718 rod. This large concentration of PCs (TiC etc.) was actually due to higher titanium content (1.25 weight %) in the rod. Thus composition plays critical role in determining mechanical properties of a material. By regulating the titanium, niobium and aluminum content, the final mechanical properties of Inconel718 alloy can be controlled to a great extent.

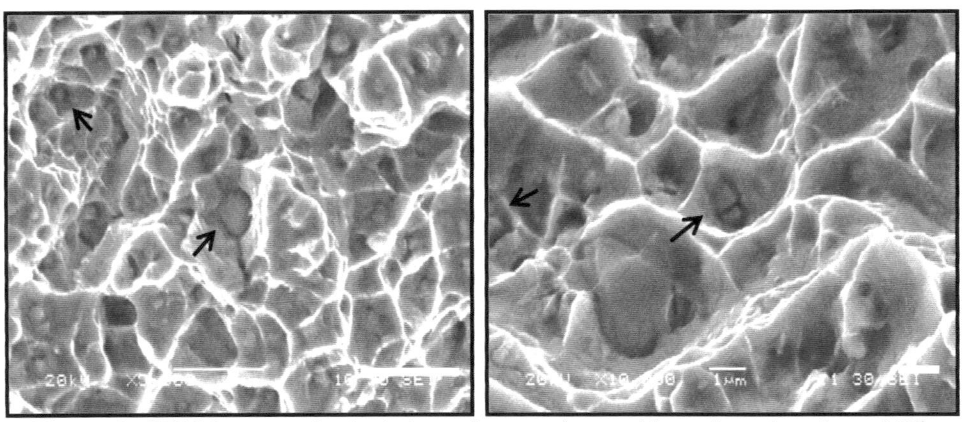

Figure 3. SEM fractographs depicting primary micro-voids nucleated at sites of (Ti, Nb)C

3.4 Precipitated phases

Scanning Transmission Electron Microscope (STEM) was utilized in order to study the microstructure (precipitated phases) in Inconel718. Objective was to understand the effect o f heat treatment on these precipitated phases. For this purpose material was subjected to: aging cycle and annealing treatment Details of heat treatment are given in Table 4.

Table 4. Aging cycle and annealing treatment for Inconel718 alloy

Aging Cycle	Annealing Treatment
Solution Treatment at 980°C/1hr/air cool	Heat at 960°C for 1 hr.
Heat at 720°C for 8hrs.	
Heat at 620°C for 8hrs.	Air cool
Air cool	

Figure 4 shows γ′ and γ″ precipitates in peak aged inconel718. These precipitates are responsible for imparting strength to the super alloy. Ni_3Nb (γ″) was identified as the dominant hardening phase. Both γ′ and γ″ are metastable in nature. Figure 5 shows the Selected Area Diffraction (SAD) patterns of γ″ precipitates obtained using the diffraction mode. Habit plane of γ′ and γ″ is {100} at [$\bar{1}$00] orientation. It was observed that γ″ precipitates grew perpendicular to [$\bar{1}$00]. Fully coherent relationship was observed γ″ phase and austenitic matrix. Ellipsoidal disc shaped γ″ phase was observed. Theses precipitates existed on (100), (010) and (001) planes at [$\bar{1}$00] orientation.

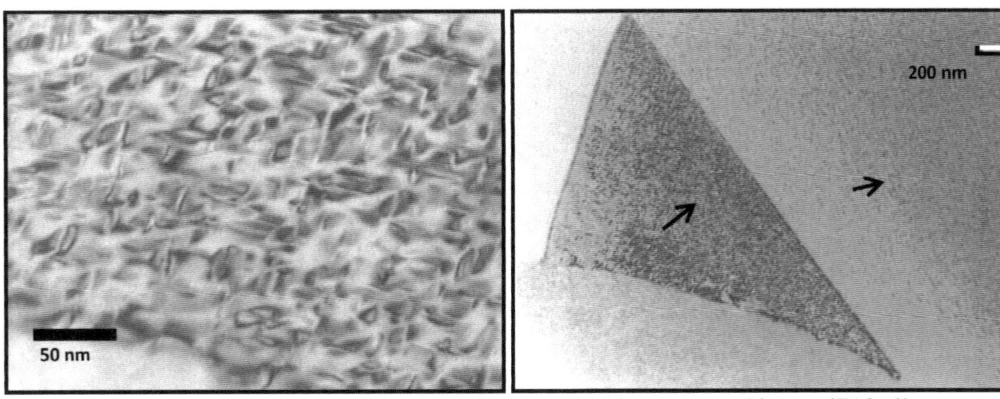

Figure 4. STEM micrographs depicting γ′ and γ″ in peak aged inconel718 alloy

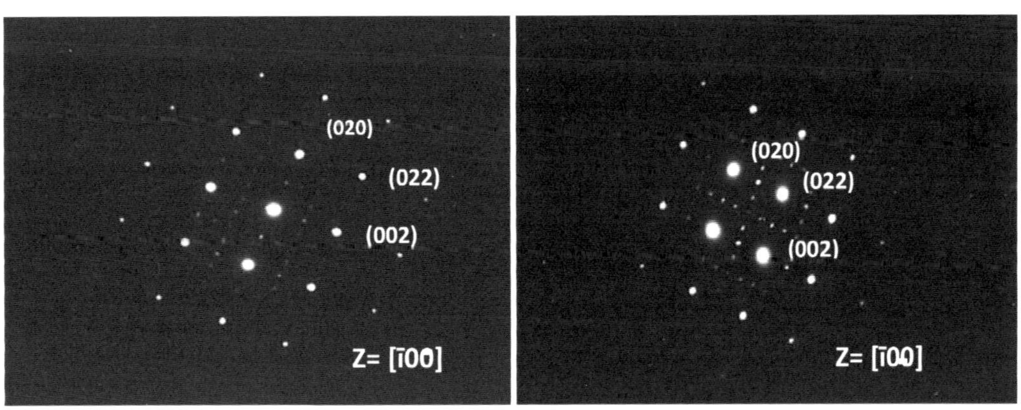

Figure 5. SAD pattern of γ″ precipitates

Figure 6 shows STEM micrographs of annealed Inconel718 alloy. In annealed condition needle like δ phase was observed. This δ phase resulted due to transformation of metastable γ″ to stable δ. Yun Z et al. reported that δ equilibrium phase nucleates at γ″ at temperature< 1173K [10]. δ can also precipitate directly in γ matrix at higher aging temperatures. Cubic/blocky niobium carbides were also observed in annealed samples. Figure 7 shows selected area diffraction pattern of annealed sample. Zone axis was calculated as Z: [2$\bar{1}$1].

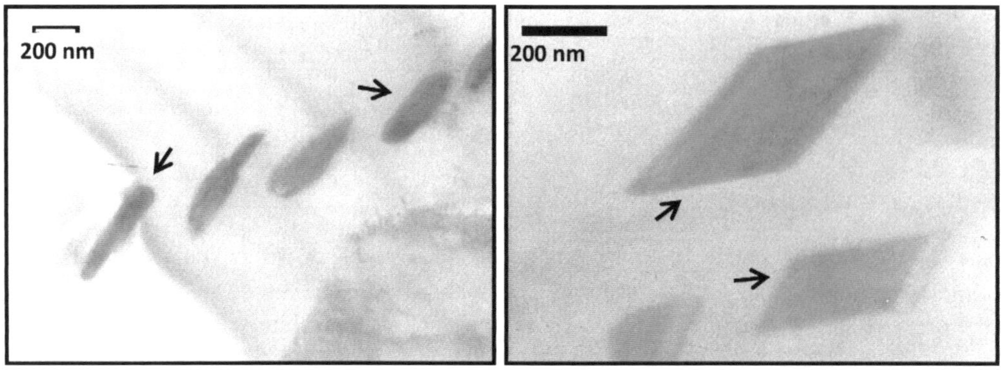

Figure 6. STEM micrographs depicting δ phase (left) and blocky NbC (right) in annealed state

Figure 7. SAD pattern of annealed inconel718 samples

4. Conclusion

i. Ti content has a profound effect on high temperature rupture ductility of Inconel718.

ii. (Al+Ti+Nb) at% was found to be 5.98. However for typical Inconel718 with 1 wt. % Ti, (Al+Ti+Nb)at% = 5.5. Thus high temperature stress rupture properties of Inconel718 sensitively depend on (Al + Ti + Nb) at%.

iii. Ti content above 1% leads to formation of TiC in large quantities. 5-6 μm primary carbides (TiC) were found at grain boundaries in samples subjected to creep testing (649 °C). Hence, presence of MCs (TiC & NbC) in large concentrations at grain boundaries deteriorated creep properties.

iv. Fractographs of fractured samples during creep testing (649 °C/ 690MPA) revealed that large primary microvoids nucleated at (Nb, Ti) C.

v. $\gamma' + \gamma''$ were observed to be primary strengthening phases in as-received Inconel718. Habit plane of γ' and γ'' is {100} at [$\bar{1}$00] orientation. γ'' precipitates grow perpendicular to [$\bar{1}$00]. Upon annealing metastable γ'' transformed to stable acicular δ phase (Ni_3Nb).

vi. Fully coherent relationship between γ'' precipitates and matrix was found. Ellipsoidal disc shaped γ'' precipitates lie on (100), (010) & (001) planes at [$\bar{1}$00] orientation.

5. References

[1] Dehmas M, Lacaze J, Niang N, Viguier B. TEM study of high temperature precipitation of delta phase in Inconel 718 alloy. Adv Mater Sci Eng 2011;2011:1-9

[2] Ghosh S, Yadav S, Das G. Study o standard heat treatment on mechanical properties of Inconel 718 using ball indentation technique. Mater Lett 2008;62:2619-22
[3] Niang A, Viguier B, Lacaze J. Some features of anisothermal solid-state transformations in alloy 718. Mater Charact 2010;61:525-34
[4] Chamanfar A, Sarrat L, Jahazi M, Asadi M, Weck A, Koul A.K. Microstructural characteristics of forged and heat treated Inconel 718 disks. Materials and Design. 2013;52:791-800
[5] Nowotnik A, Sieniawski J, Mrowka-Nowotnik G. indentification of dynamically precipitated phases in hotworking Inconel 718 alloy. Journal of achievements in materials and manufactuiring engineering. 2008;31(2):275-80
[6] Preuss M, Withers PJ, Baxter GJ. A comparison of inertia friction welds in three nickel base superalloys. Mater Saci Eng A 2006;437:38-45
[7] Salama C, Abdellaouib M. structural characterization of the aged Inconel 718. J Alloy Compd 2000;306:277-84
[8] Xie X, Dong J, Wang G, Wei Y. the effect of Nb, Ti, Al on precipitation and strengthening behavior of 718 type superalloys. The Minerals, Metals and Materials Society. 2005:287-98
[9] Xie X, Wang G, Dong J, Xu C, Cao W.D, Kennedy R. Structure stabilioty stdy on a newly developed Nickle-base superalloy-ALLVAC 718PLUS™. The minerals, Metals and Materials Society. 2005:179-91
[10] Yun Z, Xuebing H, Yong WWeicheng Y, Zhuangqi H. delta phase and deformation fracture behavior of Inconel718 alloy. The Minerals, Metals and Materials Society. 1997;229-36

Scheduling job shop – *A case study*

M. Abbas, A. Abbas and W. A. Khan
Faculty of Mechanical Engineering, GIK Institute of Engineering Sciences and Technology, Topi-23640; KPK; Pakistan

E-mail: rahmanabbas@live.com,ahmad@giki.edu.pk,wasim@giki.edu.pk

Abstract. The scheduling in job shop is important for efficient utilization of machines in the manufacturing industry. There are number of algorithms available for scheduling of jobs which depend on machines tools, indirect consumables and jobs which are to be processed. In this paper a case study is presented for scheduling of jobs when parts are treated on available machines. Through time and motion study setup time and operation time are measured as total processing time for variety of products having different manufacturing processes. Based on due dates different level of priority are assigned to the jobs and the jobs are scheduled on the basis of priority. In view of the measured processing time, the times for processing of some new jobs are estimated and for efficient utilization of the machines available an algorithm is proposed and validated.

1. Introduction

In manufacturing industries discrete products are manufactured through processing of raw materials on different machines. Different methodologies are used to organize the processing of jobs in different industries. The ideal handling areas are known as job shop, cellular manufacturing, project shop, continuous manufacturing, flow line, flexible manufacturing and reconfigurable manufacturing systems. The efficient use of the manufacturing facility depends upon the proper scheduling of jobs which is dependent upon product's nature, its lot size, the transfer of energy and material and implementation of intangible functions like production planning, process planning, inventory control and supply chain management. The scheduling of jobs in industries is mostly done by experienced workers with pen, paper and graphical approach and also through some industrial databases [1-3]. For immediate solution of problems, simple dispatching rules are typically used like sequencing at a level of work center [2-3]. This can create confusions in scheduling, where it is difficult to predict the completion dates, also some time the jobs with higher priority are chased down by the management [4].

Many rules are found in literature for scheduling of jobs based on due dates, criticality of operation, processing time, and resource utilization and critical ratio [5]. The ratio of processing time of a job and its due date is known as critical ratio. Rules on the basis of processing time are shortest processing time (SPT) and longest processing time (LPT). The jobs having shorter time are processed first in SPT, while the jobs having longer time are processed first in the LPT. The jobs having earlier due dates are processed first in due date. The other dispatching rules are first come first serve (FCFS) where the jobs are processed in their incoming order and last come first serve (LCFS) where the last jobs are first processed [6].

Content from this work may be used under the terms of the Creative Commons Attribution 3.0 licence. Any further distribution of this work must maintain attribution to the author(s) and the title of the work, journal citation and DOI.
Published under licence by IOP Publishing Ltd

In manufacturing industries single machine scheduling, flow shop scheduling and job shop scheduling are most famous. In single machine all jobs are processed on a single machine and for best results the jobs are scheduled in ascending order of time. In flow shop every job must be processed through each machine and the processing time for each job is different on every machine. In flow shop the jobs are arranged in different orders and the best order is picked for which the completion time for a single batch is minimum. In job shop each job may or may not be processed on all machines and also the time of processing of each job is different in different machine. It is one of the most complex problem and optimum scheduling is also hard to calculate [7]. For ideal scheduling of job shop many algorithms have been developed from time to time [7-9].

A job shop typically consists of machines with identical material processing capabilities. The scheduling problem aims to sequence the manufacturing decisions with low volume, high variety systems where the route for one part is fixed while it vary from part to part [10]. For instance, in a specific case involving 10 jobs and 10 machines posed by Muth and Thompson [12] in 1963 was stayed unsolved for 20 years, untilit was solved by Carlier and Pinson [13] in 1985.

Neural network for JSPs was first time applied by Takefuji [14]. Later on different neural architectures had developed and applied successfully to solve JSP [15-17]. Genetic algorithms are developed and used for job shop scheduling [18-19].

Many heuristic methods have also been developed for job shop problems. Well known are Johnson [20], Gupta [21], bottle neck [22], Palmer [23] branch and bound [24] etc. with the passage of time these methods were improved further but modifications are still needed. The major problem with these techniques are that some are based on processing time ignoring the due date while other considers the due date and ignoring the processing time.

This research intends to develop a new algorithm for job shop problem, considering the jobs having multiple machine operations with corresponding due dates, hence accounting for lead time.

2. The problem formulation

Consider a typical job shop with n jobs and m machines. Each job will be processed through one machine only once, and it is not necessary that all jobs will be processed through all machines. Every job will be having a particular order of processing. The processing time for each job is fixed and known. [25]

To schedule jobs the following parameters have to be identified first:

I. The available facility on shop floor.
II. Total jobs in system.
III. Setup and operation time for each job on each machines.
IV. Time to complete all jobs.

After identification of the parameters the scheduling of the jobs is performed on the basis of following constraints.

I. Every job will be processed through respective machines (j = 1, 2, 3, ... m).
II. One machine can process single job at a time.
III. The time for setup for different operations are not dependent of the sequence
IV. The setup times for different operations are independent of the sequences and are incorporated in processing time.
V. The scheduling of jobs will be done on priority basis (due dates). Jobs will be assigned with three levels of priority. Priority 1 being the Top priority, priority 2 being the medium priority and priority 3 the lowest priority. The top priority jobs will be scheduled first and the lowest priority at the last.

3. Analysis of the manufacturing facility and the jobs processing

The manufacturing facility is analyzed through the availability of machines and respective operations to be performed on each machine. On similarity basis the manufacturing facility is divided into Shaper (S), Lathe (L), Milling (M), Boring (B), Welding (W), and Grinding (G) machine centres.

After analysing the facility, jobs to be processed are selected. The operation time for each job on each machine center observed through time motion study with a stop watch and the operation is performed by the best worker.

Table 1 displays the order of operations and processing time for individual jobs on respective machine centers with their order of priority.

Table1. Processing time for jobs (1, 2 … m-1, m) on machines (1, 2… n-1, n)

Job Name	**Jobs No**	**S**	**L**	**M**	**B**	**W**	**G**	**Priority**
Nut	1	0	17	0	0	0	0	2
Bolt	2	0	19.5	0	0	0	0	2
Gear	3	60	18.5	30	15	0	4	1
Hexagonal Gear	4	65	17.5	45	4	0	6	1
Flange Pipe	5	0	50	35	35	27	25	2
Screw	6	62	58	15	0	0	0	1
Lead Screw	7	60	34.5	20	0	0	0	2
Universal Coupling	8	60	17.5	20	10	0	6	3
Lathe Center	9	63	29.5	0	0	0	5	3
Piston Connecting Rod	10	0	0	17	11	0	7	3
Sprocket	11	0	0	15	55	0	5	2
Tie Rod End	12	0	26	27.5	27	0	6.5	3
Gear Pump Flange	13	0	27	15	31	17	5	3

4. The Solution Methodology

In the proposed algorithm the jobs are scheduled on the basis of priority. The algorithm is described as following flow chart: Figure 1

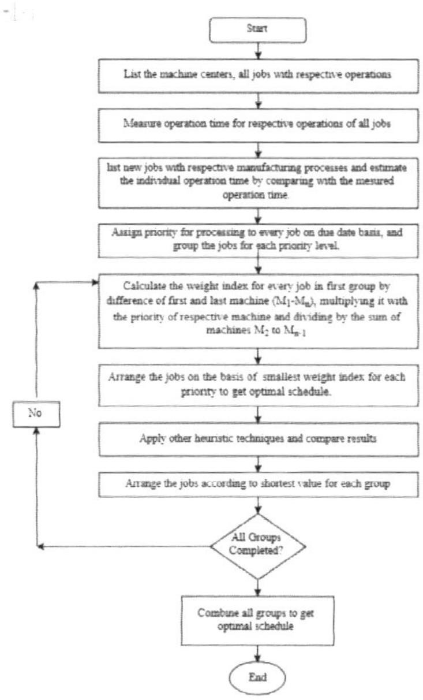

Figure 1. Algorithm flow chart

5. Results and discussions

In order to check that proposed heuristic give optimal results different well known heuristics of job shop scheduling is applied to 13 jobs and 6 machines problem in Table 1.The result obtained by applying various heuristics are shown in Table 2 with calculated sequence and total make span.

Table 2. Comparison of algorithms

Algorithm	Sequence													Make Span (Minutes)
SPT	1	2	10	11	12	13	9	8	7	3	6	4	5	559.5
LPT	5	4	6	3	7	8	9	13	12	11	10	2	1	580
Johnsons	1	2	10	11	12	13	5	6	4	3	7	8	9	404.5
Gupta	1	2	10	11	12	13	5	6	4	3	7	8	9	404.5
CDS	1	2	11	13	12	10	8	4	5	9	3	6	7	424.5
Palmer	11	5	10	13	12	1	2	8	3	9	4	7	6	443
WI	10	5	12	11	13	1	2	6	3	4	7	8	9	404.5

The comparison of different heuristics in Table 2 shows that WI method can give optimal results, which is comparable to well-known heuristic techniques like Johnson, Gupta, Palmer, and CDS. Effect of Lead time is neglected for the sequences acquired based on different algorithm in Table 2. Therefore this study is intended to achieve optimal sequence by considering the lead time as well as processing time.

Now considering the lead time, the steps discussed in solution methodology are used. Jobs with lead time of one day are given prime importance compare to jobs having lead time of 2 and 3 days. So first take the jobs having lead time of "1 day", calculate their WI and arrange it in ascending order of WI value. Similarly apply same procedure to group "2" and "3"as shown in combine form in Table 3:

Table 3. Sequence based on WI

Machines/Jobs	S	L	M	B	W	G	Lead Time	WI
6	62	58	15	0	0	0	1	0.849315
3	60	18.5	30	15	0	4	1	0.88189
4	65	17.5	45	4	0	6	1	0.887218
5	0	50	35	35	27	25	2	-0.13605
1	0	17	0	0	0	0	2	0.058824
7	60	34.5	20	0	0	0	2	0.061135
11	0	0	15	55	0	5	2	0.085714
2	0	19.5	0	0	0	0	2	0.102564
10	0	0	17	11	0	7	3	-0.25
12	0	26	27.5	27	0	6.5	3	-0.08075
13	0	27	15	31	17	5	3	-0.05556
8	60	17.5	20	10	0	6	3	1.136842
9	63	29.5	0	0	0	5	3	1.966102

Make span calculated for the schedule obtained in Table 3 is 498 minutes. By matching the results to, LPT, SPT, Gupta, Palmer, and Johnson rule it gives optimum results as shown in Table 4. If the priority of the jobs were changed the total makeup span will also vary.

Table 4. Comparison of results obtained from various algorithms

Algorithms	Total Make Span (Minutes)
SPT	503.5
LPT	525
Johnson	503.5
Gupta	503.5
CDS	522
Palmer	510
WI	498

The improved technique separates jobs based on lead time. Jobs with shorter lead time are schedule first and the sequence is obtained by computing WI method. Similarly the jobs with medium lead time are arranged after shorter lead time jobs and their sequence is achieved by WI method. Lastly the jobs with long lead time are arranged after medium lead time and their sequence is achieved using WI method. Thus a new optimal sequence is obtained by considering the lead time.

Processing time is estimated for new products based on available data. So a new problem having 6 jobs and 6 machines is created with their desired lead time. The purpose of it is to check whether the new heuristic give optimal result by applying it to other set of problem or not. The results are presented in Table 5:

Table 5. Processing time with priorities for new jobs

Machines / New Jobs	S	L	M	B	W	G	Priority
P1	65	10	0	0	5	0	1
P2	0	65	60	55	0	20	2
P3	85	60	40	60	25	0	3
P4	80	55	30	20	0	20	3
P5	80	20	30	0	10	10	2
P6	65	15	0	20	10	0	1

Neglecting lead time effect make span is calculated based on SPT, LPT, Johnson, Gupta, CDS, Palmer, heuristic as shown in Table 6:

Table 6. Comparison of algorithms

Algorithm	Sequence						Make Span
SPT	P1	P6	P5	P2	P4	P3	560
LPT	P3	P4	P2	P5	P6	P1	435
Johnsons	P2	P3	P4	P5	P6	P1	390
Gupta	P2	P3	P4	P5	P6	P1	390
CDS	P2	P6	P5	P1	P6	P3	545
Palmer	P3	P4	P5	P1	P6	P2	560
WI	P2	P3	P4	P5	P6	P1	390

As shown in Table 6 the make span calculated based on WI method give optimal result, comparable to Johnson and Gupta heuristic.

Now considering the lead time, apply the steps discussed in methodology section to get optimal sequence. The Table 7 shows optimal sequence of jobs on available machines:

Table 7. Sequence based on WI method

Machines / Jobs	S	L	M	B	W	G	Priority	WI
P6	65	15	0	20	10	0	1	1.444444
P1	65	10	0	0	5	0	1	4.333333
P2	0	65	60	55	0	20	2	-0.11111
P5	80	20	30	0	10	10	2	1.166667
P3	85	60	40	60	25	0	3	0.459459
P4	80	55	30	20	0	20	3	0.571429

Make span calculated is 500 minutes for sequence obtained based on WI method. The comparison of WI to other algorithms is shown in Table 8:

Table 8. Comparison of Algorithms

Algorithms	Total Make Span (Minutes)
SPT	560
LPT	500
Johnson	505
Gupta	505
CDS	500
Palmer	560
WI	500

6. Conclusion

Scheduling techniques namely LPT, SPT, Johnson, Gupta and Palmer are studied for job shop scheduling. A new heuristic "weighted index (WI)" method is developed which give optimal results comparable to these four techniques. The new WI method is easy and has less computational cost compared to other heuristics. New heuristic and other four heuristic are improved by addition of assumption. The assumption is based on grouping of jobs based on lead time. The earlier methods were completely based on processing time ignoring the aspect of lead time. These new modified methods include both processing time and lead time aspects.

7. References

[1] S.C. Graves, "A review of production scheduling" *Operation research*, vol. 18, pp. 841-852, 1981

[2] K. N. McKay, F.R. Safayen, and J. A. Buzacott, "Job-shop scheduling theory: What is relevant?"*Interfaces*, vol. 18, pp. 84-90, July/Aug. 1988

[3] K. P. White, Jr., "Advances in the theory and practice of scheduling," *Advances in Industrial Systems, Control and Dynamic Systems*, vol. 37, C. T. Leondes, Ed. San Diego, Ca: Academic, 1990, pp. 115-158

[4] J. E. Ashton, and F. W. Cook, Jr., "Time to reform job shop manufacturing," *Harvard Bus, Rev*,pp. 106-111, Mar,-Apr. 1989

[5] J. H. Blackstone, D. t. Phillips, and G. L. Hogg, "A state of the art survey of dispatching rules for manufacturing job shop operations," *Int. J. Production Res.* Vol. 20, pp. 27-47, Jan. 1982

[6] Chase, R. B. & Aquilano, N. J. (1995). Operations Scheduling. In Hercher, R. T., Korosa, G., Britt,S. M. (Eds.), *Production and Operations Management Manufacturing and Services* (pp. 637-647). Chicago, IL: Irwin.

[7] Fuqing Zhao, Jianlin Zhang, Chuck Zhang, Junbiao Wang "An improved shuffled complex evolution algorithm with sequence mapping mechanism for job shop scheduling problems" *Expert Systems with Applications* 42 (2015) 3953–3966

[8] Rui Zhang, Cheng Wu, "An Artificial Bee Colony Algorithm for the Job Shop Scheduling Problem with Random Processing Times" *Entropy* 2011, 13, 1708-1729; doi:10.3390/e13091708

[9] Miguel A. González, Camino R. Vela, Ramiro Varela "Scatter search with path relinking for the flexible job shop scheduling problem" *European Journal of Operational Research* 245 (2015) 35–45

[10] TamerF.Abdelmaguid "A neighborhood search function for flexible job shop scheduling with separable sequence dependent setup times" *Applied Mathematics and Computation* 260 (2015) 188–203

[11] Meeran, S., &Morshed, M. (2014). "Evaluation of a hybrid genetic tabu search framework on job Shop scheduling benchmark problems" *International Journal of Production Research*, 52(9), 5780-5798

[12] J.F. Muth and G.L. Thompson. *Industrial Scheduling* Prentice-Hall, Englewood Cliffs, N.J.,1963.

[13] J. Carlier and E. Pinson, An algorithm for solving the job-shop problem, *Manag. Sci.* 35 (1985),164-176

[14] S. Y. Foo and Y. Takefuji, "Neural networks for solving job-shop scheduling: Part 1. Problem representation," *Proc. IEEE IJCNN,* vol. 2, pp. 275–282, 1988

[15] T. M. Willems, "Neural networks for job-shop scheduling," *Control Engg. Prac.*, vol. 2, no. 1, pp. 31–39, 1994.

[16] S. Yang and D. Wang, "Constraint satisfaction adaptive neural network and heuristics combined approaches for generalized job-shop scheduling," *IEEE Trans. on Neural Networks*, vol. 11, no. 2, pp. 474-486, 2000

[17] S. Yang and D. Wang, "A new adaptive neural network and heuristics hybrid approach for job-shop scheduling,"*Comp. &Oper.Res*, vol. 28, no. 10, pp. 955–971, 2001

[18] Pei Chann Chang, Shih HsinChen , V. Mani , "A hybrid genetic algorithm with dominance properties for single machine scheduling with dependent penalties" , *Applied Mathematical Modeling* 33 (2009) 579–596

[19] T.C.E. Cheng , B.M.T. Lin , "Johnson's rule, composite jobs and the relocation problem" , *European Journal of Operational Research* 192 (2009) 1008–1013

[20] Johnson, S.M. (1954), "Optimal two- and three-stage production schedules with setup times included", *Naval Research Logistics Quarterly* 1, 61-68

[21] Gupta, J.N.D. and Reddi, S.S. (1978), "Improved dominance conditions for the three-machine flowshop scheduling problem," *Operations Research* 26, 200-203

[22] Adams J, Balas E, Zawack D. (1988) The shifting bottleneck procedure for job shop scheduling. *Management Science* 34:391– 401

[23] Palmer, D.S. (1965), "Sequencing jobs through a multi-stage process in the minimum total time-A quick method of obtaining a near optimum", *Operational Research Quarterly* 16, 101-107.

[24] Bmcker, P, Jurisch, B. and Sievers, B. (1994), "A branch and bound algorithm for the job-shopscheduling problem," *Discrete Applied Mathematics* 49, 107-127

[25] P.K. Gupta, D.S. Hira. (2003), "Sequencing models and related problems".*"Operations Research"* S. Chand & Company LTD India. 257-258

IOP Conf. Series: Materials Science and Engineering **146** (2016) 012053 doi:10.1088/1757-899X/146/1/012053

Forecasting of indirect consumables for a Job Shop

M. Shakeel, S. Khan and W. A. Khan
Faculty of Mechanical Engineering, G.I.K Institute of Engineering Sciences and Technology, Topi 23640, Swabi, KPK, Pakistan

E-mail: gme1510@giki.edu.pk, gme1512@giki.edu.pk, wasim@giki.edu.pk

Abstract. A job shop has an arrangement where similar machines (Direct consumables) are grouped together and use indirect consumables to produce a product. The indirect consumables include hack saw blades, emery paper, painting brush etc. The job shop is serving various orders at a particular time for the optimal operation of job shop. Forecasting is required to predict the demand of direct and indirect consumables in a job shop. Forecasting is also needed to manage lead time, optimize inventory cost and stock outs. The objective of this research is to obtain the forecast for indirect consumables. The paper shows how job shop can manage their indirect consumables more accurately by establishing a new technique of forecasting. This results in profitable use of job shop by multiple users.

1. Introduction

The objective of forecasting is to satisfy the need of users of the job shop efficiently by providing them all the direct and indirect consumables required. The main objective proposed in this research is to develop a forecasting method for the selected indirect consumables that are used in a job shop in order to determine the demand quantity for the ordering. If the forecasting estimates are not predicted, the figure may appear to be prone to errors. [1] Therefore in order to make it error free and more precise, forecasting is done so that their order to the supplier is appropriate in term of quantity and delivery time. Appropriate ordering results in good lead time management, inventory cost and stock outs.

Direct consumables include machines and usually the demand rises as the number of orders increases in a job shop [1-2].

Indirect consumables are the materials required for these machines to be used by the users. These materials play a supporting role in performing manufacturing processes. The demand of indirect consumables must be specified as they may cause hurdles during operation of machine tools while executing an order. Several methods for forecasting the use of indirect consumables are available.[3] Quantitative techniques are employed for forecasting of indirect consumables. The forecasting can be time-series forecasts or they may be based on other associative models. Time series data must be identified by the forecaster. These behaviors may show pattern or random variation. [3-5]

In this study the quantitative techniques weighted average and exponential smoothing were employed to express new technique to reduce the error due to limited data for analysis.

2. Proposed methodology

The methodology used is to first get the previous data from a job shop. The data used was for a period of 4 months. The estimated value of the indirect consumable was calculated through feedback from stakeholders. The forecast was estimated using averaging, weighted average and exponential smoothing (with different values of alpha) [3-4]. The results are presented in Table1, the difference

Content from this work may be used under the terms of the Creative Commons Attribution 3.0 licence. Any further distribution of this work must maintain attribution to the author(s) and the title of the work, journal citation and DOI.
Published under licence by IOP Publishing Ltd

between the actual value and the exponential smoothing decreases as the value of alpha increases but there is still a large difference. A new method is introduced whereby the weighted average and exponential smoothing are combined together. For this new method the formula is, Forecasted value = Weighted average + α (Actual-Weighted average). The results are listed in the figure below. The results show a big difference in the traditional exponential method vs the new method. The results are presented in Table 1.

Table 1. Forecasting of indirect consumables using different available techniques

Items	**Actual number**	**Estim ated**	**Aver age**	**Weighted Average**	**ES 0.1**	**ES 0.2**	**ES 0.3**	**ES 0.4**	**ES 0.5**	**ES 0.6**
Hack Saw blades	1000	1200	1100	1080	1180	1160	1140	1120	1100	1080
Fine emery paper	24	32	28	27.2	31.2	30.4	29.6	28.8	28	27.2
Coarseemer y paper	24	32	28	27.2	31.2	30.4	29.6	28.8	28	27.2
Painting brush	20	14	17	17.6	14.6	15.2	15.8	16.4	17	17.6
Welding rods	5	8	6.5	6.2	7.7	7.4	7.1	6.8	6.5	6.2
Aluminium rods	0.5	2	1.25	1.1	1.85	1.7	1.55	1.4	1.25	1.1
Copper Rods	0.5	2	1.25	1.1	1.85	1.7	1.55	1.4	1.25	1.1
Brass rods	0.5	2	1.25	1.1	1.85	1.7	1.55	1.4	1.25	1.1
Mild steel rods	1	3	2	1.8	2.8	2.6	2.4	2.2	2	1.8
Gas cylinders	1	1	1	1	1	1	1	1	1	1
Kerosine oil	20	20	20	20	20	20	20	20	20	20
Mobile oil	60	60	60	60	60	60	60	60	60	60
Soluble oil	40	40	40	40	40	40	40	40	40	40
Cotton rag	40	50	45	44	49	48	47	46	45	44
Shoes	55	55	55	55	55	55	55	55	55	55
Coats	55	55	55	55	55	55	55	55	55	55
Goggles	40	40	40	40	40	40	40	40	40	40
Lathe tools	20	21	20.5	20.4	20.9	20.8	20.7	20.6	20.5	20.4
End mill cutters	5	7	6	5.8	6.8	6.6	6.4	6.2	6	5.8
Gear module	2	3	2.5	2.4	2.9	2.8	2.7	2.6	2.5	2.4

Table 1 presents the values for average and weighted average however it requires more data. The proposed methodology is applied at different value of alpha and at alpha 0.6 it is showing very good results as compared with the actual.

Table2. Forecasting of indirect consumables using proposed methodology

Items	Actual number	Estim ated	Aver age	Weighted Average	New ES 0.4	New ES 0.5	New ES 0.6	New ES 0.7
Hack Saw blades	1000	1200	1100	1080	1048	1040	1032	1024
Fine emery paper	24	32	28	27.2	25.92	25.6	25.28	24.96
Coarse emery paper	24	32	28	27.2	25.92	25.6	25.28	24.96
Painting brush	20	14	17	17.6	18.56	18.8	19.04	19.28
Welding rods	5	8	6.5	6.2	5.72	5.6	5.48	5.36
Aluminium rods	0.5	2	1.25	1.1	0.86	0.8	0.74	0.68
Copper Rods	0.5	2	1.25	1.1	0.86	0.8	0.74	0.68
Brass rods	0.5	2	1.25	1.1	0.86	0.8	0.74	0.68
Mild steel rods	1	3	2	1.8	1.48	1.4	1.32	1.24
Gas cylinders	1	1	1	1	1	1	1	1
Kerosine oil	20	20	20	20	20	20	20	20
Mobile oil	60	60	60	60	60	60	60	60
Soluble oil	40	40	40	40	40	40	40	40
Cotton rag	40	50	45	44	42.4	42	41.6	41.2
Shoes	55	55	55	55	55	55	55	55
Coats	55	55	55	55	55	55	55	55
Goggles	40	40	40	40	40	40	40	40
Lathe tools	20	21	20.5	20.4	20.24	20.2	20.16	20.12
End mill cutters	5	7	6	5.8	5.48	5.4	5.32	5.24
Gear module	2	3	2.5	2.4	2.24	2.2	2.16	2.12

3. Results and discussion

Above calculations shows that selection of an appropriate method of forecasting requires availability of past data. Averaging methods can be used when enough data is present. Forecasting using small data for averaging technique may result in large errors. Although exponential smoothing is the method that can be used with small amount of data and gives a good results even if more variables are applied. But the new method introduced gives even better result than traditional exponential smoothing method although it is a derivative of the same method.

Figures 1-8 shows the graphs related to Table 1. Figures 9-15 shows graphs related to Table 2. Figure 16 shows the comparison between actual forecast, forecast using exponential smoothing (alpha=0.4) and forecast using proposed methodology (alpha=0.4).

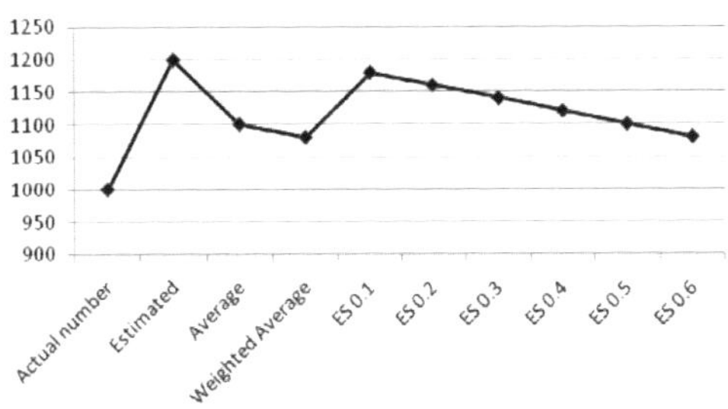

Figure 1. Forecasting for hack saw blades

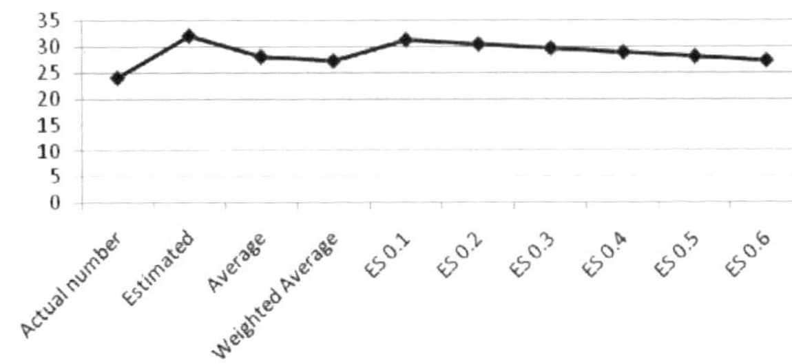

Figure 2. Forecasting for emery paper

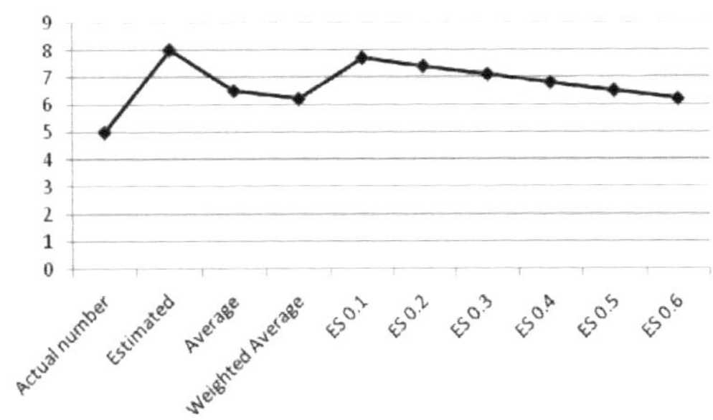

Figure 3. Forecasting for welding rods

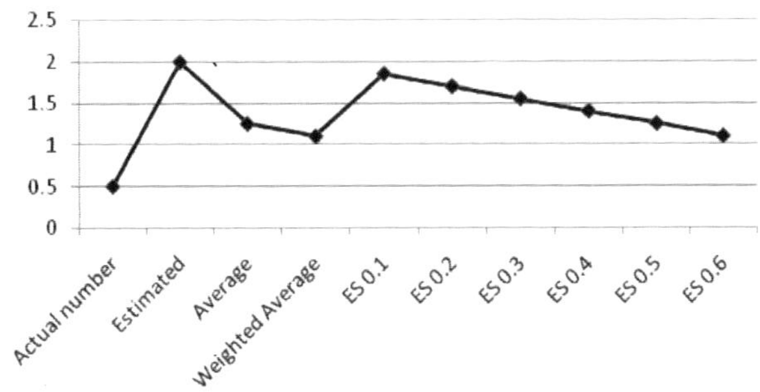

Figure 4. Forecasting for gas welding rods

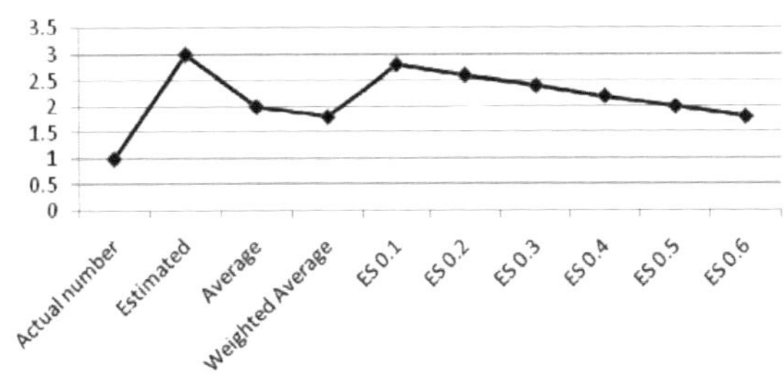

Figure 5. Forecasting for MS rods

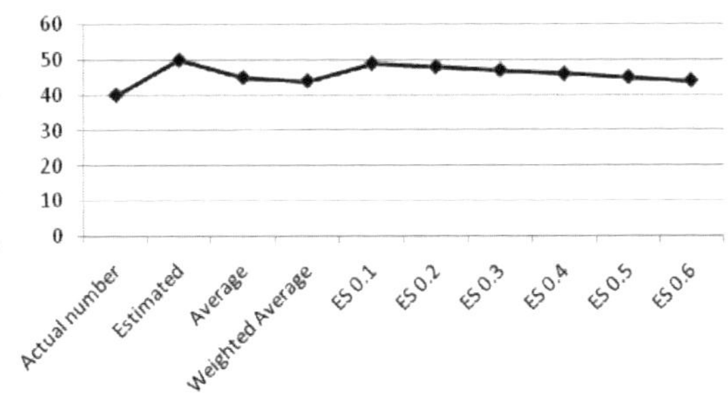

Figure 6. Forecasting for cotton rig

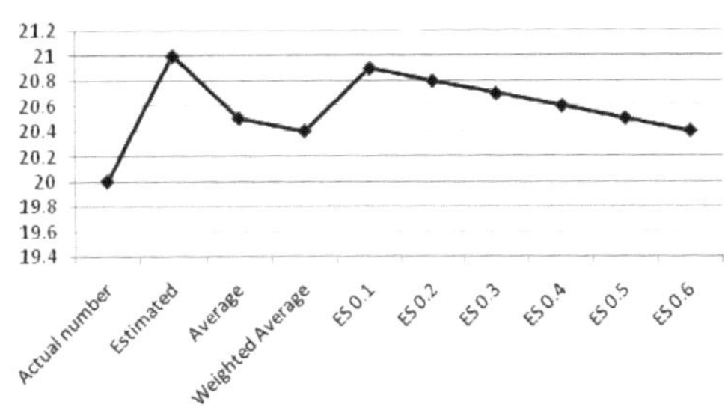

Figure 7. Forecasting for lathe tools

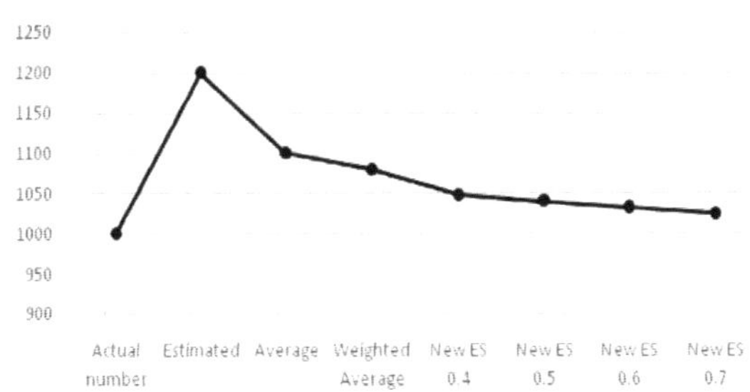

Figure 8. Forecasting for end mill cutter

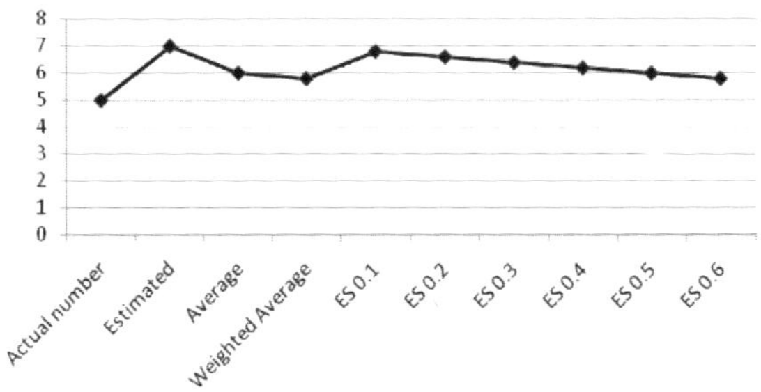

Figure 9. Forecasting for hack saw blades

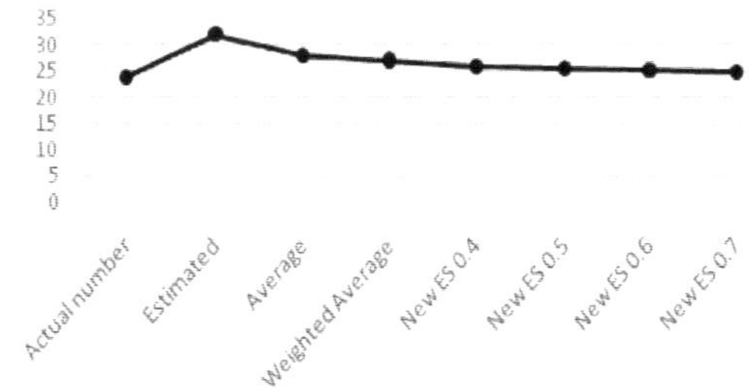

Figure 10. Forecasting for emery paper

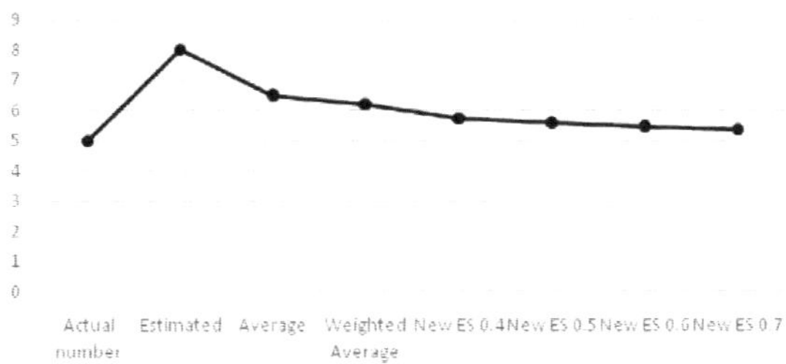

Figure 11. Forecasting for welding rods

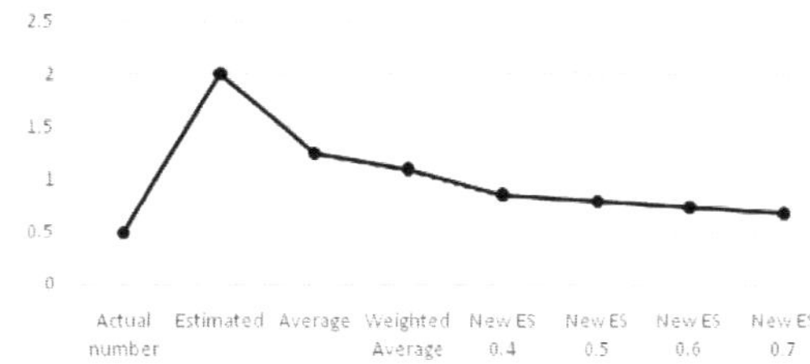

Figure 12. Forecasting for gas welding rods

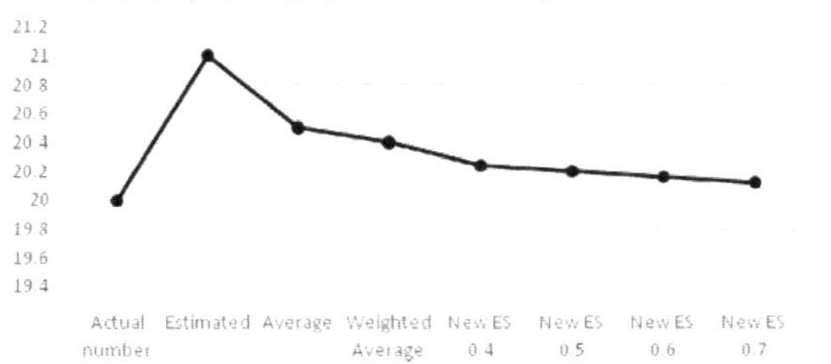

Figure 13. Forecasting for lathe tools

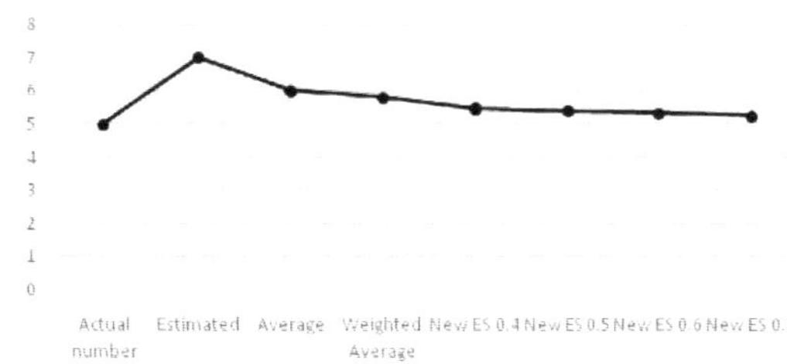

Figure 14. Forecasting for end mill cutter

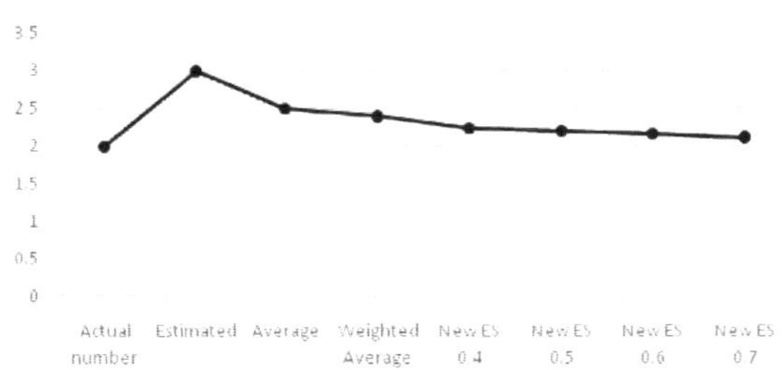

Figure 15. Forecasting for gear module

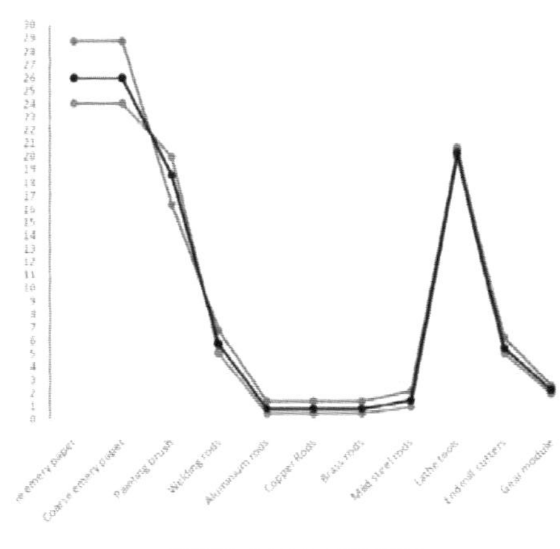

Figure 16. Comparison b/w actual, exponential smoothing and proposed methodology

4. Conclusion

The method proposed can be used for forecasting rather than older methods because it is more accurate. In fact new method is just a derivative of some old techniques together. The results show a good trend and are comparable at any level. For less critical values any method can be used but for critical problems the proposed method is the solution.

5. References

[1] *A.I. Khan; M. Karim; S. Imtia; (2008);*"Forecasting of indirect consumables in automobile industry", *MBA (MIS) Project report* Institute of Business Administration Karachi Pakistan

[2] *Richard Jones* Estimating for furniture makers 96 Monkswood, Leeds, LS14 1DP, Great Britain

[3] *Paul Savory* (June 2010); Estimation of cellular manufacturing cost components using simulation and activity-based costing Journal of Industrial Engineering and Management

[4] *M. A. Driessen, J. J. Arts, G.J. van Houtum, W.D. Rustenburg,B. Huisman;*(Sep-2010); "Maintenance spare parts planning and control:A framework for control and agenda for future research"; BETA working paper series 325

[5] *Gerald Davis;* (April 19, 2013); "Job shop estimating: Expenses driven by events in the fabrication sequence; The Fabricator

Electromagnetic flat sheet forming by spiral type actuator coil

S. Akbar, M.A. Aleem, M. N. Sarwar, A. U. Zillohu, M. S. Awan, A. Haider, Z. Ahmad, S. Akhtar and M. Farooque
Ibn-e-Sina Institute of Technology H-11/4 Islamabad, Pakistan

sam_qau@yahoo.com,

Abstract. Focus of present work is to develop a setup for high strain rate electromagnetic forming of thin aluminum sheets (0.5, 1.0, 1.5 and 2.0 mm) and optimization of forming parameters. Flat spiral coil of 99.9% pure Cu strip (2.5x8.0 mm) with self-inductance 11 μH, 13 no. of turns and resultant outer diameter of 130mm has been fabricated and was coupled to a capacitor bank of energy, voltage and capacitance of 9 kJ, 900 V and 22.8 mF, respectively. To optimize the coil design, a commercially available software FEMM-4.2 was used to simulate the electromagnetic field profile generated by the coils of different pitch but same number of turns. Results of electromagnetic field intensity proposed by simulation agree in close proximity with those of theoretical as well as experimental data. The calculation of electromagnetic force and magnetic couplings between the coil and metal sheet are made. Forming parameters were optimized for different sheet thicknesses. Electromagnetic field intensity's profile plays a principal role in forming of typical shapes and patterns in sheets.

1. Introduction

The commercial application of this process has existed since the 1960s. A recent interest in understanding the electromagnetic forming (EMF) of metals has been stimulated by the desire to use more aluminum in automobiles. The large majority of applications have involved either the expansion or compression of cylinders (tubes). Sheet materials with particular geometry are difficult to form as compared to flat sheets.

Electromagnetic forming (EMF) systems are based on the principle of mutual induction between actuator coil and a conductive workpiece [1]. In the EMF process, a transient electric pulse of high magnitude is sent through a specially designed forming coil using a low-inductance electric circuit. During the current pulse, the coil is surrounded by a strong transient magnetic field. The transient nature of the magnetic field induces current in a nearby conductive workpiece that flows opposite to the current in the coil. The coil and the workpiece act as parallel currents through two conductors to repel each other. The force of repulsion can be very high, equivalent to surface pressures of the order of tens of thousands of pounds per square inch. Thin metal sheets can be accelerated to high velocity in fractions of a millisecond [2-3]. High forming speed facilitates the formability of metals. Also, the dynamics of contact with the forming die can help mitigate springback, an undesired effect that cannot be avoided in other forming techniques such as stamping [4]. These benefits would result in increased use of aluminum in fuel-efficient (light weight) vehicles. Several studies have started from this premise; most of them involve specific situations like deformation of tubular parts by solenoid coils, while few studies have analyzed sheet metal forming by planar coils [5-8].

Content from this work may be used under the terms of the Creative Commons Attribution 3.0 licence. Any further distribution of this work must maintain attribution to the author(s) and the title of the work, journal citation and DOI.
Published under licence by IOP Publishing Ltd

The objective of this work is to develop a setup for high strain rate electromagnetic forming of thin aluminum sheets and optimization of forming parameters. For this purpose, an in house EMF facility was developed and aluminum sheets were formed successfully.

2. Work descriptions

A system of electromagnetic forming has been developed for forming of thin circular metal sheets by using a flat spiral coil. Electromagnetic forming (EMF) systems consist of two major parts: a capacitor bank couple to an electromagnetic coil termed as 'actuator'. Workpiece placed above the flat actuator coil as shown in Figure1.The capacitor bank provides stored energy in the form of high current transient electric pulse, while during discharge through a low inductive reactance actuator coil.Transient magnetic field thus generated in the coil induces a reverse field in the workpiece. Interaction of these fields provides a magnetic pressure, which in turn forms the work piece.

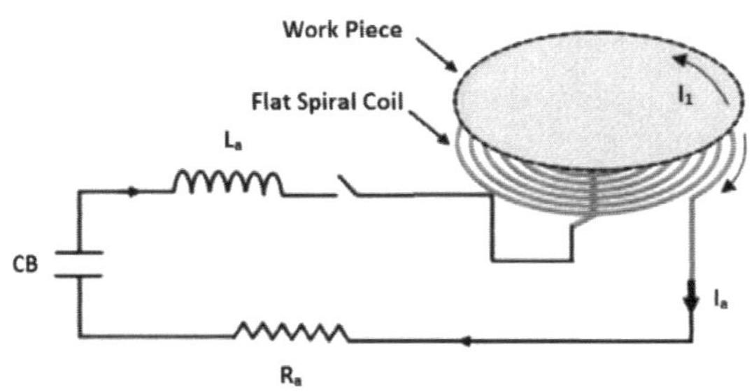

Figure 1. Schematic diagram of the electromagnetic forming system

An in-house developed capacitor bank of energy 9 kJ, voltage 900 V and capacitance 22.8 mF was used. The capacitor bank has provision to operate in the voltage range of 100 to 900 V and current limit exceeding 10 kA, for the mentioned coil. The cycle time of capacitor bank is 45 seconds.

The forming fixture is shown in Figure 2, while detailed characteristics of coil are given in Table 1. During EMF process, high-intensity electromagnetic force is applied to the coil. The coil, insulators and support structure must be capable of bearing this force, as well as thermal shock, within the elastic limit. Thus, actuator coil's design and material selection play vital role in system performance. Material selections are made by taking electrical resistance and mechanical properties into consideration. Moreover, for use in automotive industry, coil systems should be low-cost, modular, and have high durability (~100,000 cycles) [9]. Thus, the coil has been cycled several times at high current levels to confirm its durability.

Figure 2. Assembly photograph of the actuator coil and forming fixture

Table 1. Description of parameters system used in the current study

Equipment	Parameter	Value
Actuator Coil of 99.9% pure Copper	Numbers of turns	13
	Outer diameter	140mm
	Inner diameter	45mm
	Pitch	3.15mm
	Section of the wire	2.5 x 8mm
	Self inductance (La)	11μH
Capacitors Bank	Capacitance	22.8mF
	Maximum voltage	900V
	Maximum energy	8.9kJ
Metal Sheet	Material	Aluminum
	Thickness analyzed	0.5, 0.8, 1.0, 1.2, 1.5, 2.0 mm
	Diameter of metal sheet	135mm
	Gap between actuator and metal sheet	3mm

The schematic model of the system is shown in Figure 1, which shows a circularmetal sheet placed on flat spiral coil connected to a charged capacitor. Electromagnetic forming system represented by Figure 1 is governed by a set of coupled differential equations. The transient electromagnetic problem can be separated in an RLC primary circuit coupled with secondary RL circuit [10-11]. The discharge of the capacitor in the primary circuit as:

$$\frac{d}{dt}(L_a \cdot I_a + M \cdot I_1) + R_a \cdot I_a + V_C = 0 \quad (1)$$

Where L_a, R_a and V_c are the self-inductance, resistance of actuator coil and charging voltage of capacitor bank. M is the mutual inductance between the actuator coil and work piece. I_a and I_1 are discharge current in actuator coil and the induced current in work piece, respectively. For secondary RL circuit the differential equation is:

$$\frac{d}{dt}(L_1 \cdot I_1 + M \cdot I_a) + R_1 \cdot I_1 = 0 \quad (2)$$

Where L_1 and R_1 are the self inductance and resistance of work piece.

FEMM 4.2 software was used for modeling of actuator coil and the magnetic field generated while discharge through capacitor bank. It works on the principle of Biot-Savart law and Maxwell's equation. Magnetic field produced by the coil was thus simulated as shown in Figure 3 and the results have been verified experimentally using gauss meter with accuracy of more than 95 %.

3. Results and discussions

Flat spiral coil has been used for thin metals sheet forming. For this purpose, design of actuator coil was first optimized using FEMM-4.2 software. Coils with approximately same outer diameter, same no of turn but with different pitch, cross sectional area have been designed and magnetic field thus produced was modeled as shown in Figure 3a. Simulation results show that the coil with pitch 3.15 mm (bottom coil (d)) produced a uniform field (3.2 T $\pm$ 0.1) throughout the winding area as shown in Figure 3b. The other three coils i.e. a, b, c have different pitch and less Lorentz force value given in Table 2. Although the coil a has greater value of magnetic field than the coil d, but it has lesser effective length. Hence, a coil with same specifications as the coil'd' was fabricated and used in the further experiments.

For efficient operation, the fitting of the work piece to the coil also plays an important role. It has been found that minimum gap between the works piece and actuator is favorable for efficient forming. However, sufficient clearance must be provided to allow space for insulating materials and to facilitate the loading and unloading operation.

Table 2: Description of parameters of simulated actuator coils.

Coil	Lorentz Force (kN)	H_{radial} (T)
a	15.7	3.5
b	12.5	2.9
C	7.9	2.8
d	26.3	3.1

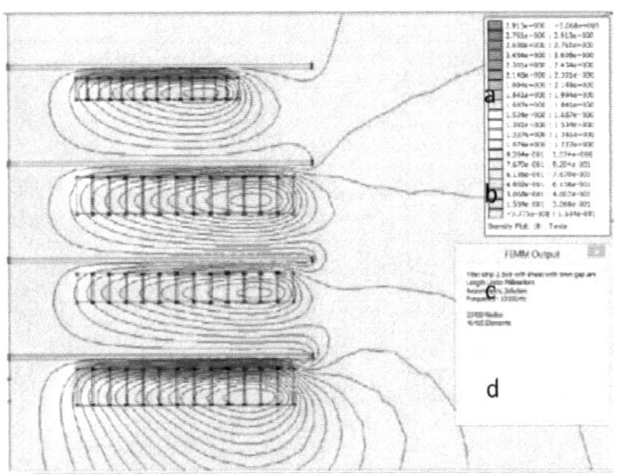

(a)

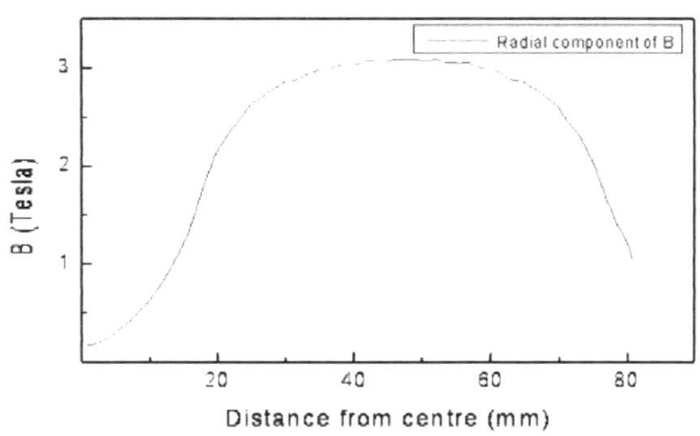

(b)

Figure 3. (a) Magnetic field profile taken by software is shown for four different actuators with circular geometry carrying current "I" under the same discharge voltage **(b)** quantitative radial component of field of bottom coil.

Limited work is available on the role of pitch in the design of flat spiral coil. Kamalet al [7] worked on a new type of flat coil to generate uniform pressure on the work piece. He argued that the spatial distribution of forming pressure in electromagnetic forming was controlled by the configuration of the coil. Golovashchenkoet al [11] also proposed a new generation coil design along with coil failure modes and its remedies. Field shaper is another tool that is used to modify the electromagnetic field to have desired distribution of the magnetic field and magnetic pressure. Yu et al [12] studied the effect of field shaper on magnetic pressure on tubular work piece and found that a greater radial magnetic pressure can be achieved with field shaper than the case without it. In the present study, the coil have been designed by optimizing the pitch as well as the cross sectional area to achieve a thoroughly uniform magnetic field.

Forming is accomplished with no physical contact on the work piece. The pressure on the work piece is applied by magnetic field and translation of force requires no medium**.** As the magnetic field can pass through non-conductive materials, it is possible to form a coated work piece without damaging the coating.

Sheet forming experiments were conducted for different sheet thicknesses and at two different energy levels (4kJ and 8.9kJ) of capacitor bank are shown in Figure 4. At low energy level the sheets with thickness 0.5 & 0.8 mm showed poor forming, however sheets with thicknesses 1.0 to 2.0 mm did not form. At the energy level of 8.9 kJ all of sheets having thicknesses 0.5, 0.8, 1.0, 1.2, 1.5 and

2.0 mm have been formed well. It is attributed to the fact that thicker sheets need more force than the thinner ones which is provided through the force generated by coupling of applied and induced fields and it is always higher in thicker sheets than that of relatively thinner sheets. Outer diameter of all sheets is 135 mm while the height of formed doom is 25-28 mm. The forming profile at high energy (9 kJ) is not symmetric in the thin sheets (0.5 and 0.8 mm), however, in sheets with thickness 1 to 2 mm, forming profile have been found symmetrical. Ali M et al [14] and Takatsu N et al [1] calculate the magnetic pressure produce by the actuator coil between the work piece. They use the coil having turns five in case of Ali M et al and six in case of Takatsu N, but we use the coil having 13 turns. Design of the coil was modified with batter understanding so that approximately same magnitude value of the force got using 900V instead of 6kV.The results are also in modest agreement with the previously reported results [1,13,14].

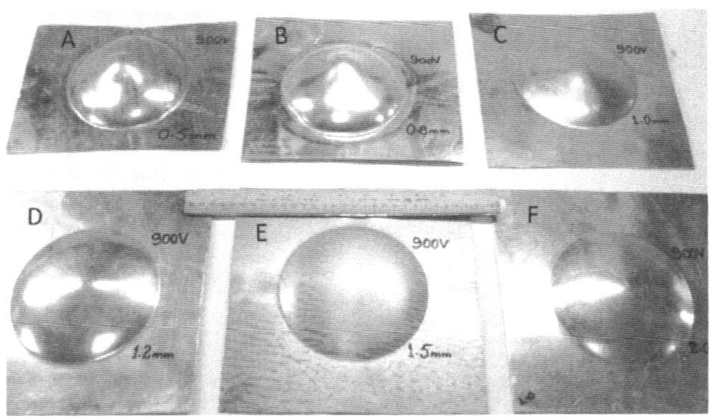

Figure 4. Different formed sheets at 900v, 9 kJ; top row from left to right 0.5, 0.8 and 1.0mm thickness, top row from left to right 1.2, 1.5 and 2.0 mm thickness.

4. Conclusion

A system for electromagnetic forming was constructed by coupling an in-house developed capacitor bank to an indigenously designed and developed actuator coil. To obtain uniform magnetic field, actuator coil was modeled in FEMM-4.2 to optimize the design parameters like pitch and cross sectional area. Results were then used to develop the coil for further use in forming experiments. Forming experiments were performed with aluminum sheets of varying thickness at different forming energies. Forming was conducted successfully and it has been thus established that the forming of thicker sheets require high forming energies.

5. Acknowledgement

Authors would like to thank Mr. Naveed Hassan Tahir, Mr. Faisal Pirandad and Mr. Abid Shazad for their support.

6. References

[1] Takatsu N., Kato M., Sato K., and Tobe T. (1988): High speed forming of metal sheets by electromagnetic force. J.S.M.E. International Journal, Vol. 31, no. 1,pp 142-148,doi: 10.1299/jsmec1988.31.142

[2] Mamalis A.G., Manolakos D.E., Kladas A.G., Koumoutsos A.K., Ovchinnikov S.G. (2006): Electromagnetic forming of aluminum alloy sheet using a grooved die: numerical modeling, The Physics of Metals and Metallography, Vol. 102, Suppl. 1, pp S90–S93, doi: 10.1134/S0031918X06140237

[3] A. El Azab, M. Garnich and A. Kapoor (2003): Modeling of the Electromagnetic Forming of Sheet Metals: state of the art and future needs, Journal of materials Processing Technology 142, pp 744-754, doi:10.1016/S0924-0136(03)00615-0

[4] P. Jimbert, A. Arroyo, I. Eguia, J.I. Fernández, E. Silveira, I. Garuz and G.S. Daehn, Efficiency Improvement and Analysis of Changes in Microstructure Associated to a Uniform Pressure Actuator, 2nd International Congress on High Speed Forming, Dortmund, Germany, 2006

[5] Zhipeng Lai, Quanliang Cao, Xiaotao Han, Zhongyu Zhou, Qi Xiong, Xiao Zhang, Qi Chen and Liang Li, Procedia Engineering (2014): Radial-axial Force Controlled Electromagnetic Sheet Deep Drawing: Electromagnetic Analysis,Vol. 81, pp. 2505 – 2511, doi:10.1016/ J. proeng. 2014.10.358

[6] Zhipeng Lai, Xiaotao Han, Quanliang Cao, Li Qiu, Zhongyu Zhou, and Liang Li (2014): The Electromagnetic Flanging of a Large-Scale Sheet Workpiece, IEEE transactions on Applied Superconductivity, Vol. 24 (3), pp. 1-5, doi: 10.1109/TASC.2013.2285443

[7] M. Kamal and G. Daehn (2006): A Uniform Pressure Electromagnetic Actuator for Forming Flat Sheets, Journal of Manufacturing Science and Engineering, Vol. 129 (2), pp. 369–379, doi:10.1115/1.2515481

[8] Li Qiu, Xiaotao Han, Tao Peng, Hongfa Ding, Qi Xiong, Zhongyu Zhou, Chengxi Jiang, Yiliang Lv, and Liang Li (2012) Design and Experiments of a High Field Electromagnetic Forming System, IEEE Transactions on Applied Superconductivity, Vol. 22 (3), doi: 10.1109/ TASC.2011.2178223

[9] R. W. Davies, F. Stavehaug and S. Golovashchenko, Automotive Light weighting Materials, FY 2005 Progress Report

[10] Shang Jianhui: Electromagnetically assisted sheet metal stamping, Ph. D. dissertation, The Ohio State University, 2006

[11] S. Golovashchenko (2007): Material formability and coil design in electromagnetic forming, Journal of Materials Engineering and Performance, 16(3), pp 314-320, doi: 10.1007/s11665-007-9058-7

[12] Yu H, Li C, Zhao Z, Li Z. (2005): Effect of field shaper on magnetic pressure in electromagnetic forming, Journal of Materials Processing Technology, 168 (2) pp 245–249, doi: 10.1016/ J. Jmatprotec. 2005.01.001

[13] E. Paese, P.A.R. Rosa, M. Geier, R.P. Homrich and R. Rossi (2014): An Analysis of Electromagnetic Sheet Metal Forming Process, Applied Mechanics and Materials, Vol. 526, pp-9-14, doi: 10.4028/www.scientific.net/AMM.526.9

[14] Ali M. Abdelhafeez, M.M. Nemat-Alla and M.G. El-Sebaie (2014): FEA of electromagnetic Forming Using a New Coupling Algorithm: effects of strain hardening properties and anisotropy, international Journal of Scientific and Engineering Research, Vol. 5, Issue 12 pp-1069-1075, doi: ISSN 2229-5518 /www.ijser.org

IOP Conf. Series: Materials Science and Engineering 146 (2016) 012055 doi:10.1088/1757-899X/146/1/012055

Effect of welding parameters on the mechanical and micro-structural properties of friction stir welded AA- 2014 joints

R. Khan, M. B. Bhatty[1], F. Iqbal, H. Zaigham[1] and I. Salam[1]
National Institute of Vacuum Science and Technology, Pakistan
[1]Institute of Industrial Control Systems, P. O. Box 1398, Rawalpindi, Pakistan

Abstract. In this study, the effect of processing parameters on the mechanical and microstructural properties of aluminum AA2014-T6 joints produced by friction stir welding was analyzed. Friction stir welding was carried out on a milling machine. Different samples were produced by varying the tool rotational rates (700, 1000 rpm) and travel speeds (45-105 mm/min). Tensile tests performed at room temperature were used to evaluate the mechanical properties of the joints. In order to analyze the microstructural evolution of the material, the welds' cross-sections were observed under optical microscope. The results shows that the resulting microstructure is free of defects and tensile strength of the welded joints is upto 75% of the base metal strength.

1. Introduction

The Friction Stir Welding (FSW), is a solid state welding process invented by The Welding Institutein 1991 [1-3]. Since then its industrial use has been increasing. For aerospace applications, this process is attractive to replace the traditional riveting process in fuselage panels, reducing the production costs (being faster) and weight of airframes. Further, it can produce high quality joints in most of the high strength and damage tolerant 2XXX and 7XXX series of aluminum alloys. Such alloys are not suitable for conventional fusion welding processes due to the poor solidification microstructure, resulting in low strength when comparing to the base metal. Further, such alloys have high propensity to develop solidification cracks [4-5].

In FSW process, the joint is obtained by a non-consumable rotating tool, containing a pin and a shoulder. The friction between the shoulder and the surface of the work-piece generates most of the heat that plasticises the material at the joining surfaces [6]. Simultaneously, the pin penetrates the work-piece, stirring the material at the interface and promoting joint. Further, the shoulder imposes forging forces to the stirred material, resulting in a defect free weld. The combination of the tool rotation and its translation through the interface is responsible for the formation of a high quality welding line.

A review from Mishra and Ma [3] summarizes several studies in FSW of aluminum alloys that converge to the identification of three distinct regions in the welding line: the stirring zone, also known as nugget, the thermomechanically affected zone (TMAZ) and the heat affected zone (HAZ). The grain size in the stir zone is fine and equiaxed, resulting in higher mechanical strength and ductility [7-10]. The thermomechanically affected zone (TMAZ) surrounds the stirring zone and is formed by high deformed non-recrystallized grains. The interface between the highly deformed and the stirred zones is not well defined in the retreating side of the welding line, but it is very well defined in the advancing side. The third zone of the FSW bead, known as the heat affected zone (HAZ),

Content from this work may be used under the terms of the Creative Commons Attribution 3.0 licence. Any further distribution of this work must maintain attribution to the author(s) and the title of the work, journal citation and DOI.
Published under licence by IOP Publishing Ltd

undergoes a thermal cycle without any plastic deformation associated. For aluminum alloys, the HAZ can be defined as the region starting at the TMAZ and extending up to regions where temperatures above 250° C are reached [6]. Below this temperature the microstructure seems to be unaffected, as the exposition time is very short.The 2014-T6 aluminum alloy is one of the 2xxx-series heat-treatable aluminum alloys and it has rarely been friction stir welded up to now. This paper aims to demonstrate friction stir weldability of AA-2014 alloy and the emphasis is laid on the relations of the tensile properties and microstructure of the joints in order to optimize the FSW parameters.

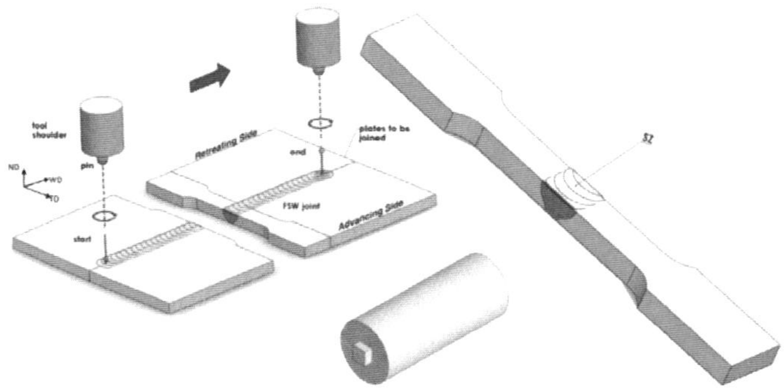

Figure 1. The schematic of FSW process, tool and tensile specimen used in present work

2. Experimental procedure

The material used in this study is 2014-T6 alloy, with the chemical compositions and mechanical properties listed in Table 1. The AA-2014 plates (600×1000×3mm) were cut and machined into rectangular welding samples 200mmx65mmx2.5mm.The test pieces were first ground using steel brush and sandpaper to remove the oxide film, and then cleaned with acetone to remove oil.

Welding was performed on a conventional milling machine (specially adopted for FSW by changing its mechanism in over arm spindle) driven by a 3hp motor with a 4A current rating using a square pin FSW tool. The schematic of friction stir welding process and tool employed for welding is shown in figure 1. There is an angle adjustment in the spindle, which allows setting of the tool tilt angle necessary for FSW. The feed speeds are gear driven with discrete speed options for both the specimen table travel and the spindle RPM. The welding parameters and welding tool size used in the experiments are shown in Table 2.

After welding, the welded joints were visually examined, were cross-sectioned transverse to the welding direction for metallographic analyses and tensile tests using Wire Cut machine. The cross-sections of the metallographic specimens were polished with a diamond paste, etched with Keller's reagent and observed under optical microscope. Room-temperature tensile tests were carried out at a crosshead speed of 1 mm/sec using an Instron testing machine. The gauge length of each specimen was 25 mm, and the tensile properties of each joint were evaluated using three tensile specimens cut from the same joint.

Table 1. Chemical compositions and mechanical properties of 2014-T6 aluminum alloy

Chemical compositions (wt. %)									Mechanical properties		
Al	Si	Fe	Cu	Mg	Mn	Ti	Zn	Cr	Tensile Strength (MPa)	Yield strength (MPa)	Elongation (%)
Bal	0.55	0.16	2.06	0.33	0.34	0.02	0.08	0.02	470	407.3	9.9

IOP Conf. Series: Materials Science and Engineering **146** (2016) 012055 doi:10.1088/1757-899X/146/1/012055

Table 2. Tool size and welding process parameters used in the experiments

Tool size (mm)			Welding parameters		
Shoulder diameter	Pin dimension	Pin length	Rotation speed (rpm)	Welding speed (mm/min)	Tool tilt (°)
20	5X5 square	2.3	700/1000	45–105	2.5

3. Results and discussions

Microstructure

Figure 2 shows the microstructure of base metal plate. It constitutes large grains; typical feature of rolled structures with almost uniform distribution of second phase particles.

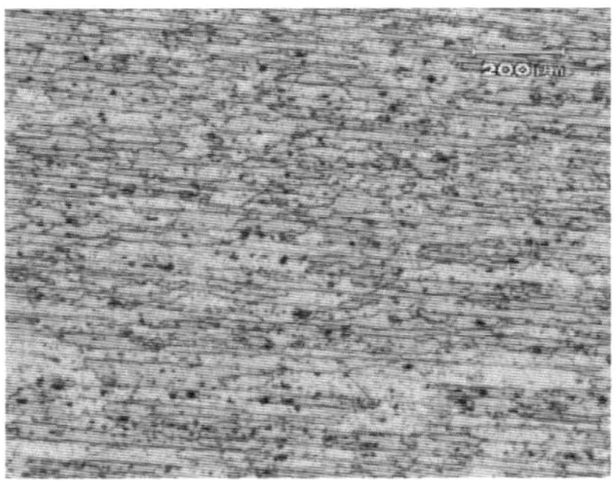

Figure 2. Microstructure of AA-2014 base plate

Figure 3 shows the cross section of joint welded at 1000 rpm and 105mm/min with low magnification. A complete penetration of the welding up to the root region of the sample was observed and no typical FSW defects like kissing bonds, voids, cracks etc. were observed.

Figure 3. Microstructure of Welded Joints at low magnification

The microstructures of distinct zones of welded joints at different travel speeds at 1000 rpm is shown in Figure 4. The grain size is remarkably smaller in the nugget zone as this region has experienced the high temperature and extensive deformation and is characterized by dynamically recrystallized grains. A clear interface is seen between the stirred (nugget) and TMAZ zone in the joint welded at lower travel speed (upto 65mm/min) indicating the insufficient flow of material. It is also obvious that region between the nugget and TMAZ zones always exhibit the poor property zone of the joint. The lower mechanical properties of the joints welded at lower travel speeds may be due to this reason. As the speed is increased the interface between nugget and TMAZ zone becoming diffused showing uniform flow of material and ultimately better mechanical properties.

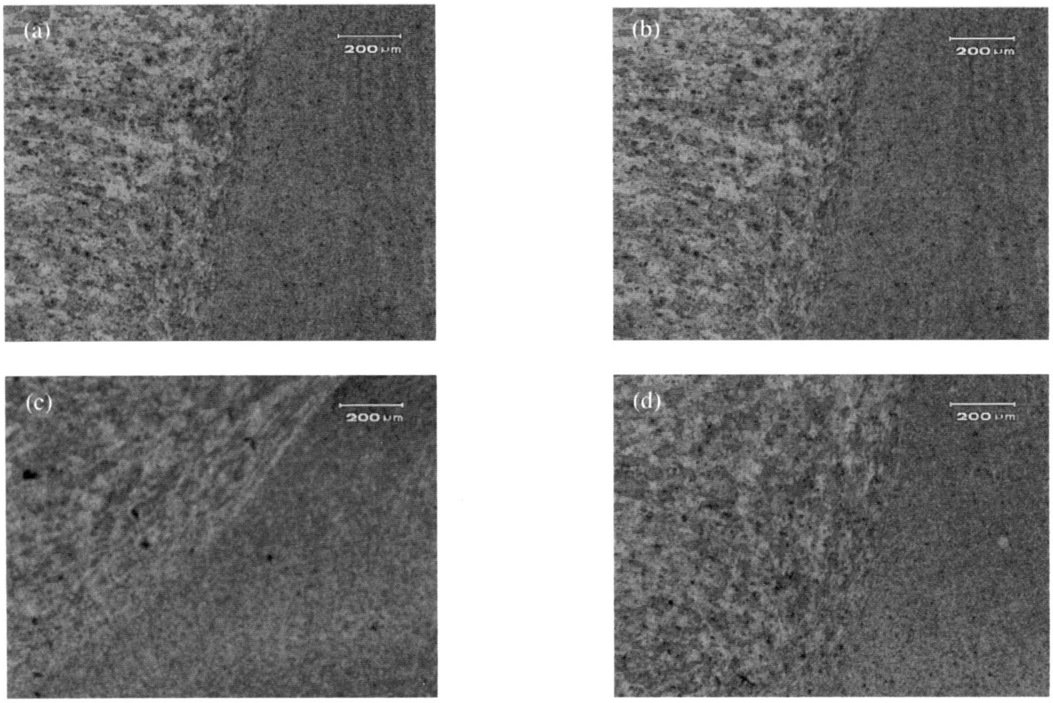

Figure 4. Microstructures in Weld Nugget -TMAZ Zone at 700 rpm. (a) 45mm/min (b) 65mm/min (c) 85mm/min (d) 105mm/min

4. Tensile properties of joints

The transverse tensile properties such as yield strength, tensile strength, percentage of elongation and joint efficiency of friction stir welded AA-2014 alloy joints were evaluated. In each condition, three specimens were tested and the average of results is presented in Figure 5. It is observed that ultimate tensile strength of the joint increases with increasing travel speed irrespective of the tool rpm reaching about 75.6 of the base metal strength and then decreases. These values are amongst the highest reported in the literature for friction stir welded AA-2014 T6 alloys, which ranges typically 60-70 % of base metal strength. [6].

The elongation of the joint reached about 66.6% of the elongation of the base metal. Nevertheless it remain around 6.6% showing good ductility. The joint strength is mainly due to microstructural aspects which in turn depends on the temperature generated during welding. As can be seen from Figure 4, difference in microstructures appear by varying the travel speeds. The decrease in joint efficiency at lower travel speeds may be due to improper heat input and non-uniform flow of material as is obvious in micrographs due to high speed.

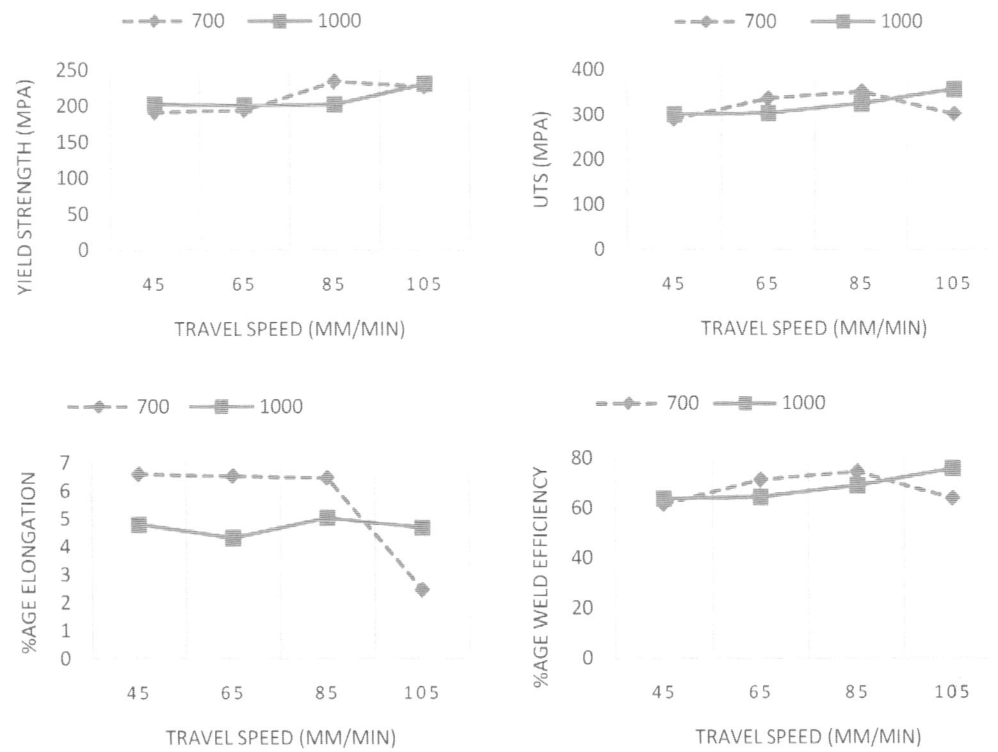

Figure 5. Graph between mechanical properties and weld efficiency of joint vs travel speeds at 700 and 1000 rpm

5. Conclusion

The maximum joint/weld efficiency is achieved at 1000rpm and 105mm/min travel speed. This may due to the diffused region between the stirred and TMAZ zone indicating the better flow of material during the welding.

6. Acknowledgement

The author like to acknowledge the contribution of Mr. Amir Siddique and Muhammad Arslan in carry out this research work.

7. References

[1] Thomas WM, Nicholas ED, Needham JC, Murch MG, Temple-Smith P and Dawes CJ European *Patent Specification* 0615480B1 1991
[2] Leonard AJ TWI Members Report 7400.01/99/1027.03 July 1999
[3] Mishra RS, Ma ZY Mater Sci Eng. R 2005 501
[4] AndrezaS, Fernando F and Dilermando N 2011*Mat& design*32 4684-88
[5] Su JQ, Nelson TW, Mishra RS AND Mahoney M. 2003 Acta Mater 51 713
[6] Norman AF, Brough I and Prangnell PB 2000 Mater Sci Forum1713 331–337
[7] Sato YS, Kokawa H, Enomoto M and Jogan S.1999 Metall Mater Trans A 30A 2429.
[8] Jata KV, Sankaran KK and RuscheauJj 2000 Metall MaterTrans A 31A 2181.
[9] H.J. Liu, H. Fujii, M. Maeda and K. Nogi 2003 *J. Materials Processing tech.142* 692-696
[10] Yan-hua Z, San-bao Lin, Lin Wu and Fu-xingQu 2005 *Materials Letter* 59